Optical Properties of Semiconductor Nanostructures

NATO Science Series

A Series presenting the results of activities sponsored by the NATO Science Committee. The Series is published by IOS Press and Kluwer Academic Publishers, in conjunction with the NATO Scientific Affairs Division.

A. Life Sciences	IOS Press
B. Physics	Kluwer Academic Publishers
C. Mathematical and Physical Sciences	Kluwer Academic Publishers
D. Behavioural and Social Sciences	Kluwer Academic Publishers
E. Applied Sciences	Kluwer Academic Publishers
F. Computer and Systems Sciences	IOS Press

1. Disarmament Technologies	Kluwer Academic Publishers
2. Environmental Security	Kluwer Academic Publishers
3. High Technology	Kluwer Academic Publishers
4. Science and Technology Policy	IOS Press
5. Computer Networking	IOS Press

NATO-PCO-DATA BASE

The NATO Science Series continues the series of books published formerly in the NATO ASI Series. An electronic index to the NATO ASI Series provides full bibliographical references (with keywords and/or abstracts) to more than 50000 contributions from international scientists published in all sections of the NATO ASI Series.
Access to the NATO-PCO-DATA BASE is possible via CD-ROM "NATO-PCO-DATA BASE" with user-friendly retrieval software in English, French and German (WTV GmbH and DATAWARE Technologies Inc. 1989).

The CD-ROM of the NATO ASI Series can be ordered from: PCO, Overijse, Belgium

Series 3. High Technology – Vol. 81

Optical Properties of Semiconductor Nanostructures

edited by

Marcin L. Sadowski

Groupe d'Etude des Semi-Conducteurs,
Université Montpellier II,
Montpellier, France

Marek Potemski

Grenoble High Magnetic Field Laboratory,
Max-Planck-Institut für Festkörperforschung and
Centre National de la Recherche Scientifique,
Grenoble, France

and

Marian Grynberg

Institute of Experimental Physics,
Warsaw University,
Warszawa, Poland

Kluwer Academic Publishers

Dordrecht / Boston / London

Published in cooperation with NATO Scientific Affairs Division

Proceedings of the NATO Advanced Research Workshop on
Optical Properties of Semiconductor Nanostructures
Jaszowiec, Poland
12-16 June 1999

A C.I.P. Catalogue record for this book is available from the Library of Congress.

ISBN 0-7923-6316-7

Published by Kluwer Academic Publishers,
P.O. Box 17, 3300 AA Dordrecht, The Netherlands.

Sold and distributed in North, Central and South America
by Kluwer Academic Publishers,
101 Philip Drive, Norwell, MA 02061, U.S.A.

In all other countries, sold and distributed
by Kluwer Academic Publishers,
P.O. Box 322, 3300 AH Dordrecht, The Netherlands.

Printed on acid-free paper

Table of contents

II   *Excitons and Polaritons*

III   Diluted magnetic semiconductors and other II-VI structures

IV   *Quantum dots*

PREFACE

The NATO Advanced Research Workshop on the "Optical properties of semiconductor nanostructures" was held in Jaszowiec, Poland, in June 1999. We are pleased to present, in this volume, the Proceedings of the Workshop.

Optical methods of investigating semiconductors, and the theoretical description of optical processes, have always been an important part of semiconductor physics. The emphasis placed on different materials and different structures changes with time. A Workshop such as this, bringing together major specialists in the field, gives an overview of the state of the art and current trends of development. We had the privilege of listening to 30 invited lectures, while the poster session, which lasted well into the evening, was a nice mixture of hard work and a pleasant social event.

A large number of papers was devoted to quantum dots. Theory, spectroscopic investigations and methods of producing such structures were presented. Another major part of the contributions reflected the growing interest in diluted semiconductors and II-VI nanosystems in general. Fascinating developments in the field of photonic crystals were presented. "Classical" low dimensional systems, such as GaAs/GaAlAs quantum wells and heterostructures, still occupy a significant part of the presented results and serve as model systems for new phenomena. New materials are being sought for; new experimental techniques - combinations of different spectroscopies - are appearing.

We take the opportunity of acknowledging the financial help received from the NATO Science Programme, and the Polish Committee for Scientific Research (KBN). We would also like to express our gratitude to our friends from the Department of Physics of the University of Warsaw whose efficient and unstinting help made the smooth functioning of the Workshop possible. Finally, it should also be mentioned that the weather, tailor-made on demand to suit the requirements for a fruitful Workshop, was remarkably consistent throughout.

Winter 1999

Marian Grynberg
Marek Potemski
Marcin L. Sadowski

NEAR-FIELD SPECTROSCOPY OF A GATED ELECTRON GAS

G. EYTAN,   Y. YAYON,   M. RAPPAPORT,
H. SHTRIKMAN, AND I. BAR-JOSEPH
Department of Condensed Matter Physics,
The Weizmann Institute of Science,
Rehovot 76100, Israel

Abstract. We study the spatial distribution of the photoluminescence of a gated two-dimensional electron gas with sub-wavelength resolution. This is done by scanning a tapered optical fibre tip with an aperture of 250 nm in the near field region of the sample surface, and collecting the photoluminescence. The spectral line of the negatively charged exciton, formed by binding of a photo-excited electron-hole pair to an electron, serves as an indicator for the local presence of charge. The local luminescence intensity of this line is directly proportional to the number of electrons under the tip. We observe large spatial fluctuations in this intensity in the gate voltage range, where the electron conductivity exhibits a sharp drop. The amplitude of these fluctuations increases and the Fourier spectrum extends to lower spatial frequencies as the gate voltage becomes more negative. We show that the fluctuations are due to the statistical distribution of localised electrons in the random potential of the remote ionised donors. We use these fluctuations to image the electron and donor distribution in the plane.

1. Introduction

The spatial separation between the donors and the two-dimensional electron gas (2DEG) in a modulation-doped semiconductor quantum well strongly inhibits electron scattering and allows obtaining high electron mobility. Applying a gate voltage enables to alter the 2DEG state from a high density mobile metallic state to a low density insulating one. The low temperature behaviour of this system has been a subject of intensive experimental

1

M.L. Sadowski et al. (eds.), Optical Properties of Semiconductor Nanostructures, 1–18.
© *2000 Kluwer Academic Publishers. Printed in the Netherlands.*

and theoretical studies [1, 2]. It was found that while the density changes linearly with gate voltage the conductivity exhibits a much stronger dependence. Below a certain critical voltage it exhibits a very sharp drop, of a few orders of magnitude. This large drop of conductivity is accompanied by a relatively small change in the electron density, typically less than an order of magnitude. It has been realised that the remote ionised donors, which provide the electrons to the 2DEG, play an important role in this drop of conductivity [1, 2]. Spatial variations in the density of these donors are manifested as random potential fluctuations in the plane of the 2DEG. At small negative gate voltages, when the conductivity is high, these fluctuations are effectively screened by the 2DEG. As the electron density is decreased at larger negative gate voltages, the screening becomes less effective, and the fluctuations grow, giving rise to localisation of the electrons.

These changes in electron density and screening properties which are induced by the gate voltage are clearly manifested in the absorption and emission spectra [3, 4]. It was found that the optical spectrum abruptly changes, from a broad line, which characterises a Fermi sea of free electrons, to two narrow excitonic resonances.[5] The higher energy resonance was shown to be associated with a neutral exciton (X) and the low energy one with the negatively charged exciton (X^-). The X^- is a bound complex, which consists of two electrons and a hole, and is a semiconductor analogue of the hydrogen ion H^-. The existence of the X^- was predicted nearly 50 years ago [6], but it was only recently observed in CdTe/CdZnTe quantum wells (QW)[7], and subsequently in GaAs/AlGaAs QW [5, 8, 9, 10]. It is clear that an excess density of electrons, which could bind to the photo-excited electron-hole pairs, is needed for the X^- to be formed. Indeed, X^- is observed in structures where extra electrons are created by modulation doping [5, 7, 8], tunnelling [9], or optical excitation [10]. The electron density, in which the X - X^- doublet is observed, is typically of the order of $\sim 10^{10}$ cm^{-2}, but can be as high as 10^{11} cm^{-2} in modulation doped samples with a thin spacer [5]. Since a free electron gas with such high density would quench the excitonic interaction [11], it was concluded that the electrons are localised [5]. Indeed, the appearance of the X and X^- is correlated with the sharp drop in the electron conductivity.

These findings have led us to propose that the X^- is formed in these structures by binding of a photo-excited electron-hole pair to a localised electron. The electrons are localised, separately one from another, in the minima of a random potential induced by the remote ionised donors. Our purpose in this work is to prove the localised nature of the X^- and use it as *a probe for the presence of electrons*. By measuring the intensity distribution of the X^- luminescence in the range of gate voltages where the conductivity drops, we can image the electron distribution in the plane. This is done using near-field scanning optical microscopy (NSOM), which provides subwavelength spatial resolution [12],[13]. By collecting the PL through a small aperture, which is scanned in close proximity to the surface, and analysing its spectrum, we map the charge distribution in the plane. We show that in the range of gate voltages where the conductivity drops, the electrons are localised in the potential fluctuations of the remote ionised donors.

A short presentation of some of the results has recently appeared in a letter format [14]. Here we wish to present a more complete account of that work. The paper is organised as follows: In section II we describe the microscope and present measurements that demonstrate the system performance. We also discuss the structure of the sample we study. Section III is devoted to the experimental results. We show the existence of large spatial fluctuations of the X^- PL intensity, and we analyse the Fourier spectrum and the dependence on gate voltage of these fluctuations. We also present our observations of temporal fluctuations at high excitation intensity. In section IV we discuss the origin of the spatial fluctuations and relate them to the electron localisation.

2. The near-field microscope

We have built a low temperature near-field scanning optical microscope (NSOM), which collects the emitted PL through a tapered optical fibre tip. The microscope operates in a storage Dewar and has a very stable mechanical structure so scanning is conducted while the Dewar is standing on the laboratory floor. The cool-down procedure is simple: there is a large (0.7 mm) Z range, which enables positioning the tip at a safe distance from the sample at the beginning of the cooling process. The NSOM is operated inside a single-walled, exchange gas-cooled cryostat with 36 mm ID which fits inside the storage Dewar. The body of the microscope was made extremely rigid by machining it out of solid stainless steel, giving rise to a relatively high (a few kHz) resonance frequency. It is supported by a long stainless steel tube, with a low horizontal resonance frequency ($\sim$ 6 Hz). The large difference between the two resonances provides isolation from horizontal vibrations.

The tip-sample distance regulation is done by shear force detection, using a commercial quartz tuning fork, which is bonded to a ϕ 2 mm x 10 mm long piezoelectric tube [15]. The fibre tip is glued to one of the prongs of the fork, and oscillates at an amplitude of less than 1 nm as the piezoelectric tube is driven at the fork resonance frequency ($\sim$ 33 kHz). Since the fork is made of quartz this oscillation generates an oscillating piezoelectric signal at its electrodes. This signal is a direct measure of the oscillation amplitude of the tip. As the tip approaches the sample the oscillations are damped by the interaction with the surface, causing the amplitude and phase of the piezoelectric signal to change. Thus, continuous measurement of the fork signal, together with an electronic feedback loop, can be used to control the height of the tip above the sample surface (typically at ~ 10 nm).

The sample is mounted on a ϕ 1/4 inch x 2 inch long piezoelectric tube, giving a scan range of 11x11 μm^2 at 4.2 K. A coarse X-Y movement assembly translates the sample piezo and enables movement to different regions of the sample within an area of 2x2 mm^2, with ¡1 μm resolution. The tip, which is manufactured by Nanonics, is coated with aluminium and has a clear aperture of ~ 250 nm and a transmission of $5x10^{-3}$ at 810 nm. The transmission factor was determined by illuminating the tip and measuring the light intensity from the other end of the fibre. The tip diameter and the continuity of the metallic coating were determined using SEM imaging.

The excitation source for the PL measurements was either a Ti-sapphire or a He-Ne laser. The collected PL was measured by a 0.5 m spectrometer and a thermoelectrically cooled, back-illuminated CCD detector with matrix of 1100x330, 24 μm x 24 μm pixels. The gain of the CCD is 4, namely, four incoming photons give one count. The system spectral resolution, determined by the spectrometer and the pixel width of the CCD, is 0.04 nm.

An essential point in the operation of the microscope is related to the way the photo-excitation and the PL collection were performed. Operation in the so-called illumination mode, where the sample is photo-excited through the tip and the PL is collected by a broad lens, is not appropriate [16]. The resolution in this mode is limited by diffusion of the photo-excited carriers (typically ~ 1 μm in GaAs samples). The alternative method of operation is the collection-mode, where the tip is used to collect the emitted PL. In applying this method there are two possible means of excitation. One popular technique is to use the tip itself for excitation [17]. This is a simple and straight forward technique, but it creates background optical noise due to Raman scattering and fluorescence from the fibre itself [18] and from the tip [17]. Another disadvantage of this technique is the non-uniform excitation that is created. The second technique is to use a separate single-mode fibre oriented such that the light is nearly parallel to the surface. This form of excitation gives uniform excitation over an area of ~ 1 mm^2 and a background-free PL signal. In this work we have used the collection mode and a separate fibre for excitation throughout our measurements. For an excitation intensity of 50 mW/cm^2 at a wavelength of 632.8 nm we measure 30 - 40 counts/sec at the exciton peak of a single 20 nm GaAs quantum well.

An indication of a near-field measurement is an abrupt increase in the

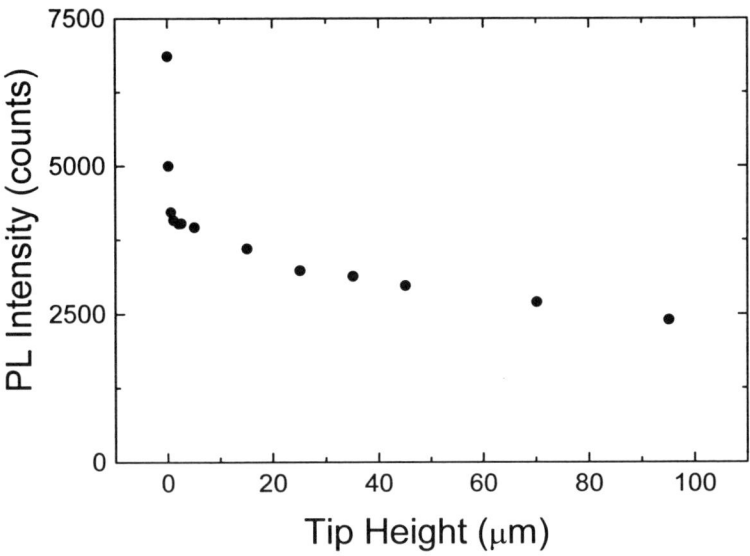

Figure 1. The PL intensity versus the tip-sample separation. Note the sharp rise at the near field region.

collected signal as the tip enters the near-field zone (another indication is a loss of polarisation in this zone [19]). In Fig. 1 the PL intensity versus the tip-sample separation is shown. It is seen that the PL intensity increases gradually as the tip approaches the sample. However, in the last micron, the signal rises abruptly by a factor of ~ 2. This factor is tip dependent: the smaller the tip, the larger the increase. A factor of ~ 6 is reported in [20] for tip diameter of 100nm.

To test the optical spatial resolution of the system a calibration sample has been prepared, consisting of seven GaAs quantum wells, 10 nm each, separated by 20 nm AlGaAs barriers. A grating of 30 nm thick opaque Pd/Au strips with 1.5 μm width and 3.5 μm pitch was evaporated on the sample surface. The optical spatial resolution is determined by measuring the change in the PL intensity as the tip is scanned across the metal strips. Defining the resolution as the distance between the points at which the PL intensity rises from 10% to 90% of its maximum, we get 250 nm. This is in a good agreement with the SEM measurements of the tip aperture diameter.

It is important to take into account a significant factor determining the spatial resolution of a signal from a QW buried below the surface. The PL is emitted at some wavelength λ from a layer which is ~ 100 nm below the surface. PL emission carrying spatial frequencies smaller than n/λ, where n is the refractive index, can propagate in the medium to the surface. There

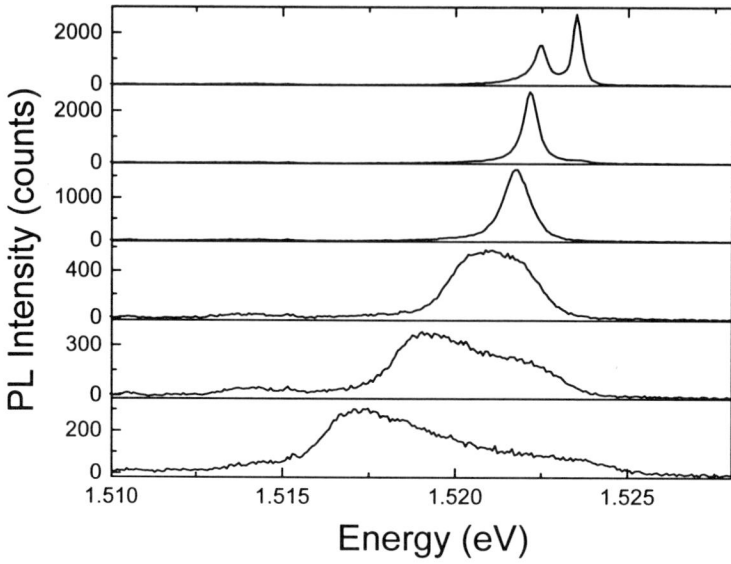

Figure 2. The evolution of the far-field PL of a 2DEG with gate voltage (upper curves correspond to more negative gate voltage).

they can be detected by the tip, which must be in the near-field region. PL emission carrying higher spatial frequencies decays substantially over the distance to the sample surface [21]. For n = 3.5 and λ = 815 nm this spatial frequency cut-off is 230^{-1} nm^{-1}.

3. Experimental results

The sample measured in this work consists of a 20 nm GaAs quantum well followed by a 37.5 nm $Al_{0.37}Ga_{0.63}As$ spacer layer and a 10 nm $Al_{0.37}Ga_{0.63}As$ layer doped with silicon at a concentration of 3.5×10^{18} cm^{-3}. The structure is capped by a 20 nm undoped $Al_{0.37}Ga_{0.63}As$ layer and a 10 nm GaAs layer. A 2x2 mm^2 mesa was etched, and ohmic contacts were alloyed into the 2DEG layer. A 4 nm Pd/Au semi-transparent gate was evaporated on top. The continuity and uniformity of the gate was verified by a high power optical microscope and SEM imaging. The 2DEG concentration was 4×10^{11} cm^{-2}, and the mobility 1.3×10^{6} cm^2/V sec, both measured at 4.2 K. Upon excitation with the laser the electron density is reduced by an amount that depends on the laser intensity. Consequently, the gate voltage needed to deplete the sample is intensity dependent.

In Fig. 2 a series of PL spectra of the gated 2DEG is shown. These spectra were taken with the tip in the far-field zone at different gate voltages. We observe an evolution from a broad spectrum of free electrons to an excitonic spectrum characterised by two peaks. The high and low energy peaks are due to recombination of neutral (X) and negatively charged (X^-) excitons, respectively. Our focus in this work is the gate voltage regime where the spectrum becomes excitonic. The underlying idea is to scan the sample with the tip at the near-field zone and measure the PL emission at a constant gate voltage. Since the X^- is a localised object associated with a localised electron, a change in the PL intensity of the X^- should be observed as the tip moves between regions with different electron concentrations.

3.1. NON-UNIFORMITY OF THE LOCAL SPECTRA

Figure 3a shows near-field spectra measured at four different positions and the same gate voltage, V=−0.135 Volts. At this gate voltage the spectrum is excitonic, consisting of X and X^- peaks. It can be seen from the figure that the near-field spectrum is different from one location to another: while the height of the X peak exhibits small changes, the height of the X^- peak varies substantially. Since the X^- is an indicator for the presence of electrons, these variations in its intensity show that the electrons are non-uniformly distributed.

In Fig. 3b the intensity under the X and X^- peaks of each local PL spectrum is integrated and is shown as a function of the tip position along an 11 μm line with 100 nm steps. It is evident that there are large fluctuations in the X^- intensity. The fluctuations occur throughout the scanned region and on any length scale, down to the resolution limit. It should be emphasised that at low excitation levels (< 50 mW/cm^{-2}) these fluctuations are stable over time: repeating the measurement over and over again reproduces the line scan.

To further prove the local nature of the X^- intensity fluctuations we performed line scan measurements of the PL spectrum, first with the tip in the near-field region (~ 10 nm above the sample surface), and then at a height of 1 μm above the sample. The intensities of the X and the X^- peaks were integrated as in Fig. 3b, and the result is shown at Fig. 4. It is seen that at a height of 1 μm the fast spatial frequencies disappear, and only frequencies smaller than $\lesssim (3\ \mu m)^{-1}$ survive. This is also seen in the X line. Thus, the far-field spectrum in the gate voltage regime where the spectrum is excitonic is an average of local spectra that are different from each other.

Figure 5 demonstrates this statement. It is constructed by the following procedure: first a square of 1x1 μm^2 is scanned and the PL spectra from all the points are summed up (Fig. 5, solid line). Then the tip is withdrawn to a distance of 1 μm from the sample and the PL is measured above the

8

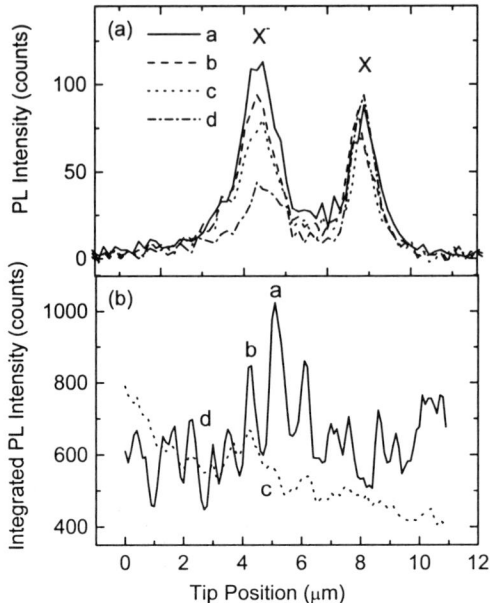

Figure 3. (a) Four near-field spectra taken at $V_g = -0.135$ V and at different tip locations. (b). The integrated X^- (solid line) and X (dotted line) intensity along an 11 μm-long line scan. The letters denote the location of the spectra shown in (a).

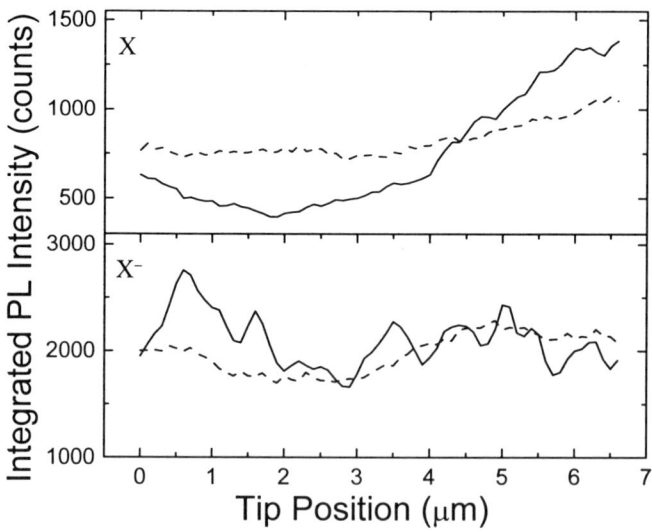

Figure 4. The integrated intensity of the X and X^- peaks along a 7 μm-long line scan. The solid and dashed curves are measured at a tip height of 10 nm and 1 μm, respectively.

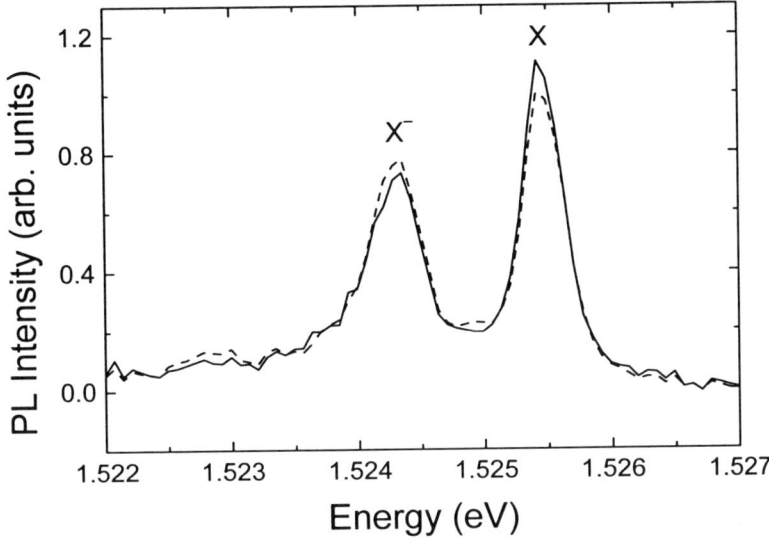

Figure 5. A sum of near-field spectra from 1x1 μm^2 region (solid) and a far-field spectrum taken at 1 μm above this region (dotted).

centre of this square (Fig. 5 dashed line). It can be seen that the two curves nearly coincide.

3.2. FOURIER ANALYSIS OF THE FLUCTUATIONS

Figure 6 describes the dependence of the fluctuations on the gate voltage. In Fig. 6a we show the integrated X^- PL intensity along an 11 μm line scan for three gate voltages (the average spectrum at each scan is shown in Fig. 6b). It is seen that at small negative gate voltages the intensity along the line is relatively smooth. As the gate voltage becomes more negative, the X^- intensity exhibits large spatial fluctuations. It can also be seen that these fluctuations evolve adiabatically with the gate voltage: as the gate voltage becomes more negative local maxima become higher and local minima lower.

To determine the evolution of the spatial frequencies with gate voltage we calculated the Fourier transforms of these line scans. In Fig. 6c we present the wavelengths of the Fourier spectrum in units of μm. To average out the effect of temporal noise we repeated the scan 8 times at each gate voltage, and performed the Fourier transform on the sum of the spectra. It is evident that as the 2DEG is depleted (more negative gate voltage) the intensity of all the Fourier components (long and short) increases. This

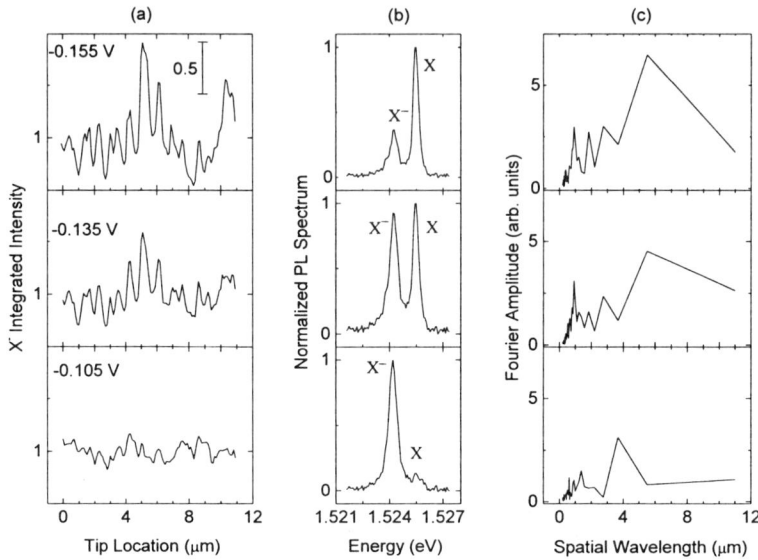

Figure 6. (a) The normalised X^- integrated intensity along an 11 μm-long line at three gate voltages. The curves are normalised by dividing the value at each point by the average of the whole line. (b) An average spectrum for each gate voltage. (c) The Fourier spectrum of the line scans of (a).

change in amplitude will be further discussed in the next section. However, it is also seen that the relative intensity of the short wavelength components decreases as compared to the longer ones. The most noticeable peak in the dense 2DEG (Vg = −0.105 V) is at 3.6 μm, and is shifted to 5.5 μm in the dilute 2DEG (Vg = −0.135 V and Vg = −0.155 V). This behaviour is a result of the fact that long wavelength components are the easiest to screen [24]. Hence, as the 2DEG is depleted these components get unscreened and become pronounced.

The shortest component in all spectra that is still pronounced is at 0.9 μm, which is ∼3 times the tip diameter. Below this cut-off wavelength the amplitude decreases abruptly. This is better seen in the two top spectra. This abrupt jump in the amplitude is due to the tip diameter that averages out spatial components smaller than three tip diameters. It should be noticed, however, that Fourier components exist at all possible wavelengths down to the FFT resolution.

3.3. FLUCTUATION AMPLITUDE

Figure 7 summarises the dependence of the fluctuation amplitude on gate voltage (note that this measurement is taken at a lower illumination in-

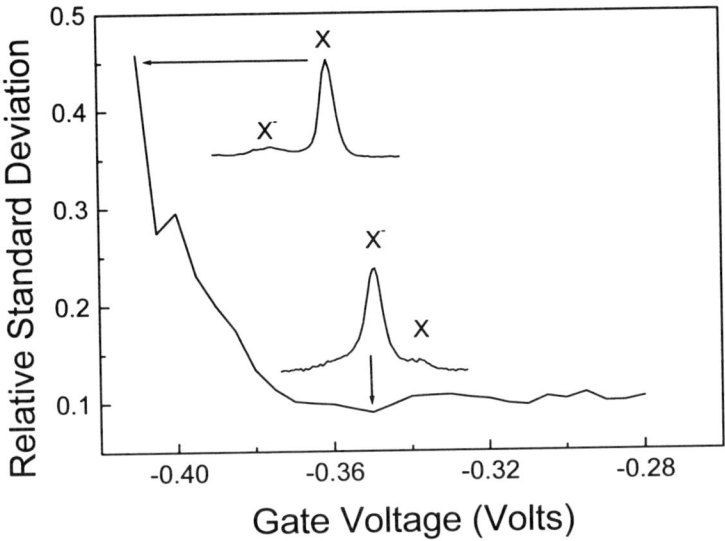

Figure 7. The standard deviation of the relative fluctuations amplitude of the X⁻ intensity along an 11 μm-long line, as a function of gate voltage. The averaged spectra at $V_g = -0.35$ V and $V_g = -0.41$ V are also shown.

tensity than the measurements of Fig. 6, hence the more negative gate voltages). The graph describes the dependence of the normalised standard deviation of the X⁻ fluctuation amplitude on gate voltage. The two spectra are typical spectra at the gate voltages denoted by the arrows. It can be clearly seen that the fluctuation amplitude is constant over a large range of gate voltages and starts to increase at the voltage at which the spectrum becomes excitonic. This correlation between the rise of the fluctuation amplitude and the appearance of the excitonic spectrum is not accidental. The appearance of large fluctuations in the X⁻ intensity is a clear sign of a non-uniform distribution of the 2DEG, or in other words − strong localisation of the electrons. This localisation gives rise to a change in the screening properties of the 2DEG: it becomes ineffective in screening the electron-hole interaction, and hence excitons appear. The weak PL intensity fluctuations, which are observed before the appearance of the exciton, are of different character, and are probably due to local fluctuations in the density of the photo-excited holes that are also influenced by the potential landscape.

To analyse the localised electronic system let us assume a simplified model, where the localising potential is periodic in the plane and the electrons are randomly distributed in it. In such a model the tip samples a region in which there are N sites with a probability p to be occupied by electrons. The average number of electrons occupying the area under the tip is $n_{av} = pN$, and the corresponding standard deviation is $\sigma_n = \sqrt{pN(1-p)}$. The average PL intensity of the X$^-$ line that is measured by the tip can be written as $I_{av} = \alpha n_{av}$, where α is a proportionality factor. Hence, the maximum intensity, I_{max}, when all the sites are occupied, can be written as $I_{max} = \alpha N$. The standard deviation, σ, in the X$^-$ intensity is proportional to σ_n with the same proportionality factor α. Thus α can be expressed in terms of the measurable quantities I_{av}, I_{max}, and σ as:

$$\alpha = \frac{\sigma^2}{I_{av}(1 - \frac{I_{av}}{I_{max}})} \tag{1}$$

By substituting α, N can be calculated to be:

$$N = \frac{I_{max} I_{av}(1 - \frac{I_{av}}{I_{max}})}{\sigma^2} \tag{2}$$

Since I_{av} and σ depend on the gate voltage, the consistency of this model can be checked by calculating N for each gate voltage. Figure 8a shows N for the measurement in Fig. 7. It can indeed be seen that a relatively small scatter around an average value of $N \approx 30$ is obtained. Since the tip diameter is 250 nm this implies an average fluctuation size of ~ 40 nm.

Figure 8b shows the distribution of the X$^-$ intensities over a scanned area of 11x11 μm^2. The dashed curve describes a Gaussian fit to this distribution. The nice fit shows that the X$^-$ distribution is indeed Gaussian, in agreement with our simple model. It is important to note that a gate voltage range in which the spectrum is excitonic in one area and 2DEG-like in another area has not been found. Such a spectrum might represent an isolated puddle of free electrons. The absence of such regions indicates that there are no large clusters of free electrons, and the localisation is of single electrons.

3.4. TEMPORAL FLUCTUATIONS

In addition to spatial fluctuations of the X$^-$ and X intensities, it is found that there are also temporal intensity fluctuations. This phenomenon occurs while exciting the sample with a high enough power (above 50mW/cm^2). Figure 9 shows the integrated intensities of the X$^-$ and X peaks during a one hour measurement (480 points) at the same tip position at a high excitation level of 60 mW/cm^2. Large temporal oscillations, with a typical period ranging from a few to hundreds of seconds, are clearly observed. It is evident from the figure that the fluctuations of the X and X$^-$ have opposite phases. Figure 10a shows the relative standard deviation of the temporal intensity

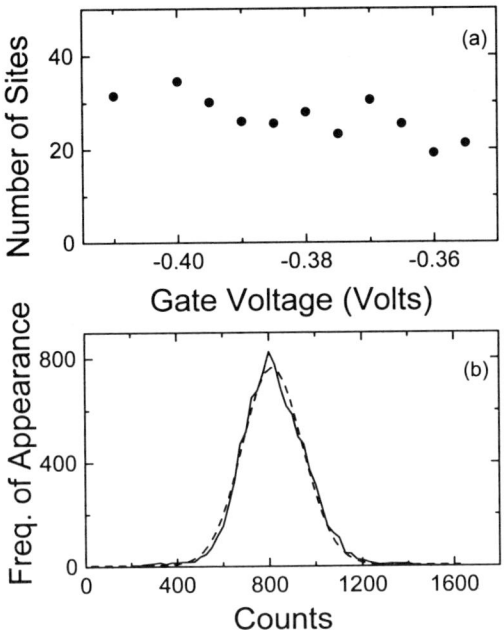

Figure 8. (a) The number of sites N under the tip, calculated for the gate voltages of Fig. 7. The calculation method is explained in the text. (b) The distribution of the X^- intensities over a scanned area of 11x11 μm^2. The dashed curve describes a Gaussian fit to this distribution.

fluctuations (calculated by dividing the absolute standard deviation by the average intensity) as a function of excitation intensity. It is seen that the fluctuations of the X intensity are always larger than those of the X^-. This is in contrast to the spatial fluctuations, which are much more pronounced in the X^- intensity. The dotted line represents the limit of the statistical shot noise of the X^-, which is the square root of the average intensity. It is seen that at low intensities the statistical noise becomes the dominant noise source in the X^- line; however, temporal fluctuations in the X intensity persist to the lowest measured excitation intensity. Figure 10b shows the relative standard deviation of the temporal fluctuations as a function of the tip height above the sample. It is seen that the temporal fluctuations decay very slowly as the tip is withdrawn from the sample, and even at a height of ~ 30 μm the fluctuations are clearly observed.

The effect of high excitation intensity on the temporal behaviour of the PL spectrum is very similar to that of a fluctuating gate potential. A negative gate voltage causes a growth of the X peak at the expense of X^-. Since the oscillator strength of X is larger than that of X^-, the changes in the X intensity are larger than those in the X^-. The experimental result also indicate that the changes occur on a relatively large scale, a few tens

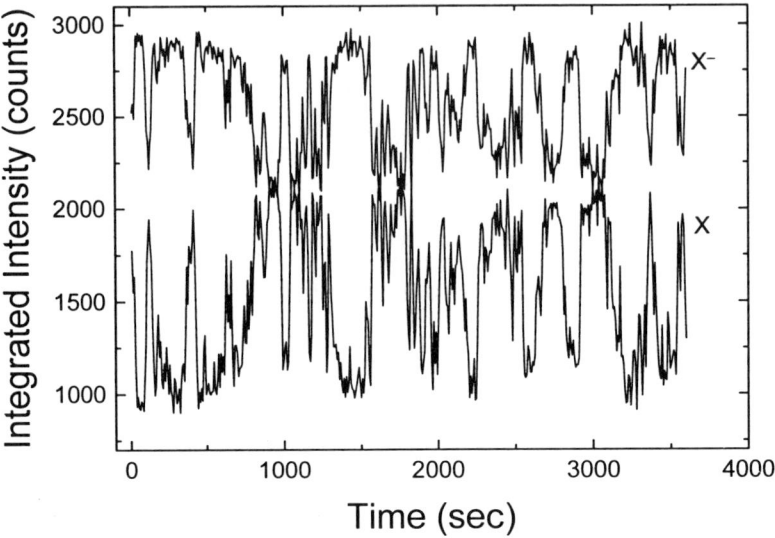

Figure 9. The integrated intensity of the X and X⁻ peaks over a one hour scan.

of microns. We believe that these temporal fluctuations are due to charge movement in the donor layer. Electrons that are activated to the donor layer screen the gate potential, and their diffusion in the plane causes a change of the lateral potential at the 2DEG layer. Further studies of these fluctuations are needed in order to clarify the underlying mechanism.

4. Discussion

It was previously shown that the appearance of excitons in the PL spectrum of a gated 2DEG is correlated with the sharp drop of the mobility that is observed below a certain gate voltage [5]. This drop was studied in GaAs/AlGaAs heterostructures and was shown to be due to crossover from weak to strong localisation [1, 22]. The underlying mechanism is a change in the screening of the remote ionised donors. These donors are distributed randomly in the doped layer during the growth process. Hence, when ionised they induce a random potential at the 2DEG layer. A high density 2DEG screens this random potential effectively, and its properties are almost insensitive to density changes. At low 2DEG density the screening becomes less effective, and the potential fluctuations grow, limiting the 2DEG mobility. The reduction in the electron mobility causes a further decrease of the screening ability of the 2DEG, and a run-away process develops. Thus, at a critical density, a small change in the 2DEG density is

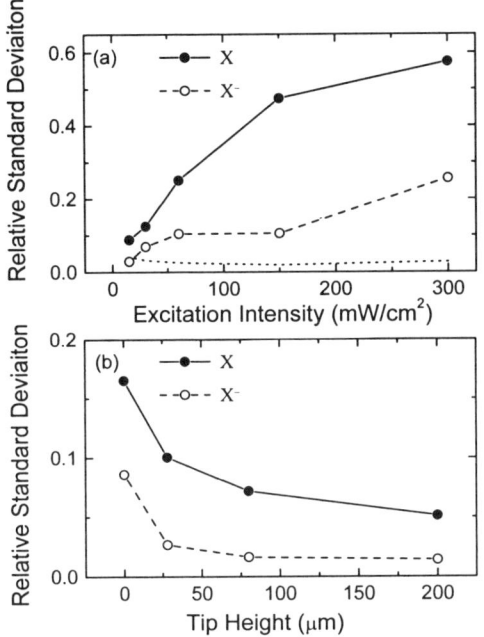

Figure 10. The relative standard deviation of the temporal fluctuations of the X and X^- as a function of (a) excitation intensity and (b) tip height.

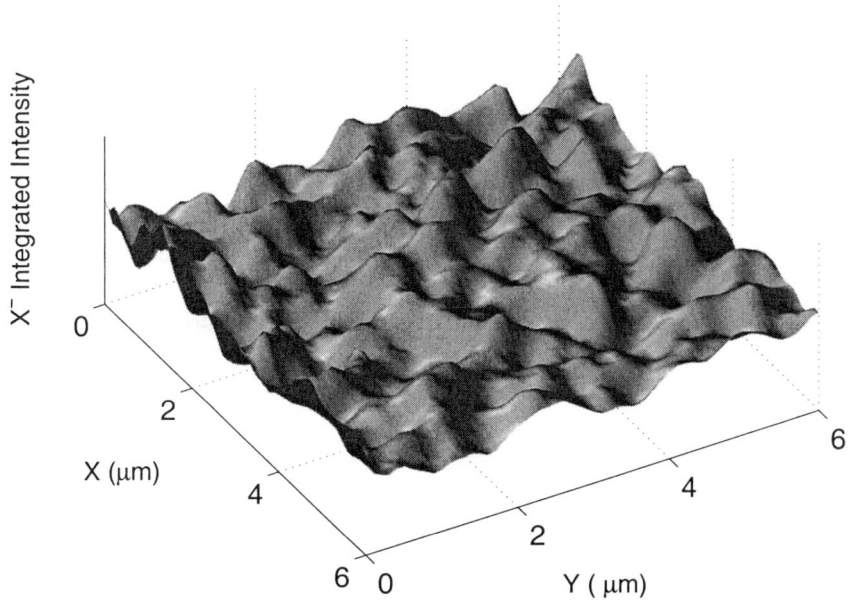

Figure 11. A two-dimensional image of the X^- integrated intensity for a 6×6 μm^2 region.

accompanied by a large change in the 2DEG mobility, and the electrons become strongly localised. The X^- is a probe of these localised electrons, and the fluctuations in its intensity mirrors the random distribution of the localised electrons in the localising potential. It should be emphasised that the question whether the system is initially weakly localised or metallic [23] can not be answered by this optical experiment.

A quantitative estimation of the potential fluctuations at high and low 2DEG density was presented in Ref. [24]. Following this derivation let us assume for simplicity that the donor layer is infinitely thin (δ doping), and that the donors are randomly distributed and uncorrelated in the plane. The RMS potential fluctuation for the case of a high density 2DEG, taking screening into account, is

$$\sqrt{< V^2 >} = W \frac{1}{2 q_s s}, \tag{3}$$

where $W = (e^2/2\varepsilon\varepsilon_0)\sqrt{N_0/2\pi}$, ε is the GaAs dielectric constant, ε_0 is the permittivity of free space, N_0 is the donor density, s is the thickness of the spacer between the doped layer and the 2DEG, and q_s is the inverse screening length of the 2DEG. Since usually $q_s s \gg 1$ we get that $\sqrt{< V^2 >} \ll W$. If, on the other hand, one assumes that there is no screening, the expression becomes

$$\sqrt{< V^2 >} = W \sqrt{\ln\left(\frac{L}{2s}\right)}, \tag{4}$$

where L is the sample size. It is readily seen that the fluctuations with no screening are significantly larger, $\sqrt{< V^2 >} \approx W$. These potential fluctuations were also calculated numerically, including the screening by the metal gate [25]. For the particular choice of parameters the amplitude of the potential fluctuations was shown to increase from 2.6 to 18 meV as the electron density decreased.

An important result of this derivation is that the spatial frequencies, which characterise the potential, are below $1/2s$, and higher frequencies are exponentially small. Thus, the typical size of a potential fluctuation is the spacer width (37.5 nm in our sample), in good agreement with our finding that N, the number of potential fluctuations under our 250 nm tip, is ≈ 30. Another finding is that at high 2DEG density the low frequencies are screened and the spectrum consists of a narrow range below $1/2s$. As the 2DEG becomes more localised, lower spatial frequencies appear in the spectrum. This behaviour is indeed found in our measurements (Fig. 6c).

Finally, the fact that the X^- are localised in the donor potential can be used to image the donor distribution. It was shown in [24] that the number of localised electrons in a given area is proportional to the square root of the local donor density. Thus, the fluctuations in the X^- intensity mirror the fluctuations in the donor density. By fixing the gate voltage at a convenient value and scanning the sample, we obtain a two dimensional map of the X^- intensity, which is directly related to the donor distribution. Figure 11 shows such a map for a region of 6x6 μm^2.

Acknowledgement. This research was supported by the Minerva Foundation and the Israel Academy of Science.

References

1. Jiang, C., Tsui, D.C., and Weimann, G. (1988), Threshold transport of high-mobility two-dimensional electron gas in GaAs/AlGaAs heterostructures, *Appl. Phys. Lett.* **53**, 1533 -1535.
2. Efros, A. L. (1988), Density of states of 2D electron gas and width of the plateau of IQHE, *Solid State Commun.* **65**, 1281-1284 and (1989), Metal-non-metal transition in heterostructures with thick spacer layers, *ibid* **70**, 253-256.
3. Bar-Joseph, I., Kuo, J. M., Klingshirn, C., Livescu, G., Miller, D.A.B., Chang, T.Y., and Chemla, D. S. (1987), Absorption spectroscopy of the continuous transition from low to high electron density in a single modulation doped InGaAs quantum well, *Phys. Rev. Lett.* **59**, 1357-1360.
4. Delalande, C., Orgonasi, J., Brum, J.A., Bastard, G., and Voos, M. (1987), Optical studies of GaAs quantum well based field-effect transistor, *Appl. Phys. Lett.* **51**, 1346-1348.
5. Finkelstein, G., Shtrikman, H., and Bar-Joseph, I. (1995), Optical spectroscopy of a two-dimensional electron gas near the metal-insulator transition, *Phys. Rev. Lett.* **74**, 976-979.
6. Lampert, M.A. *Phys. Rev. Lett.* **1**, 450 (1950).
7. Kheng, K., Cox, R.T., Merle d'Aubigne, Y., Bassani, F., Saminadayar., K, and Tatarenko, S. (1993), Observation of negatively charged excitons X^- in semiconductor quantum wells, *Phys. Rev. Lett.* **71**, 1752-1755.
8. Shields, A.J., Pepper, M., Ritchie, D.A., Simmons, M.Y., and Jones, G.A.C. (1995), Quenching of excitonic optical transitions by excess electrons in GaAs quantum wells, *Phys. Rev. B* **51**, 18049-18052.
9. Buhmann, H., Mansouri, L., Wang, J., Beton, P.H., Mori, N., Eaves, L., Henini, M., and Potemski, M. (1995), Electron-concentration-dependent quantum-well luminescence: Evidence for a negatively charged exciton, *Phys. Rev. B* **51**, 7969-7972.
10. Ron, A., Yoon, H.W., Sturge, M.D., Manassen, A., Cohen, E., and Pfeiffer, L.N. (1996), Thermodynamics of free trions in mixed type GaAs/AlAs quantum wells, *Solid State Commun.* **97**, 741-745.
11. Schmitt-Rink, S., Chemla, D.S., and Miller, D.A.B. (1989), Linear and nonlinear optical properties of semiconductor quantum wells, *Advances in Physics* **38**, 89-188.
12. Betzig, E., and Trautman, J. K. (1992), Near-field optics: microscopy, spectroscopy, and surface modification beyond the diffraction limit, *Science* **257**, 189-195.
13. Paesler, M.A., and Moyer, P.J. *Near Field Optics* (Wiley, New York, 1996).
14. Eytan, G., Yayon, Y., Rappaport, M., Shtrikman, H., and Bar-Joseph, I. (1998), Near-field spectroscopy of a gated electron gas: a direct evidence for electron localization, *Phys. Rev. Lett.* **81** , 1666-1669.
15. Karrai, K. and Grober, R. D. (1995), Piezoelectric tip-sample distance control for near field optical microscopes, *Appl. Phys. Lett.* **66**, 1842-1844.
16. Hess, H.F., Betzig, E., Harris, T.D., Pfeiffer, L.N., and West, K.W. (1994), Near-field spectroscopy of the quantum constituents of a luminescent system, *Science*

18

264, 1740-1745.

17. Harris, T.D., Gershoni, D., Grober, R.D., Pfeiffer, L., West, K., and Chand, N. (1996), Near-field optical spectroscopy of single quantum wires, *Appl. Phys. Lett.* **68,** 988-990.

18. Yayon, Y. M.Sc thesis work, The Feinberg graduate school, Weizmann Institute, Rehovot, Israel (1997).

19. Obermuller, C., Karrai, K., Kolb, G., and Abstreiter, G. (1995), Transmitted radiation through a subwavelength-sized tapered optical fiber tip in near-field scanning optical microscopy, *Ultramicroscopy* **61**, 171-177.

20. Levy, J., Nikitin, V., Kikkawa, J.M., Cohen, A., Samarth, N., Garcia, R., and Awschalom, D. D. (1996), Spatiotemporal near-field spin microscopy in patterned magnetic heterostructures, *Phys. Rev. Lett.* **76,** 1948-1951 .

21. G. A. Massey (1984), Microscopy and pattern generation with scanned evanescent waves, *Appl. Opt.* **23,** 658-660.

22. Sajoto, T., Suen, Y.W., Engel, L.W., Santos, M.B., and Shayegan, M. (1990), Fractional quantum Hall effect in very-low-density $GaAs/Al_xGa_{1-x}As$ heterostructures, *Phys. Rev. B* **41**, 8449-8460.

23. Kravchenko, S.V., Simonian, D., Sarachik, M., Mason, P., and W. Furneaux, J.E. (1996), Electric field scaling at a B=0 metal-insulator transition in two dimensions, *Phys. Rev. Lett.* **77**, 4938-4941.

24. Efros, A.L., Pikus, F.G., and Burnett, V.G. (1993), Density of states of a two-dimensional electron gas in a long-range random potential, *Phys. Rev. B* **47**, 2233-2243.

25. Nixon, J.A. and Davies, J.H. (1990), Potential fluctuations in heterostructure devices, *Phys. Rev. B* **41**, 7929.

TWO-DIMENSIONAL ELECTRON-HOLE SYSTEMS IN A STRONG MAGNETIC FIELD

JOHN J. QUINN[1], ARKADIUSZ WÓJS[1,2],
IZABELA SZLUFARSKA[1,2], AND KYUNG-SOO YI[1,3]
[1]*Department of Physics, University of Tennessee, Knoxville,
Tennessee 37996, USA*
[2]*Institute of Physics, Wroclaw University of Technology,
Wroclaw 50-370, Poland*
[3]*Physics Department, Pusan National University,
Pusan 609-735, Korea*

Abstract. Two-dimensional systems containing N_e electrons and N_h holes ($N_e > N_h$) strongly correlated through Coulomb interactions in the presence of a large magnetic field are studied by exact numerical diagonalisation. Low lying states are found to contain neutral (X^0) and negatively charged (X^-) excitons and higher charged exciton complexes (X_k^-, a bound state of k neutral excitons and an electron). Representing these states in terms of angular momenta and binding energies of the different exciton complexes, and the pseudopotentials describing their interactions with electrons and with one another, permits numerical studies of systems that are too large to investigate in terms of individual electrons and holes. Laughlin incompressible ground states of such a multi-component plasma are found. A generalised composite Fermion picture based on Laughlin type correlations is proposed. It is shown to correctly predict the lowest band of angular momentum multiplets for different charge configurations of the system for any value of the magnetic field.

1. Introduction

In two-dimensional electron-hole systems in the presence of a strong magnetic field, neutral excitons X^0 and spin-polarised charged excitonic ions X_k^- (X_k^- consists of k neutral excitons bound to an electron) can occur[1, 2, 3, 4]. The complexes X_k^- should be distinguished from spin-

M.L. Sadowski et al. (eds.), Optical Properties of Semiconductor Nanostructures, 19–31.

unpolarised ones (e.g. spin-singlet biexciton or charged exciton) that are found at lower magnetic fields[5] but unbind at very high fields as predicted by hidden symmetry arguments[6]. The excitonic ions X_k^- are long lived Fermions whose energy spectra display Landau level structure[2, 3, 4]. In this work we investigate, by exact numerical diagonalisation within the lowest Landau level, small systems containing N_e electrons and N_h holes ($N_e > N_h$) confined to the surface of a Haldane sphere [7, 8]. For $N_h = 1$ these systems serve as simple guides to understanding photoluminescence[1, 2, 3, 5, 9, 10, 11, 12]. For larger values of N_h it is possible to form a multi-component plasma containing electrons and X_k^- complexes[4]. We propose a model[13] for determining the incompressible quantum fluid states[14] of such plasmas, and confirm the validity of the model by numerical calculations. In addition, we introduce a new generalised composite Fermion (CF) picture[15] for the multi-component plasma and use it to predict the low lying bands of angular momentum multiplets for any value of the magnetic field.

The single particle states of an electron confined to a spherical surface of radius R containing at its centre a magnetic monopole of strength $2S\phi_0$, where $\phi_0 = hc/e$ is the flux quantum and $2S$ is an integer, are denoted by $|S, l, m\rangle$ and are called monopole harmonics [7, 8]. They are eigenstates of $\hat{l}^2$, the square of the angular momentum operator, with an eigenvalue $\hbar^2 l(l+1)$, and of $\hat{l}_z$, the z component of the angular momentum, with an eigenvalue $\hbar m$. The energy eigenvalue is given by $(\hbar\omega_c/2S)[l(l+1) - S^2]$, where $\hbar\omega_c$ is the cyclotron energy. The $(2l+1)$-fold degenerate Landau levels (or angular momentum shells) are labelled by $n = l - S = 0, 1, 2, \ldots$

2. Four electron–two hole system

In Fig. 1 we display the energy spectrum obtained by numerical diagonalisation of the Coulomb interaction of a system of four electrons and two holes at $2S = 17$. The states marked by open and solid circles are multiplicative [6] (containing one or more decoupled X^0's) and non-multiplicative states, respectively. For $L < 12$ there are four rather well defined low lying bands. Two of them begin at $L = 0$. The lower of these consists of two X^- ions interacting through a pseudopotential $V_{X^- - X^-}(L)$. The upper band consists of states containing two decoupled X^0's plus two electrons interacting through $V_{e^- - e^-}(L)$. The band that begins at $L = 1$ consists of one X^0 plus an X^- and an electron interacting through $V_{e^- - X^-}(L)$, while the band which starts at $L = 2$ consists of an X_2^- interacting with a free electron.

Knowing that the angular momentum of an electron is $l_{e^-} = S$, we can see that $l_{X_k^-} = S - k$, and that decoupled excitons do not carry angular momentum ($l_{X^0} = 0$). For a pair of identical Fermions of angular momentum l

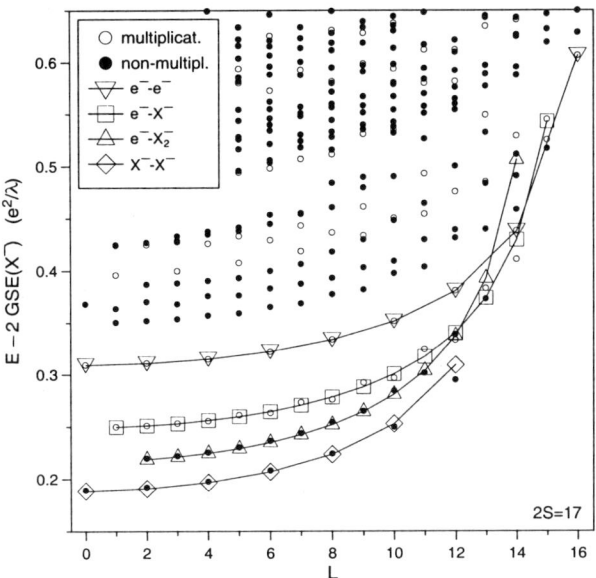

Figure 1. Energy spectrum of four electrons and two holes at $2S = 17$. Open circles – multiplicative states; solid circles – non-multiplicative states; triangles, squares, and diamonds – approximate pseudopotentials.

the allowed values of the pair angular momentum are $L = 2l - j$, where j is an odd integer. For a pair of distinguishable particles with angular momenta l_A and l_B, the total angular momentum satisfies $|l_A - l_B| \leq L \leq l_A + l_B$. The states containing two free electrons and two decoupled neutral excitons fit exactly the pseudopotential for a pair of electrons at $2S = 17$; the maximum pair angular momentum is $L^{\text{MAX}} = 16$ as expected. The states containing two X^-'s terminate at $L = 12$. Since the X^-'s are Fermions, one would have expected a state at $L^{\text{MAX}} = 2l_{X^-} - 1 = 14$. This state is missing in Fig. 1. By studying two X^- states for low values of S, we surmise that the state with $L = L^{\text{MAX}}$ does not occur because of the finite size of the X^-. Large pair angular momentum corresponds to the small average separation, and two X^-'s in the state with L^{MAX} would be too close to one another for the bound X^-'s to remain stable. We can think of this as a "hard core" repulsion for $L = L^{\text{MAX}}$. Effectively, the corresponding pseudopotential parameter, $V_{X^- - X^-}(L^{\text{MAX}})$ is infinite. In a similar way, $V_{e^- - X^-}(L)$ is effectively infinite for $L = L^{\text{MAX}} = 16$, and $V_{e^- - X_2^-}(L)$ is infinite for $L = L^{\text{MAX}} = 15$.

Once the maximum allowed angular momenta for all four pairings AB are established, all four bands in Fig. 1 can be roughly approximated by the pseudopotentials of a pair of electrons (point charges) with angular momenta l_A and l_B, shifted by the binding energies of appropriate composite particles. For example, the X^-–X^- band is approximated by the e^-–e^- pseudopotential for $l = l_{X^-} = S - 1$ plus twice the X^- energy. The agreement is demonstrated in Fig. 1, where the squares, diamonds, and two kinds of triangles approximate the four bands in the four-electron–two-hole spectrum. The fit of the diamonds to the actual X^-–X^- spectrum is quite good for $L < 12$. The fit of the e^-–X^- squares to the open circle multiplicative states is reasonably good for $L < 14$, and the e^-–X_2^- triangles fit their solid circle non-multiplicative states rather well for $L < 13$. At sufficiently large separation (low L), the repulsion between ions is weaker than their binding and the bands for distinct charge configurations do not overlap. There are two important differences between the pseudopotentials $V_{AB}(L)$ involving composite particles and those involving point particles.

The main difference is the hard core discussed above. If we define the relative angular momentum $\mathcal{R} = l_A + l_B - L$ for a pair of particles with angular momentum l_A and l_B, then the minimum allowed relative angular momentum (which avoids the hard core) is found to be given by

$$\mathcal{R}_{AB}^{\min} = 2\min(k_A, k_B) + 1, \tag{1}$$

where $A = X_{k_A}^-$ and $B = X_{k_B}^-$. The other difference involves polarisation of the composite particle. A dipole moment is induced on the composite particle by the electric field of the charged particles with which it is interacting. By associating an "ionic polarisability" with the excitonic ion X_k^-, the polarisation contribution to the pseudopotential can easily be estimated. When a number of charges interact with a given composite particle, the polarisation effect is reduced from that caused by a single charge, because the total electric field at the position of the excitonic ion is the vector sum of contributions from all the other charges, and there is usually some cancellation. We will ignore this effect in the present work and simply use the pseudopotentials $V_{AB}(L)$ obtained from Fig. 1 to describe the effective interaction.

3. Eight electron–two hole system

As an illustration, we first present the results of exact numerical diagonalisation performed on the ten particle system ($8e^-$ and $2h^+$). We expect low-lying bands of states containing the following combinations of complexes: (i) $4e^- + 2X^-$, (ii) $5e^- + X_2^-$, (iii) $5e^- + X^- + X^0$, and (iv) $6e^- + 2X^0$. The total binding energies of these configurations are: $\varepsilon_{\mathrm{i}} = 2\varepsilon_0 + 2\varepsilon_1$, $\varepsilon_{\mathrm{ii}} = 2\varepsilon_0 + \varepsilon_1 + \varepsilon_2$, $\varepsilon_{\mathrm{iii}} = 2\varepsilon_0 + \varepsilon_1$, and $\varepsilon_{\mathrm{iv}} = 2\varepsilon_0$. Here ε_0 is the binding energy of an X^0, ε_1 is the binding energy of an X^0 to an electron to form an X^-, and ε_k is the binding energy of an X^0 to an X_{k-1}^- to form an X_k^-. Some estimates of these binding energies (in magnetic units e^2/λ where λ is the magnetic

TABLE 1. Binding energies ε_0, ε_1, ε_2, and ε_3 of X^0, X^-, X_2^-, and X_3^-, respectively, in the units of e^2/λ.

$2S$	ε_0	ε_1	ε_2	ε_3
10	1.3295043	0.0728357	0.0411069	0.0252268
15	1.3045679	0.0677108	0.0395282	0.0262927
20	1.2919313	0.0647886	0.0381324	0.0260328

length) as a function of $2S$ are given in Tab. 1. Clearly, $\varepsilon_0 > \varepsilon_1 > \varepsilon_2 > \varepsilon_3$. The total energy depends upon not only the total binding energy, but the interactions between all the charged complexes in the system as well. All groupings (i)–(iv) contain an equal number of $N = N_e - N_h$ singly charged complexes. However, both angular momenta of involved complexes and the relevant hard cores are different. Which of the groupings has a state with the lowest total repulsion and binding energy, i.e. the absolute (possibly incompressible) ground state of the electron-hole system, depends on $2S$. It follows from the mapping between electron-hole and spin-unpolarised electron systems[6] that the multiplicative state $Ne^- + N_h X^0$ in which all holes are bound into decoupled X^0's is the absolute ground state (only) at the values of $2S$ corresponding to the filling factor $\nu = 1 - 1/m = 2/3$, $4/5$, ... of N (excess) electrons. At other values of $2S$, a non-multiplicative state containing an X^- is likely to have lower energy.

In Fig. 2, we show the low energy spectra of the $8e+2h$ system at $2S = 9$ (a), $2S = 13$ (c), and $2S = 14$ (e). Filled circles mark the non-multiplicative states, and the open circles and squares mark the multiplicative states with one and two decoupled excitons, respectively. In frames (b), (d) and (f) we plot the low energy spectra of different charge complexes interacting through appropriate pseudopotentials (see Fig. 1), corresponding to four possible groupings (i)–(iv). As marked with lines, by comparing left and right frames, we can identify low lying states of type (i)–(iv) in the electron-hole spectra.

The fitting of energies in left and right frames at $2S = 13$ and 14 is noticeably worse than at $2S = 9$. It is also much worse than almost a perfect fit obtained for the three charge system ($6e + 3h$ vs. $3X^-$, $e^- + X^- + X_2^-$, etc.)[4]. This is almost certainly due to treating the polarisation effect of the six charged particle system improperly by using the pseudopotential obtained from the two charged particle system (Fig. 1). A better fit is obtained by ignoring the polarisation effect, and only including the hard core effect on the pseudopotentials of a pair of point charges with angular momentum l_A and l_B.

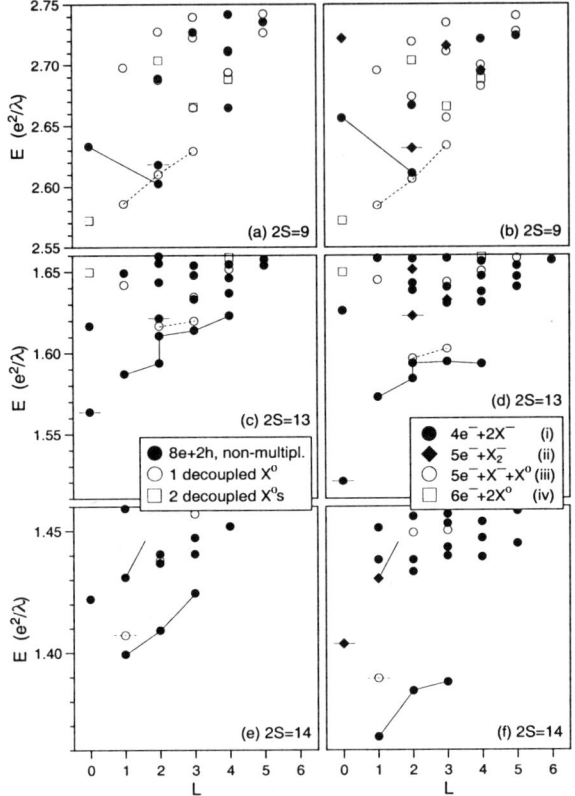

Figure 2. Left: low energy spectra of the $8e + 2h$ system on a Haldane sphere at $2S = 9$ (a), $2S = 13$ (c), and $2S = 14$ (e). Right: approximate spectra calculated for all possible groupings containing excitons (charged composite particles interacting through pseudopotentials as in Fig. 1). Lines connect corresponding states in left and right frames.

4. Larger systems

It is unlikely that a system containing a large number of different species (e.g. e^-, X^-, X_2^-, etc.) will form the absolute ground state of the electron-hole system. However, different charge configurations can form low lying excited bands. An interesting example is the $12e + 6h$ system at $2S = 17$. The $6X^-$ grouping (v) has the maximum total binding energy $\varepsilon_v = 6\varepsilon_0 + 6\varepsilon_1$. Other expected low lying bands correspond to the following groupings: (vi) $e^- + 5X^- + X^0$ with $\varepsilon_{vi} = 6\varepsilon_0 + 5\varepsilon_1$ and (vii) $e^- + 4X^- + X_2^-$ with $\varepsilon_{vii} = 6\varepsilon_0 + 5\varepsilon_1 + \varepsilon_2$.

Although we are unable to perform an exact diagonalisation for the $12e + 6h$ system in terms individual electrons and holes, we can use appropriate pseudopotentials and binding energies of groupings (v)–(vii) to

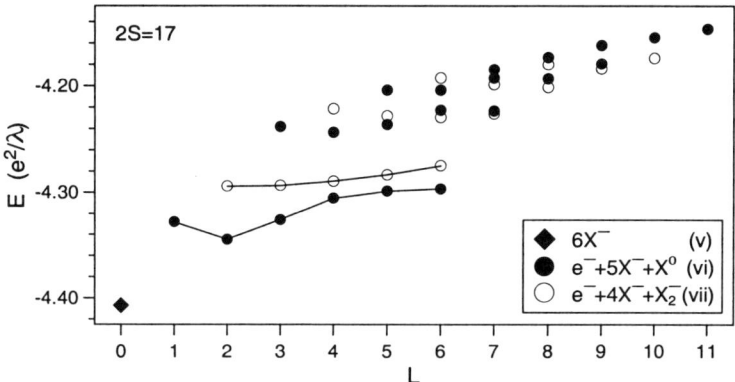

Figure 3. Low energy spectra of different charge configurations of the $12e + 6h$ system on a Haldane sphere at $2S = 17$: $6X^-$ (diamonds), $e^- + 5X^- + X^0$ (filled circles), and $e^- + 4X^- + X_2^-$ (open circles).

obtain the low lying states in the spectrum. The results are presented in Fig. 3. There is only one $6X^-$ state (the $L = 0$ Laughlin $\nu_{X^-} = 1/3$ state[4]) and two bands of states in each of groupings (vi) and (vii). A gap of 0.0626 e^2/λ separates the $L = 0$ ground state from the lowest excited state.

In Fig. 4 we present the spectra of the $6X^-$ charge configurations for $2S = 21$, 23, 25, and 27. The dashed lines are obtained by adding to the ground state energy the binding energy difference appropriate for the next lowest charge configuration; no states other than the plotted six X^- states are expected below these lines. The $L = 0$ ground states observed at different $2S$ correspond to $\nu_{X^-} = 2/7$, 2/9, 6/29, and 1/5. The $\nu_{X^-} = 1/5$ state is a Laughlin state and $\nu_{X^-} = 2/7$ and 2/9 are states in Jain sequence. The $\nu_{X^-} = 6/29$ state is a CF hierarchy state[7] corresponding to two quasiparticles (QP's) of the $\nu_{X^-} = 1/5$ state forming a $\nu_{QP} = 1/5$ state at the next level of the CF hierarchy. Without knowing the nature of the QP-QP interaction vs. pair angular momentum L, there is no guarantee that the CF hierarchy picture (which assumes the validity of the mean field approximation) is valid. Fig. 4c seems to indicate that it is, since the $L = 0$ state has a lower energy than the other two states at $L = 0$ and 4, predicted for two QP's each with $l_{QP} = 5/2$. Our study of the pseudopotential of QP's in the Laughlin $\nu = 1/5$ state at $\nu_{QP} = 1/5$ very strongly suggests that it behaves like the Coulomb pseudopotential, so that the MFCF picture should work.

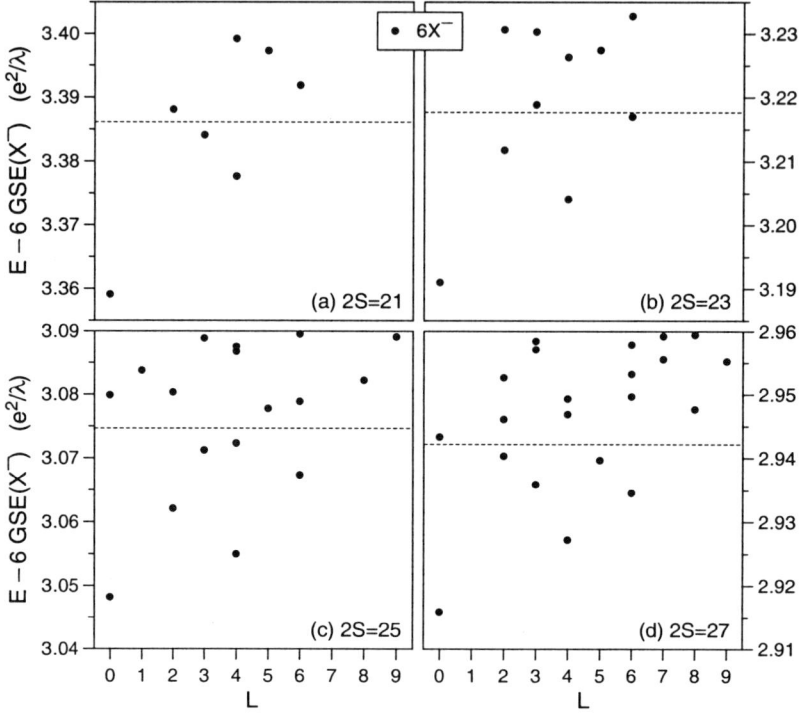

Figure 4. Low energy spectra of the $6X^-$ charge configuration of the $12e + 6h$ system on a Haldane sphere at $2S = 21$, 23, 25, and 27. Dashed lines – estimated lower bounds of higher bands.

5. Generalised Laughlin wavefunction

It is known that if the pseudopotential $V(\mathcal{R})$ decreases quickly with increasing $\mathcal{R}$, the low lying multiplets avoid (strongly repulsive) pair states with one or more of the smallest values of $\mathcal{R}$[17]. For the (one-component) electron gas on a plane, avoiding pair states with $\mathcal{R} < m$ is achieved with the factor $\prod_{i<j}(x_i - x_j)^m$ in the Laughlin $\nu = 1/m$ wavefunction. For a system containing a number of distinguishable types of Fermions interacting through Coulomb-like pseudopotentials, the appropriate generalisation of the Laughlin wavefunction will contain a factor $\prod(x_i^{(a)} - x_j^{(b)})^{m_{ab}}$, where $x_i^{(a)}$ is the complex coordinate for the position of ith particle of type a, and the product is taken over all pairs. For each type of particle one power of $(x_i^{(a)} - x_j^{(a)})$ results from the antisymmetrisation required for indistinguishable Fermions and the other factors describe Jastrow type correlations between the interacting particles. Such a wavefunction guarantees

that $\mathcal{R}_{ab} \geq m_{ab}$, for all pairings of various types of particles, thereby avoiding large pair repulsion[13, 16]. Fermi statistics of particles of each type requires that all m_{aa} are odd, and the hard cores defined by Eq. (1) require that $m_{ab} \geq \mathcal{R}_{ab}^{\min}$ for all pairs.

6. Generalised composite fermion picture

In order to understand the numerical results obtained in the spherical geometry (Figs. 2 and 3), it is useful to introduce a generalised CF picture by attaching to each particle fictitious flux tubes carrying an integral number of flux quanta ϕ_0. In the multi-component system, each a-particle carries flux $(m_{aa} - 1)\phi_0$ that couples only to charges on all other a-particles and fluxes $m_{ab}\phi_0$ that couple only to charges on all b-particles, where a and b are any of the types of Fermions. The effective monopole strength[10, 15, 17, 18] seen by a CF of type a (CF-a) is

$$2S_a^* = 2S - \sum_b (m_{ab} - \delta_{ab})(N_b - \delta_{ab}) \qquad (2)$$

For different multi-component systems we expect generalised Laughlin incompressible states (for two components denoted as $[m_{aa}, m_{bb}, m_{ab}]$) when all the hard core pseudopotentials are avoided and CF's of each kind fill completely an integral number of their CF shells (e.g. $N_a = 2l_a^* + 1$ for the lowest shell). In other cases, the low lying multiplets are expected to contain different kinds of quasiparticles (QP-a, QP-b, ...) or quasiholes (QH-a, QH-b, ...) in neighbouring filled shells.

Our multi-component CF picture can be applied to the system of excitonic ions, where the CF angular momenta are given by $l_{X_k^-}^* = |S_{X_k^-}^*| - k$. As an example, let us first analyse the low lying $8e + 2h$ states in Fig. 2. At $2S = 9$, for $m_{e^- e^-} = m_{X^- X^-} = 3$ and $m_{e^- X^-} = 1$ we predict the following low lying multiplets in each grouping: (i) $2S_{e^-}^* = 1$ and $2S_{X^-}^* = 3$ gives $l_{e^-}^* = l_{X^-}^* = 1/2$. Two CF-$X^-$'s fill their lowest shell ($L_{X^-} = 0$) and we have two QP-e^-'s in their first excited shell, each with angular momentum $l_{e^-}^* + 1 = 3/2$ ($L_{e^-} = 0$ and 2). Addition of L_{e^-} and L_{X^-} gives total angular momenta $L = 0$ and 2. We interpret these states as those of two QP-e^-'s in the incompressible [331] state. Similarly, for other groupings we obtain: (ii) $L = 2$; (iii) $L = 1, 2$, and 3; and (iv) $L = 0$ ($\nu = 2/3$ state of six electrons).

At $2S = 13$ and 14 we set $m_{e^- e^-} = m_{X^- X^-} = 3$ and $m_{e^- X^-} = 2$ and obtain the following predictions. First, at $2S = 13$: (i) The ground state is the incompressible [332] state at $L = 0$; the first excited band should therefore contain states with one QP-QH pair of either kind. For the e^- excitations, the QP-e^- and QH-e^- angular momenta are $l_{e^-}^* = 3/2$ and $l_{e^-}^* + 1 = 5/2$, respectively, and the allowed pair states have $L_{e^-} = 1, 2, 3,$

and 4. However, the $L = 1$ state has to be discarded, as it is known to have high energy in the one-component (four electron) spectrum[18]. For the X^- excitations, we have $l^*_{X^-} = 1/2$ and pair states can have $L_{X^-} = 1$ or 2. The first excited band is therefore expected to contain multiplets at $L = 1$, 2^2, 3, and 4. The low lying multiplets for other groupings are expected at: (ii) $L = 2$ and 3; (iii) $2S^*_{X_2^-} = 3$ gives no bound X_2^- state; setting $m_{e-X^-} = 1$ we obtain $L = 2$; and (iv) $L = 0$, 2, and 4. Finally, at $2S = 14$ we obtain: (i) $L = 1$, 2, and 3; (ii) incompressible [3*2] state at $L = 0$ (m_{X-X^-} is irrelevant for one X^-) and the first excited band at $L = 1$, 2, 3, 4, and 5; (iii) $L = 1$; and (iv) $L = 3$.

For the $12e + 6h$ spectrum in Fig. 3 the following CF predictions are obtained: (v) For $m_{X-X^-} = 3$ we obtain the Laughlin $\nu = 1/3$ state with $L = 0$. Because of the hard core of V_{X-X^-}, this is the only state of this grouping. (vi) We set $m_{X-X^-} = 3$ and $m_{e-X^-} = 1$, 2, and 3. For $m_{e-X^-} = 1$ we obtain $L = 1$, 2, 3^2, 4^2, 5^3, 6^3, 7^3, 8^2, 9^2, 10, and 11. For $m_{e-X^-} = 2$ we obtain $L = 1$, 2, 3, 4, 5, and 6. For $m_{e-X^-} = 3$ we obtain $L = 1$. (vii) We set $m_{X-X^-} = 3$, $m_{e-X_2^-} = 1$, $m_{X-X_2^-} = 3$, and $m_{e-X^-} = 1$, 2, or 3. For $m_{e-X^-} = 1$ we obtain $L = 2$, 3, 4^2, 5^2, 6^3, 7^2, 8^2, 9, and 10. For $m_{e-X^-} = 2$ we obtain $L = 2$, 3, 4, 5, and 6. For $m_{e-X^-} = 3$ we obtain $L = 2$. In groupings (vi) and (vii), the sets of multiplets obtained for higher values of m_{e-X^-} are subsets of the sets obtained for lower values, and we would expect them to form lower energy bands since they avoid additional small values of $\mathcal{R}_{e-X^-}$. However, note that the (vi) and (vii) states predicted for $m_{e-X^-} = 3$ (at $L = 1$ and 2, respectively) do not form separate bands in Fig. 3. This is because the V_{e-X^-} pseudopotential increases more slowly than linearly as a function of $L(L + 1)$ in the vicinity of $\mathcal{R}_{e-X^-} = 3$. In such case the CF picture fails[17].

The agreement of our CF predictions with the data in Figs. 2 and 3 (marked with lines) is really quite remarkable and strongly indicates that our multi-component CF picture is correct. We were indeed able to confirm predicted Jastrow type correlations in the low lying states by calculating their coefficients of fractional parentage[17, 19]. We have also verified the CF predictions for other systems that we were able to treat numerically. If exponents m_{ab} are chosen correctly, the CF picture works well in all cases.

7. Special case: many electron–one hole systems

In an investigation of photoluminescence, the eigenstates of a system containing up to $N_e = 7$ electrons and a single hole have been studied as a function of d, the separation between the surfaces on which electrons and the hole are confined[10, 11, 12]. For d larger than a few magnetic lengths λ, the low energy spectra can be understood quite simply[10] in terms of the lowest band of multiplets of N_e electrons weakly coupled to the hole. There is clear evidence for bound states of the hole to one or more Laughlin[14] quasielectrons. For $d < \lambda$ there has been no convincing interpretation of the low lying states, although Apalkov et al.[12] suggested an explanation in terms of "dressed" X^0 excitons.

At $d = 0$ there are two types of states which contain excitons, viz. multiplicative states containing $N_e - 1$ electrons and one X^0, and non-multiplicative states containing $N_e - 2$ electrons and one X^-. The multiplicative states are particularly simple; their energies are simply the energies of $N_e - 1$ interacting electrons less the binding energy ε_0 of an X^0. The non-multiplicative states are an example of a two-component plasma and can be understood in our generalised CF picture.

For $N_e = 7$, the $6e^- + X^0$ and $5e^- + X^-$ states can be found in the $8e+2h$ spectra shown in Fig. 2, where they correspond to the $6e^-+2X^0$ and $5e^- + X^- + X^0$ multiplicative states marked with open symbols. We have shown that the predictions of our model work very well for this system. In particular, it is clear from Fig.2ab that while the $7e + 1h$ ground state at $2S = 9$ is the (multiplicative) incompressible $\nu = 2/3$ state of six electrons, the low lying states at $L = 1$, 2, and 3 all contain an X^- and thus their nature is very different.

Similarly, at $2S = 15$, the pseudopotential calculation for the $5e^- + X^-$ grouping (as in Fig.2bdf) as well as the CF prediction for $m_{e-e-} = 3$ and $m_{e-X-} = 2$ undoubtedly preclude the interpretation of the low energy band at $L = 1$, 2, 3, and 4 (see figures in Refs. [10, 12]) in terms of a "dressed" exciton in favour of the $5e^- + 1X^-$ configuration. In the CF picture of those states, one electron binds to the X^0 forming an X^- and leaving behind a quasihole (QH-e^-) in the Laughlin $\nu = 1/3$ state. The X^- (with $l^*_{X^-} = 3/2$) and the QH-e^- (with $l^*_{e-} = 5/2$) have opposite charges and attract one another; what results in their excitonic dispersion. We have checked that the present interpretation remains valid at inter-layer separations d up the order of λ, when X^-'s unbind (detailed analysis of spatially separated systems will be presented elsewhere).

8. Photoluminescence

A single X^- cannot emit a photon by $e - h$ recombination and leave behind a free electron. In the simplest terms, this is because the luminescence operator conserves total angular momentum, and an X^- has $l_{X-} = S - 1$, while the electron has $l_{e-} = S$. For separated electron and hole planes, the hidden symmetry theorem does not hold, and it is possible to have weak luminescence from an X^- interacting with other charged particles. However, the luminescence intensity is much weaker than the fundamental luminescence line due to a neutral X^0. The existence of free X_k^- complexes appears to act as a trap that inhibits a strong luminescence intensity from X^0's. Observation of a strong X^- luminescence signal seems likely to be associated with excitons bound to an impurity and/or mixing of higher Landau levels. This might break the selection rule that forbids luminescence for a free X^-.

9. Speculation

The generalised CF picture will be of value if it can make predictions for systems which at the moment are too large to evaluate numerically. An example that we have not been able to study numerically is that of $N_e = 14$ and $N_h = 5$. The configuration with the largest binding energy is (viii) $4e^- + 5X^-$, but the configuration (ix) $5e^- + 3X^- + X_2^-$ is only slightly smaller in binding energy. Which of these configurations has the lowest energy at a given value of $2S$ will depend on both the binding energy and the interparticle interactions. For $2S = 36$, we can choose $m_{e-e-} = m_{X-X-} = 5$ and $m_{e-X-} = m_{e-X_2^-} = m_{X-X_2^-} = 4$. This choice satisfies all the requirements imposed by the Pauli principle and by the hard cores of the different pseudopotentials. ¿From Eq. (2), we find $2S^*_{e-} = 2S^*_{X-} = 2S^*_{X_2^-} = 4$ so that $2l^*_{e-} = 4$, $2l^*_{X-} = 2$, and $2l^*_{X_2^-} = 0$. This leads to an $L = 0$ state of configuration (ix). If it is lower in energy than the lowest state of configuration (viii), it is very probably a Laughlin incompressible state. For configuration (viii), we find that there is a quasihole in the electron shell of angular momentum $l^*_{e-} = 2$ and a pair of quasiparticles in the X^- shell of angular momentum $l^*_{X-} = 2$. This gives $L_{e-} = 2$, $L_{X-} = 1$, 3, and thus $L = 1$, 2^2, 3^2, and 4. It seems likely that the quasiparticle energy in configuration (viii) more than compensates its slightly higher binding energy and that configuration (ix) is an incompressible quantum fluid state. It is unlikely that one will be able to diagonalise the nineteen particle electron-hole system at $2S = 36$, but the nine particle systems ($4e^- + 5X^-$ and $5e^- + 3X^- + X_2^-$) might be possible.

10. Summary

Charged excitons and excitonic complexes play an important role in determining the low energy spectra of electron-hole systems in a strong magnetic field. We have introduced general Laughlin type correlations into the wavefunctions, and proposed a generalised CF picture to elucidate the angular momentum multiplets forming the lowest energy bands for different charge configurations occurring in the electron-hole system. We have found Laughlin incompressible fluid states of multi-component plasmas at particular values of the magnetic field, and the lowest bands of multiplets for various charge configurations at any value of the magnetic field. It is noteworthy that the fictitious Chern–Simons fluxes and charges of different types or colours are needed in the generalised CF model. This strongly suggests that the effective magnetic field seen by the CF's does not physically exist and that the CF picture should be regarded as a mathematical convenience rather than physical reality.

Acknowledgements. We thank P. Hawrylak and M. Potemski for helpful discussions. AW and JJQ acknowledge partial support from the Materials Research Program of Basic Energy Sciences, US Department of Energy. KSY acknowledges support from the Korea Research Foundation (Project No. 1998-001-D00305).

References

1. A. J. Shields, M. Pepper, M. Y. Simmons, and D. A. Ritchie, Phys. Rev. B **52**, 7841 (1995); G. Finkelstein, H. Shtrikman, and I. Bar-Joseph, Phys. Rev. B **53** 1709, (1996).
2. A. Wójs and P. Hawrylak, Phys. Rev. B **51** 10 880, (1995).
3. J. J. Palacios, D. Yoshioka, and A. H. MacDonald, Phys. Rev. B **54**, 2296 (1996).
4. A. Wójs, P. Hawrylak, and J. J. Quinn, Physica B **256–258**, 490 (1998); Phys. Rev. Lett. (submitted, cond-mat/9810082); A. Wójs, I. Szlufarska, K.-S. Yi, and J. J. Quinn, Phys. Rev. Lett. (submitted, cond-mat/9904395).
5. K. Kheng, R. T. Cox, Y. Merle d'Aubigne, F. Bassani, K. Saminadayar, and S. Tatarenko, Phys. Rev. Lett. **71**, 1752 (1993); H. Buhmann, L. Mansouri, J. Wang, P. H. Beton, N. Mori, M. Heini, and M. Potemski, Phys. Rev. B **51**, 7969 (1995).
6. I. V. Lerner and Yu. E. Lozovik, Sov. Phys. JETP **53**, 763 (1981); A. H. MacDonald and E. H. Rezayi, Phys. Rev. B **42**, 3224 (1990).
7. F. D. M. Haldane, Phys. Rev. Lett. **51**, 605 (1983).
8. T. T. Wu and C. N. Yang, Nucl. Phys. B **107**, 365 (1976).
9. A. H. MacDonald, E. H. Rezayi, and D. Keller, Phys. Rev. Lett. **68**, 1939 (1992).
10. X. M. Chen and J. J. Quinn, Phys. Rev. B **50**, 2354 (1994); ibid. **51**, 5578 (1995).
11. E. I. Rashba and M. E. Portnoi, Phys. Rev. Lett. **70**, 3315 (1993).
12. V. M. Apalkov, F. G. Pikus, and E. I. Rashba, Phys. Rev. B **52**, 6111 (1995).
13. B. I. Halperin, Helv. Phys. Acta **56**, 75 (1983).
14. R. B. Laughlin, Phys. Rev. Lett. **50**, 1395 (1983).
15. J. K. Jain, Phys. Rev. Lett. **63**, 199 (1989).
16. F. D. M. Haldane and E. H. Rezayi, Phys. Rev. Lett. **60**, 956 (1988).
17. A. Wójs and J. J. Quinn, Solid State Commun. **108**, 493 (1998); ibid. **110**, 45 (1999); Phys. Rev. B (submitted, cond-mat/9903145).
18. P. Sitko, S. N. Yi, K.-S. Yi, and J. J. Quinn, Phys. Rev. Lett. **76**, 3396 (1996).
19. A. de Shalit and I. Talmi, *Nuclear Shell Theory*, Academic Press, New York and London 1963.

A FAR INFRARED MODULATED PHOTOLUMINESCENCE (FIRM-PL) STUDY OF A 2-D ELECTRON GAS IN GaAs/Al$_x$Ga$_{1-x}$As HETEROJUNCTIONS AND QUANTUM WELLS

R.J. NICHOLAS, C.C. CHANG, AND V. ZHITOMIRSKIY
Physics Dept., Oxford University, Clarendon Laboratory, Parks Rd., Oxford, OX1 3PU, U.K.

Abstract. The new technique of Far InfraRed Modulated PhotoLuminescence (FIRM-PL) has been used to study the properties of a high mobility 2-D electron gas in a series of GaAs/AlGaAs heterojunctions and quantum wells. When the occupancy of the lowest Landau level is between 1 and 2 very large transfers of PL intensity are observed at the cyclotron resonance condition. The PL intensity is transferred from the E$_0$ line to E$_1$ corresponding to emission from the first two quantised electric subbands. A study of the power dependence of this signal allows us to deduce that the relative recombination efficiencies of the two lines are around 5 orders of magnitude different in such structures. This difference leads to a very high sensitivity for this technique for the observation of internal excitations within the electron system. In quantum wells signals are seen for all occupancies due to transfer between PL from higher Landau levels.

1. Introduction

Double resonance spectroscopy is a very sensitive method for the selective study of specific carriers and states in semiconductors. A good example of this is Optically Detected Cyclotron Resonance (ODCR) in which the cyclotron (or other) resonances of carriers in semiconductors is detected by the modulation produced to the interband photoluminescence (PL) at one energy by Far InfraRed (FIR) radiation [1,2,3,4]. The first reports are now appearing (Herold et al [5], Nicholas et al [6]) of a new variation on this technique based on measurement of the total PL spectrum, known as FIRM-PL (Far Infrared Modulated-PhotoLuminescence). In this paper we report a comprehensive study of this phenomenon in high mobility GaAs/AlGaAs heterojunctions, which shows that substantial modulations to PL can occur, with both enhancement and suppression and that the behaviour is strongly linked to the Landau level occupancy.

Recently, there has been intense interest in the photoluminescence properties of the two-dimensional electron systems (2DES) as a function of the occupancy of Landau levels [7-14] since the spectra of radiative recombination between the 2DES and photo-excited holes are sensitive probes of correlation in the electron ground states. In this paper, FIRM-PL has been studied on a series of GaAs/Al$_x$Ga$_{1-x}$As heterostructures with

M.L. Sadowski et al. (eds.), Optical Properties of Semiconductor Nanostructures, 33–44.

different electron concentrations. The Landau level filling factor (ν) and 2DES density are found to be very important in determining the response and modulation of the PL under illumination by FIR radiation.

2. Experimental

The samples measured in this study are a series of ultra-high-mobility GaAs/Al$_x$Ga$_{1-x}$As heterojunctions grown by MBE at the Philips Research Laboratory Redhill by C. T. Foxon and J. J. Harris. The Al fraction x was kept at 33±1%, keeping the barrier height constant about 330 meV. The GaAs substrates are followed by a 50Å period GaAs/Al$_x$Ga$_{1-x}$As superlattice to trap any residual impurities, principally carbon. 5100 Å of GaAs is grown on top of the superlattice followed by an Al$_x$Ga$_{1-x}$As spacer layer. Then a lightly doped region of Al$_x$Ga$_{1-x}$As spread over 2000 Å is grown on top of the spacer layer. The low doping level helps reduce the remote impurity scattering in the sample. Due to the spatial separation between the 2DES and the dopants, an ultra high mobility can be achieved of up to 4.6×10^6 cm^{2}V^{-1}s^{-1} at 4K. The 2DES carrier density is dependent on the thickness of the spacer layers. Photo-luminescence has already been extensively studied in this series of samples [7-10].
TABLE 1 summarises the properties of all the samples employed in this study.

TABLE 1. Sample characteristics.

Sample number	G650	G640	G646
Spacer Layer (Å)	400	1200	2400
Carrier density (cm^{-2})	2.26×10^{11}	1.17×10^{11}	7.0×10^{10}

The precise form of the potential distribution is dependent on the illumination conditions as was shown by Michels et al. [15]. Experiments were performed under conditions of low continuous above band gap illumination corresponding to the classification by Michels et al as metastable equilibrium. In this configuration the depletion field is flattened by the minimal charge remaining in the GaAs. In the Al$_x$Ga$_{1-x}$As, the conduction band is lower than the Fermi level which indicates a full occupation of the donors. Only one subband, E$_0$, is populated at all fields but the Fermi level is very close to the E$_1$ subband at zero field.

Photoluminescence was measured using a sorption pumped ^{3}He system which gave temperatures down to 330 mK in a superconducting magnet giving fields up to 14 T. A calibrated ruthenium oxide resistor is mounted at 2mm away from the sample on the mount to monitor the temperature during the measurement and give an indication of the total FIR power load on the system. The insert allows us to apply far infrared and visible light to the sample at the same time. A hole was drilled at the side of the focusing light cone to let two 600 μm fibres through. The collecting fibre is 2mm above the sample holder at an angle to eliminate the strong laser beam reflected from the sample surface. The end of excitation fibber is around 2mm further back to give a uniform illumination on the sample. PL excitation is by a diode laser with a wavelength of 690 nm, attenuated to give a typical illumination intensity of approximately 10 μW/cm^2. The luminescence signal is dispersed by a one-meter spectrometer and detected by a nitrogen cooled Si CCD camera. Typical exposure times are of order 60 seconds with automated magnet sweeps between set field values. Series of spectra are taken around the expected resonance position, and then repeated with an FIR laser applied. The FIR radiation is produced from a CO_2 pumped molecular gas laser as used previously for ODCR measurements on bulk GaAs [15].

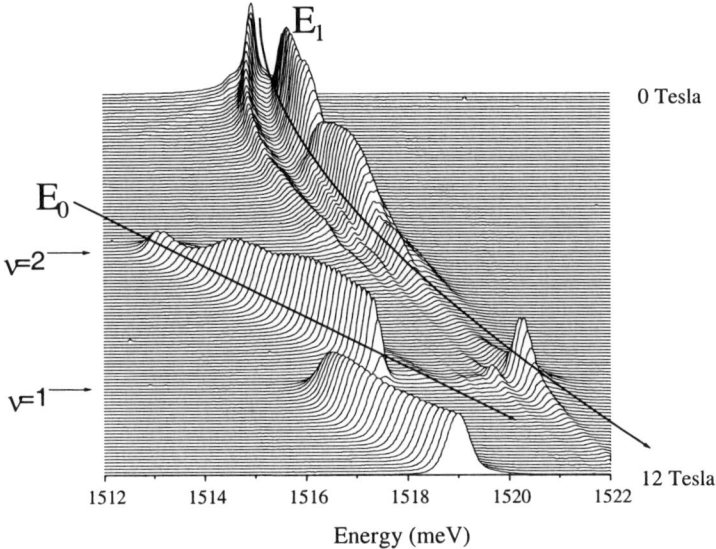

Figure 1: A waterfall plot of the photoluminescence spectra of sample G650at 340 mK for every 0.1T from 0 to 12T.

3. Overview of the photoluminescence behaviour

Fig. 1 shows a typical waterfall plot of the conventional photoluminescence spectra at 340 mK for sample G650. The very long carrier lifetime in the bulk GaAs allows the emission from the 2-D electron gas to dominate. At zero field, a single strong peak labelled E_1 can be seen at 1515 meV. At around $v=2$ (~4.5 Tesla)(v: filling factor), the luminescence begins getting weaker and is indiscernible by the time the field reaches 5 Tesla. At the same time the E_0 line appears 5 meV lower and strengthens with increasing field. At $v=1$ (~9 Tesla) a second discontinuity occurs accompanied by a brief transfer of intensity to E_1. This behaviour has been extensively studied [6],[7], and is now attributed mainly to the change over in luminescence of the 2DES from excitonic to free hole character at fields above $v=1$ [6],[11],[16]. The E_1 feature results from excitonic recombination from the higher subbands and is very close in energy to the Bulk GaAs free exciton. The luminescence transfer between E_0 and E_1 is due to the incompressible nature of the 2DES when the Landau levels are filled [7].

The dominance of E_1 at $v=2+$ is due to the fact that the lowest two Landau levels of the E_0 subband are completely filled and incompressible. Only the higher partly filled Landau levels can contribute to luminescence, but due to the fast hole relaxation process most of the photo-excited holes are in the lowest Landau level where recombination is only weakly allowed to higher electron levels [17]. For the lowest density sample G646 the separation between the electron and hole is smaller, and under these conditions the recombination is always excitonic [11].

4. Infrared modulation of the luminescence

We now discuss the influence of far infrared absorption on the photoluminescence spectrum. The perturbation can either cause simple heating effects on the luminescence - or induce resonances, or both. In both cases the infrared power level is expected to play a major role in the modulation strength. The temperature measured by the RuO resistor close to the sample is used to ensure that a similar low level of infrared power is incident on the samples for different spectral scans and FIR laser wavelengths.

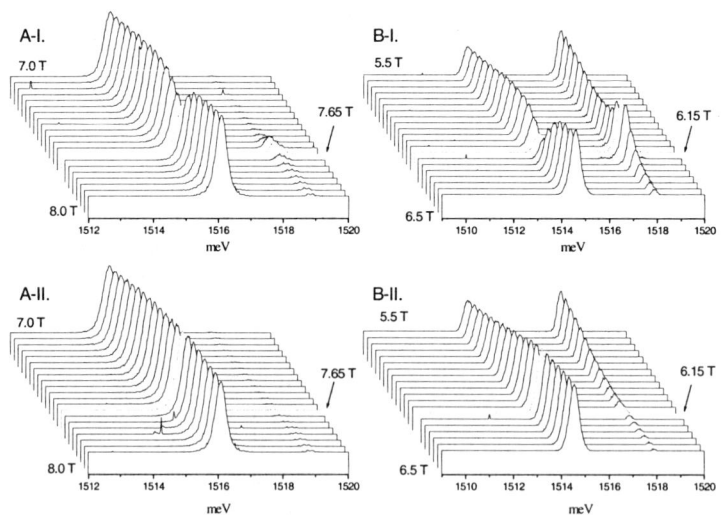

Figure 2. Photoluminescence spectra of Sample G650 with and without infrared modulation over a range of 1T around the cyclotron resonance condition for two different infrared wavelengths. A-I and A-II are measured with and without 96.52 μm radiation respectively, and B-I and B-II with and without 118.88μm.

Fig. 2. shows two series of photoluminescence spectra over a range of 1T around the energy of the electron cyclotron resonance for sample G650. PL spectra are shown both with and without infrared radiation of wavelength 96.52 μm and 118.88 μm. Significant luminescence transfers from E_0 to E_1 can be seen under the influence of the far infrared. The central magnetic field of these transfers corresponds to the cyclotron resonance field which rules out the possibility that these are just global infrared heating. The effective mass deduced from the resonances is around 0.068 m_0, in good agreement with direct cyclotron resonance [15] experiments. The resonance is very sharp with a full width of around 0.15 Tesla.

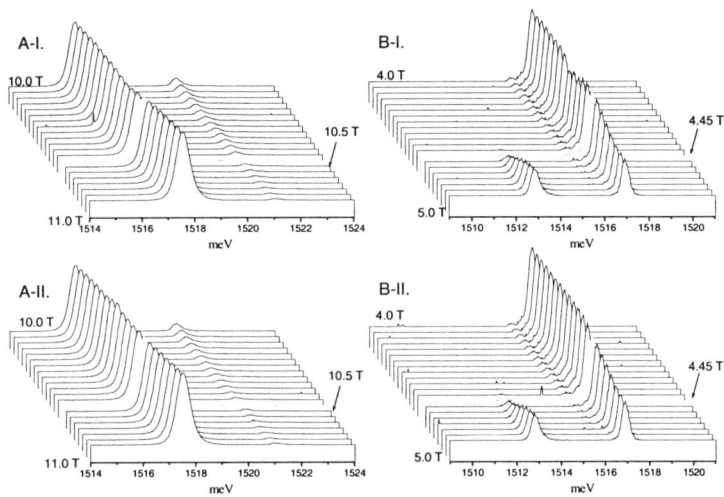

Figure 3: Photoluminescence spectra of Sample G650 as in fig. 3 for two different wavelengths. A-I and A-II are measured with and without 70.51 μm radiation respectively, and B-I and B-II with and without 163.03 μm radiation

Fig. 3 shows two further sets of photoluminescence spectra measured in the same sample with and without infrared radiation of similar intensity at wavelengths of 70.51 μm and 163.03 μm. For 70.51 μm, there is no luminescence transfer from E_0 to E_1 and only very weak modulation of the E_0 luminescence is observed. For 163.03 μm, where the E_1 luminescence dominates at its resonance field, the E_1 feature is slightly suppressed by the infrared beam without any extra feature appearing on the high energy side. Fig. 4 summarises the far infrared modulated photoluminescence of G650 in contour plots. Outside this field range, there are only weak suppressions of the major features. To confirm the dependence on filling factor rather than density alone, we then investigated sample G640 using the same set of infrared lines. With roughly the same intensity illuminating the samples, the modulations produced by the infrared are different both qualitatively and quantitatively. For 118.88 μm for example there is no luminescence transfer from E_0 to E_1 in G640 and the resonance is much weaker. This corresponds to the FIRM-PL spectra from G650 when the filling factor is less than 1.

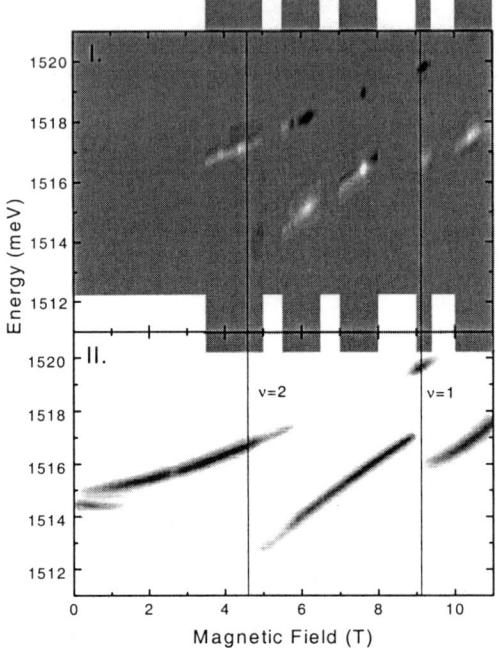

Figure 4. Contour plots of the PL intensity for sample G650, (I) shows the intensity change produced by infrared radiation at 186 μm, 163.03 μm, 118.88 μm, 96.52 μm, 80.0 μm and 70.51 μm each taken in a 1T scan about the cyclotron resonance position, mid grey represents no change, with black representing enhancement and white suppression respectively. (II) the photoluminescence intensity without infrared modulation.

Further investigation with other wavelengths confirms that the systematic dependence is based almost entirely on filling factor. This can be seen from an analysis of the absolute magnitude of the modulation amplitude for samples G650 and G640, which is shown in Fig. 5. The normalised suppression or enhancement of the intensity ($\frac{\Delta I_{E_i}}{I_{E_1} + I_{E_2}}$) is presented to give a picture of the modulation independent of the absolute PL intensity.

We can summarise the influence of the modulation made by the infrared on the photoluminescence of sample G650 and G640 with respect to filling factor.

1) For $\nu=1-$, the free hole recombination feature is weakly suppressed by applying infrared but no enhancement of the E_1 feature can be observed.
2) For $\nu=2+$, the E_1 feature is suppressed.
3) For $2 > \nu > 1$, a large transfer of intensity occurs from the E_0 excitonic feature to the E_1 feature which is strongest close to $\nu=1$.

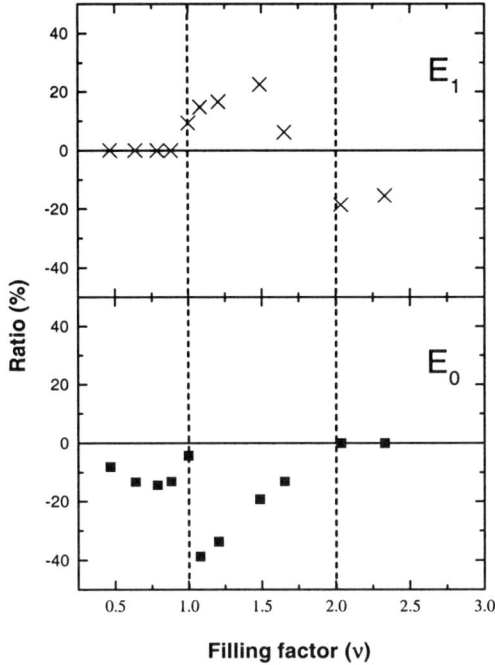

Figure 5. A summary of the far infrared modulation of the photoluminescence from samples G650 and G640 as a function of the Landau level occupancy. The plot shows the percentage change in PL intensity for each line as a function of the total luminescence at the resonance field.

Under cyclotron resonance excitation, the electrons will redistribute themselves between different Landau levels and three things can be expected from the redistribution. First, the excited electrons in higher Landau levels may recombine directly with holes and produce luminescence with higher energy than E_0. No evidence for this is found. Second, these electrons may promote E_1 luminescence by relaxing to the E_1 subband and forming excitons. Third, the lattice can be heated by absorbing the infrared directly or scattering with the hot electrons and hence promote non-radiative recombination and eventually reduce the number of holes capable of radiative recombination. Direct emission from higher Landau levels will not happen due to the short electron lifetime in higher levels. The second case is very much dependent on the relaxation process and finally the infrared heating is likely to be a function of absolute power but not individual wavelengths.

Fig. 6 illustrates the possible cyclotron resonance transitions as a function of magnetic field showing the typical filling factors and the angular momentum quantum numbers of the different levels. At $v=1-$, the only possible transition is from $n=0$ to $n=1$. The relaxation from the $n=1$ state to either $n=0$ or $n=-1$ of E_1 is an intersubband process involving both losing angular momentum and parity flipping, which is slower than the intrasubband relaxation back to the ground state of E_0 [17]. No enhancement of E_1 is therefore expected and so we expect only a weak suppression of E_0 luminescence due to the lower E_0 ground level population giving holes a better chance to recombine non-radiatively. For filling factors between 2 and 1 (2>v>1), two cyclotron transitions are possible – one from a completely filled $n=0$ level and one from a partly filled $n=-1$ level.

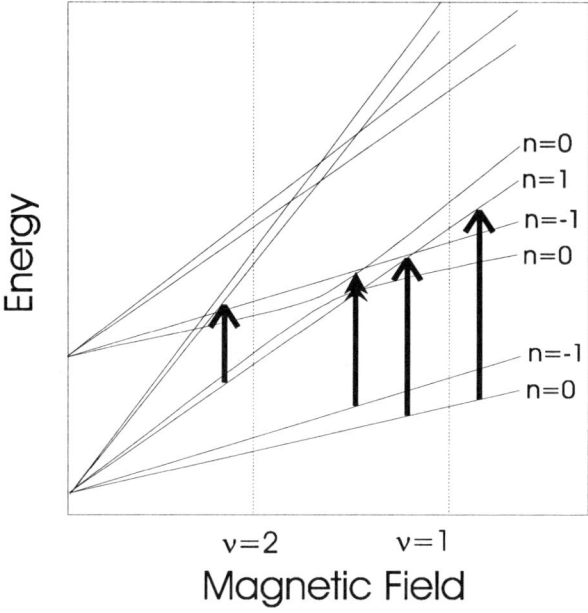

Figure 6. A schematic diagram to illustrate the possible cyclotron resonance transitions for both the E_0 and E_1 electron subbands. n shows the total angular momentum quantum number for the spin resolved levels.

Since the spin-down $n=0$ upper state in E_0 has the same parity as the spin-up $n=0$ ground state in E_1, there is likely to be substantial mixing of these states caused by any misorientation of the field relative to the surface normal and coupling to hole levels. The signature of this band mixing is the anti-crossing between the Landau fans as shown in Fig. 6. With a fixed number of holes available for recombination, it is the ratio between the E_1 and E_0 recombination probability that matters, which is a function of both wavefunction overlap and carrier density. A *k.p* calculation of G650 at 118.88 μm with a subband separation (E_1–E_0) of 15 meV gives a band mixing around 0.5% even for perfect magnetic field alignment due to the presence of far band mixing terms [18]. In other words, 0.5% of the electrons in the spin-down $n=0$ upper state in E_0 can be counted as E_1 electrons. Although this mixing is small, it can still make a significant contribution to the E_1 luminescence due to the larger wavefunction overlap with holes than that of E_0. At v=2+, the final states of the cyclotron transitions have quantum numbers n= 1 and 2 and mixing with the E_1 levels is again weak. The main contributions to luminescence in this regime are electrons photo-excited by the visible laser but not in the 2DES. These electrons are pumped to higher Landau levels of the E_1 subband when in resonance with the infrared photons leading to fewer electrons in the ground state of the E_1 subband and a suppression of luminescence. In summary therefore we suggest that the large PL transfer between 2>v>1 is due to the combination of subband mixing and the specific transitions which are excited in this regime.

The lowest density sample G646 was also studied. The photoluminescence spectrum of this sample is always excitonic due to the small electron-hole separation caused by the lower surface electric field [11]. A small decrease in intensity is observed (<5%) at all filling factors. This is similar to what happens for G650 and G640 at v=2+ where the PL is dominated by excitonic recombination from E_1.

5. Power dependence of the FIRM-PL signal

The relation between infrared power and modulation was studied in sample G650 at resonance with a wavelength 118.88 μm, corresponding to an occupancy v =1.49. This is the strongest line available from our far infrared laser with a maximum output power around 10 mW. After the attenuation by optical windows and light pipes, we estimate a transmission of 10% giving ~1mW incident on the sample. The dependence on power is studied using calibrated absorbers.

Fig. 7 illustrates the PL spectra at the resonance field (6.15 T) under different levels of infrared illumination. With the full power from the infrared laser the luminescence transfer from E_0 to E_1 is almost total, leaving very little signal from E_0. The modulation then decreases as less power is used. With the full power from the infrared laser the sample temperature also rises to 580mK, however separate studies of the temperature dependence of the PL show that this amount of heating alone only causes a very small (a few %) shift in PL from E_0 to E_1. To understand the dependence on FIR power quantitatively, the modulation has been normalised against the total luminescence without far infrared, as shown in Fig. 8(A). The suppression or enhancement $\dfrac{\Delta I_{E_i}(W)}{I_{E_1}(0) + I_{E_2}(0)}$ is highly non-linear with infrared power. The modulation

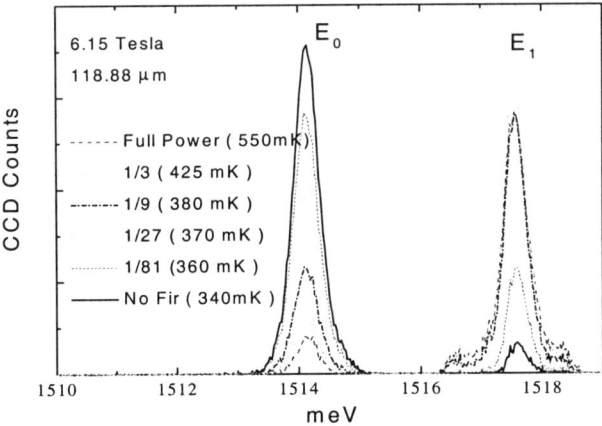

Figure 7. A series of PL spectra for sample G650 at the magnetic field corresponding to cyclotron resonance with 118.88 μm for a series of different FIR illumination powers.

reaches its plateau at around 25% of full power. At this point the E_1 luminescence is five times stronger than the E_0 luminescence. Fig.8(B) shows the relative contributions from the E_0 and E_1 lines against power, which also shows a saturation close to 25% of full power. Since the number of holes is determined by the PL photo-excitation process and is almost independent of the cyclotron resonance, the E_1/E_0 ratio results from a competition between the two recombination processes. The recombination strength is proportional to (transition probability) and (population) so we may write the

ratio $\dfrac{I_{E_1}}{I_{E_0} + I_{E_1}}$ as

$$\frac{I_{E_1}}{I_{E_0} + I_{E_1}} = \frac{1}{1 + \dfrac{n_0}{(n_1 + n_{HL}X)}\dfrac{P_0}{P_1}} \tag{1}$$

where n_0 and P_0 represent the carrier density and transition probability for E_0, n_1 and P_1 are the E_1 carrier density without infrared pumping and the transition probability. n_{HL} is the carrier density of the higher Landau level under cyclotron excitation in the E_0 subband and the factor X is defined as the proportion of n_{HL} being mixed into the E_1 subband due to coupling. With the full power of the infrared laser illuminating an area of 4mm^2, n_{HL} is about 3×10^9 cm^{-2} assuming a lifetime of around 1 ns [19] and 50% of the photons are absorbed.

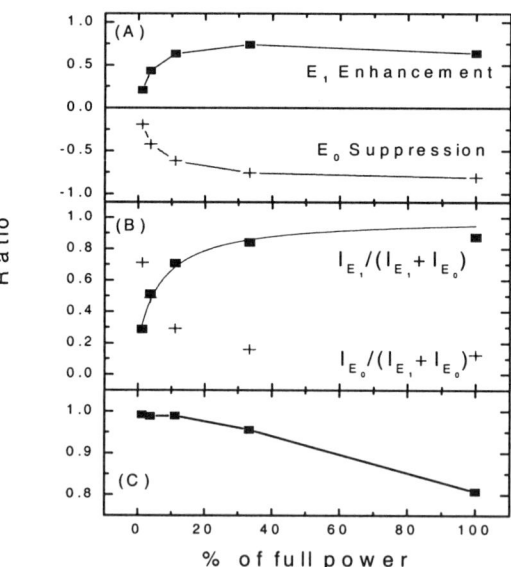

Figure 8. The PL modulation as a function of the FIR power level at 6.15T. (A) shows the ratio of the intensity change of the PL from the E_1 and E_0 levels compared to the total PL intensity without infrared, (B) shows the relative proportions of PL coming from E_1 and E_0 as a function of FIR intensity. The solid line is a fit using the calculation described in the text and (C) shows the total PL intensity change as a function of FIR intensity.

If n_{HL} is linear with the infrared power, which is a sensible assumption, we can then fit the experimental data with Eq. (6.4). Fig. 8 (B) shows the curve that fits best the $\frac{I_{E_1}}{I_{E_0}+I_{E_1}}$ trace, giving $\frac{XP_1}{P_0}$ around 1.7×10^3. Assuming a mixing (X) of order 0.5% gives a ration P_1/P_0 of order 3×10^5. This demonstrates that the efficiency of the E_1 recombination is very substantially larger than for E_0 due to the much greater electron-hole overlap. It is the large difference in the two recombination rates which gives rise to the very high sensitivity of the FIRM-PL signal.

The small decrease in the total photoluminescence level shown in fig. 8(C) ($I_{E1}+I_{E0}$) is caused by an increase in the non-radiative recombination rate. This is probably due to the increased total population of E1 which is very close to the bulk band edge. In consequence the probability of bulk non-radiative recombination increases.

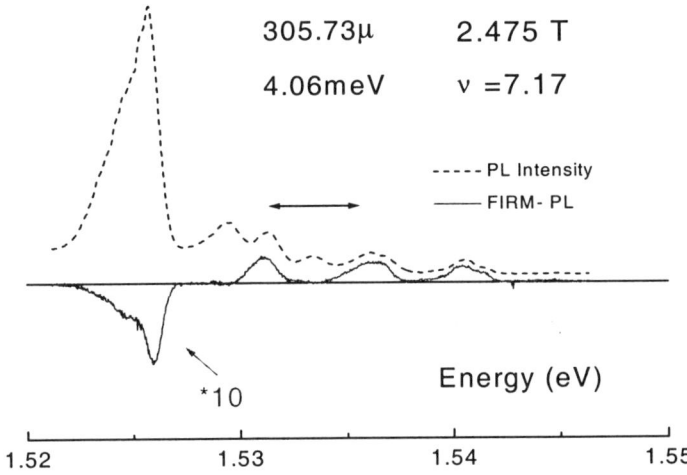

Figure 9. PL and FIRM-PL signals from a GaAs-GaAlAs single quantum well at the cyclotron resonance magnetic field for 305 μm radiation. The length of the arrow shows the cyclotron energy

Further evidence for the concept that the sensitivity of the FIRM-PL signal is strongly enhanced when there are competing recombination mechanisms comes from preliminary measurements on single side doped quantum wells, shown in Fig. 9. This shows a PL and FIRM-PL trace for a low field (high occupancy). The PL is all from the lowest quantum well level but transitions can be seen from many higher Landau levels as well as the main line from the lowest Landau level. In this case there is a clear transfer of intensity into the higher Landau levels associated with the cyclotron resonance transition.

6. Conclusions

We have demonstrated that FIRM-PL is an interesting new tool to investigate the properties of a 2-D electron gas in high magnetic fields. We have shown that the modulation of the PL is very strongly dependent on factors such as the level occupancy and the relative recombination rates for different PL channels. We expect that the selectivity and enhanced sensitivity of this technique will offer a new range of possibilities for investigating the internal transitions and excitations of semiconductor heterostructures.

Acknowledgements. We would like to thank J.J. Harris and C.T. Foxon for the provision of the high mobility GaAs/AlGaAs heterojunction samples, and EPSRC for financial support.

References

1. M.G. Wright et al, Semicond. Sci. and Technol. **5** (1990) 438
2. R.J. Warburton, J.G. Michels, R.J. Nicholas, J.J. Harris and C.T. Foxon, Phys. Rev. B **46,** 13394 (1992).
3. J.G. Michels, R.J. Warburton, R.J. Nicholas and C.R. Stanley Semicon. Sci. and Technol., **9** , 198 (1994).
4. J. Kono et al, Phys. Rev. B **52,** R8654 (1995).
5. G.S. Herold, H.A. Nickel, J.G. Tischler, B.A. Weinstein and B.D. McCombe, Physica E **2** 39 (1998).
6. R.J. Nicholas, D. Kinder, A.N. Priest, C.C. Chang, H.H. Cheng, J.J. Harris and C.T. Foxon, Physica B **249-251,** 553 (1998).
7. A. J. Turberfield, S. R. Haynes, P. A. Wright, R. A. Ford, R. G. Clark, J. F. Ryan, J. J. Harris and C. T. Foxon, Phys. Rev. Lett. **65,** 637 (1990).
8. A. J. Turberfield, R. A. Ford, I. N. Harris, J. F. Ryan, C. T. Foxon and J. J. Harris, Phys. Rev. B **47,** 4794 (1993).
9. H.D.M. Davies, J.C. Harris, J.F. Ryan, and A.J. Turberfield, Phys. Rev. Lett. **78** 4095 (1998).
10. R.J. Nicholas, H.H. Cheng and A.N. Priest, in High Magnetic Fields in the Physics of Semiconductors, (World Scientific, Singapore, 1997), 527
11. J. L. Osbourne, A. J. Shields, M. Y. Simmons, N. R. Cooper, D. A. Ritchie and M. Pepper, Phys. Rev. B **58,** R4227-R4230, (1998)
12. E. H. Aifer, B. B. Goldberg and D. A. Broido, Phys. Rev. Lett. **76,** 680 (1996).
13. B. B. Goldberg, D. Heimann, A. Pinczuk, L. Pfeiffer and K. West, Phys. Rev. Lett. **65,** 641 (1990).
14. D. Heiman, B. B. Goldberg, A. Pinczuk, C. W. Tu, A. C. Gossard and J. H. English, Phys. Rev. Lett. **61,** 605 (1988).
15. J.G. Michels, R. J. Nicholas, G. M. Summers, D. M. Symons, C. T. Foxon and J. J. Harris, Phys. Rev. B **52,** 2688 (1995).
16. N. R. Cooper and D. B. Chklovskii, Phys. Rev. B **55,** 2436 (1997).
17. R. Ferreira and G. Bastard, Phys. Rev. B **40,** 1074 (1989).
18. R.J. Nicholas, T.A. Vaughan, C.C. Chang and A.J.L. Poulter, to be published (1999)
19. I. Maran, W. Seidenbusch, E. Gornik, G. Weimann and M. Shayegan, Semicond. Sci. Technol. **9,** 700 (1994).

MAGNETO-OPTICAL PROPERTIES AND POTENTIAL FLUCTUATIONS IN HIGH MOBILITY 2D ELECTRON GAS

GERARD MARTINEZ
Grenoble High Magnetic Field Laboratory, MPI-FKF and CNRS
B.P. 166, Grenoble, France

Abstract. I address the question of the relation between potential fluctuations and optical properties in high mobility 2D electron gas under high magnetic fields. In view of recent results on the magnetisation of these systems, I first try to evaluate the density of localised states attached to Landau levels and show that it is significantly lower than usually assumed. It will then be demonstrated that they however play a major role in the relaxation mechanisms between Landau levels and between different electric subbands. Finally it will be shown that there are still some unexplained observations in the intensity and linewidths of the magneto-optical recombination of a 2D gas: their possible relation with the presence of disorder is discussed.

1. Potential fluctuations in high mobility 2D gas without magnetic field

Let us first try to evaluate how potential fluctuations could affect the density of states of a 2D gas. In a pure 2D gas the density of states $G(E)$ is constant and the energy dispersion is parabolic (Fig. 1). Classically, one gets the relations: $G(E) = G_0 = \dfrac{m^*}{\pi \hbar^2}$

and $E(k) = \dfrac{\hbar^2 k^2}{2m^*}$. For a 2D carrier density n_S, the Fermi level E_F is such that

$n_S = E_F\, G_0 \times 2$. If one takes, as an example, the case of GaAs with $m^* = 0.067\ m_0$, $G_0 = 0.278 \times 10^{11}$ meV/cm^2 and for $n_S = 2 \times 10^{11}$/cm^2 one gets $E_F = 3.6$ meV and $k_F = 0.73\ 10^4$ cm^{-1} whereas for $n_S = 2 \times 10^{12}$/cm^2 one gets $E_F = 36$ meV and $k_F = 2.31\ 10^4$ cm^{-1}.

In high mobility samples the only source of potential fluctuations is that due to the remote ionised donors. Assuming a δ doped structure at a distance (spacer) d (Fig. 2) from the 2D layer with a concentration N_D^+ one has $N_D^+ = n_S$. The distribution of these centres is assumed to be Gaussian and they give rise to fluctuating potentials $\Phi(r)$, which are screened by the 2D electrons. Following Ando et al. [1], the screening vector q_S is given by:

M.L. Sadowski et al. (eds.), Optical Properties of Semiconductor Nanostructures, 45–63.
© 2000 *Kluwer Academic Publishers. Printed in the Netherlands.*

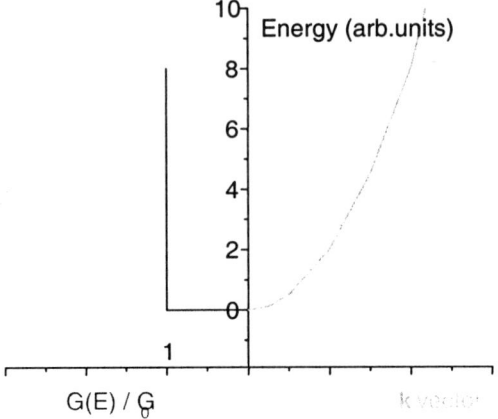

Figure 1. Schematic diagram of the density of states (left) and of the dispersion (right) of a 2D electron gas.

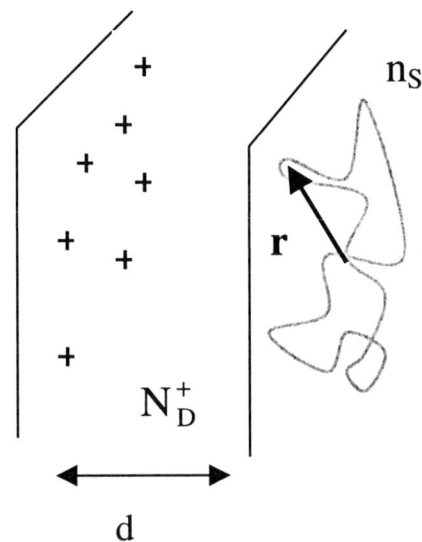

Figure 2. Diagram for the remote impurity scattering.

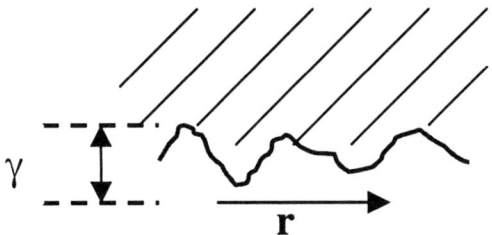

Figure 3. Energy scheme for scattering in the linear scattering regime.

$$q_S = \frac{2\pi e^2}{\kappa} \frac{dn_S}{dE_F} = \frac{2\pi e^2}{\kappa} G_0 \times 2$$ and does not depend on the energy. For GaAs where κ,

the dielectric constant is equal to 12, $q_S = 4.198 \times 10^6$ cm^{-1} and $q_S^{-1} = 23.8 \times 10^8$ cm. Therefore for a typical spacer width d = 40 nm one gets $q_S d = 16.8 \gg 1$.

For high mobility samples, the case to be considered is the one corresponding to the linear screening theory [2]. This corresponds to a mean fluctuating potential energy $\gamma \ll E_F$ (Fig. 3). In this case $nS(r) = G_0 \times (E_F + V(r))$ with $V(r) = -e\Phi(r)$ and $\gamma^2 = \langle V(r)V(r') \rangle$ is averaged over the impurity distribution (assumed to be Gaussian). Following Efros [3],

$$V(r) = \frac{2\pi e^2}{\kappa} \int C_q \exp(iq.r - qd) \frac{dq}{q + q_S} \text{ with } \langle C_q C_{q'} \rangle \propto \delta(q - q').$$

When $q_S d \gg 1$, one gets [3] $\quad \gamma = \frac{\sqrt{2\pi} \, e^2 \sqrt{N_D^+}}{2\kappa (q_S d)}$

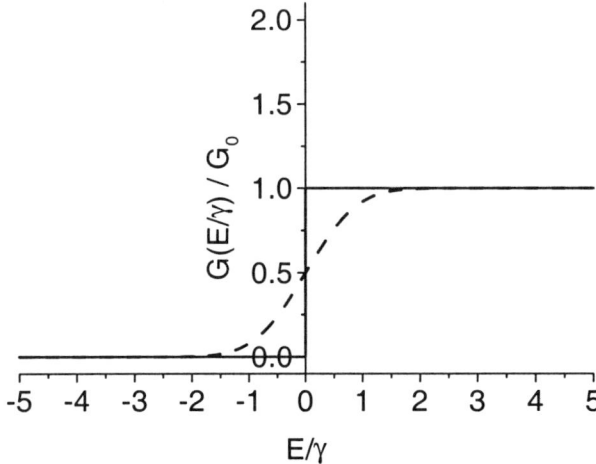

Figure 4. Density of scattering states for a 2D system in the linear scattering regime.

For instance in GaAs (d = 40 nm) this corresponds to $\gamma = 0.4$ meV ($E_F/\gamma = 9.1$) for $n_S = 2 \times 10^{11}/cm^2$ and to $\gamma = 1.26$ meV ($E_F/\gamma = 28.6$) for $n_S = 2 \times 10^{12}/cm^2$.

Therefore the linear screening theory is only valid for electron concentrations $\geq 10^{11}$ cm^{-2} which will be the case considered in the following. When this holds the density of states is given by [2]:

$$G(E) = G_0 \int\limits_{E > V(r)} \frac{dV(r)}{dE} dr = G_0 \int\limits_{-\infty}^{E} F(V)dV$$

with $F(V) = \dfrac{1}{\gamma\sqrt{\pi}} \exp(-\dfrac{V^2}{\gamma^2})$ a consequence of the Gaussian distribution of impurities.

As a result, one obtains $G(\dfrac{E}{\gamma}) = G_0 \dfrac{1}{2} (1 + erf(\dfrac{E}{\gamma}))$. This result, only valid when the linear screening is effective, gives rise to a softening of the density of states around the threshold as illustrated in Fig. 4. One clearly sees that the fluctuations are essentially affecting the energy range where $E/\gamma < 2$. This corresponds to a k-vector $k_{imp} \approx k_F$ and characteristic sizes of fluctuations $\lambda \gg a_H$ (the magnetic length) in general.

In GaAs when $n_S = 2 \times 10^{11}/cm^2$, $\gamma = 0.4$ meV; $k_{imp} = 1.1 \times 10^4$ cm^{-1}; $\lambda \geq 4.1 \times 10^{-4}$ cm and when $n_S = 2 \times 10^{12}/cm^2$; $\gamma = 1.26$ meV; $k_{imp} = 6.4 \times 10^4$ cm^{-1}; $\lambda \geq 2.3 \times 10^{-4}$ cm.

2. Potential fluctuations in high mobility 2D gas with magnetic field

The question to address in this case is: what is the expected disorder-induced density of states attached to each Landau level?

One usually represents the effect of magnetic field on the 2D density of states in the way sketched in Fig. 5. This picture is correct as far as the degeneracy ($G_B = G_0\hbar\omega_C = 2B/\Phi_0$) of the Landau level (LL) is concerned but it is misleading in the sense that the wavefunctions of the LL are not made with those states which are visualised. This can be quantified in the following way: in the symmetric gauge the LL wavefunction writes:

$$\Phi_{n,m}(r,\varphi) = \left[\frac{n!}{2\pi a_H^2\, 2^m (n + |m|)!} \right]^{\frac{1}{2}} e^{-im\varphi}\, e^{-\left(\frac{r}{2a_H}\right)^2} \left(\frac{r}{a_H}\right)^m L_n^{|m|}\left(\frac{r^2}{2a_H^2}\right)$$

where $a_H = \sqrt{\hbar c /(eB)}$ is the magnetic length. One can project this function on the basis $\dfrac{1}{\sqrt{S}} e^{i\,k.r}$ such that $\Phi_{n,m}(r,\varphi) = \dfrac{1}{\sqrt{S}} \sum\limits_k b_{n,m}(k) e^{i\,k.r}$ expression in which $b_{n,m}(k)$ measures the "weight" of each plane wave $e^{i\,k.r}$ for each LL(n,m).

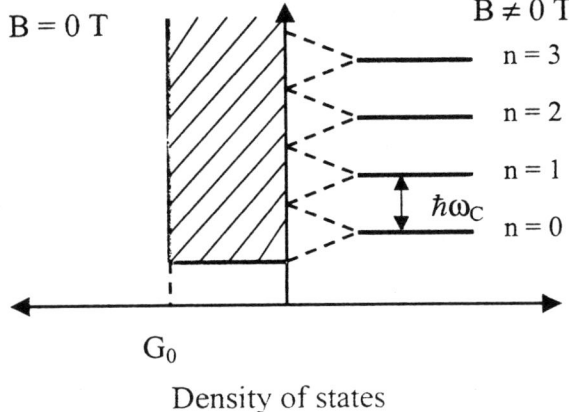

$B = 0\ T$ $B \neq 0\ T$

$n = 3$

$n = 2$

$n = 1$

$\hbar\omega_C$ $n = 0$

G_0

Density of states

Figure 5:Density of states for a 2D gas without (left) and with (right) magnetic field.

Setting $x = k\, a_H$ and taking S as unity, one finds for instance for the first LL (n=0):

$$b_{0,m}(k) = \varepsilon \left[\frac{2\pi\, 2^{m+2}}{m!} \right]^{\frac{1}{2}} a_H\, x^m\, e^{-x^2} \qquad \text{where}$$

$$\varepsilon = \begin{cases} (-1)^{m/2} & \text{if } m = 2p \\ i\,(-1)^{(m+1)/2} & \text{if } m = 2p+1 \end{cases}$$

One has of course the relations $\sum_k \left| b_{n,m}(k) \right|^2 = 1$ (normalisation of the

wavefunction) and $\sum_{k,m} \left| b_{n,m}(k) \right|^2 = G_B = \dfrac{1}{\pi a_H^2}$.

For a single impurity it is easy to show that the maximum of $b_{0,m}$ occurs for $x^2 = k^2 a_H^2 = m/2$ which implies that states influenced by the potential fluctuations ($k_{imp} \geq k$) correspond to small values of m and that in general the higher the LL index the larger is the relative contribution of perturbed states. This conclusion is not necessarily valid in real cases where the mean distance between potential fluctuations $d_f \approx 1/\sqrt{N_D^+}$ is not much bigger than $a_H\,(cm) = 2.56 \times 10^{-6} / \sqrt{B(T)}$. For instance in GaAs, for $n_S = 2$ x $10^{11}/cm^2$, $d_f = 2$ x 10^{-5} cm and for $n_S = 2$ x $10^{12}/cm^2$, $d_f = 7$ x 10^{-6} cm. More elaborate theories are therefore necessary to take into account the actual distribution of impurities but they should result in a large proportion of free states even at moderate fields for the lower LL which is not apparently the case [4]. Therefore one needs experiments to clarify the situation.

Wiegers et al. [5] have recently reported on magnetisation experiments performed on single heterojunctions with high mobility 2D gases using a very sensitive torque-meter. The results are displayed in Fig. 6 and Fig. 7. In both samples, one observes a saw-tooth behaviour of the magnetisation, with abrupt jumps at even filling factors when the Fermi level scans the localised states between Landau levels (Fig. 7). This is the expected variation of the magnetisation as predicted a long time ago by Peierls [6]. In such experiments, when the field is increased by δB, the number of states which are swept is $\nu\delta(G_B)$. Therefore the number of states scanned in the cyclotron gap $N_{Gap} = \nu\delta(B)G_B/B = G_B\delta(\nu)$, where $\delta(\nu)$ is the increment of ν during the jump of the Fermi level at even values of ν. One can then deduce the relative density of localised states N_{Gap}/G_B. This is represented in Fig. 8 for both samples. In the lower part of the figure is also plotted the relative concentration of gap-states with respect to the total carrier density. This variation reveals a rather constant and weak impurity density of states between LL, which is of course sample dependent. The lower value found for the higher doped sample is due to its particular structure [7]. For all samples we find a relative density of states much weaker than that predicted by current theories [4]. Therefore, one expects in the quantum limit to have a number of localised states not larger than a few percent of the carrier concentration, which implies that the major part of 2D electrons are essentially free!

Despite their relative small number, are these potential fluctuations playing a significant role in the optical properties of 2D gases?

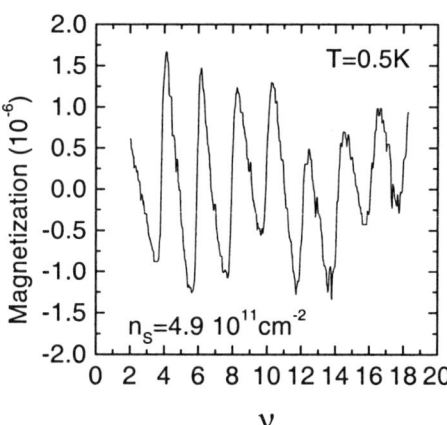

Figure 6. Magnetisation of a 2D gas in a single heterojunction.

3. Phonon–impurity assisted relaxation mechanism in 2D gases

There is at least one known case where the potential fluctuations play a key role in the optical properties of 2D gas. This was reported by Hartmann et al. [8] and concerns the relaxation mechanism inside a heterojunction under illumination. The structure of the sample is shown in Fig. 9. The sample contains a δ-doped layer of acceptors inside the

GaAs wide well, far away from the interface, and this system is known to be well adapted for optical measurements of all the correlated properties of the 2D gas [9]. For constant laser power illumination the quasi equilibrium electron concentration is constant. The initial state of the recombination, which is strongly spin polarised, involves an impurity in the neutral state and this recombination corresponds to a time scale much longer than that of other relaxation processes. Typical luminescence spectra are displayed in Fig. 10.

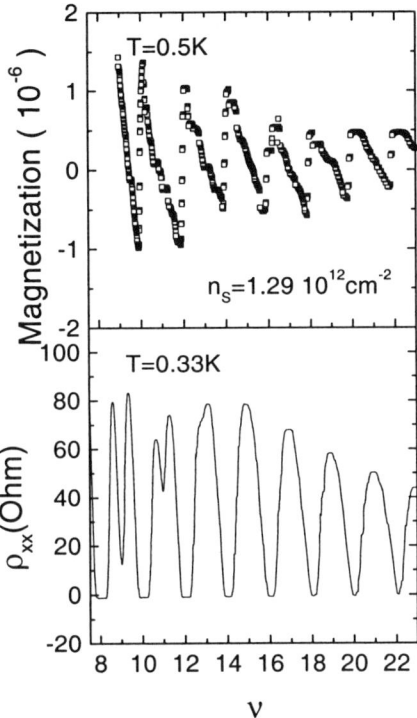

Figure 7. Magnetisation (upper part) and resistivity (lower part) of a 2D gas in a single quantum well.

One observes a contribution of the second electric subband with an intensity that fluctuates strongly with the magnetic field. We want to understand the origin of these fluctuations and in particular to interpret the variation of the ratio of intensities of the two contributions originating from the two electric subbands at N=0. This variation is displayed in Fig. 11. The ratio of intensities is written as:

$$\frac{I_{20}}{I_{10}} = \frac{n_{20}}{n_{10}} \times \frac{|M_{20}|^2}{|M_{10}|^2}$$

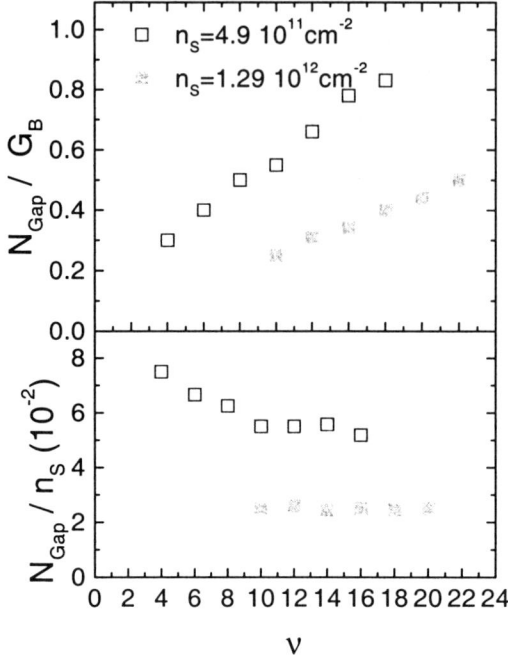

Figure 8. Relative magnitude of localised states with respect to the degeneracy of the Landau level (upper part) and with respect to the total number of carriers (lower part) as a function of the filling factor.

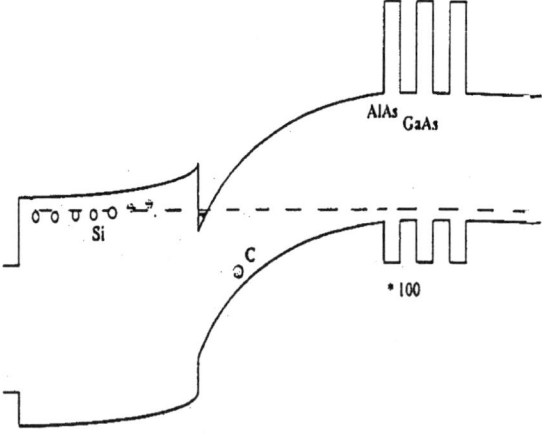

Figure 9. Schematic level structure of the sample for the study of the electron-bound hole recombination.

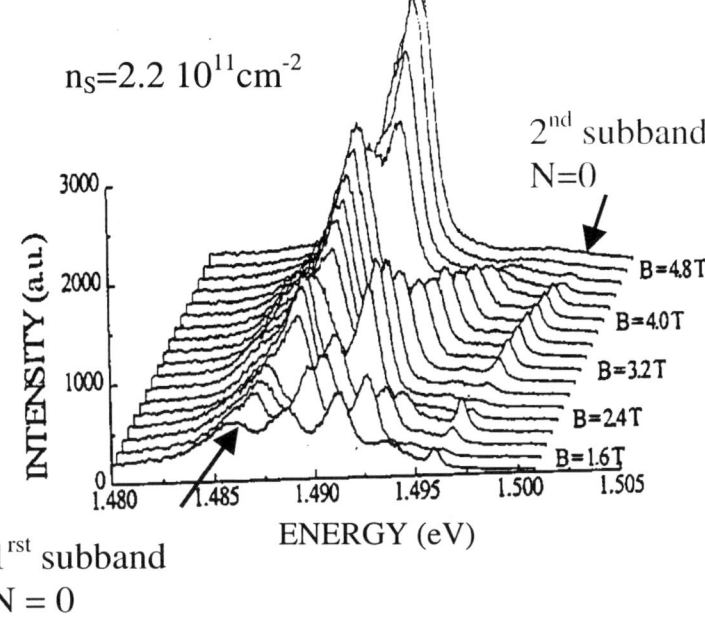

Figure 10. Luminescence spectra for different values of the magnetic field.

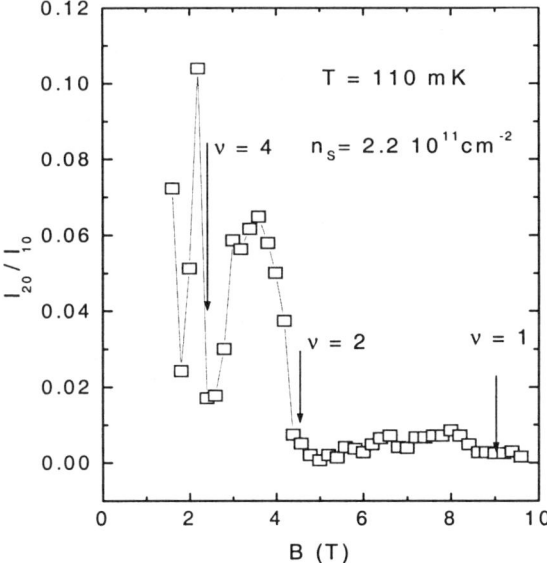

Figure 11. Ratio of intensities relative to the N=0 Landau level of the two lower electric subbands.

54

where n_{20} and n_{10} are the population of the Landau levels N=0 and the ratio of optical matrix elements is easily calculated with the Fang-Howard model for the electron wavefunction [10]. Knowing that for $v > 1$ n_{10} is equal to the degeneracy of the LL and for $v < 1$, $n_{10} = n_S$, one can get directly the variation of n_{20} with B (Fig. 12). As expected $n_{20} \ll n_S$ and fluctuates not only for large values of υ but also for $\upsilon < 2$. Knowing the absolute values of n_{20} allows us to solve exactly the rate equations provided one is able to evaluate the different relaxation rates in the system.

The situation is complicated because one has to take into account all relaxation mechanisms inside the Landau levels as schematically shown in Fig. 13. One has to deal with spin relaxation times $\tau_{S\,ij}$ within LL and phonon assisted relaxation times between LL. For instance the solution of the rate equations at $v < 2$ is [10]:

$$n_{20} = \frac{n_S}{2\tau_{R1}} \frac{(2-v)/\tau_{20-10} + 1/\tau_{20-11} + 2/\tau_{S20}}{(2-v)/\tau_{20-10} + 1/\tau_{20-11} + 1/\tau_{S20}} \frac{1}{\alpha/\tau_{20-10} + 1/\tau_{20-11} + 1/\tau_{R2}}$$

where τ_{R1} and τ_{R2} are the radiative recombination times towards the neutral acceptor of electrons in the N_{10} and N_{20} states respectively. $\alpha = 1 - v_{10}$ is the measure of the number of places made available for the relaxation in N_{10}. τ_{R1} ($\approx 200-300$ ns) and τ_{R2} ($\approx 20-30$ ns) are calculated [10] and results correspond very well to measured values obtained on similar structures by Dite et al.[11]. It remains to calculate spin and phonon assisted relaxation times.

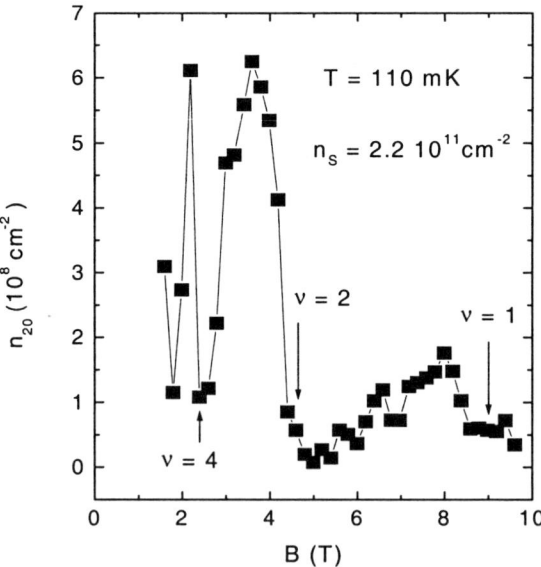

Figure 12. Variation of the population of the N=0 Landau level of the second electric subband.

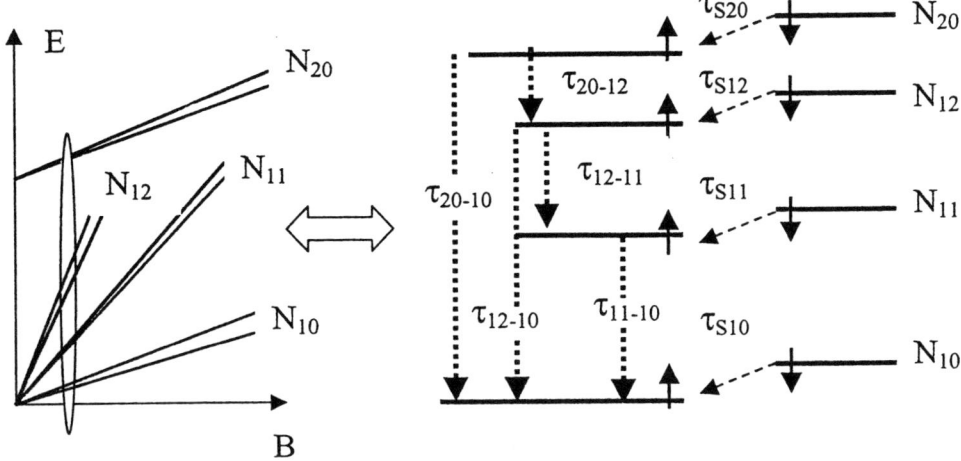

Figure 13. Diagram illustrating the different relaxation mechanisms governing the relaxation of excited carriers.

3.1. SPIN RELAXATION RATES

The spin relaxation times can be evaluated using the Dyakonov-Perel mechanism with all known parameters for GaAs along the lines developed by Bastard [12]. Results in the present case [10] are shown in Fig. 14. The spin flip relaxation times are of the order of nanoseconds and decrease as expected with the increasing Zeeman splitting at high fields.

The comparison with experimental results reported by Timofeev et al. [13] at 5T on similar structures shows that this mechanism is probably realistic.

3.2. PHONON ASSISTED RELAXATION TIMES

In GaAs and for energies lower than 10 meV, there are two kinds of processes that favour the acoustic phonon assisted relaxation between the LLs. It turns out that the deformation potential mechanism is more efficient than the piezoelectric potential one [10]. In any case they are exponentially dependent on the energy difference between the levels as clearly shown in Fig. 15. There is an abrupt decrease of W_{20-12} and W_{20-11} when the LL attached to the lower electric subband crosses the N=0 LL of the second electric subband. This explains the oscillations observed at low fields.

Results obtained at fields higher than 6 T (Fig. 12) clearly show that we need a relaxation rate in the range of 10^9 s^{-1}, at least three orders of magnitude higher than the calculated value of W_{20-10}.

One clearly needs a new mechanism to explain the results.

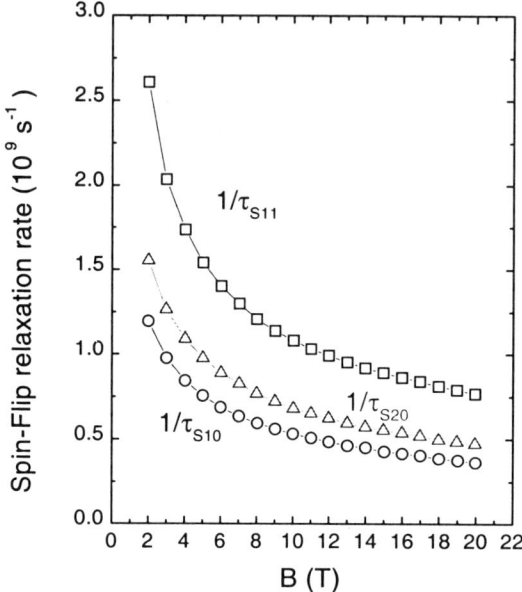

Figure 14. Spin-flip relaxation rates as a function of the magnetic field. The black hexagon corresponds to the experimental result (see text).

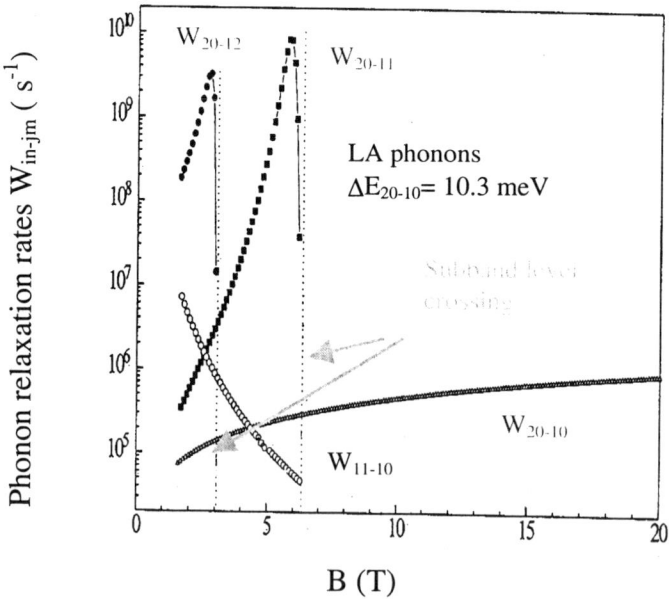

Figure 15. Phonon relaxation rates calculated with the deformation potential mechanism.

3.3 PHONON-IMPURITY RELAXATION MECHANISM

The deficiency observed in the previous analysis is due to the fact that the large energy (about 10 meV) required to relax at high fields corresponds to acoustic phonons with a large q vector, which the intrinsic system cannot provide. On the other hand potential fluctuations can provide this q and the idea is to analyse the possibility of having a combined relaxation implying both acoustical phonons and potential fluctuations [8] in a process sketched in Fig. 16.

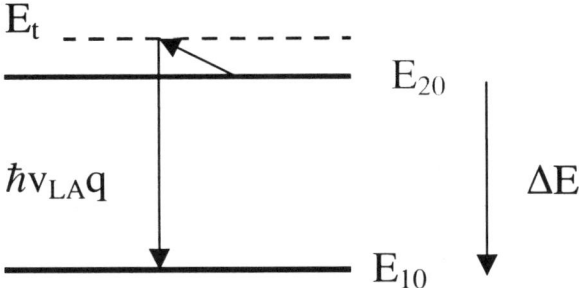

Figure 16. Schematic diagram for the phonon-impurity assisted relaxation mechanism.

The relaxation rate of such a process can be evaluated within the second order perturbation theory:

$$W_{20-10}^{i-p} = \frac{2\pi}{\hbar} \int \frac{d^2k}{(2\pi)^2} \int \frac{d^3q}{(2\pi)^3} \sum_{\mu} \left| M^{(2)} \right|^2 \delta \left(\Delta E - \hbar v_{LA} q \pm \Gamma \right)$$

where $M^{(2)}$ is the second order matrix element given by:

$$M^{(2)} = A_\mu (k) \sqrt{B(q)} \sum_t \frac{\langle i | e^{-ikz} e^{ik.r} | t \rangle \langle t | e^{iq.R} | f \rangle}{E_i - E_t}$$

with $A_\mu(k)$ and $B(q)$ being the Fourier transforms of the screened impurity potential (within the RPA) and the electron-phonon deformation potential, respectively [14].

$A_\mu (k) = \frac{2\pi e^2}{\varepsilon_0} \left[k + \tilde{F}(k) \right]^{-1} e^{kz_\mu}$ with $\tilde{F}(k)$ having a known analytic expression in the

Fang-Howard model [1], and z_μ denotes the position of the ionised impurities assumed to be located in a single plane. This distance corresponds to the spacer thickness d as defined above. In the spirit of the self consistent Born approximation [1], an empirical approach consists in assuming a mean value Γ for E_i - E_t which will have to be fitted.

The result is the following:

$$W_{20-10}^{i-p} = \frac{3(2\pi)^4 a_{nk}^2 N_\mu}{2\hbar\rho v_{LA}^2} \frac{1}{\Gamma^2} \left(\frac{e^2}{\varepsilon_0}\right)^2 \left(\frac{\omega}{v_{LA}^2}\right)^5 \int\limits_0^\infty dy \frac{y e^{2bz_\mu y}}{\left[y + \tilde{F}(y)\right]^2} \frac{\left(1+3y^2\right)^2}{(1+y)^{10}} e^{-\frac{b^2 a_H^2 y}{2}}$$

$$\times \int\limits_0^1 du\, e^{-\left(\frac{\omega}{v_{LA}}\right)^2 \frac{a_H^2 u}{2}} \frac{\sqrt{1-u}}{\left[1+\left(\omega/(bv_{LA})\right)^2 (1-u)\right]^4}$$

In this expression, all parameters are tabulated for GaAs except Γ which remains to be fitted.

If one chooses a value of Γ equal to the half width of the luminescence line related to the N_{20} level (0.5 meV) one gets results for n_{20} which can be compared (Fig. 17) to the experimental values.

One notes that the agreement is correct on an absolute scale and reproduces the observed structures. These structures correspond indeed to the experimental variation of ΔE (Fig. 16) with magnetic field. These variations are in turn dependent of n_S which can be monitored with the laser power. The same good qualitative and quantitative agreement is found for different experimental conditions and different samples [10].

Therefore one believes that this new phonon-impurity relaxation mechanism is indeed the dominant one in this range of fields and demonstrates the importance of the potential fluctuations.

It is therefore natural to ask if these potential fluctuations could play a key role in other unexplained observations.

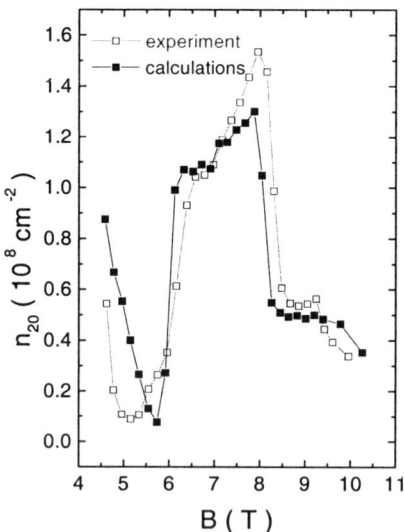

Figure 17. Comparison of the experimental (open squares) and calculated (full squares) data.

4. Open questions

There are different anomalies observed in the optical properties of 2D gas which cannot be explained within the one electron picture. Rather than listing them in an exhaustive way it is more convenient to take a specific example.

We consider the case of the photoluminescence of a highly doped single quantum well (9 nm width) of GaInAs doped in the middle with Be acceptors. This quantum well has a carrier concentration $n_S = 2.7 \times 10^{12}$ cm^{-2} corresponding to $E_F = 130$ meV. The density of acceptors $n_{Be} = 10^{10}$ cm^{-2}. The luminescence spectra [15] as a function of the magnetic field are displayed in Fig. 18.

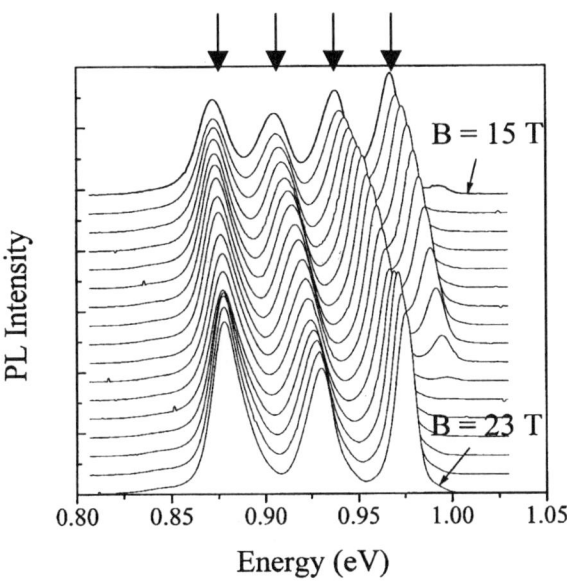

Figure 18. Electron-bound hole luminescence spectra of a GaInAs quantum well for different values of the magnetic field.

The linewidth of the LL structures is much larger (20-25 meV) than that usually observed in GaAs based structures. This is due to the fact that these structures have a much lower mobility, in part because of the presence of acceptors... One can nevertheless easily follow the variation of the LL energies with B which are shown in Fig. 19. This variation is non-linear with B and this can be very well explained [15] using known effective mass values including the non-parabolicity (continuous lines in Fig. 19). However, the relative Landau level photoluminescence intensities are not easy to understand.

These intensities I_N are directly proportional to the optical matrix element:

$$M_N^{opt} = |M_N|^2 = \sum_m \left| \langle \chi(z)\Phi_{N,m}(r,\varphi) | \Psi_{1s}(r,z_i) \rangle \right|^2$$

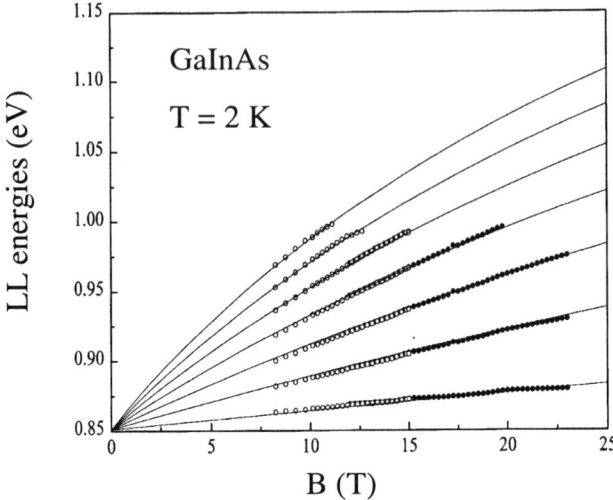

Figure 19. Variation of the energies of the different Landau levels (points) and comparison with theory (continuous lines) for the GaInAs QW magneto luminescence.

where z_i, the location of the impurity along the z axis, is equal to zero in the present case. Using here the appropriate model for χ (z) one obtains [15] results displayed in Fig. 20.

It is clear that if the intensity of the N=0 level is correctly reproduced, this is not the case for higher Landau levels. One obtains currently discrepancies of at least one order of magnitude between experimental results and predictions. The fact that the agreement between theory and experiments is only obtained for the LL N=0 is not specific for this sample but is also observed in high mobility ones, like the one discussed in the preceding section. Clear evidence is obtained for highly doped samples where the higher LL are populated in high fields. One possible reason for this discrepancy is that, in a one electron picture, the symmetry of the impurity wave function imposes a selection rule $\delta_{0,m}$ for M^{opt}. We have tried to invoke here a mechanism where the optical recombination is assisted by potential fluctuations, which also mix the LL wavefunctions. We did not succeed to explain the observations with this mechanism. Therefore, it is very likely that electron-electron and electron-hole interactions play a major role in this problem.....Before suggesting a more complete explanation, it is instructive to discuss the linewidth problem.

Indeed the variation of the linewidth also raises questions: it is difficult to discuss its large value because of the poor mobility of the sample and the fact that the extrinsic imperfections are not easy to quantify, but the major contribution to this linewidth cannot a priori originate from the acceptors since their number is much smaller than the degeneracy of LL at high fields. But whatever is the origin of the large linewidth, it is also difficult to understand the way the LL line disappears when it empties as a function of B (see Fig. 18, LL N=3). In particular the picture of a Landau

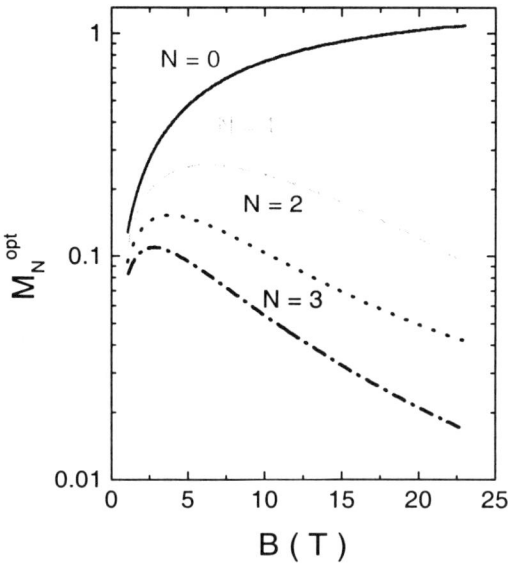

Figure 20. Variation of the optical matrix element for different Landau levels.

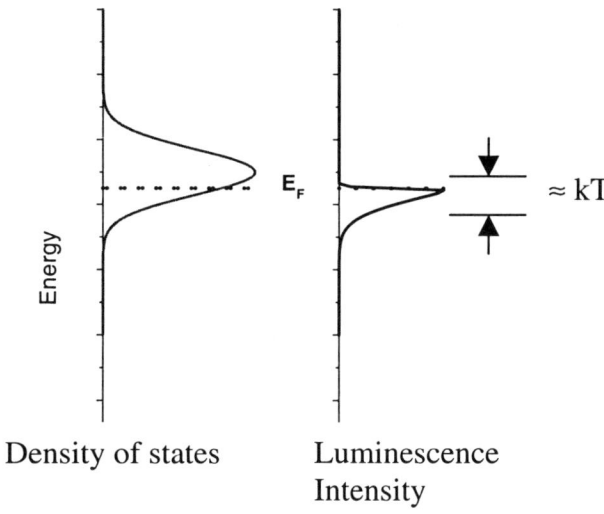

Density of states

Luminescence
Intensity

Figure 21. Schematic line-shape expected for a luminescence originating from a Landau level broadened by impurity scattering.

level broadened by impurities cannot be used to fit the data. In this picture (Fig. 21), when the Fermi level lies between two LL, the lineshape of the electron-bound hole luminescence should reflect the density of states whereas when E_F is inside this density of states the high energy side of the PL line should have a width of the order of kT. This is NEVER observed, in any sample, including those which have a very high mobility. Instead the lineshape remains more or less symmetric with a value for the width oscillating with the filling factor due to the well-known dependence of the screened dielectric constant on ν.

It is very likely that the intensity should reflect the effect of electron-electron or electron-hole interactions although, in part, they are already included in the calculation of the dielectric constant. All these effects are reminiscent of the Fermi Edge singularity problem. In its standard formulation this requires finite q-dependent density of states values around E_F , which, in principle, a pure 2D electron gas has difficulties with providing in high magnetic fields.... However, if the shake-up process is assisted by potential fluctuations that could provide values of q necessary to compensate for the hole recoil, it is not unlikely that one could explain the different observations. This remains to be formulated.

5. Conclusions

The influence of potential fluctuations on the optical properties of a 2D gas in high magnetic fields appear to be important. Though the relative density does not exceed a few percent of the degeneracy of the Landau levels they play a major role in the dynamic properties of the system each time the observed process requires conservation of the momentum. This is clearly the case for relaxation inside the Landau levels and may also be true for processes implying electron-bound hole recombination.

Acknowledgements. I would like to thank Israel Vagner and Dung-Hai Lee for many helpful discussions. The Grenoble High Magnetic Field Laboratory is associated to the Université Joseph Fourier de Grenoble et L'Institut National Polytechnique de Grenoble.

References

1.   Ando, T. Fowler, A. B., and Stern, F. (1982) "Electronic properties of two-dimensional systems", *Rev. Mod. Physics*, **54**, 437-672.
2.   Shklovskii, B. I., Efros, A. L. (1984) "Electronic properties of doped semiconductors", in M. Cardonna, P. Fulde and H. –J. Queisser (eds), Springer Series in Solid-State Sciences, Springer-Verlag, pp 1- 386.
3.   Efros, A. L. (1988) "Density of states od 2D electron gas and width of the plateau of IQHE", *Sol. State Com.* **65**, 1281-1284.
4.   Avishai, Y., Azbel, M. Ya., Gredeskul, S. A. (1993) "Electron in a magnetic field interacting with point impurities", *Phys. Rev. B* **48**, 17280-17295.
5.   Wiegers, S. A. J., Specht, M., Levy, L. P., Simmons, M. Y., Ritchie, D. A., Cavanna, A., Etienne, B., Martinez, G., and Wyder, P. (1997) " Magnetisation and energy gaps of a high-mobility 2D electron gas in the quantum limit", *Phys. Rev. Lett.*, **79**, 3238-3241.
6.   Peierls, R. (1933) " Zur Theorie des Diamagnetismus von Leitungselektronen", *Z. Phys.*, **81**, 186-194.

7. Friedland, K.-J., Hey, R., Kostial, H., Klann, R. and Ploog, K. (1996) "New concept for the reduction of the impurity scattering in remotely doped GaAs Quantum wells", *Phys. Rev. Lett.*, **77**, 4616-4619.
8. Hartmann, C., Martinez, G., Fischer, A., Braun, W., and Ploog. K. (1998) " Intersubband relaxation rates of a two-dimensional electron gas under high magnetic fields", *Phys. Rev. Lett.*, **80**, 810-813.
9. Kukushkin, I. V., Pulsford, N. J., von Klitzing, K., Haug, R. J., Ploog, K., Buhmann, H., Potemski, M., Martinez, G., and Timofeev, V. B. (1993) "Thermal collapse of the fractional-quantum-Hall-effect energy gaps", *Europhys. Lett.*, **22**, 287-292.
10. Hartmann, C. (1997) "Magneto-optical study of the two-dimensional electron gas by electron-bound hole recombination", Thesis of the University of Konstanz, Hartung-gorre Verlag Konstanz.
11. Dite, A. F., Kukushkin, I. V., Timofeev, V. B., Filin, A. I., von Klitzing, K. (1991) "Magnetooscillations in the luminescence decay time of a 2D electron gas at a single GaAs-AlGaAs heterojunction with a monolayer of acceptors", *JETP Lett.*, **54**, 642-645.
12. Bastard, G. (1992) "Spin-flip relaxation time of quantum-well electrons in a strong magnetic field", *Phys. Rev. B* **46**, 4253-4256.
13. Timofeev, V. B., Kirpichev, V. E., Kukushkin, I. V., Filin, A. I., Zhitomirkii,V. E., (1995) "Spin relaxation of photoexcited electrons and holes in GaAs/AlGaAs single heterojunction ", in D. J. Lockwood (ed) Proc. of the 22nd ICPS, Vancouver, World Scientific, vol.2, pp 1320-1324.
14. Yu, P. Y., Cardona, M. (1996) *"Fundamentals of Semiconductors, Physics and Material Properties"* Springer, Berlin.
15. Mordenti, S. (1997) "Etude magnéto-optique dans le proche infra-rouge de systèmes bidimensionels élaborés à partir d'alliages II-VI ou III-V", Thesis of L'Université Joseph Fourier de Grenoble.

SPECTRUM SHAPE ANALYSIS OF MAGNETOREFLECTIVITY AND MAGNETOROTATION OF POLARISATION PLANE OF LIGHT REFLECTED FROM GaAs/AlGaAs MQW'S IN THE QUANTUM HALL REGIME[1].

J. BOŻYM, E. DUDZIAK, D. PRUCHNIK
Institute of Physics, University of Technology of Wrocław
Wybrzeże Wyspiańskiego 27, 50-370 Wrocław, Poland

Z.R. WASILEWSKI
Institute for Microstructural Sciences, National Research Council
K1A OR6 Ottawa, Canada

Abstract. The magnetooptic Kerr effect (MOKE) in GaAs/AlGaAs multiple quantum wells was investigated in the integer quantum Hall regime. The shapes of the spectra are explained using transfer matrix calculations. Evidence of the magneto-polariton in GaAs (cap or bulk) has been shown. A large reduction of spin polarisation of the 2DEG measured by means of the MOKE in the region of Landau filling factors $1 \leq v < 4$, has been found.

Interband magnetooptical spectroscopy in strong magnetic fields is a sensitive method in studies of two-dimensional electron gas (2DEG). It allows us to study the correlations between the optical properties of a 2DEG and the integer and fractional quantum Hall effects (IQHE and FQHE). Many experimental and theoretical papers are devoted to quantum Hall state with filling factor $v=1$, when all electrons occupy the lowest spin-split Landau level [1-4]. Our report concerns the magnetooptical properties of the 2DEG at higher filling factors $v \geq 1$.

In most cases the methods of photoluminescence (PL), photoluminescence excitation (PLE) and reflectance were used. We propose measurements of the magnetorotation (MR) of linearly polarised light reflected from samples with multiple quantum wells (MQWs) of GaAs/AlGaAs [5]. The proposed method combines the advantages of magnetoabsorption and modulation spectroscopy. Magnetorotation has also been measured by other authors in the InGaAs/GaAs [6] and CdTe/CdMnTe quantum well structures [7,8].

When linearly polarised light is incident on a sample, the polarisation of both the transmitted and reflected beams in a direction parallel to the external magnetic field is in general elliptical. The angle between the major axis and the incident polarisation

---

[1] The measurements were performed at the International Laboratory of High Magnetic Fields and Low Temperatures, Wrocław, Poland

M.L. Sadowski et al. (eds.), Optical Properties of Semiconductor Nanostructures, 65–70.

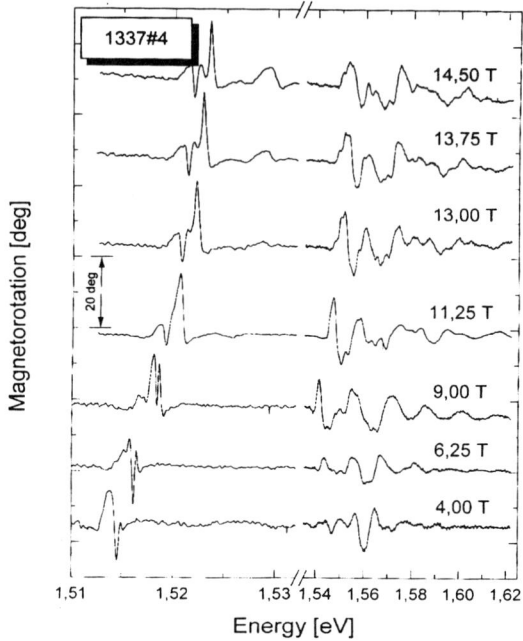

Figure 1. Magnetorotation spectra for sample 1337#4 at chosen magnetic fields. The sharp lines in the region of low energy show development of magneto-polariton in the sample. The lines at higher energies correspond to interband transitions and excitons in the quantum wells.

direction defines the Faraday rotation (FR) in transmission and the magneto-optic Kerr rotation (MOKE) in reflection. Both effects can be phenomenologically described by the off-diagonal $\widetilde{\varepsilon}_{xy}$ component of the dielectric tensor in the presence of a longitudinal magnetic field: $i\widetilde{\varepsilon}_{xy} = \Delta\widetilde{\varepsilon}\big/2$ with $\Delta\widetilde{\varepsilon} = \widetilde{\varepsilon}^{-} - \widetilde{\varepsilon}^{+}$, where $\widetilde{\varepsilon}^{\mp}$ are the complex dielectric functions for right and left circularly polarised light ($\sigma^{\mp}$) [9]. When the sample is a thin film or contains a multilayer structure, as in MQWs, multiple internal reflections cause the superposition of the FR and KR in the reflection configuration and we can talk about an effective magnetorotation (MR).

Following Manfra et al. [3], we assume that the integrated absorption peak in each polarisation is proportional to the available density of empty states $N_{A\uparrow(\downarrow)}$ in the spin-up (down) band of the lowest Landau level. Because the Kerr rotation Φ_K results from the difference of the absorption for σ^- and σ^+ polarisation, Φ_K will be the measure of the difference $N_{A\downarrow} - N_{A\uparrow}$ and finally the measure of the spin polarisation per particle

$$S_Z = \frac{N_\uparrow - N_\downarrow}{N} = \frac{N_{A\downarrow} - N_{A\uparrow}}{N}$$ ($N_{\uparrow(\downarrow)}$ is the number of spin-up (down) electrons in the

Landau level). The integrated Kerr rotation is therefore a measure of sample magnetisation. It is worth noticing here that MOKE is often applied to the measurements of surface magnetisation in different magnetic materials.

The MQWs were grown by molecular beam epitaxy (MBE). Two types of samples have been measured. The samples of the first type were composed of 30 GaAs wells of about 200 Å separated by about 400 Å wide $Al_{0.312}Ga_{0.688}As$ barriers. The barriers were asymmetrically δ-doped with Si atoms at a distance of 75 Å from one side of the barrier with concentrations (4÷9) x $10^{11}cm^{-2}$. The design of the samples of the second type was similar, but they contained 60 GaAs narrow wells of 50Å. The sample structure was always ended by a GaAs cap of thickness equal to the well width.

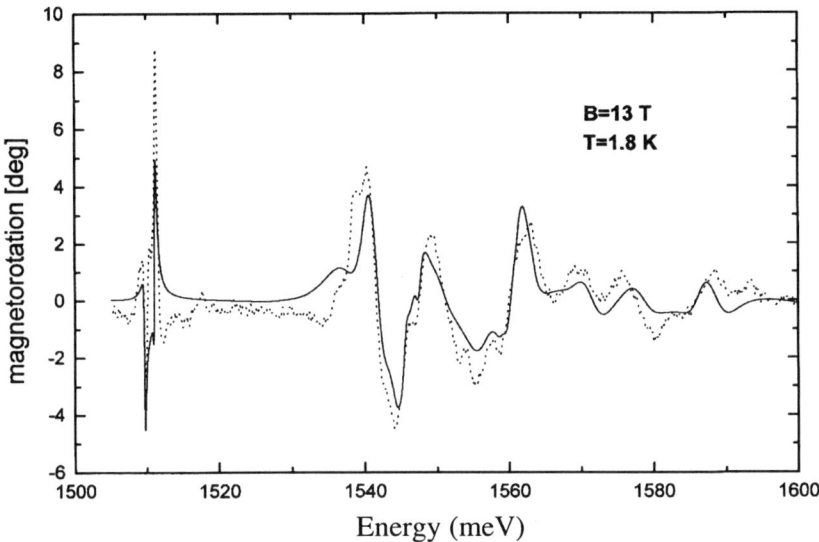

Figure 2. Juxtaposition of the measured (dashed line) and calculated (solid line) MOKE spectra for sample 1337#4 at a magnetic field of 13 T and a temperature of 1.8K.

The measurements were performed in the reflection configuration, at a temperature of 1.8K. The magnetic field (up to 15T) of a superconductive coil was perpendicular to the interfaces of the MQWs. Linearly polarised monochromatic light was incident perpendicular to the sample. The polarisation state of the light reflected from the sample was analysed by a rotating polariser placed before the detector [10].

Figure 1 shows the measured magnetorotation spectra Θ(E) at several magnetic fields (the magnetic field was changed by steps of 0,25T). The results are shown for the sample containing 30 periods of GaAs wells and $Al_{0.312}Ga_{0.688}As$ barriers of thickness 183 Å and 402 Å, respectively. Two different kinds of structures can be distinguished in the Θ(E) spectra. On the higher energy side of the spectra a rich structure of relatively broad lines can be seen. We attribute this part of the spectrum to transitions involving Landau Levels (LL) in the quantum wells. The transition energies agree quite well with energy distances between the electron and hole energy levels calculated in the local density approximation (LDA). The electron sheet density was determined to be 3,51 x $10^{11}cm^{-2}$, derived from the sharp discontinuity (red shift) at Landau filling factor ν=2 on the curve of the transition energy vs. magnetic field. The nominal δ-doping level was 4 x $10^{11}cm^{-2}$.

The sharp, exciton-polariton-like structure on the low energy side of the spectra in Fig.1 (note the enlarged energy scale) we attribute to the GaAs with zero free carrier concentration in the cap or the bulk (we cannot unambiguously resolve this question in numerical simulations). At magnetic fields B>3T exciton Zeeman splitting is seen. This splitting increases to 1.65 meV at B=15T. At magnetic fields higher than 11.5 T in magnetorotation spectra and higher than 13T in reflectivity spectra, a relatively broad upper polariton (UP) can be seen. Our experimental data concerning the exciton-polariton structure in GaAs show a striking resemblance to the experimental results obtained in [11] for $In_{0.13}Ga_{0.87}As$ QW's located centrally in a microcavity in high magnetic field. This may support the view that our exciton-polariton spectra are also emitted in the cap quantum well.

The fitting of the spectra shape was performed by means of transfer matrix calculations taking the dielectric response function from [12], where the inhomogeneous broadening is taken into account by the exciton resonance frequency ω_0:

$$w(z) = e^{-z^2} erfc(-iz) \quad \text{where} \quad z = \frac{\omega - \omega_0 + i\gamma}{\Delta}. \text{ The parameters } \gamma, \ \Delta > 0 \quad \text{are the}$$

nonradiative homogeneous broadening and inhomogeneous broadening of the oscillators, respectively. For small inhomogeneous broadening Δ the response function reduces to a Lorentzian oscillator. An example of a fit of the magnetorotation spectra, with the exciton-oscillator placed in the cap, is shown in Fig 2.

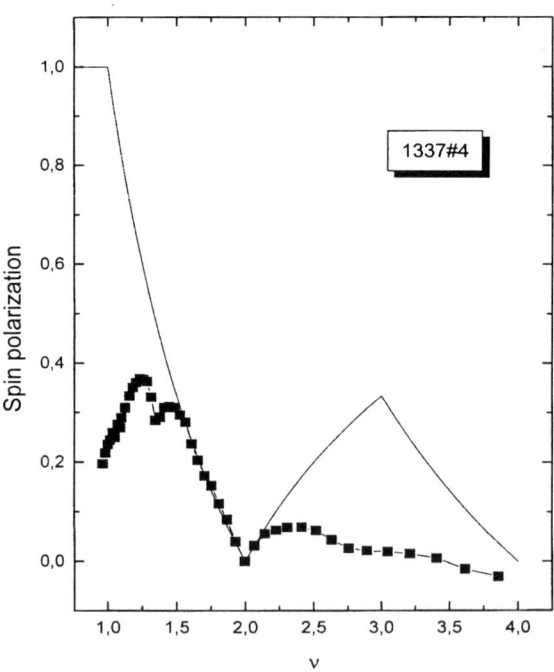

Figure 3. The ν dependence of the value proportional to spin polarisation calculated from experimental MOKE spectra as a strength of the lowest transition in the MQWs. The solid line is the ν dependence of spin polarisation for a noninteracting electron gas model.

Experiments on 2DEG magnetisation are rather rare because of the small number of electrons giving rise to the spin polarisation. A highly sensitive magnetometer is needed (about 10^{-14}J/T, which corresponds to ≈ 1 x $10^9 \mu_B$). Optionally, nuclear magnetic resonance (NMR) spectroscopy can be used [13,14]. The frequency Knight shift of the [71]Ga NMR signal due to coupling with the conduction electrons is proportional to the electron spin polarisation. Polarisation-resolved magnetoabsorption spectroscopy has also been applied to study the spin polarisation of a 2DEG [15,16] (our MOKE technique may be considered as a modulated magnetoabsorption method). The results reported in Refs. [13-16] concerned the spin polarisation of a 2DEG in the vicinity of filling factor $v = 1$. They showed a surprisingly sharp reduction of the electron spin polarisation on either side of filling factor one [13-16]. These observations have changed our understanding of the $v = 1$ quantum Hall state, showing that this state is not well described in the framework of a single-particle model. The sharp spin-depolarisation on either sides of $v = 1$ confirmed the theoretical prediction that charged excitations at $v = 1$ involve more than one spin flip. It was established that on either side of $v = 1$, the ground state of a 2DEG exhibits a collective character, in agreement with the theoretical prediction of the formation of spin-textures, called Skyrmions, around $v = 1$ [17,18].

Figure 3 shows the results of our measurements of the magnetisation of a 2DEG in the region $1 \leq v < 4$. The solid line represents the dependence expected for a noninteracting electron gas model. In agreement with this model, the experimentally derived magnetisation of a 2DEG vanishes in the close vicinity of $v = 2$. For other filling factors, the measured magnetisation is appreciably reduced as compared to theoretical expectations based on a one-particle model. This discrepancy can be hardly explained by the effects of disorder. These effects act in a way similar to the increase of temperature [19] and do not explain our experimental findings. Figure 3 shows that there is a broad maximum of the spin polarisation at $v=2,5$ instead of the one expected at $v =3$. Our result may be supported by the theoretical work [20]. Quite recently the spin polarisation near $v =3$ was measured by means of NMR [14]. The authors suggest the existence of Skyrmion excitations for $v=3$ although in their experimental conditions, the Zeeman energy exceeds the critical theoretical value postulated in [21].

We would like to notice that peaking of the spin polarisation a little away from $v=1$ has been found also in other experiments (see, for example, the data for sample C in Fig.3 of Ref. 22). It seems that this peak shift is apparent and originate from questionable assumption of field independent density of 2D electron density in a quantum well. In real 2DEG structures, the surface states should be taken into account. They form a reservoir of electrons and can partly pin the Fermi level. This leads to possible transfer of electrons into and out of the well with changing magnetic field. As a result, 2D electron density exhibits some oscillations as a function of magnetic field as it has been shown in Ref. 23.

In conclusion, measurements of the MOKE in GaAs/AlGaAs multiple quantum wells have been performed in the integer quantum Hall regime. The spectra shapes have been fitted using transfer matrix calculations. Evidence of the magneto-polariton in GaAs (cap or bulk) has been shown. A large reduction of spin polarisation of the 2DEG, measured by means of the MOKE in the region $1 \leq v < 4$, was found.

70

References

1. E. H. Aifer, B.B. Goldberg and D.A. Broido, Phys. Rev.Lett. **76**, 680 (1996).
2. I.V. Kukushkin, K.V. Klitzing and K. Eberl, Phys. Rev. B **55**, 10607 (1997).
3. M.J. Manfra, B.B. Goldberg, L. Pfeiffer and K. West, Phys. Rev. B **54**, R17327 (1996) also Acta Phys. Pol. A **92**, 621 (1997).
4. F. Plentz, D. Heiman, L.N. Pfeiffer and K.W. West, Phys. Rev. B **57**, 1370 (1998).
5. E. Dudziak, J. Bożym, D. Pruchnik and Z.R. Wasilewski, Acta Phys. Pol. A **90**, 1022 (1996) and Acta Phys. Pol. A **94**, 291 (1998).
6. A. V. Kavokin, M. R. Vladimirova, M. A. Kaliteevski, O.Lyngues, J. D. Berger, H. M. Gibbs, and G. Klitrova, Phys. Rev.B **56**, 1087 (1997).
7. C. Testelin, C. Rigaux, and J. Cibert, Phys. Rev.B **55**,2360 (1997).
8. A. Lemaitre, C. Testelin, C. Rigaux, S. Maćkowski, J. A. Gaj, G. Karczewski, T. Wojtowicz, and J. Kossut, to be published in Phys. Rev.B.
9. F.R. Kessler and J. Metzdorf „Landau Level Spectroscopy: Interband Effects and Faraday Rotation" ed. by G. Landwehr and E. I. Rashba, Elsevier Science Publishers B.V. 1991.
10. J.C. Suits, Rev. Sci. Instruments, **42**, 19(1971).
11. T.A. Fisher, A.M. Afshar, M.S. Skolnik, D.M. Whittaker and J.S. Roberts, Phys. Rev. B **53**, R10469 (1996).
12. L.C. Andreani, G. Panzarini, A.V. Kavokin and M. R. Vladimirova, Phys. Rev. B **57**, 4670 (1998)
13. S.E. Barret, G. Dabbagh, L.N. Pfeiffer, K.W. West and R. Tycko, Phys. Rev. Lett. **74**, 5112 (1995).
14. Y.-Q. Song, B.M. Goodson, K. Maranowski and A.C. Gossard, Phys. Rev. Lett., **82**, 2768 (1999).
15. M.J. Manfra, B.B. Goldberg, L.Pfeiffer and K. West, Phys. Rev. B **54**, R17327 (1996).
16. B.B. Goldberg, M.J. Manfra, L. Pfeiffer and K. West, Physica B 249-251, 7 (1998).
17. S.L. Sondhi, A. Karlheck, S.A. Kivelson and E.H. Rezayi, Phys. Rev. B **47**, 16419 (1993).
18. H.A. Fertig, L. Brey, R. Cote and A.H. MacDonald, Phys Rev. B **50**, 11018 (1994).
19. A.J. Nederveen and Y.V. Nazarov, Phys. Rev. Lett. **82**, 406 (1999).
20. R.H. Morf, Phys. Rev.Lett. **80**, 1505 (1998).
21. N.R. Cooper, Phys. Rev. B **55**, 1934 (1997).
22. B.B. Goldberg , M. J. Manfra, L. Pfeifer, K. West, Physica B **249-251**, 7 (1998)
23. A. Raymond, S.Juillagnet, I. Elmezouar, W. Zawadzki, M.L. Sadowski, M. Kamal-Saadi, and B. Etienne, Semicond. Sci. Technol. **14**, 915 (1999)

MAGNETIC FIELD INDUCED INSULATING PHASE IN TWO-DIMENSIONAL ELECTRON GAS IN InGaAs/InP WITH STRONG DISORDER

B. PŐDÖR[1], G. KOVÁCS[2], G. REMÉNYI[3], I. G. SAVEL`EV[4],
G. KISS[1,3], AND S. V. NOVIKOV[4]
[1] Hungarian Academy of Sciences,
Research Institute for Technical Physics and Materials Science,
Budapest, Hungary
[2] Department of General Physics, Eötvös Loránd University, Budapest,
Hungary
[3] CNRS Centre de Recherches sur les Très Basses Températures
et Laboratoire des Champs Magnétiques Intenses, Grenoble, France
[4] A. F. Ioffe Physical Technical Institute, St. Petersburg, Russia

Abstract. Here we report on the results of our study of the properties of the magnetic field-induced insulating phase in a two-dimensional electron gas in InGaAs/InP heterostructures. These structures are characterised by a much greater disorder potential than AlGaAs/GaAs heterostructures due to the inherent alloy disorder in the InGaAs layer. At millikelvin temperatures, below a Landau level filling factor of 0.35-0.45, signs of a magnetic field-induced insulating phase (Wigner solid?) were seen in this system with large disorder: divergent resistivity, nonlinear current-voltage characteristics with a threshold, and a transition from non-activated to activated transport were observed. A model is proposed for the ordering of electrons in spite of the strong disorder.

1. Introduction

In recent years the occurrence and nature of insulating phases in two-dimensional electron systems (2DES) at semiconductor interfaces in high magnetic field have been the subject of intensive research in connection with such exotic phases as the Wigner solid [1,2] and Hall insulator [3], to mention but a few. At low temperature and small Landau level filling factor ($\nu = n_s h/eB$), the ground state of a 2DES with small disorder is expected to be an electron solid (Wigner crystal) [1,2]. The Wigner transition has been observed in nearly perfect GaAlAs/GaAs heterostructures [4-7], and recently also in an InGaAs/InP heterostructure [8,9]. In GaAlAs/GaAs the critical filling factor ν_C at T = 0 K, below which Wigner crystallisation occurs, was about 0.2 [4,5] and 0.28 [6], while in InGaAs/InP a higher value of $\nu_C = 0.49$ was obtained [9]. Experiments on the GaAlAs/GaAs system, where the ratio of the disorder potential to the Coulomb energy

M.L. Sadowski et al. (eds.), Optical Properties of Semiconductor Nanostructures, 71–76.
© 2000 Kluwer Academic Publishers. Printed in the Netherlands.

is $e\langle V\rangle/U \langle\langle 1$, indicated that the solid exhibits activated [10,11] and non-linear transport properties [4,5,7], which can be attributed to a pinning on the random potential introduced by defects and impurities. Similar transport properties were also observed by the present authors [8,9,12] in an InGaAs/InP system with large disorder potential, where $e\langle V\rangle/U \approx 1$, when the ground state of the 2DES might be expected to be a glass-like phase [13].

Here we report experiments on 2DEG in InGaAs/InP heterostructures with strong disorder aimed at the study of the magnetic field induced insulating phase.

2. Experimental

The samples were modulation-doped liquid phase epitaxial $In_{0.53}Ga_{0.47}As/InP$ heterostructures [8,9]. The electron density and mobility were $(0.3-1.5) \times 10^{11}$ cm^{-2} and $(1-5) \times 10^4$ cm^2/Vs, respectively. The low temperature mobility of a 2DEG in such structures is inherently limited by alloy scattering. Magnetotransport measurements were carried out down to 40 mK in a He^3-He^4 dilution refrigerator fitted into the bore of a resistive magnet capable of reaching 22 Tesla. Persistent photoconductivity was used to control the 2DEG density. Both conventional dc and (8 Hz) lock-in measuring techniques were used, and both two- and four-terminal resistivity measurements were performed.

3. Experimental results

Below filling factors $v \leq 0.5$ the longitudinal resistivity ρ_{xx} displayed an activated behaviour for temperatures above about 200 mK (Fig. 1) [12], attributed to the emergence of an insulating phase (IP). The activation energies E_A, determined according to $\rho_{xx} = \rho_0 exp(E_A/2kT)$, when plotted as a function of the filling factor (Fig. 2) decrease roughly linearly with v, with similar slopes for all the three samples. The intercepts with the horizontal axis (v_C) mark the onset of the region of activated transport. The higher the disorder (as indicated by the value of the 2DEG mobility), the larger is the extrapolated activation energy at $v = 0$, and the critical filling factor v_C for the onset of the activated transport.

The data for the onset for various samples at various magnetic fields were extrapolated to infinite magnetic field values (assumed in all theoretical models) using $v = n_S/eB$ as suggested in [10] (Fig. 3). A limiting value of $v(B \rightarrow \infty) = 0.37$ was obtained for infinite B, which is about twice as much as the corresponding value ($v = 0.19$) obtained for high mobility (low-disorder) GaAlAs/GaAs in [10].

In the IP, two-terminal I-V characteristics (Figs. 4 and 5) revealed a non-linearity with a filling factor dependent 'plateau' and 'threshold', as well as a characteristic shift in the phase between current and voltage in the measurements of the longitudinal resistivity ρ_{xx}. For higher filling factors these features were absent.

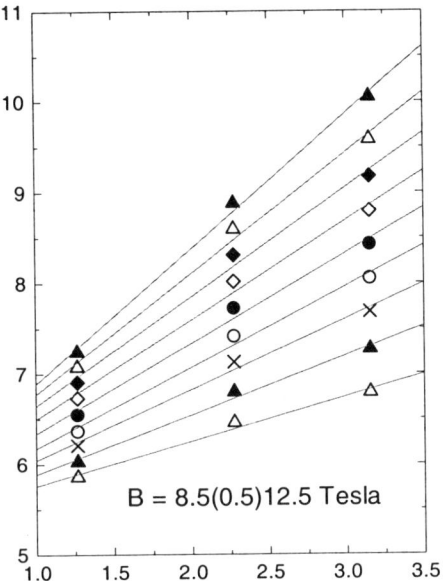

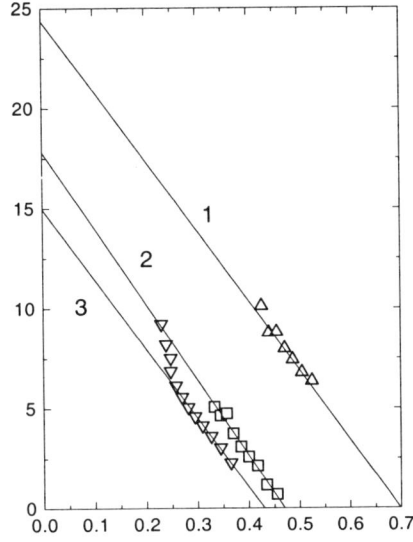

Figure 1. Longitudinal resistance ρ_{xx} versus the reciprocal temperature for a sample with $n_S = 7.5 \times 10^{10}$ cm^{-2}. The magnetic fields correspond to range of filling factors from 0.25 to 0.37.

Figure 2. Dependence of the activation energy E_A on the filling factor v. 2DEG mobility and density for samples 1 through 3 are 1.6 x 10^4, 2.9 x 10^4, and 5.2 x 10^4 cm^2/Vs, and 3 x 10^{10}, 1.2 x 10^{11}, and 7.5 x 10^{10} cm^{-2} respectively. Linear fits indicated correspond to $E_A = E_{A0} - \alpha v$.

4. Discussion

These results, as in the case of the GaAlAs/GaAs system in the literature [4-7], were interpreted as an indication of the formation of a magnetic-field-induced insulating phase below filling factors of about 0.5. The insulating phase is thought to be a Wigner solid pinned by the strong disorder. Its conduction mechanism is governed by the pinning-depinning phenomena.

The threshold electric fields are much larger than expected for an ordinary Wigner crystal. This may be explained by assuming the pinning of the crystalline electron solid by the large disorder. The strength of the disorder is about e<V> = 8-10 meV, comparable in magnitude to the Coulomb energy U, e<V>/U ≈ 1. This disorder is much stronger that in GaAlAs/GaAs, where this ratio can be smaller than 1/100. The incompressible fluid state of the 2DEG is fully destroyed and a glass-like condensate pinned by the disorder is formed. The stronger the disorder, the more glassy the Wigner solid pinned by the disorder. In low mobility samples a strongly pinned Wigner solid is expected with the onset of activated transport at higher values of v than in the high mobility case. Indeed, the extrapolated activation energies in the IP show a surprisingly good correlation with the 2DEG mobility (see Fig. 6), with a logarithmic slope of -0.5. This means a direct correlation with the disorder, which is quantified by the Landau level width $\Gamma \sim (B/\mu)^{1/2}$.

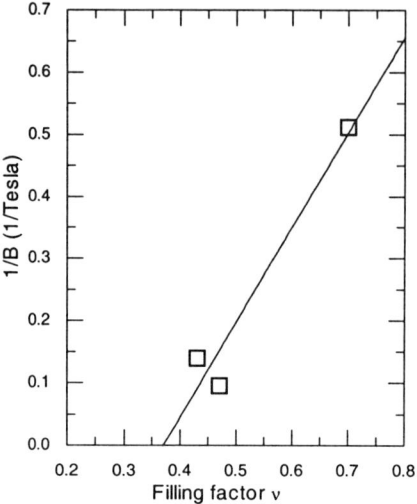

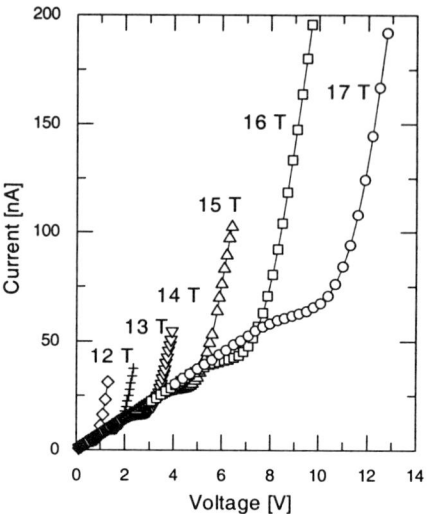

Figure 3. Extrapolation of the onset $\nu(E_A)$ to infinite B using $\nu = n_S h/eB$.

Figure 4. Two-terminal I-V characteristics at 70 mK in a sample with $n_S = 1.2 \times 10^{11}$ cm^{-2}. The filling factors are $\nu = 0.414, 0.382, 0.354, .331, 0.310,$ and 0.292 respectively.

We propose the following model for a Wigner crystal-like ordering of electrons in the presence of strong disorder. In the InGaAs/InP system studied here, the strong disorder is due to the inherent alloy disorder and to the roughness of the heterointerface. Both disorder potentials are characterised by a short range. The range of the alloy disorder potential is comparable to the lattice constant. Previous investigations have also shown [14] that in our material the characteristic period of the heterointerface roughness is L = 5 to 7 nm. Both characteristic distances are less than the lattice constant of the Wigner crystal, i.e. $L < a_W = (\pi n_S)^{-0.5} = 15\text{-}30$ nm in our case. The following scenario is thus possible: long range disorder ($L > a_W$) prevents Wigner crystallisation and leads to the usual single particle localisation, while in the case of short range disorder ($L < a_W$), when electrons are localised by strong short range potentials, they can also retain a Wigner solid-like long range order. The distinction between Wigner solidification, magnetic freeze-out, and the emergence of other IPs like the Hall insulator [15], has not yet been solved in most studies. We think that the hypothesis of Wigner solidification is a plausible one in our samples; however, an alternative interpretation based on the concept of the Hall insulator, which might occur in our samples at the lowest 2DEG concentrations, cannot be ruled out either.

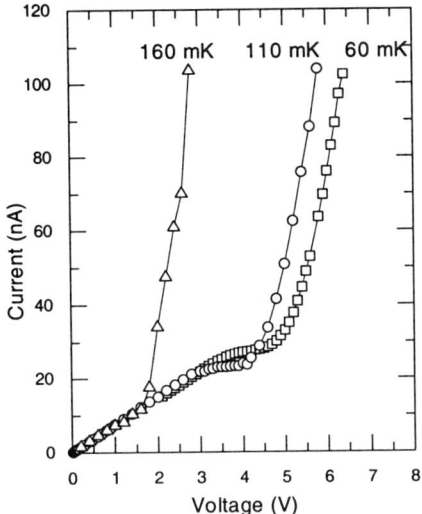

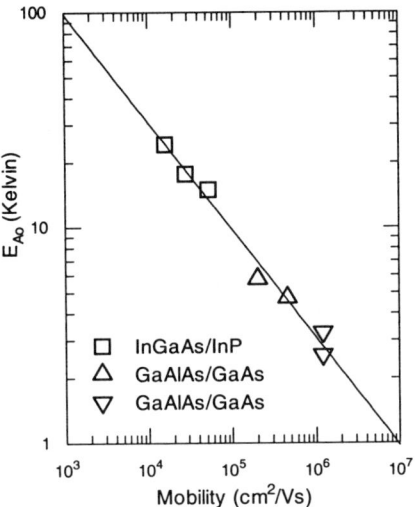

Figure 5. I-V characteristics at different temperatures, at a fixed value of the magnetic field B = 15 T (n_S = 1.2 x 10^{11} cm^{-2}).

Figure 6. Plot of the extrapolated activation energies E_{A0} versus the 2DEG mobility. Squares - our data, down triangles - data from [10], up triangles - data from [11]. The slope of the fitted line is -0.5.

5. Conclusions

To sum up, we observed diverging resistance, transition from non-activated to activated transport, and non-linear I-V characteristics with filling factor-dependent thresholds in 2DEG in InGaAs/InP in high magnetic fields at fractional filling factors. The observations were interpreted as the formation of a glass-like Wigner solid, and a model has been proposed to for the ordering of electrons in the presence of a strong short-range disorder.

Acknowledgements. The high magnetic field measurements were performed in the Grenoble High Magnetic Field Laboratory under the projects SE1295 and SE2198. We express our gratitude for the hospitality of the CNRS. Partial support was also made available from the Hungarian National Research Fund project No. OTKA/22894.

References

1. Wigner, E. P. (1934) On the interaction of electrons in metals, *Phys. Rev.* **46**, 1002-1011.
2. Lozovik, Y.E. and Yudson, V. I. (1975) Kristallizatsiya dvumernogo elektronnogo gaza v magnitnom pole, *Pis`ma v ZhETF* **22**, 26-28. (*JETP Lett.* **22**, 11 (1975).)
3. Kivelson, S., Lee, D. H., and Zhang S. C. (1992) Global phase diagram in the quantum Hall effect, *Phys. Rev. B* **46**, 2223-2238.

4. Andrei, E. Y., Deville, G., Glattli, D. C., Williams, F. I. B., Paris, E., and Etienne, B. (1988) Observation of magnetically induced Wigner solid, *Phys. Rev. Lett.* **60**, 2765-2768.
5. Goldman, V. J., Santos, M., Shayegan, M., and Cunningham, J. E. (1990) Evidence for two-dimensional quantum Wigner crystal, *Phys. Rev. Lett.* **65**, 2189-2192.
6. Buhmann, H., Joss, W., von Klitzing, K., Kukushkin, I. V., Plaut, A. S., Martinez, G., Ploog, K., and Timofeev, V. B. (1991) Novel magneto-optical behavior in the Wigner-solid regime, *Phys. Rev. Lett.* **66**, 926-929.
7. Andrei, E. Y., Williams, F. I. B., Glattli, D. C., and Deville G. (1993) Experiments on two-dimensional Wigner crystals, in P. Butcher (ed.) *The Physics of Low-Dimensional Semiconductor Structures*, Plenum Press, New York, pp. 499-537.
8. Kovács, Gy., Remenyi, G., Gombos, G., Savel`ev, I. G., Kreshchuk, A. M., Hegman, N., and Pődör, B. (1995) Wigner crystallization in InGaAs/InP heterostructures with strong disorder, *Acta Physica Polonica A* **88**, 783-786.
9. Remenyi, G., Kovács, Gy., Savel`ev, I. G., Gombos, G., and Pődör, B. (1996) Wigner-like phase transition in the strong disorder limit, (1996), *Czechoslovak J. Phys.* **46**, 2527-2528. (*Proc. 21ˢᵗ Int. Conf. Low Temperature Physics, 8-14 August 1966, Prague.*)
10. Willet, R. L., Stormer, H. L., Tsui, D. C., Pfeiffer, L. N., West, K. W., and Baldwin, K. W. (1988) Termination of the series of fractional quantum Hall states at small filling factors, *Phys. Rev. B* **38**, 7881-7884.
11. Fukano, A., Kawaji, S., Wakabayashi, J., Hirakawa, K., Sakaki, H., Goto, T., Koike, Y., and Fukase, T. (1996) Inter-electron Coulomb interaction and disoder in the activated transport in magnetic-field-induced insulating phase in 2D systems, in D. Heiman (ed.), *High Magnetic Fields in the Physics of Semiconductors*, World Scientific, Singapore, pp. 440-443.
12. Pődör, B., Gombos, G., Reményi, G., Kovács, Gy., Savel`ev, I. G., and Novikov, S.V. (1997) Activated transport in magnetic-field induced insulating phase in two-dimensional electron gas in InGaAs/InP heterostructures, *Acta Physica Polonica A* **92**, 945-949.
13. Aoki, H. (1979) Effect of coexistence of random potential and electron-electron interaction in two-dimensional systems: Wigner glass, *J. Phys. C: Solid State Phys.* **12**, 633-645.
14. Kreshchuk, A. M., Novikov, S. V., Polyanskaya, T. A., Savel`ev, I. G., and Shik, A. Y. (1995) Quantum transport effects in a two-dimensional electron gas as a tool for the investigation of heterointerfaces, *J. Crystal Growth* **146**, 153-158.
15. Manoharan, H. C. and Shayegan, M. (1994) Wigner crystal versus Hall insulator, *Phys. Rev. B* **50**, 17662-17665.

OPTICAL PROPERTIES OF BOUND POLARONS IN QUANTUM WELLS

S. BEDNAREK, B. SZAFRAN, J. ADAMOWSKI
Faculty of Physics and Nuclear Techniques,
University of Mining and Metallurgy (AGH),
Kraków, Poland

I. ESSAOUDI
Department of Physics, University Moulay Ismail,
Meknes, Morocco

AND

B. STÉBÉ
Institut de Physique et d'Electronique,
Université de Metz, France

Abstract. The influence of spatial confinement on the bound-polaron energy spectrum is studied. We show that the polaron effects for donors in quantum wells considerably enhance the donor binding energy and the donor energy levels can be splitted due to a resonance with phonons.

In ionic semiconductors, the electron-phonon coupling changes the properties of donors leading to the formation of bound polarons [1]. In semiconductor quantum wells (QWs), the polaron properties of donors are additionally modified by two effects. First, there appear new phonon modes: interface, confined, and half-space modes [2]. A modification of polaron properties, which results from the different phonon modes, was discussed by Mori and Ando [2]. Second, the confinement of electrons in the QW and the change of electron localisation around a donor center essentially modify the energies of optical transitions between the donor states [3]. If the QW heterostructure is fabricated from semiconductors of similar dielectric properties, i.e. the phonon modes and interaction amplitudes differ only slightly (as in $CdTe/Zn_xCd_{1-x}Te$), the first modification of bound-polaron states can be neglected. Then, the polaron effects of the second type are of substantial importance. The present paper is devoted to these effects

M.L. Sadowski et al. (eds.), Optical Properties of Semiconductor Nanostructures, 77–80.

and deals with the joint influence of the spatial confinement and electron-phonon coupling on the binding and transition energy for donors in QWs. For the purposes of clarity of the description and uniqueness of argumentation, we assume the same electron-LO phonon interaction as in the bulk crystal, which corresponds to an averaging over the heterostructure-phonon modes and is compatible with the sum rule [2].

We consider a system which consists of a donor impurity located at the center of a QW, an excess conduction-band electron, and a field of LO phonons. We assume that the confinement potential of the QW has the form of a quantum well of depth V_0 and thickness L measured in the growth (z) direction. In ionic heterostructures, the electron-LO phonon coupling can be approximated by the Fröhlich interaction [2]. In order to solve the eigenvalue problem we adopt the properly modified version of the optimised canonical transformation method, which was elaborated for excitons [4] and bound polarons [1] in bulk crystals. Accordingly, the trial electron-phonon state is proposed in the form: $|\psi\rangle = \exp(S)\varphi(\mathbf{r})|0\rangle$, where

$$S = \sum_{\mathbf{k}} \left[F_{\mathbf{k}}(\mathbf{r})a_{\mathbf{k}}^+ - H.c. \right] , \tag{1}$$

$\varphi(\mathbf{r})$ is the electron wave function, $|0\rangle$ is the phonon vacuum state, and $a_{\mathbf{k}}^+$ is the phonon creation operator. In order to choose the best displacement amplitude $F_{\mathbf{k}}(\mathbf{r})$, we have performed test calculations using the simple trial wave function of the form $\varphi(\mathbf{r}) = \exp[-\alpha(x^2+y^2)-\beta z^2]$ with two variational parameters α and β, and the following amplitudes: $F_{\mathbf{k}}^{(1)}(\mathbf{r}) = f_{\mathbf{k}}^{(1)}$, $F_{\mathbf{k}}^{(2)}(\mathbf{r}) = f_{\mathbf{k}}^{(2)} \exp(-i\mathbf{k} \cdot \mathbf{r})$, $F_{\mathbf{k}}^{(3)}(\mathbf{r}) = f_{\mathbf{k}}^{(3)} \exp(-i\mathbf{k}_\perp\cdot\mathbf{r})$, $F_{\mathbf{k}}^{(4)}(\mathbf{r}) = f_{\mathbf{k}}^{(4)} \exp(-i\mathbf{k}_\parallel\cdot\mathbf{r})$, as well as their linear combinations, e.g., $F_{\mathbf{k}}^{(123)} = F_{\mathbf{k}}^{(1)} + F_{\mathbf{k}}^{(2)} + F_{\mathbf{k}}^{(3)}$. Here, $\mathbf{k}_\perp$ and $\mathbf{k}_\parallel$ denote the in-plane and z wave-vector components, respectively. Phonon amplitudes $f_{\mathbf{k}}^{(i)}$ are determined by minimisation of the expectation value of the Hamiltonian of the system.

In Fig. 1 (a), we display the lattice relaxation energy (the expectation value of the Hamiltonian of the phonon field and electron-phonon interaction) calculated with the use of different displacement amplitudes for the material parameters of a GaAs/Al$_{0.3}$Ga$_{0.7}$As nanostructure. The binding energy of the bound polaron is an increasing function of the lattice relaxation energy. Therefore, due to the variational character of the present approach, the larger values of the lattice relaxation energy correspond to the better variational estimates of the bound polaron energy. The best estimates are obtained with the linear combination of all the four displacement amplitudes $F_{\mathbf{k}}^{(i)}$, but the results obtained with the use of three amplitudes, i.e. with $F_{\mathbf{k}}^{(123)}$, are almost indistinguishable from these results.

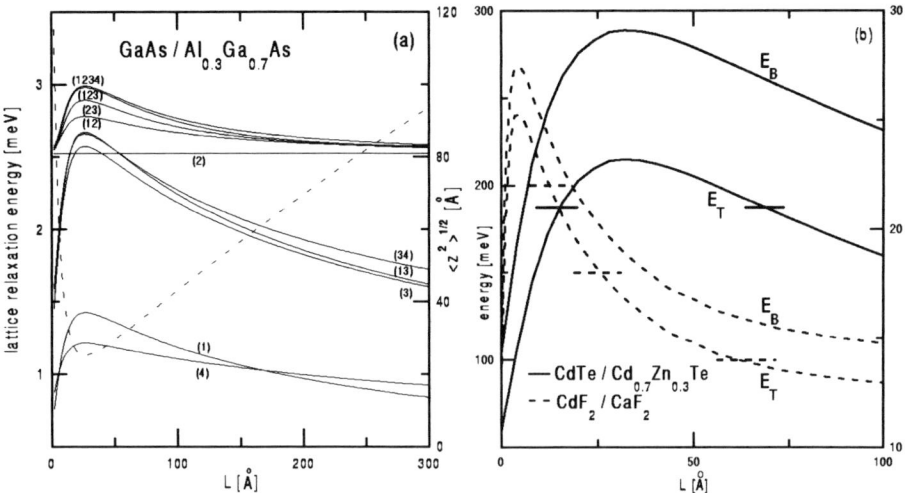

Figure 1. Results of calculations for bound polarons in QWs of thickness L. (a) Lattice relaxation energy (solid curves) obtained with different displacement amplitudes (see text) and square-root mean value of electron z coordinate (dashed curve) for GaAs/Al$_{0.3}$Ga$_{0.7}$As. (b) Binding energy (E_B) and $1s - 2p$ transition energy (E_T) for the bound polaron in CdTe/Zn$_{0.3}$Cd$_{0.7}$Te (right energy scale) and CdF$_2$/CaF$_2$ (left energy scale). Short horizontal lines show the positions of phonon resonances.

Therefore, we have chosen amplitude $F_{\mathbf{k}}^{(123)}$ for the more extended calculations of the bound-polaron energy levels. We note that the application of only two amplitudes, e.g., $F_{\mathbf{k}}^{(12)}$ or $F_{\mathbf{k}}^{(23)}$, is not sufficient for thin QWs. The maximum lattice relaxation energy corresponds to the minimum electron wave-function extension in the z direction [cf. dashed curve in Fig. 1 (a)], i.e., maximum electron localisation in the QW.

Using the amplitude $F_{\mathbf{k}}^{(123)}$ and 15-element Gaussian basis, we have calculated the ground-state ($1s$) binding energy E_B and $1s - 2p$ transition energy E_T for the donor located at the center of the QW. The results of our calculations show that the polaron effects are negligibly small in the weakly ionic GaAs/AlGaAs structure, which results from an almost exact cancellation of lattice relaxation energies for the two states used in the calculation of E_B and E_T. In Fig. 1 (b), we report the results for the ionic heterostructures made of CdTe/Zn$_{0.7}$Cd$_{0.3}$Te ($V_0 = 0.146$ eV) and CdF$_2$/CaF$_2$ ($V_0 = 3$ eV). For donors in these structures, the polaron effects are large. For example, the joint effect of the electron-phonon coupling and confinement of electrons enlarges by a factor of two the binding energy of donors in CdF$_2$/CaF$_2$ QW with $L \simeq 20$Å in comparison to that in bulk CdF$_2$ crystals. The experiments with donors in these structures have just begun [5]. The maxima of the binding and transition energy appear for the QW thickness, which corresponds to the largest electron localisation in the

z direction shown in Fig. 1 (a).

For the ionic heterostructures, the calculated $1s - 2p$ transition energy exceeds the LO-phonon energy, which means that phonon resonances are expected in the optical spectra of donors. These resonances should appear for the $2p$ and $1s + N$-phonon energy-level crossing, which leads to a level repulsion (level anticrossing). The results displayed in Fig. 1 (b) allow us to predict that in $CdTe/Zn_{0.3}Cd_{0.7}Te$ the phonon resonances involving one LO phonon with an energy of 21 meV can be observed for $L \simeq 10\text{Å} \div 20\text{Å}$ and $60\text{Å} \div 70\text{Å}$. In CdF_2/CaF_2, the phonon resonances with $N = 2, 3$, and 4 phonons involved should appear at the transition energies 100 meV, 150 meV, and 200 meV for $L \simeq 60\text{Å}$, 25Å, and 15Å, respectively. In the QWs with these parameters, a splitting of the $1s - 2p$ donor transition energy should be observed. The present variational approach does not allow us to reproduce the anticrossing of energy levels at the resonance. However, we can predict the values of the QW parameters, for which this effect appears. Until now, similar phonon resonances were observed for polarons in GaAs/AlGaAs multiple QW structures in magnetic field [6]. The results of the present paper show that in QWs made of ionic materials the phonon resonances in donor spectra can be detected even without an external magnetic field.

In summary, we have shown that the polaron effects in QWs are enhanced in comparison to those in bulk crystals, since the confinement potential reduces the penetration of electrons in the growth direction, which considerably changes the lattice deformation around the donor centre. This leads – in the case of ionic materials – to a considerable increase of the donor binding energy and the appearance of phonon resonances in the optical spectra of donors.

Acknowledgements. This work is performed in the frame of the French-Polish scientific cooperation programme POLONIUM and is partially supported by the Polish State Scientific Committee (KBN Grant No. 2 P03B 5613).

References

1. J. Adamowski, Phys. Rev. B **32**, 2588 (1985).
2. N. Mori and T. Ando, Phys. Rev. B **40**, 6175 (1989).
3. N.C. Jarosik, B.D. McCombe, B.V. Shanabrook, J. Comas, J. Ralston, and G. Wicks, Phys. Rev. Lett. **54**, 1283 (1985).
4. S. Bednarek, J. Adamowski, and M. Suffczyński, Solid State Commun. **21**, 1 (1977).
5. N.S. Sokolov, S.V. Gastev, A.Yu. Khilko, S.M. Suturin, I.N. Yassievich, J.M. Langer, and A. Kozanecki, Phys. Rev. B **59**, R2525 (1999).
6. Y.J. Wang, H.A. Nickel, B.D. McCombe, F.M. Peeters, J.M. Shi, G.Q. Hai, X.-G. Wu, T.J. Eustis, and W. Schaff, Phys. Rev. Lett. **79**, 3226 (1997).

SECOND HARMONIC GENERATION IN ASYMMETRIC QUANTUM WELLS: THE ROLE OF DEPOLARISATION AND EXCITONIC-LIKE EFFECTS

V. BONDARENKO
Institute of Physics National Academy of Sciences of Ukraine,
pr. Nauki 46, 252650 Kiev-28, Ukraine

AND

M. ZAŁUŻNY
Institute of Physics, M. Curie-Skłodowska University, 20-031,
Lublin, pl. M. Curie-Skłodowskiej 1, Poland

In this paper we discuss theoretically the influence of the depolarisation effect (DE) and the exciton-like effect (EE) on the second harmonic generation (SHG) in two level asymmetric quantum well structures.

Our approach is based on the TDLDA and the density matrix formulation used in our previous paper [1]. In calculating contributions associated with the indirect Coulomb interaction we take into account the additional "driving" term resulting from a nonlinear dependence of the exchange-correlation potential on the electron density [2].

As in most papers we assume that the external perturbation $\triangle V_{\text{ext}}(z,t)$ $[= ez\mathcal{E}\exp(-i\omega t)+\text{c.c.}]$ is small ($\mathcal{E}$ is the amplitude of the external electric field) [1, 2, 3]. Then, the change in the electron concentration $\triangle N(z,t)$ can be expanded in powers of $\triangle V_{\text{ext}}(z,t)$ as

$$\triangle N(z,t) = \sum_{n=1,2} N^{(n)}(z,n\omega) \exp(-in\omega t). \tag{1}$$

(A similar expansion can also be used in the case of the self-consistent perturbing potential $[V(z,t)]$ and the density matrix.)

The nth order self-consistent perturbing potential $V^{(n)}(z,n\omega)$ takes the form

$$V^{(n)}(z,n\omega) = ez\mathcal{E}\delta_{k1} + V_{\text{H}}^{(k)}(z,k\omega) + V_{\text{XC}}^{(k)}(z,k\omega), \tag{2}$$

where

$$V_{\text{H}}^{(n)}(z,n\omega) = -\frac{e^2}{\epsilon_0\epsilon} \int_{-\infty}^{z} dz \int_{-\infty}^{z'} dz'' N^{(n)}(z'',n\omega), \tag{3}$$

M.L. Sadowski et al. (eds.), Optical Properties of Semiconductor Nanostructures, 81–84.
© 2000 *Kluwer Academic Publishers. Printed in the Netherlands.*

describes the depolarisation effect and

$$V_{XC}^{(n)}(z, n\omega) = V'_{XC}(z)N^{(n)}(z, n\omega) + \frac{1}{2}V''_{XC}(z)[N^{(1)}(z, \omega)]^2\delta_{n2}, \quad (4)$$

describes the exciton-like effect. $V'_{XC}(z)$, $V''_{XC}(z)$ are the first and second functional derivatives of the exchange-correlation potential with respect to the equilibrium density distribution of the electrons, ϵ_0 is the dielectric permittivity of the vacuum, and ϵ is an average dielectric constant.

The SHG surface susceptibility (in two subband asymmetric QWs) is connected with the second order density matrix by relation [1]

$$\chi_{zz}^{(2)}(2\omega) = \frac{-e2}{\epsilon_0\mathcal{E}2} \sum_{i,j} \sum_{\mathbf{k}_\parallel} \rho_{ij}^{(2)}(2\omega)z_{ji} \quad (5)$$

where $\mathbf{k}_\parallel$ is the parallel component of the electron wavevector, $z_{ij} = \int_{-\infty}^{\infty} \varphi_i(z)z\varphi_j(z)dz$ and $\varphi_i(z)$ is the electron envelope function.

The matrix elements of the density matrix obey equation [1]

$$\mp in\omega\rho_{ij}^{(n)}(\pm n\omega) = \frac{1}{i\hbar}[\hat{H}_0, \rho^{(n)}(\pm n\omega)]_{ij}$$
$$+ \frac{1}{i\hbar}\sum_{k>0}^{n}[V^{(k)}, \rho^{(n-k)}(\pm n\omega)]_{ij} - \frac{\rho_{ij}^{(n)}(\pm n\omega)}{\tau_{ij}}, \quad (6)$$

where $\rho_{ij}^{(n)}(-n\omega) = [\rho_{ij}^{(n)}(n\omega)]^*$ and $\hat{H}_0$ is the unperturbed Hamiltonian.

The final expression for $\chi^{(2)}(2\omega)$ resulting from Eqs. (1)-(6) can be written as a sum of three terms

$$\chi^{(2)}(2\omega) = \chi_A^{(2)}(2\omega) + \chi_B^{(2)}(2\omega) + \chi_C^{(2)}(2\omega), \quad (7)$$

where

$$\chi_A^{(2)}(2\omega) = \frac{-2e^3N_{12}z_{12}^2(z_{22} - z_{11})[E_{21}^2 - (\hbar\omega + i\Gamma)^2]}{\epsilon_0[\tilde{E}_{21}^2 - (\hbar\omega + i\Gamma)^2]^2} \times$$
$$\left[\frac{E_{21}^2 + (2\hbar\omega + i\Gamma)(\hbar\omega + i\Gamma)}{\tilde{E}_{21}^2 - (2\hbar\omega + i\Gamma)^2}\frac{\tilde{E}_{21}^2 - (\hbar\omega + i\Gamma)^2}{E_{21}^2 - (\hbar\omega + i\Gamma)^2} + \frac{\hbar\omega + i\Gamma}{2\hbar\omega + i\bar{\Gamma}}\right], \quad (8)$$

$$\chi_B^{(2)}(2\omega) = \frac{2e^3N_{12}^2z_{12}^3\bar{\alpha}_{21}E_{21}[E_{21}^2 - (\hbar\omega + i\Gamma)^2]}{\epsilon_0[\tilde{E}_{21}^2 - (\hbar\omega + i\Gamma)^2]^2[\tilde{E}_{21}^2 - (2\hbar\omega + i\Gamma)^2]}$$
$$\times \left[\frac{E_{21}^2 + (2\hbar\omega + i\Gamma)(\hbar\omega + i\Gamma)}{E_{21}^2 - (\hbar\omega + i\Gamma)^2} + \frac{\hbar\omega + i\Gamma}{2\hbar\omega + i\bar{\Gamma}}\right], \quad (9)$$

$$\chi_C^{(2)}(2\omega) = \frac{8e^3 N_{12}^3 z_{12}^3 \bar{Z}_{12} E_{21}^3}{\epsilon_0 [\tilde{E}_{21}^2 - (\hbar\omega + i\Gamma)^2]^2 [\tilde{E}_{21}^2 - (2\hbar\omega + i\Gamma)^2]^2}. \tag{10}$$

Here $E_{21} = E_2 - E_1$ is the intersubband energy, $N_{12} = N_1 - N_2$ is the difference in the electron concentration in the ground (1) and excited (2) subband, $\Gamma = \hbar/\tau_{12}$, $\bar{\Gamma} = \hbar/\tau_{11} = \hbar/\tau_{22}$, τ_{ii}^{-1} is the relaxation rate from the ith subband, $\tau_{12}^{-1} = \tau_{21}^{-1}$ is the off-diagonal elastic dephasing rate connected with $1 \rightarrow 2$ transitions,

$$\tilde{E}_{21} = E_{21}[1 + 2\bar{\alpha}(1,2;1,2)N_{12}/E_{21}]^{1/2}$$

is the resonant intersubband energy shifted by DE and EE [4],

$$\bar{\alpha}_{21} = \bar{\alpha}(1,2;2,2) - \bar{\alpha}(1,2;1,1),$$

$$\bar{\alpha}(i,j;k,l) = \frac{e^2}{\epsilon_0\epsilon} L_{ijkl} + Y_{ijkl},$$

with

$$L_{ijkl} = \int_{-\infty}^{\infty} dz \left[\int_{-\infty}^{z} dz' F_{ij}(z') \right] \left[\int_{-\infty}^{z} dz' F_{kl}(z') \right],$$

$$Y_{ijkl} = \int_{-\infty}^{\infty} dz F_{ij}(z) F_{kl}(z) V'_{XC}(z)$$

and

$$\bar{Z}_{12} = \frac{1}{2} \int_{-\infty}^{\infty} dz F_{12}^3(z) V''_{XC}(z),$$

where $F_{ij}(z) = \varphi_i(z)\varphi_j(z)$.

For comparison, the authors of Ref.[3] give (without presenting the details of the calculations) the following expression for $\chi^{(2)}(2\omega)$:

$$\chi^{(2)}(2\omega) = \frac{-e^3 N_{12} 3 z_{21}^2 (z_{22} - z_{11})}{\epsilon_0}$$

$$\times \frac{[(E_{21} + i\Gamma)^2 - (\hbar\omega)^2](E_{21} + i\Gamma)^2}{[(\tilde{E}_{21} + i\Gamma)^2 - (2\hbar\omega)^2][(\tilde{E}_{21} + i\Gamma)^2 - (\hbar\omega)^2]^2}. \tag{11}$$

We want to note that, in the one electron limit, only our result reduces to that derived by Tsang et al. [5]

$$\chi^{(2)}(2\omega) = \frac{-2e^3 N_{12} z_{12}^2 (z_{22} - z_{11})}{\epsilon_0 [E_{21}^2 - (\hbar\omega + i\Gamma)^2]}$$

$$\times \left[\frac{E_{21}^2 + (2\hbar\omega + i\Gamma)(\hbar\omega + i\Gamma)}{E_{21}^2 - (2\hbar\omega + i\Gamma)^2} + \frac{\hbar\omega + i\Gamma}{2\hbar\omega + i\bar{\Gamma}} \right]. \tag{12}$$

It is also interesting to note that, in the one electron limit $(\tilde{E}_{21} = E_{21})$ the difference between numerical values of $|\chi^{(2)}(2\omega)|$ resulting from Heyman's and Tsang's expressions is negligibly small. Unfortunately, Heyman's expression gives a different sign for $\mathrm{Im}[\chi^{(2)}(2\omega)]$ and contradicts the well known rule saying that the poles of the susceptibility should lie in the lower half of the complex-frequency plane [6]. This indicates that in Ref. [6] the damping factor was not introduced correctly.

Since the expression for $\chi^{(2)}(2\omega)$ is complex we have performed appropriate numerical calculations for the n-doped coupled asymmetric double quantum wells (similar to those studied experimentally in Ref. [3]). A detailed inspection of the analytical and numerical results shows that two peaks appear in the SHG spectrum; lower at $\omega = \omega_{res}^l$ and upper at $\omega = \omega_{res}^u$. When the electron concentration is not too small $(N_1 \geq 10^{11} \mathrm{cm}^{-2})$, the difference between $\hbar\omega_{res}^l$ and $\tilde{E}_{21}/2$ ($\hbar\omega_{res}^u$ and $\tilde{E}_{21}$) is negligible. Thus, as in the case of linear response, the peaks in the SHG spectrum are shifted by DE and EE. (The coincidence of $\hbar\omega_{res}^l$ with $\tilde{E}_{21}/2$ is supported by experimental results reported in Ref. [3].) However, in contrast with the linear case, the peak heights are also strongly affected by Coulomb interaction. This interaction reduces (enhances) the lower (upper) peak by more than one order of magnitude when the Coulomb correction to the resonant intersubband energy is comparable with E_{21}. We have checked numerically that the above mentioned peak height modification is not described correctly (it is substantially underestimated) by Heyman's expression.

References

1. Załużny, M. and Bondarenko, V. (1997) Doubly resonant second-harmonic generation in quantum well with anisotropic and tilted valleys: resonant screening effects, *J. Appl. Phys.* **81**, 3276-3280.
2. Gusmao, M.S, and Mahan, G.D. (1996) Field-effect transistors as tunable infrared detectors, *J. Appl. Phys.* **79**, 2752-2754.
3. Heyman, J.N. , Craig, K., Galdrikian, B., Sherwin, M. S., Capman, K., Hopkins, P.F., Fafard, S., and Gossard, A.C. (1994) Resonant harmonic generation and dynamical screening in a double quantum well, *Phys. Rev. Lett.* **72**, 2183-2186.
4. Ando, T. (1976) Lineshape of intersubband optical transitions in space charge layers, *Z. Phys. B* **24** 33-39.
5. Tsang, L., Ahn, D., and Chuang, S.L. (1988) Electric field control of optical second-harmonic generation in a quantum well, *Appl. Phys. Lett.* **52**, 697-699.
6. Butcher, P.N. and Cotter, D. (1993) *The elements of nonlinear optics* (Cambridge University Press 1993) p.92

OPTICAL ABSORPTION AND CARRIER TRANSMISSION IN HETEROSTRUCTURES

R. OSZWAŁDOWSKI

Instytut Fizyki, UMK, Grudziądzka 5, Toruń, Poland

Abstract. Bound to continuum optical intraband transitions in semiconductor heterostructures are investigated. A variational method is formulated and applied for calculating the optical absorption and electron transmission coefficient profiles. This method can be applied for any (asymptotically flat) band profile. The resulting optical transition matrix elements and transmission coefficient profiles are juxtaposed for the case of the conduction band. Calculations of absorption are performed both for conduction band (AlGaAs) and for valence band (Si/SiGe, 6-band Hamiltonian) quantum well type structures.

1. Introduction

Semiconductor heterostructures absorbing infrared radiation via electronic transitions within one band are attracting much attention, due to the potential application as infrared detectors [1, 2, 3]. Some of these devices are based on transitions from a bound (localised) state to a delocalised (free) one, both states being derived from the same band. Transitions of this kind may take place both in conduction and valence bands. An electron, having absorbed a photon, will contribute to electrical current if an external bias removes it from the QW region (perpendicular transport) before deexcitation. Thus a knowledge of both the optical absorption $\alpha(\hbar\omega)$ and electronic transmission $T(E)$ coefficients is necessary to determine a given heterostructure's optical properties. The aim of this paper is to present a unified *large box* description for these two characteristics.

The *large box* method for determining an averaged absorption coefficient has been put forward in an earlier article [4]. Here we concentrate mainly on a comparison of the transmission coefficient and optical transition matrix elements in $Al_xGa_{1-x}As/GaAs$ structures. Some further results of the intravalence band absorption of strained $Si_{1-x}Ge_x/Si$ QW's are also presented and briefly discussed.

2. Theoretical approach

The results presented here are obtained using the one band effective mass Hamiltonian for electrons and 6 band Luttinger-Kohn Hamiltonian (with terms due to strain) for

M.L. Sadowski et al. (eds.), Optical Properties of Semiconductor Nanostructures, 85–90.

holes.

A variant of the *large box* approach, developed originally for bound states by Fishman [5], is employed. Thus, the structures are considered to be enclosed between two distant impenetrable barriers, which are parallel to the interfaces. In our calculation this is achieved by using a variational basis, all the functions of which vanish at the points $z = \pm L$, where $2L$ is the *large box* width.

2.1. TRANSMISSION COEFFICIENT

The electronic quantum transmission coefficients T of the considered structure's layers are necessary to determine the current density for a given bias voltage, both with and without illumination. The standard definition of T requires two degenerate solutions of the effective mass equation to exist, both of them having plane wave asymptotic form (flat bands beyond the QW structure are assumed). One obtains only non-degenerate states when using the *large box* approach with a fixed width L. Therefore to obtain T for a given energy E_0, we are led to use two different widths (L_S and L_C) such that for each width there exists a state at the same energy E_0. By forming linear combinations of the two corresponding envelopes in the smaller L *large box*, we obtain typical plane wave solutions on the right (wave vector k_R) and left (k_L) sides of the structure. The final formula for T reads:

$$T = \frac{k_R}{k_L} \frac{4a^2 \sin^2(k_R \Delta L)}{a^2 b^2 + 1 + 2ab\cos((k_R - k_L)\Delta L)}, \tag{1}$$

where a and $1/b$ are ratios of right-hand to left-hand sides' amplitudes of the two envelopes and $\Delta L = L_S - L_C$. The wave vectors k_R and k_L will be different in case of an external bias along the z axis. Derivation of (1) goes along the same lines as in ref. [6].

2.2. ABSORPTION COEFFICIENT

The formulae we use to calculate the absorption coefficient are similar for example to those used in ref. [7]. The main difference lies in the way the density of states is taken into account. Our approach, suitable for *large box* calculations, is described in ref.[4].

In the calculation of oscillator strengths, the effective masses were assumed to be constant in space. In the case of the valence band we used a simple, approximate form of transition matrix elements, which is similar to that defined in [8], but reduced to the three valence bands, i.e. it does not contain the conduction band influence. Thus only z-polarised transitions are calculated.

3. Results

3.1. CONDUCTION BAND

All structures considered in this subsection host only a single bound subband, while all other states are of the delocalised type.

We first compare profiles of the squared momentum matrix elements r_{if} and transmission coefficient T for a QW type structure which consists of a GaAs 6nm layer clad with $Al_{0.3}Ga_{0.7}As$ 4nm barriers on both sides and $Al_{0.15}Ga_{0.85}As$ contact layers; the bias voltage is zero. The calculated maxima of T and r_{if}, shown in Fig.1, occur at the same energy. This results from the existence of a resonance in the continuum of odd states. The coincidence of maxima is also found in the case of a simple QW, this time due to a virtual state.

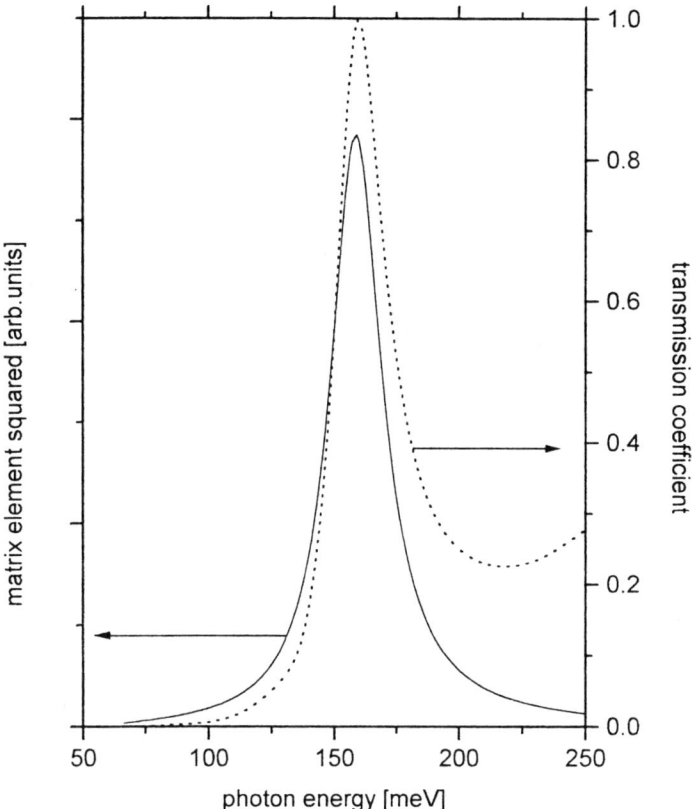

Figure 1. Transmission coefficient and matrix elements of bound to continuum transitions compared. The structure consists of a GaAs 6nm well-acting layer, clad with $Al_{0.3}Ga_{0.7}As$ 4nm barriers, and $Al_{0.15}Ga_{0.85}As$ contact layers.

The next to be considered is a biased Double Barrier Structure with 50nm undoped $Al_{0.27}Ga_{0.73}As$ barriers and a 4nm n-doped GaAs layer between them [9]. The barriers are relatively thick, so the ground subband can be considered to be bound. The voltage drop is assumed to occur only in the barriers, as both contact layers are also n-doped. Both T and r_{if} were again calculated and juxtaposed. However, due to the width of the barriers and relatively high bias, the T coefficient of the whole structure was found to be negligible in the interval of those final state energies which correspond to the region of strong transitions. Consequently, T can be used only to determine the short-circuiting dark current at high temperatures. In view of this, the collector transmission coefficient T_C must be adopted to determine the optical component of current. This was done in our calculation by retaining, as the band profile, only the collector barrier and the voltage drop between the well and the collector (i.e. half of the original bias of 350 meV). The T_C and r_{if} profiles are not parallel as in the previous case (Fig.2). For example, the maxima of r_{if} marked with squares (triangles) in Fig.2 correspond roughly to the maxima (minima) of T_C.

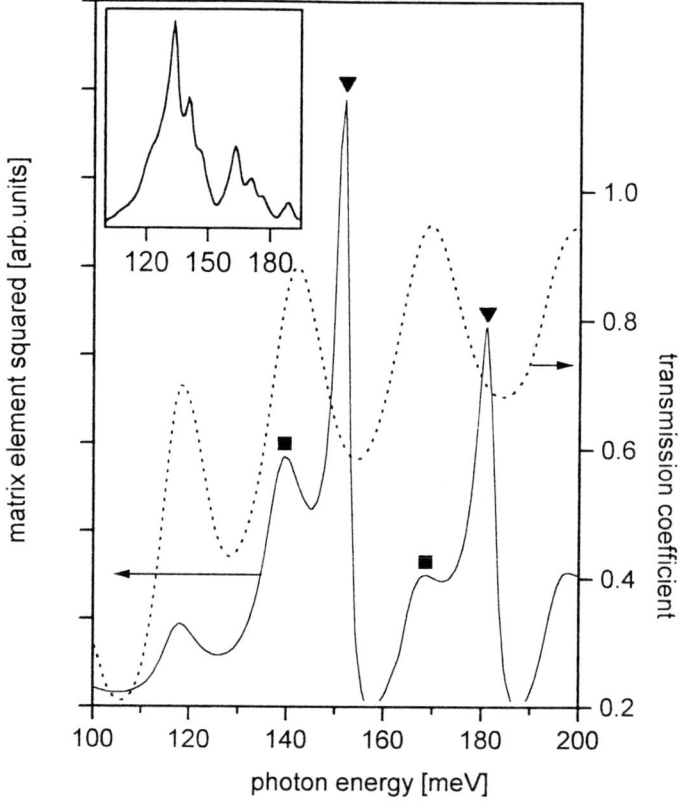

Figure 2. Collector transmission coefficient and matrix elements of bound to continuum transitions compared. The AlGaAs/GaAs Double Barrier Structure (described in the text) is biased (0.35V). The inset shows the absorption coefficient (arb.units) of this structure.

The averaged absorption coefficient profile (for $T = 300$ K and the Fermi level 20meV above the bound subband minimum), shown in the inset of Fig.2, does not repeat the sharp structure of the matrix element profile. This is due to the fact that the ground subband and the continuum onset are not parallel, which leads to shifting of r_{if} profiles with in-plane wave vector k_t values. This, in turn, results in a smearing out of the peaks when the profiles for all relevant k_t's are summed up.

3.2. VALENCE BAND

Two Si/Si$_{0.7}$Ge$_{0.3}$/Si QW's structures of 2 and 3nm width are considered in this section. We use parameters given in [10], assuming a strain induced splitting of 40meV in the central layer. Both wells contain heavy and light hole ground subbands (HH1 and LH1), of which only the first one is relevant. For absorption calculations we have taken $k_F = 0.2$ nm^{-1} and $T = 0$ K.

The above mentioned technique of determining the absorption coefficient proves convenient in the case of a valence band where few continua interact and resonances may exist. These resonances appear when the well-acting layer is relatively thin, as in the case considered here. In such a situation the SO1 (ground level of the spin-orbit derived subband) may be pushed from the interior of the well onto the background of heavy and light hole continua. The heavy and light hole components of

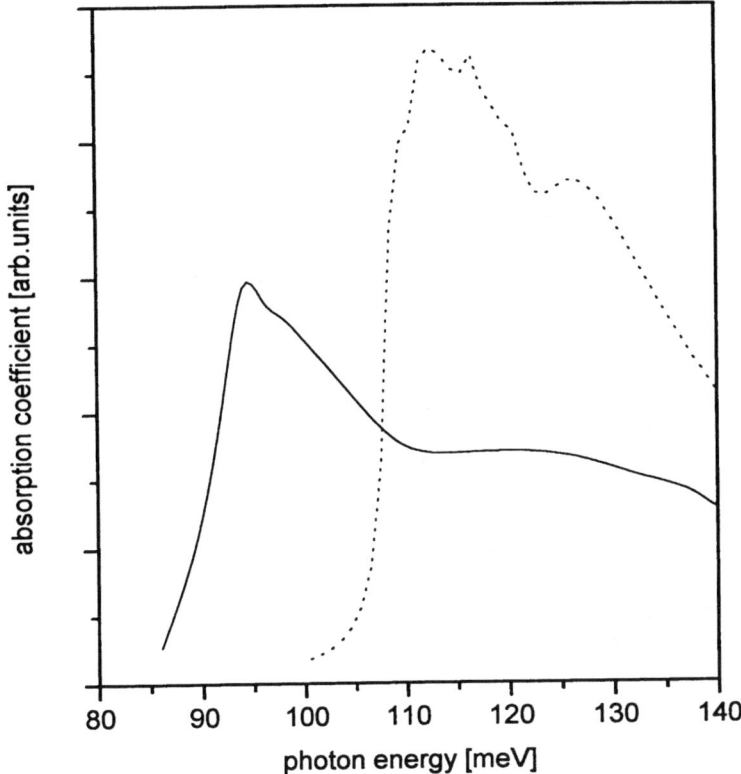

Figure 3. Absorption coefficient of Si/Si$_{0.7}$Ge$_{0.3}$/Si quantum wells of 2nm (solid line) and 3nm (dotted line) width. The low energy maximum is due to HH1-SO1 (resonant level) transitions. The high energy features are mainly HH1-HH continuum transitions.

the SO1 function will be of a delocalised character, while the spin-orbit envelope will be localised in the QW. Another complexity arises from the fact that dispersion curves of the subbands are not parabolic, therefore a level bound for $k_t = 0$ may become delocalised for $k_t \neq 0$ or vice versa. Both resonances and subbands of mixed character are easily handled in the proposed method, because there is no need to distinguish types of states while summing oscillator forces over small intervals of energy [4].

The results of calculations of the absorption coefficient are presented in Fig.3. In both cases the peak at low energies is due to the SO1 levels. However, in the 30nm QW the SO1 subband becomes bound for $k_t \geq 0.05$ nm^{-1}. The bound-bound HH1-SO1 transition is smeared out by exactly the same effect as in the case of conduction band.

4. Conclusion

The *large box* method is a convenient tool for determining absorption spectra of heterostructures where bound to continuum transitions take place. Electron transmission coefficients are also easily obtained in this model. Thus the method is a promising framework for modelling certain semiconductor device properties.

References

1. Manasreh , M. O. (ed.) (1993) *Semiconductor Quantum Wells and Superlattices for Long-Wavelength Infrared Detectors*, The Artech House Materials Science Library.
2. Levine, B. F. (1994) Quantum-well infrared photodetectors, *J. Appl. Phys.* **74**, R1.
3. Załużny, M. (1995) Infrared quantum-well devices, *Optoelectron. Rev.* **3-4**, 86.
4. Oszwałdowski, R. and Jaskólski, W. (1998) Infrared absorption of multiple quantum wells: bound to continuum transitions, *Acta Phys. Pol.* **94**, 473.
5. Fishman, G. (1995) Hole subbands in strained quantum-well semiconductors in [hhk] directions, *Phys. Rev. B* **52**, 11132.
6. Trzeciakowski, W. (1993) The density of states in the continuous spectrum of one-dimensional systems, *J. Phys. Condens. Matter* **5**, 105.
7. Tadić, M. and Ikonić, Z. (1995) Bound-free intersubband absorption in p-type doped semiconductor quantum wells, *Phys. Rev. B* **52**, 8266.
8. Oszwałdowski, R. and Fishman, G. (1999) Infrared absorption in Si-Si$_{1-x}$Ge$_x$–Si quantum wells, *Physica E* **4**,11.
9. Bandara, K.M.S.V. et al (1993) Optical and transport properties of single quantum well infrared photodetectors, *J. Appl. Phys.* **74**, 1826.
10. Dekel, E.et al. (1997) Recombination processes in SiGe/Si quantum wells measured by photoinduced absorption spectroscopy, *Phys. Rev. B* **56**, 15734.

OPTICAL INVESTIGATION OF COUPLED GaAs/Al$_{0.3}$Ga$_{0.7}$As DOUBLE QUANTUM WELLS SEPARATED BY AlAs BARRIERS

G. SĘK, K. RYCZKO, M. CIORGA, L. BRYJA, M. KUBISA,
J. MISIEWICZ
*Institute of Physics, Wrocław University of Technology, Wybrzeże
Wyspiańskiego 27, 50-370 Wrocław, Poland*

M. BAYER, J. KOETH, A. FORCHEL
*Institute of Physics, University of Würzburg, Am Hubland, D-97074
Würzburg, Germany*

Electronically symmetric coupled double quantum wells are structures where two quantum wells are separated by a thin barrier layer. Both experiment and theory show that when the barrier is so narrow that there is considerable overlap of wave functions in the two wells, the single quantum well electronic one-particle states split into symmetric and antisymmetric states with different energy levels. The splitting of energy levels is very sensitive to barrier width and barrier height and increases with decreasing barrier width and with decreasing barrier height.

We present an investigation of DQW structures with various widths of the separating barriers. We use photoreflectance (PR) spectroscopy which, as a modulation technique, is a very sensitive method for investigating optical transitions in semiconductor systems of reduced dimensionality, even at room temperature. In reduced-dimensional systems, it has been shown that PR produces a lineshape that is the first derivative of the unmodulated optical constants [1,2]. As a supplementary method we also use low temperature photoluminescence excitation (PLE) spectroscopy.

The investigated structures are undoped symmetric GaAs/Al$_{0.3}$Ga$_{0.7}$As/ Al$_{0.35}$Ga$_{0.65}$As double quantum wells separated by thin AlAs spikes grown by solid source molecular beam epitaxy along the [001] direction on semi-insulating substrates. For the two quantum wells a width of 7.5 nm was chosen in all samples, in which the width of separating barrier was varied. Four different samples were investigated with barrier widths of 1, 3 and 5 monolayers (ML) and without AlAs barrier (corresponding to a wide single quantum well with a width of 15 nm).

The PR and PLE measurements were performed with an energy step and spectral resolution below 1 meV. The low temperature measurements were carried out in a continuous-flow liquid helium cryostat at a temperature of 10 K. The details of the PR apparatus have been described elsewhere [3].

In Figure 1 the room temperature photoreflectance spectrum for a structure with a 1 ML thick AlAs barrier is shown. The whole spectrum consists of two groups of PR resonances. The first one, starting above the energy of the GaAs-related feature, is

M.L. Sadowski et al. (eds.), Optical Properties of Semiconductor Nanostructures, 91–95.
© 2000 *Kluwer Academic Publishers. Printed in the Netherlands.*

connected with transitions between confined states in the DQW. The second one is probably related to transitions between confined states and resonant states in the quasi continuum region above the band edge of the $Al_{0.35}Ga_{0.7}As$ barrier.

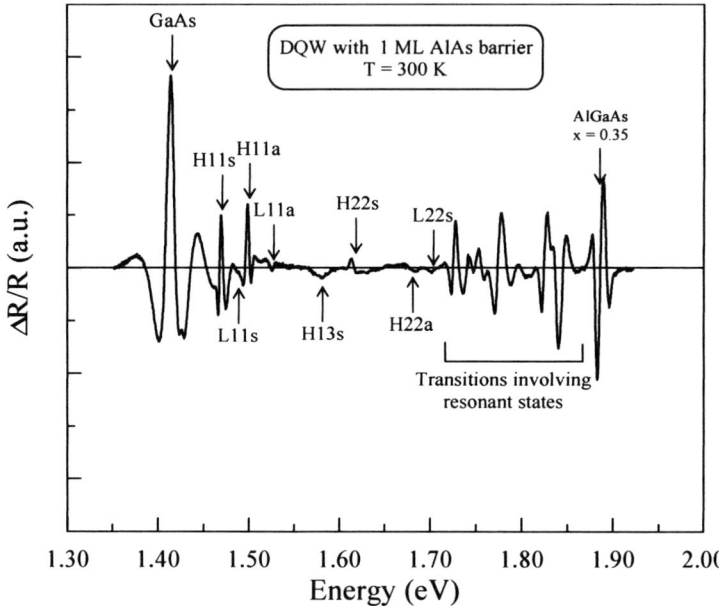

Figure 1. Room temperature photoreflectance spectrum of a GaAs/$Al_{0.3}Ga_{0.7}As$/$Al_{0.35}Ga_{0.65}As$ DQW structure with a 1 ML AlAs barrier. Arrows indicate the energies of confined transitions.

The experimental energies of the confined transitions were obtained from a least-squares fitting procedure according to the first derivative of Gaussian lineshape, the most appropriate profile of the PR signal in the case of transitions involving confined states at room temperature [1,2]. The transitions are labelled according to the notation H(L)mn, which denotes a transition between the m-th conduction and n-th valence subbands of heavy (H) or light (L) holes, respectively. The indexes s and a indicate transitions between symmetric and antisymmetric states, respectively.

The energies of the observed transitions were compared with the results of theoretical considerations. We used the transfer-matrix formalism [4,5] in the effective mass approximation to calculate the energy levels and envelope wave functions of both confined and resonant states in the investigated DQW structure. The material parameters used here were taken after reference [6].

In Figure 2 the AlAs barrier width dependence of all observed confined transitions energies is shown. The solid lines are the theoretical curves and the circles are the experimental data. It can be clearly seen that with the increase of the AlAs barrier thickness the coupling between the wells weakens and the splitting between the symmetric and antisymmetric transitions decreases. The transition energies tend toward the energies of the 7.5 nm well with infinitely thick barriers. It is also seen that the splitting energy between symmetric and antisymmetric transitions is smaller for heavy holes. This can be understood on the base of the following relation, which gives the splitting energy between the symmetric and antisymmetric single-particle states

$$\Delta_{a-s} = E_a - E_s = \frac{E}{\pi} \exp\left(-\sqrt{\frac{2mE}{\hbar^2}} L\right) \tag{1}$$

In Eq. (1) E is the energy of the electron or hole in the decoupled single quantum well, m is the electron/hole effective mass and L is the separating barrier width. We can see that the splitting energy Δ_{a-s} is strongly dependent on the barrier width, the barrier height (via E) and the carrier mass. Hence, for different masses of holes the splitting has a different value.

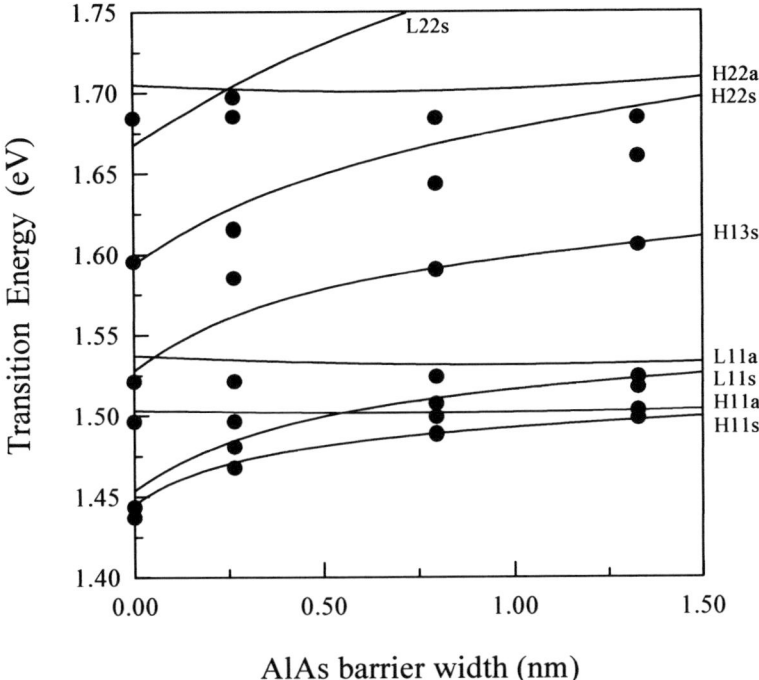

Figure 2. The energies of confined transitions as a function of the width of the separating AlAs barrier. Solid lines – calculations; circles - experimental data from the room temperature PR spectra.

The experimental results agree quite well with the calculations. However, some discrepancies can be seen. The theory slightly overestimates the energies of transitions obtained from the experiment, probably due to the exciton binding energy, which was not included into the calculations. The results are in good agreement with those reported previously for such structures by Bayer at al. [7] from photoluminescence excitation measurements.

The high-energy part of the spectrum in Fig. 1 looks like a superposition of several sharp resonances. To resolve them and to analyse them in detail we performed measurements in this region at 10 K. The comparison of low temperature PLE and PR spectra of the sample with a 1 ML AlAs barrier is presented in Fig. 3. In Table 1, we

94

compare the experimentally obtained transition energies with the calculated ones, assuming that the transitions occur between confined and resonant states. Almost all the transitions between the confined heavy hole states and the electron resonant states above the band edge of the $Al_{0.35}Ga_{0.65}As$ barrier are recognised. The discrepancy of about 6 meV between theory and experiment is mainly due to the exciton binding energy. We observe such a large number of transitions because the common selection rules for confined states are no longer valid in the case of resonant states and all possible transitions become allowed. An analogous analysis was also made for the remaining structures. Unfortunately, no significant dependence of the transitions involving resonant states on the barrier thickness was obtained.

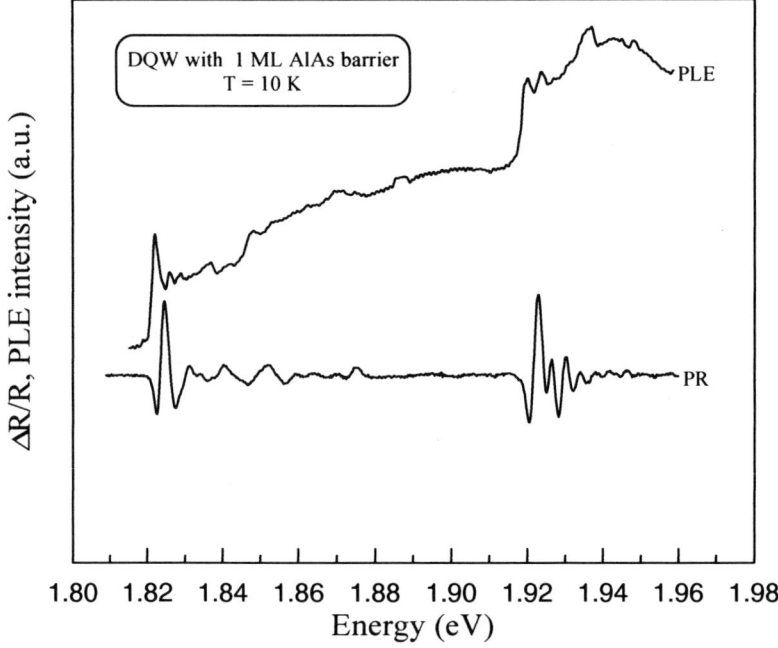

Figure 3. Low temperature PR and PLE spectra in the region of transitions involving resonant states for a DQW with a 1 ML separating barrier.

Concluding, we may say that photoreflectance spectroscopy allowed us to observe transitions between confined symmetric and antisymmetric states in DQW structure at room temperature. In addition to the usually observed confined transitions we also observed transitions involving resonant states in the quasi continuum above the band edge of the $Al_{0.35}Ga_{0.65}As$ barrier. These transitions were analysed in detail taking into account the low temperature results of PR and PLE measurements.

TABLE 1. Experimental and calculated values of the transition energies (in eV) for a $GaAs/Al_{0.3}Ga_{0.7}As$ DQW structure with a 1 ML thick AlAs separating barrier.

Transition	Calculations	PR	PLE
H1 – R1	1.829	1.824	1.822
H2 – R1	1.832	1.826	1.826
H2 – R2	1.835	1.831	1.829
H2 – R5	1.846	1.840	1.836
H2 – R7	1.854	1.848	1.848
H2 – R8	1.860	1.855	—
H3 – R3	1.870	1.866	—
H4 – R1	1.875	1.871	1.871
H4 – R3	1.881	1.875	—
H4 – R8	1.903	1.900	—
H5 – R2	1.928	1.922	1.920
H5 – R3	1.931	1.925	1.924
H5 – R4	1.934	1.928	—
H5 – R5	1.938	1.931	—
H5 – R6	1.942	1.936	1.936
H6 – R2	1.948	1.942	—
H6 – R3	1.951	1.946	—

References

1. Glembocki, O. J., Shanbrook, B. V. (1992) Photoreflectance spectroscopy of microstructures, in D. G. Seiler and C. L. Littler (eds.), *Semiconductors and Semimetals*, Academic Press, New York, vol. 36, pp. 221-292.
2. Pollak, F. H (1994) Modulation spectroscopy of semiconductors and semiconductor microstructures, in M. Balkanski (ed.), *Handbook on Semiconductors*, Elsevier Science, Amsterdam, vol. 2, pp. 527-635.
3. Misiewicz, J., Jezierski, K., Sitarek, P., Markiewicz, P., Korbutowicz, R., Panek, M., Sciana, B., and Tłaczała, M. (1995) Photoreflectance characterisation of GaAs/AlGaAs structures grown by MOCVD, *Adv. Mater. Opt. Electron.* 5, pp. 321-324.
4. Bastard, G. (1992) Wave mechanics applied to semiconductor heterostructures, *Les Editions de Physique*, Paris, pp. 31-61.
5. Worren, T., Ozanyan, K.B., Hunderi, O., Martelli, F. (1998) Above-barrier states in $In_xGa_{1-x}As/GaAs$ multiple quantum well with a thin cap layer, *Phys. Rev. B* 58, pp. 3977-3988, and references therein.
6. Adachi, S. (1985) GaAs, and $Al_xGa_{1-x}As$: Material parameters for use in research and device applications, *J.Appl.Phys.* 58, pp. R1-R29.
7. Bayer, M., Timofeev, V.B., Faller, F., Gutbrod, T., and Forchel, A. (1996) Direct and indirect excitons in coupled $GaAs/Al_{0.3}Ga_{0.7}As$ double quantum wells separated by AlAs barriers, *Phys. Rev. B* 54, pp. 8799-8808.

OPTICAL AND RELATED PROPERTIES OF THE SYNTHETIC QUASI-TWO-DIMENSIONAL SEMICONDUCTORS $K_2Cd_3S_4$, $Rb_2Cd_3S_4$ AND $Cs_2Cd_3S_4$

G. C. PAPAVASSILIOU, I. B. KOUTSELAS, G. A. MOUSDIS,
J. A. KAPOUTSIS, E. A. AXTELL III*, AND M. G. KANATZIDIS*,
Theoretical and Physical Chemistry Institute, National Hellenic Research Foundation,48, Vassileos Constantinou Ave., Athens 116/35, Greece
*Department of Chemistry, Michigan State University, East Lansing, MI 48824 – 1322, USA

During the last few years, a large number of low-dimensional (LD) semiconducting systems based on inorganic units, i.e., quantum wells (2D), quantum wires (1D), and quantum dots (0D) have been prepared and studied (for a review see [1]). It has been found that, when the dimensionality of the inorganic-units network decreases, the peak position of excitons is shifted to higher energy with increased excitonic binding energy and oscillator strength. In the case of PbI_6-based perovskites, for example, the excitonic peak positions exhibit shifts of 778, 264, and 638 meV for the "transitions" 3D→2D, 2D→1D, and 1D→0D, respectively. Similar effects have been observed for the PbX_6 based materials (X=I, Br) with fractional dimensionality (see [1-3]).

In this paper the optical and related properties of $K_2Cd_3S_4$ and similar synthetic quasi-two-dimensional (q-2D) semiconductors ($Rb_2Cd_3S_4$, and $Cs_2Cd_3S_4$) are described. The results are compared to those of the corresponding 3D system (i.e. CdS).

The preparation and crystal structure determination of $K_2Cd_3S_4$ and other CdS-based compounds are described in [4]. These compounds have been obtained in single crystal form (plates). Optical absorption (OA) spectra of thin deposits [5] were recorded on a Varian model 2390 spectrophotometer. Photoluminescence (PL) and resonance Raman (RR) spectra were recorded on a Jobin-Yvon model HG2S Raman spectrophotometer using Argon and Krypton Lasers. Resonant Raman excitation (RRE) spectra were observed from rotating single-crystal mosaics on a KBr compressed pellet containing an internal standard. The out-of-resonance Raman spectra were recorded on a Bruker model RFS 100 spectrophotometer using a Nd-YAG Laser. Far-IR spectra were recorded on a Bruker model IFS 113v spectrophotometer. Photoconductivity (PC) spectra were recorded as in [6]. All measurements were performed at room temperature.

Fig. 1 shows the crystal structures of $K_2Cd_3S_4$ and $Cs_2Cd_3S_4$ as well as the structure of CdS (hexagonal) for comparison. $Rb_2Cd_3S_4$ is isostructural with $K_2Cd_3S_4$. One can see that the compounds form $(Cd_3S_4)_n^{2n-}$ layers interspersed with K^+ (or Rb^+ or Cs^+) cations [4].

The in-plane (layers) dc-conductivity at room temperature is *ca* 10^{-4} S/cm, while the out-of-plane conductivity is almost 100 times smaller. The compounds

M.L. Sadowski et al. (eds.), Optical Properties of Semiconductor Nanostructures, 97-100.
© *2000 Kluwer Academic Publishers. Printed in the Netherlands.*

98

characterised as quasi-2D semiconductors, in which the $(Cd_3S_4)_n^{2n-}$ layers are the active part of the system and $A^+(A=K, Cs, Rb)$ or A_2S play the role of barrier.

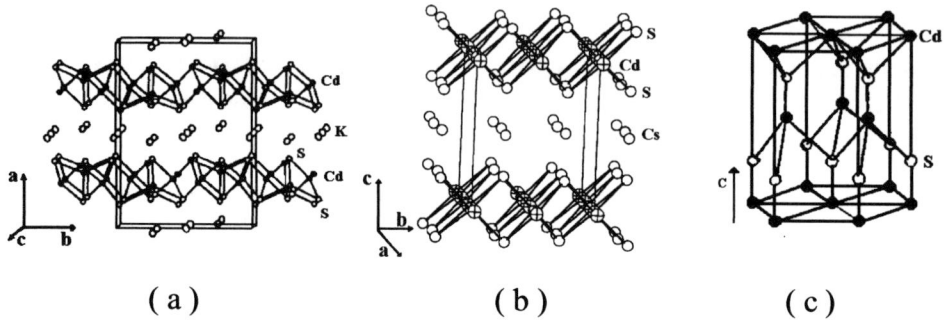

(a)　　　　　　　　(b)　　　　　　　　(c)

Figure 1. Crystal structures of $K_2Cd_3S_4$ (a), $Cs_2Cd_3S_4$ (b) and CdS (c).

Fig.2 shows the OA spectra of $K_2Cd_3S_4$ (two samples), $Rb_2Cd_3S_4$ and $Cs_2Cd_3S_4$ as well as the OA spectrum of CdS, for comparison. One can see that the OA spectra of q-2D compounds exhibit broad bands at ca 474 nm, while that of CdS at 496nm. These bands are attributed to excitons. Fig.3a and Fig. 3b show the PL and RR spectra observed from the surface of freshly cleaved single crystals of $K_2Cd_3S_4$ and CdS, respectively. The broad bands at 484 nm and 505 nm are the excitonic bands, while the sharp lines with shifts of *ca* 303, 607, 911 etc cm^{-1} are the lattice vibrational lines, i.e., the fundamental longitudinal optical mode (LO) and its overtones (2LO, 3LO, etc), arising from the exciton-phonon interaction (see also [3,7]). Figure 3a' shows the PL and RR spectra of $K_2Cd_3S_4$ after ageing. The aged samples exhibit the RR lines. They do not exhibit the excitonic PL band but a broader band (not shown in Fig. 3a') in the region 550-660 nm, which varies from crystal to crystal (see also [4,7]). The effects are similar to those obtained from other CdS -based systems (bulk, thin films and small particles or clusters) [4,8-11].

Fig. 4 and Fig. 5 show the corresponding RRE and PC spectra of $K_2Cd_3S_4$ and CdS single crystals. As in the cases of the OA and PL spectra, the RRE and PC excitonic bands of $K_2Cd_3S_4$ are blue-shifted in comparison to those of CdS. This is due to the confinement of excitons, as in the case of other synthetic LD semiconductors or the LD artificial systems based on conventional semiconductors (see for example [1]).

The spectra of freshly cleaved crystals covered with a polymer do not show ageing effects, i.e., the crystals remain stable for long time.

The PL, RR, RRE, and PC spectra of $Rb_2Cd_3S_4$ and $Cs_2Cd_3S_4$ are the same as the corresponding spectra of $K_2Cd_3S_4$ in accordance with the OA spectra (Fig. 2). However, the out-of-resonance Raman and far-IR spectra of $Rb_2Cd_3S_4$ are the same as $K_2Cd_3S_4$, but different from those of $Cs_2Cd_3S_4$ (and CdS), due to the different symmetry (Fig. 1). Fig. 6 shows the out-of-resonance Raman spectra of $K_2Cd_3S_4$, $Rb_2Cd_3S_4$, $Cs_2Cd_3S_4$ and CdS, single crystals. Fig. 7 shows the far-IR reflectance spectra of $K_2Cd_3S_4$, $Cs_2Cd_3S_4$, $Cs_2Cd_3S_4$, and CdS single-crystal mosaics.

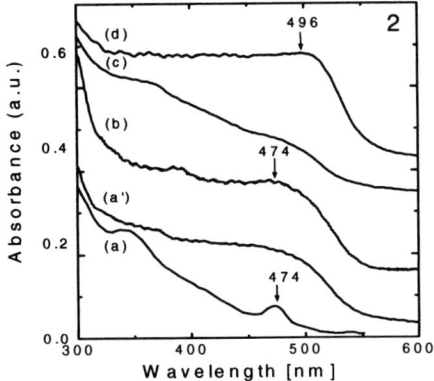

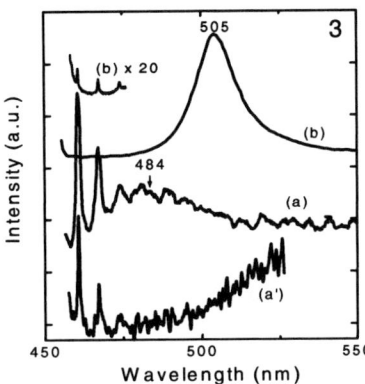

Figure 2. OA spectra of K₂Cd₃S₄ (a, a'), Rb₂Cd₃S₄ (b), Cs₂Cd₃S₄ (c) and CdS (d).

Figure 3. PL and RR spectra of K₂Cd₃S₄ (a,a') and CdS (b) freshly cleaved single crystals (a,b) and after ageing (a') (excitation line 454.5 nm).

Fig. 8 shows the $-\mathrm{Im}(1/\varepsilon)$ and ε_2 spectra of $K_2Cd_3S_4$ and CdS obtained from a Kramers-Krönig analysis of the reflectance spectra (where $\varepsilon=\varepsilon_1+i\varepsilon_2$). Two common features are the dip in the reflectance spectra and the corresponding peak in $-\mathrm{Im}(1/\varepsilon)$, i.e. the LO vibrational mode around 300 cm^{-1}, which is also observed in the RR spectra of CdS-based systems. The far-IR structures in the region *ca* 50 - *ca* 200 cm^{-1} are attributed to A_2S-vibrations. However, details on the lattice vibrational spectra of these compounds will be published elsewhere.

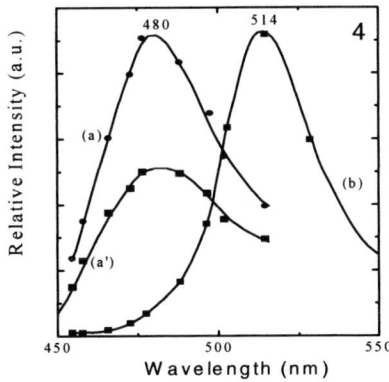

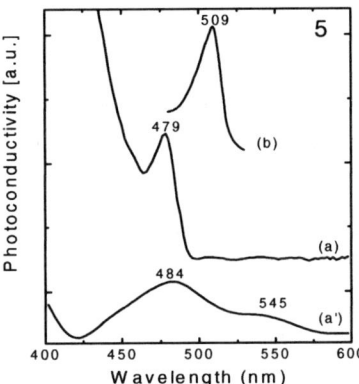

Figure 4. As in Fig. 3, but for RRE spectra of single-crystal mosaics.

Figure 5. As in Fig. 3, but for PC spectra of single-crystals.

100

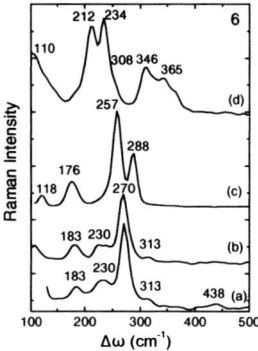

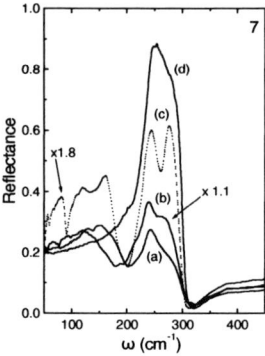

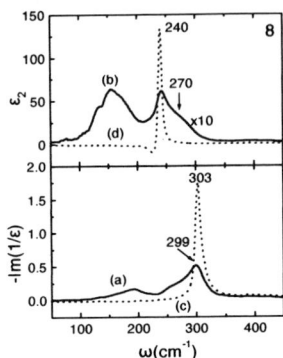

Figure 6. Out-of-resonance Raman spectra of $K_2Cd_3S_4$ (a), $Rb_2Cd_3S_4$ (b), $Cs_2Cd_3S_4$ (c) and CdS (d), single crystals (excitation line 1064 nm).

Figure 7. Far-IR reflectance spectra of $K_2Cd_3S_4$ (a), $Rb_2Cd_3S_4$, (b) $Cs_2Cd_3S_4$ (c), and CdS (d) single- crystal mosaics.

Figure 8. -lm(1/ε) and ε$_2$ spectra of $K_2Cd_3S_4$ (a,b) and CdS (c,d) obtained from Kramers-Krönig analysis of reflectance spectra.

In conclusion, the electronic spectra of the q-2D semiconductors $K_2Cd_3S_4$, $Rb_2Cd_3S_4$ and $Cs_2Cd_3S_4$ show excitonic bands at shorter wavelengths than those of CdS. These new systems as well as $K_2Cd_3S_{4-x}Se_x$, $K_2Cd_{3-y}Mn_yS_4$, etc. [4,7] can be considered as candidates for optoelectronic applications as CdS and other similar systems based on CdSe, $CdS_{1-x}Se_x$, ZnS, $Cd_{1-y}Mn_yS$, etc. (see for example [9-12]).

References

1. G. C. Papavassiliou, Progr. Solid State Chem., **25**, 125 (1997).
2. G. C. Papavassiliou, Mol. Cryst. Liq. Cryst., **286**, 231 (1996).
3. G. C. Papavassiliou, et al, Adv. Mater. Opt. Electron., **8**, 263 (1998).
4. E. A. Axtell III, J.-H. Liao, Z. Pikramenou, and M.G. Kanatzidis, Chem. Eur.J., **2**, 656 (1996); E. A. Axtell III, PhD Thesis, Michigan State University (1995).
5. Thin deposits were obtained by rubbing the materials on quartz plate or chromatographic paper and surrounding by CCl₄, see G. C. Papavassiliou, Z. Phys. Chemie, Leipzig., **258**, 174 (1977). G. C. Papavassiliou, et al, J. Chem. Soc., Faraday Trans. II, **78**, 17 (1982).
6. G. J. Papaioannou et al, J. Appl. Phys. **65**, 4864 (1989).
7. G. C. Papavassiliou et al, Synth. Metals, in press (1999); Electrochem. Soc. Proc. 98/25, 161 (1999).
8. G. C. Papavassiliou, J. Solid State Chem., **40**, 330 (1981).
9. V. S. Gurin, J. Phys.: Condens. Matter., **6**, 8691 (1994).
10. G. C. Papavassiliou et al, Mol. Cryst. Liq. Cryst., **253**, 103 (1994).
11. T. Yamashi et al, Radiat. Phys. Chem., **50**, 199 (1997); T. Miyoshi et al, Solid State Commun., **104**, 451 (1997); W. Chen et al, Ibid., **105**, 129 (1998); S. Okamoto et al, Ibid., **105**, 7 (1998); K. Tanno et al, Colloids and Surfaces A, **141**, 143 (1998); G. Counio et al, J. Phys. Chem. B, **102**, 527 (1998); A. Kasuya et al, Supramol. Sci. **5**, 235 (1988): B. Ullrich et al, Solid State Commun., **109**, 757 (1999); and references cited therein.
12. Y. Yamada et al, J. Cryst. Grouth. **138**, 570 (1994); P. Gupta et al, J. Phys. D: Appl. Phys., **26**, 1709 (1993); M. Nyman et al, Chem. Mater. **10**, 914 (19988); X. Zhang and M. Izutsu, Jpn. J. Appl. Phys., **37**, 6025 (1988); M. Grus and A. Sikorska, Physica B, **266**, 139 (1999).

INTERNAL TRANSITIONS OF NEUTRAL AND NEGATIVELY CHARGED MAGNETOEXCITONS IN GaAs NANOSTRUCTURES : THE IMPORTANCE OF SYMMETRY

B.D. McCOMBE[†], H.A. NICKEL[†], G. KIOSEOGLOU[†], Z. X. JIANG[†],
T. YEO[†], H.D. CHEONG[†], A. PETROU[†], A.B. DZYUBENKO[*],
A.YU. SIVACHENKO[#], AND D. BROIDO[‡]

[†] *Department of Physics and Center for Advanced Photonic and Electronic Materials, University at Buffalo, State University of New York, Buffalo, NY 14260, USA*
[*] *General Physics Institute, Russian Academy of Sciences, Moscow 117942, Russia*
[#] *The Weizmann Institute of Science, Rehovoth 76100, Israel*
[‡] *Department of Physics, Boston College, Chestnut Hill, MA 02167, USA*

Abstract. Cyclotron Resonance (CR) of free electrons and holes and internal transitions of their bound complexes have been studied in GaAs/AlGaAs quantum wells in high magnetic fields by Optically Detected Resonance (ODR) spectroscopy. Two nearly degenerate $1s \rightarrow 2p_+$ internal exciton transitions (IETs), as well as two well-separated $1s \rightarrow 2p_-$ IETs resulting from the two distinct heavy-hole magneto-excitons were observed. Measurements of both electron and hole CR and these IETs in the same sample evidence the predicted consequences of the symmetry of the magneto-exciton Hamiltonian. Extensions of these studies to samples that showed neutral (heavy-hole hhX) as well as negatively charged (X^-) exciton photoluminescence-lines have permitted the first observation of internal transitions of X^-. Both singlet and triplet transitions of X^- were observed, and new physics associated with translational invariance for this interesting three-particle complex was revealed. The observation of electron and hole CR and several IETs in a single sample exemplifies the promise of the ODR technique for the study of semiconductor nanostructures.

1. Introduction

The fundamental interband excitation in semiconductors is the well-known Coulomb-correlated electron-hole complex, the exciton. In bulk GaAs excitons are hydrogen-like with a Bohr radius of about 100 Å and a binding energy of about 5 meV. In quantum-well structures that confine the charge carriers strongly, a series of such excitons appears corresponding to the various interband transitions between confined electron and hole subbands; a quasi-2D exciton is associated with each of the allowed interband

M.L. Sadowski et al. (eds.), Optical Properties of Semiconductor Nanostructures, 101–116.
© 2000 *Kluwer Academic Publishers. Printed in the Netherlands.*

transitions. Neutral excitons have been studied in a number of different systems for many years.

Since the initial observation of negatively charged excitons in CdTe quantum wells [1], there has been considerable interest in <u>charged excitons</u> in quasi-2-dimensional (2D) semiconductor systems. The evolution of the exciton system in quasi-2D as a function of excess electron concentration from isolated neutral excitons to "isolated" negatively charged excitons to a few-hole/many-electron plasma (the metal-insulator transition) and the effects of a magnetic field on these transitions have constituted a focus of recent investigations [2-4]. In most of these studies the electron (or hole) density was varied in a <u>single</u> quantum well by a gate voltage or by optical means while monitoring photoluminescence (PL) or absorption features. Very recently, the interpretation of the PL feature attributed to X^- by a number of authors has been called into question [5]. These authors have suggested that it is associated with a neutral-donor- (in the barrier) bound exciton (D^0-X_b), but there is no strong evidence favoring this interpretation, and considerable evidence favoring the "conventional" interpretation [6].

There are many superficial similarities between the neutral and charged exciton systems and neutral and charged hydrogenic donors. However, there are also important differences related to the free motion of the complex in the quantum-well (QW) plane in the former case. For neutral excitons and donors the distinction appears most clearly in the difference between the cylindrical-symmetry-derived relationship equations relating the internal transition energy differences and the cyclotron resonance(s). For X^- and its negative donor ion counterpart, D^-, there are even more striking differences in the basic physics and the optical transitions; these derive from the free motion of the complex, and the continua of final states of intra-excitonic transitions.

For neutral excitons there is a series of hydrogen-like excited states with internal transitions satisfying the requisite electric-dipole selection rules, *e.g.*, s to np in the usual 3D, spectroscopic notation at zero magnetic field. At low magnetic fields these energy levels and transition energies are only slightly modified, leading to the usual Zeeman splittings. At very high magnetic fields, on the other hand, the magneto-exciton states are greatly changed and are more easily understood by considering the Coulomb interaction to be a perturbation on the high-field, free-carrier (electron and hole) Landau states. In this limit the intra-excitonic transitions can be considered to be Coulomb-shifted electron cyclotron resonance (CR) (evolving from the $1s \rightarrow 2p_+$ low field transition) and Coulomb-shifted hole CR ($1s \rightarrow 2p_-$), with $\pm$ denoting the quantum number ($m = \pm 1$) for projection of orbital angular momentum. The transition energy is blue-shifted from the corresponding bare CR energy in each case.

In QW systems with excess electrons in the well the negatively charged exciton [1,2] is the ground state. Due to the additional binding energy of the second electron, the recombination energy of this complex is lowered (by about 10-11 cm^{-1} at zero field for a 20 nm QW) with respect to the neutral, heavy-hole exciton, hhX. Internal transitions of X^- are naively expected to be similar to the internal transitions of the shallow donor analog (D^-), which have been extensively studied [7].

Study of the internal transitions offers an additional tool for understanding the excitonic state in the dilute situation and its evolution with excess electron density and magnetic field. For high exciton densities it may also provide a possibility to probe exciton-exciton interactions and, for properly designed structures, the signature of an exciton condensate.

2. Theoretical background

The Hamiltonian for a two-dimensional neutral exciton in a magnetic field can be written

$$H = \frac{p_{rel}^2}{2m_r} + \frac{e^2 B^2}{8m_r c^2} - \frac{e^2}{\varepsilon r} + \frac{1}{2}(\omega_{ce} - \omega_{ch})\hat{l}_z + \frac{e}{Mc}\vec{B} \bullet [\vec{r} \times \vec{K}] + \frac{\hbar^2 K^2}{2M}, \tag{1}$$

where m_r is the reduced electron-hole mass, $M = m_e + m_h$ with $m_{e(h)}$ the electron (hole) mass, l_z is the operator for projection of the relative angular momentum along the magnetic field direction, B is the magnetic induction, r is the relative coordinate in the plane perpendicular to B, ε is the background dielectric constant, and K is the momentum of the center of mass of the exciton. The last two terms in Eq. (1) are associated with the motion of the center of mass; the next-to-last term describes the coupling between the center-of-mass motion and the relative motion [8]. The magneto-PL lines are predominantly transitions with K = 0, and at K = 0 the solutions of the Schrödinger equation with this hamiltonian comprise a set of discrete states that evolve continuously from the usual 3D, low field hydrogenic states with increasing confinement and magnetic field. In this sense they can be labeled with the usual 3D quantum numbers at low fields (1s, $2p_{\pm1,0}$, $3d_{\pm2,\pm1,0}$, etc. in the spectroscopic notation). At high fields, the states can be labeled by the electron (hole) Landau quantum number, n_e (n_h), and the quantum number for projection of relative angular momentum, $m_z = n_e - n_h$. Allowed electric-dipole transitions satisfy the selection rule $\Delta m_z = \pm 1$. It can be shown [8] that as a result of the cylindrical symmetry leading to Eq. (1), the following important relationship is satisfied: $E_{1s\rightarrow np+} - E_{1s\rightarrow np-} = \hbar(\omega_{ce} - \omega_{ch})$, where $\omega_{ce(h)}$ is the electron(hole) cyclotron resonance frequency, and $E_{1s\rightarrow np+(-)}$ is the transition energy from the ground state to an excited state.

The three-particle problem of the negatively charged exciton is considerably more complex, and we provide here a simple, physically based, qualitative description. A detailed discussion, including an analytical demonstration of the existence of families of dark (localized) states resulting from the hidden (translational) symmetry, will be published elsewhere [9]. We consider a strictly-2D system in the limit of high magnetic fields (no Landau-level mixing). In this case the only bound state associated with the lowest two-electron/hole Landau levels is the X^--triplet; there are no bound X^--singlet states [10, 11]. This is in contrast to the B = 0 case, where the singlet is bound, and there are no bound triplet X^- states. The binding of X^- results from a delicate balance between the e-e repulsion and e-h attraction. As a result, the X^--triplet binding energy in the lowest Landau levels (LLs) is very small, $0.043 E_0$; here $E_0 = [\pi/2]^{1/2}[e^2/\varepsilon l]$ is the characteristic 2D-donor binding energy in high magnetic fields with $l = [c\hbar/eB]^{1/2}$, the characteristic magnetic length.

The X^- states can be labeled by exact quantum numbers: the total spin of the electrons, singlet (S) or triplet (T), the total angular momentum projection, $M_z = m_{ze1} + m_{ze2} + m_{zh}$, and, in high fields, by the total electron (hole) LL number $N_e(N_h)$ ($|N_e,N_h,M_z; S(T)>$), with $N_e = n_{e1} + n_{e2}$ and $N_h = n_h$. In a magnetic field the X^- states are macroscopically degenerate in M_z; this is just the LL degeneracy for the charged complex. For example, the degenerate bound X^--triplet states in zero LLs are $|00,M_z;T>$, with $M_z = -1,-2,-3 \dots$.

To discuss the internal X^- transitions, it is necessary to consider the eigenstates associated with higher LLs. We have found that in the next LL (one electron promoted to $n_e = 1$), there also exists only one bound triplet state, $|10,M_z;T>$ (also macroscopically degenerate, $M_z = 0,-1,-2\dots$) and no bound X^--singlets. The binding energy of the $|10,M_z;T>$ state is 0.086 E_0, twice that of $|00,M_z;T>$. Physically, the increased binding is due to the fact that the two electrons in the excited triplet state can occupy the 1s (zeroth electron LL) and 2s (first electron LL) single particle states. This enhances the e-h attraction relative to the ground triplet state in which the two electrons occupy an antisymmetric combination of the 1s (zeroth LL) and 2p_ (zeroth LL) single particle states; the latter is more spatially extended than the 2s state. These states are shown in Fig. 1.

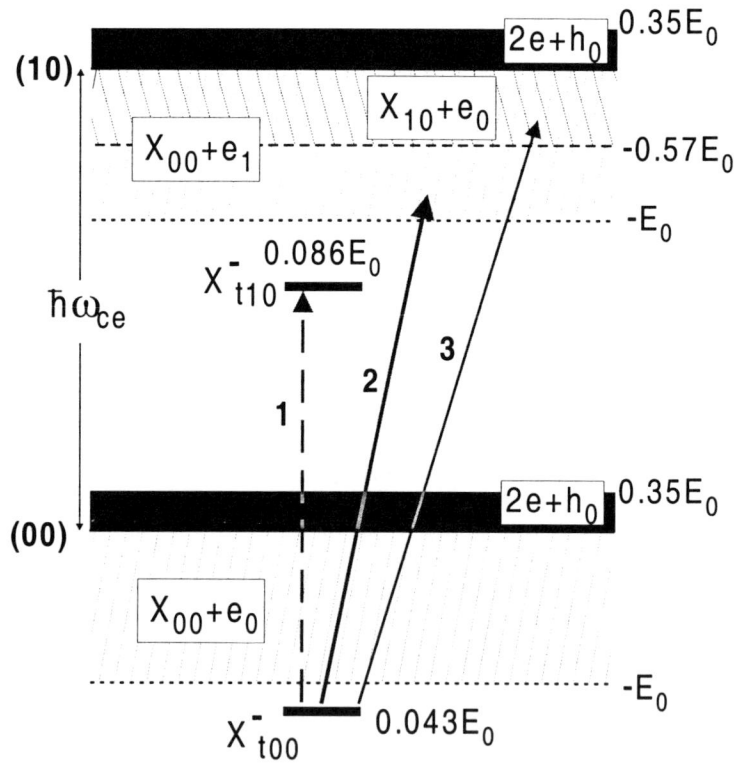

Figure 1. Schematic diagram of the triplet energy states associated with the second electron in two-electron/hole Landau levels for a strictly two-dimensional, high magnetic field limit. The notation is discussed in the text. The shaded bands labeled $2e + h_0$ represent electron-hole scattering states.

As mentioned in the introduction, there is a fundamental difference between the eigenspectra of the X^--states and D^--states in quasi-2D in a magnetic field. For the latter (and, generally, for any electron system in B) the eigenspectra are completely discrete, and all eigenstates are spatially localized. For X^- (and, generally, for any electron-hole system in B) the spectra consist of discrete bound states and a continuum. The continuum corresponds to the motion of a neutral magneto-exciton as a whole. Therefore, for the quasi-2D D^- centers in a magnetic field all internal transitions are bound-to-bound, and there are no photoionizing transitions. For X^-, however, both bound-to-bound and photoionizing inter-LL transitions to the continuum are possible.

Far infrared spectra, in general, would encompass both bound-to-bound and bound-to-continuum transitions. The bound-to-bound transition $|00,M_z;T> \rightarrow |10,M_z;T>$ shown by arrow 1 in Fig. 1 is allowed by the usual selection rules ($M_z \rightarrow M_z+1$ for σ^+ polarization, spin conserved). It lies below e-CR in energy. This is the analog of the strong T^--transition for the D^- center [12]. However, an initially surprising result of numerical calculations for a 20 nm wide quantum well was that this transition for X^- has negligibly small (10^{-4}) oscillator strength. Dzyubenko and Sivachenko have recently shown [9] that in the 2D, high field limit this is strictly forbidden due to an additional selection rule arising from a hidden symmetry of the Hamiltonian (translational invariance). Final three-particle states in the first LL can also belong to a continuum, which has a rich structure. This includes a band of energies of width E_0 below the first electron-hole LL. This band corresponds to the 1s exciton plus a scattered electron in the first LL, labeled $X_{00}+e_1$. A second band begins at the first electron-hole LL minus $0.57E_0$ and extends to the first LL. This corresponds to a $2p_+$ exciton plus a scattered electron in the ground LL, labeled $X_{10}+e_0$. These bands are shown by the hatched regions in Fig. 1.

The FIR absorption spectra reflect this rich structure of possible final states. The transitions to the $X_{00}+e_1$ continuum are dominated by a sharp onset at the edge indicated by transition 2 in Fig. 1, *i.e.*, at an energy = e-CR plus the $|0,0,M_z;T>$ binding energy. In addition, there is a broader and weaker peak corresponding to the transition to the lower edge of the $X_{10}+e_0$ magneto-exciton band indicated by arrow 3 in Fig. 1. The latter may be thought of as the $1s \rightarrow 2p_+$ internal transition of the magneto-exciton, which is shifted and broadened by the presence of the second electron. Thus the X^--triplet behaves physically in the FIR absorption as an exciton that very loosely binds an electron, and the two "parts" of the complex can absorb the FIR photon, to some extent, independently. Transition 2 corresponds to exciting the loosely bound electron to the next LL, while transition 3 corresponds approximately to exciting the "strongly bound" electron to an excited state within the X_{00} complex, while leaving the loosely bound electron in the lowest LL. At finite fields and confinement these qualitative features are preserved.

3. Experimental details

A powerful technique that circumvents some limitations associated with magneto-PL (primarily sensitive to s-like transitions), or far infrared (FIR) magneto-transmission spectroscopy (non-existent or very weak signals for undoped samples or samples with

106

low carrier densities) has recently been developed. It has been applied to the investigation of the electronic properties of semiconductors and semiconductor nanostructures [14], in particular, to the study of internal excitonic transitions in type I multiple quantum wells (MQWs) [15,16]. Optically detected resonance (ODR) spectroscopy combines the sensitivity of visible/near infrared (IR) photon detection with a far-IR laser probe of low energy excitations [14-20]. In typical implementations the sample is illuminated simultaneously by two laser beams, one visible and one FIR. The chopped FIR laser excites electronic transitions whose energies are tuned into resonance with the photon energy by an applied magnetic field. The resulting changes induced in the near IR PL simultaneously excited by a visible laser having photon energy greater than the effective bandgap of the structure under investigation are synchronously detected with a lock-in amplifier. The output signal is proportional to changes in intensity and/or shape of a particular band-edge photoluminescence (PL) feature, which have been induced by the resonant absorption of FIR radiation.

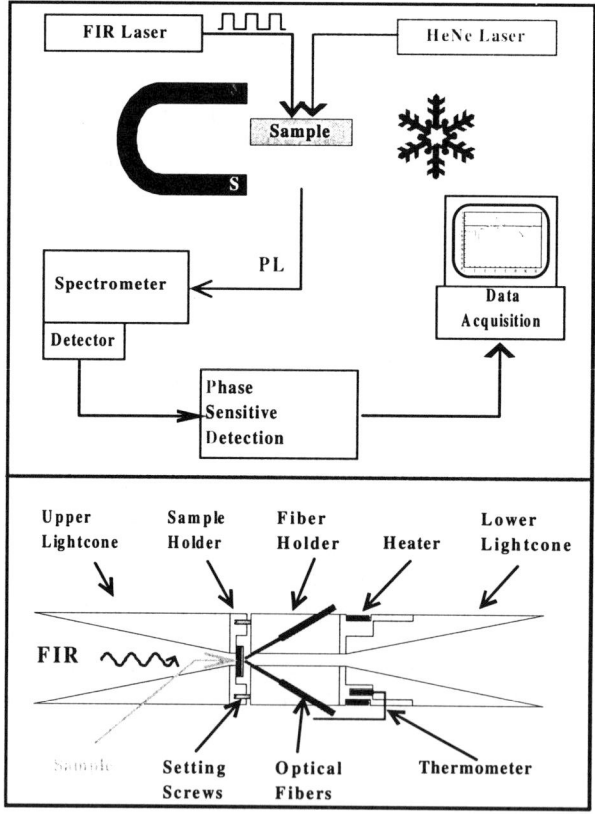

Figure 2. Top - Schematic diagram of the experimental setup for ODR. Bottom - Cross section of the sample holder showing the fibers and FIR light cones.

Among the major advantages of ODR spectroscopy are very high sensitivity and spectral specificity, both for absorption in the FIR and for the near IR PL lines. The former permits studies of undoped materials and structures (including single quantum wells and heterostructures) and detection of resonances involving very low (photoexcited) densities of free carriers, excitons and impurities. The latter offers the possibility of obtaining detailed information about mechanisms of energy transfer from the various internal transitions excited by the FIR to the different recombination channels. A new variant of this technique, involving monitoring changes in the reflectivity features induced by resonant far infrared absorption, has been developed very recently [21].

In the present low temperature experiments electrons, holes and excitons were continuously created in multiple-quantum-well (MQW) samples by optical excitation with the 6328 Å line of a He-Ne laser via an optical fiber as shown in Fig. 2. The sample holder is shown in the lower panel. The PL from the structure was collected with a second optical fiber, analyzed with a single 0.75 m, grating-monochromator and detected with a Si photodiode or a cooled photomultiplier. A CO_2-pumped molecular-gas laser was used to generate FIR radiation at a number of wavelengths ranging from 70.1 to 513 μm. The chopped FIR radiation excites various FIR transitions -- electron or hole CR, and IETs in the present case. The energies of these transitions are tuned into resonance by an applied magnetic field of up to 17 T. A lock-in amplifier referenced to the FIR laser chopper was used in conjunction with a dedicated personal computer to record the ODR signal (the difference in PL with the FIR laser on and off). The computer was also programmed to step the spectrometer drive to track the center of the peak of a particular selected PL feature as it shifted with magnetic field; spectrometer resolution was selected to incorporate most of the excitonic PL line. The sensitivity of this system corresponds to changes in PL signal of better than 0.1%.

A number of samples have been studied during the course of this work, but for the sake of simplicity and brevity the results for four are described in detail. The barrier composition for samples 1,3 and 4 is x = 0.3, and that of sample 2 is 0.15. Sample 4 is δ-doped with Si in the barriers. The other characteristics are given in Table I.

4. Neutral magnetoexcitons

TABLE I: Characteristics of the GaAs/Al$_x$Ga$_{1-x}$As MQW samples used in this study. The barrier composition of Samples 1,3 and 4 is x = 0.3; the barrier composition of Sample 2 is x = 0.15

Sample Number	Layer Width Well/Barrier (nm)	Nominal Doping Well/Barrier ($\times 10^{10}$ cm^{-2})	Number of Repetitions
1	12.5/12.5	0/0	30
2	20/15	0/0	40
3	20/60	0/0	20
4	20/40	0/2	40

Studies of internal excitonic transitions (IETs) and CR in a 12.5 nm well-width sample (sample 1 of Table I) by ODR spectroscopy have been previously reported [15]. With this technique in a single series of experiments on a single sample, electron CR, hole

CR, and several internal transitions of excitons were measured, thereby verifying and quantifying the predicted results of the symmetry inherent to such systems, as discussed above [8].

The two principal hole CR transitions originating from the different spin states of the highest heavy-hole subband (hh1) Landau level ($|0,+3/2>\rightarrow |1,+3/2>$ and $|0,-3/2>\rightarrow|1,-3/2>$) are clearly identified from comparison with theoretical calculations. Here the hole states are labeled $|n_h\ m_J>$, where $n_h = n' - m_J - 1/2$, with n' the harmonic oscillator index, and m_J the z-component of total angular momentum ($J = 3/2$). The hole wavefunction takes the closed form: $[F_{n'-2,\ 3/2},\ F_{n'-1,1/2},\ F_{n',\ -1/2},\ F_{n'+1,-3/2}]$. The following Luttinger parameters were used in the calculation: ($\gamma_1 = 6.85$, $\gamma_2 = 2.1$, $\gamma_3 = 2.9$ and $\kappa = 1.2$.

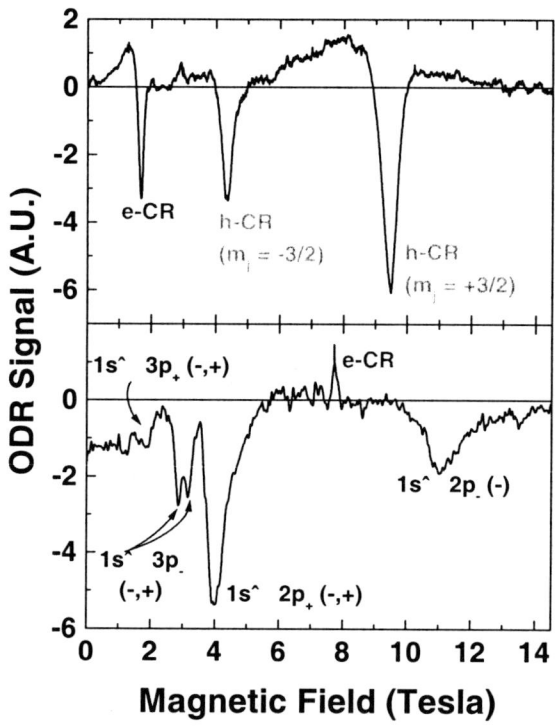

Figure 3. ODR scans from Sample 1 at 4.2 K as a function of magnetic field for FIR laser wavelengths of 432 μm (upper panel), and 96.5 μm (lower panel). IETs in the lower panel are marked with (-) and/or (+) to indicate the electron and hole spin states of their constituents, *i.e.*, (+1/2, -3/2) or (-1/2, +3/2), respectively.

An example of an ODR scan showing electron CR and well-separated $m_J = \pm 3/2$ hole CRs is shown in the upper panel of Fig. 3 at a FIR laser wavelength of 432.8 μm. The predominantly negative-going ODR signals reflect a decrease in the PL associated with X . The simultaneous observation of electron and hole CR along with the IETs (shown in the lower panel) allow the symmetry-derived relationship to be verified, and the complex spectrum of IETs to be disentangled and identified as discussed below.

The lower trace in Fig. 3 shows several of the IETs as well as electron CR at a FIR laser wavelength of 96.5 μm. The 1s→2p+ IETs, *two* resolved 1s→3p. IETs and one of the corresponding 1s→2p. IETs are identified. The two 1s→np. lines result from internal transitions of the [+1/2,-3/2] and [-1/2, +3/2] excitons (the first number denotes the spin of the electron and the second the spin of the participating hole). There is a near degeneracy of the two 1s→2p+ transitions (over this field region) due to the very small spin splitting in the conduction band. In the high field limit the 1s→2p+ transitions correspond to promotion of an electron from the lowest to the first LL. Since the energy difference between e-CR(+1/2) and e-CR(-1/2) is extremely small in this field range, the two corresponding allowed 1s→2p+ features are not resolved. On the other hand, the 1s→2p. transitions from the two different excitons correspond to promoting a hole from the lowest to the first hole LL. There is a large energy separation between the two 1s→2p. transitions, due to the large splitting between the ±3/2 h-CRs.

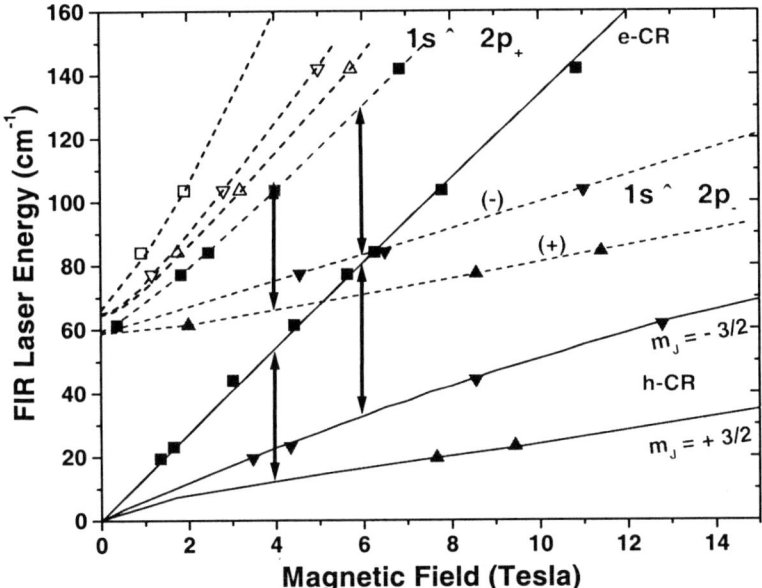

Figure 4. Summary plot of the transition energies of the observed IETs and CR vs. magnetic field for sample 1. Dashed lines are guides to the eye. Solid lines for e-CR and h-CR are calculations as described in the text. The open symbols are 1s→3pᵥ transitions. The lower arrows show the energy difference between e-CR and h-CR(+3/2) (at 4T) and h-CR(-3/2) (at 6T). The upper arrows are the lower arrows vertically displaced to originate on the 1s→2p.(+) transition (at 4T) and the 1s→2p.(−) transition (at 6T).

The highest energy excitonic feature of Fig. 3 is tentatively identified as the 1s→3p+ transitions (nearly degenerate as in the 2p+ case). Comparison of the energy difference between electron and hole CR with that of the 1s→3p. to 1s→3p+ transitions is in agreement with this assignment. However, it is difficult to make this identification unambiguously due to the sparse data set and the weakness of the higher energy IETs.. At 118.8 μm the two 1s→3p. transitions are not resolved and appear as a single feature.

Figure 4 summarizes the observed internal excitonic transitions and electron and hole CRs for different FIR laser energies. The solid lines labeled ±3/2 are calculated hole CRs for the corresponding spin states. Agreement with experiment is good. The solid line through the electron CR data points is the result of a three-band model calculation of the nonparabolicity of the conduction band. The resonant magneto-polaron interaction with LO phonons, which depresses the CR energy measurably at magnetic fields above about 10 T, is ignored.

Arrows are shown at magnetic fields of 4 and 6 T to display the electron/hole CR, 1s→np± relationship. The length of the lower double-headed arrows represents the energy difference between electron CR and the appropriate hole CR determined from the experimental data. With the bottom of the arrow placed at the respective 1s→np. transition, the top of the arrow should give the position of the 1s→np+ transition. The CR energy differences are very close to the 1s-2p± energy differences, verifying the predicted symmetry-based relationship within experimental error.

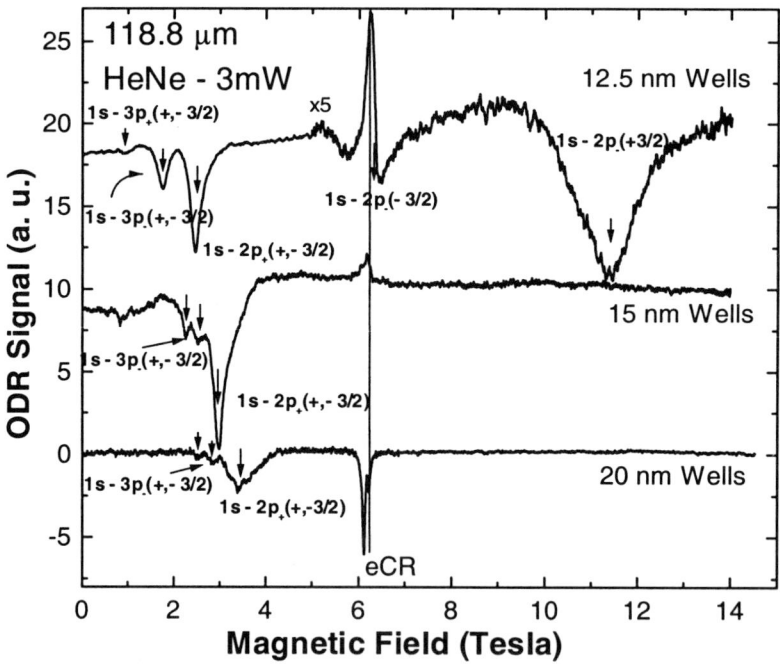

Figure 5. Comparison of ODR spectra for Samples 1, 2 and an undoped 15 nm well-width sample at a laser wavelength of 118.8 :m. The various transitions are labeled

A comparison of ODR spectra at 118.8 μm for Samples 1 and 2 (and a third sample having 15 nm wells) is given in Fig. 5. The systematic shift of the IET features to higher fields with increasing well width is apparent. The e-CR shows very little dependence on well width over this range. The shift to higher fields (lower energies) of the IETs is expected based on the fact that the reduced confinement leads to smaller binding (and IET) energies at B = 0.

Figure 6 provides a summary of ODR data for sample 2 for a number of FIR laser lines. As anticipated, the ODR features for the wider wells occur at lower energies than the corresponding transitions from the narrower-well sample. Hole CRs were observed at only two FIR laser energies, and only one CR was discernable for the m_J = -3/2 state. The calculations shown for this well width are for a barrier composition of x = 0.3, while this sample has a composition of 0.15 and a correspondingly smaller barrier height. Agreement with experiment is, not surprisingly, poorer than for sample 1. Calculations are presently underway for the reduced barrier height. The sparse data set makes it more difficult to check the symmetry-derived relationship for this sample. However, as indicated by the arrows the results in this case agree reasonably well with the prediction for the −3/2 state but do not agree at all well for the +3/2 state. Similar data for the 15 nm well-width sample agree quite well for both transitions. It is likely that the disagreement for the widest-well sample (sample 2) is related to the stronger mixing of the valence band states and the lack cylindrical symmetry.

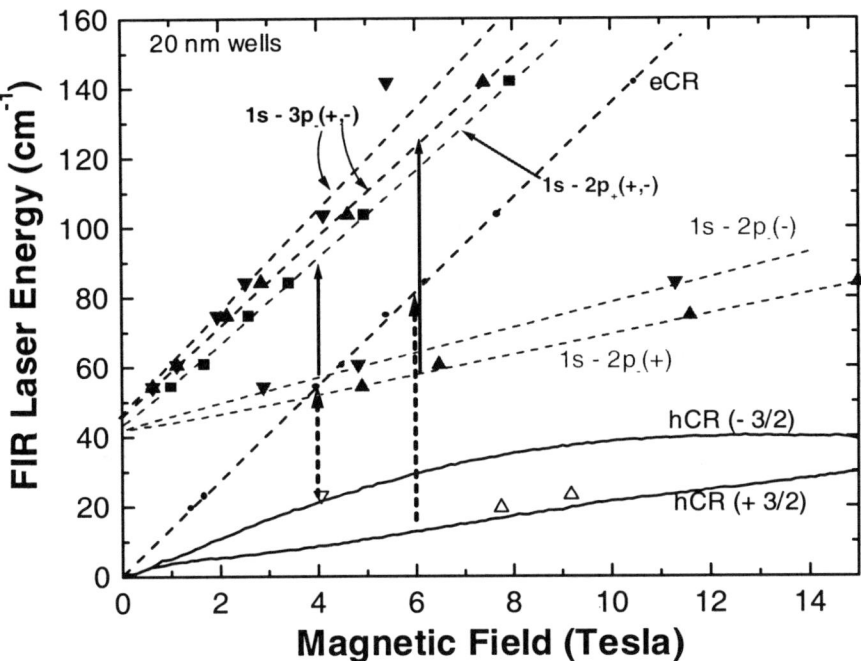

Figure 6. Summary plot of the transition energies of the observed IETs and CR vs. magnetic field for sample 2. Dashed lines are guides to the eye. Solid lines for the h-CRs are calculations described in the text; the assumed composition was x = 0.3. The lower arrows show the energy difference between e-CR and h-CR(+3/2) (at 4T) and h-CR(-3/2) (at 6T). The upper arrows are the lower arrows vertically displaced to originate on the 1s→2p_(+) transition (at 4T) and the 1s→2p_(−) transition (at 6T).

The data for the 12.5 nm wells of sample 1 extrapolate to a zero-field 1s→2p transition energy of 60 cm^{-1}, while the corresponding extrapolated zero-field energy for the 20 nm wells of sample 2 is about 43 cm^{-1}, both in reasonable agreement with calculations [22]. The data for the 15 nm well-width sample of Fig. 5 extrapolate to a 1s→2p energy of 48 cm^{-1}, also in reasonable agreement with calculations.

5. Negatively charged magnetoexcitons

Figure 7a) shows PL data obtained from samples 3 (lower panel) and 4 (upper panel) at zero magnetic field. The increase in the relative intensity of the X$^-$ peak in sample 4 results from the increased electron density in the wells due to the barrier doping. The separation between the two peaks labeled X and X$^-$ is 10.5 cm^{-1}. Assignment of the lower energy peak to X$^-$ is substantiated by the magneto-PL measurements shown in Fig. 7b). These data show the characteristic small initial decrease in recombination energy with magnetic field (the final state of the recombination is the emitted photon plus an electron in the N=0 Landau level).

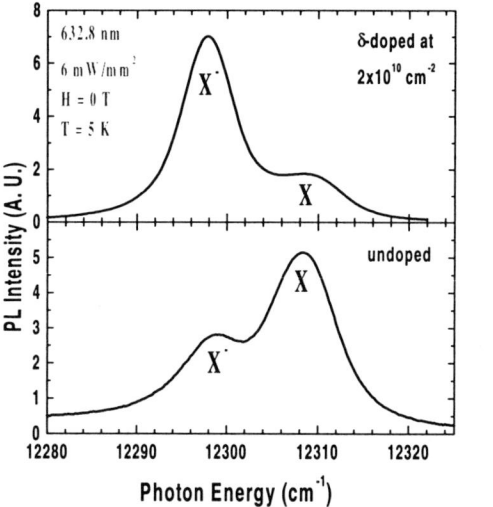
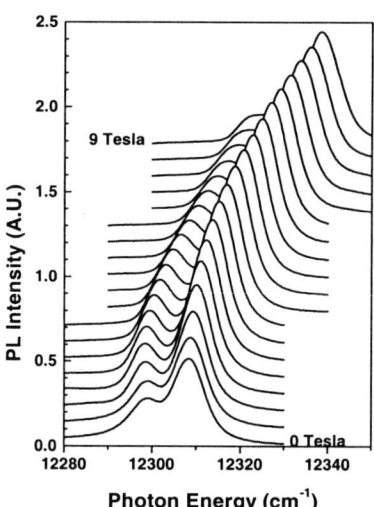

Figure 7. a) Photoluminescence spectra at B = 0 for samples 3 (lower) and 4 (upper). The neutral heavy-hole exciton (X) and the negatively charged exciton (X$^-$) features are indicated. b) Magnetic-field dependence of the X and X^t PL lines for sample 3.

Figure 8 shows examples of ODR scans from samples 3 and 4. The ODR signal obtained by tracking the X$^-$ feature is plotted vs. magnetic field at several FIR laser wavelengths, as indicated. Very similar, but positive-going, ODR spectra are obtained by tracking the X-PL line. The sharp feature at 6.3 T at 117.7 μm is e-CR. Similar sharp e-CR features can be seen in both samples at all wavelengths. Three features attributed to X$^-$-internal transitions are labeled a, b and c in both panels. Features a and c are very strongly enhanced in sample 4, clearly showing that they are related to the excess electron density. The strongest feature in sample 4 is the broad minimum at 3.76 T at 117.7 μm (labeled a); Feature b is also seen in sample 4 at elevated temperatures, shifted to slightly lower field. In sample 3 (undoped) line b is much sharper (and weaker) and occurs about 0.1 T higher in field. Another strong minimum in sample 4, asymmetrically broadened to low fields, appears just below CR. In sample 3 this is a low-field shoulder on CR.

Assignment of the various features is based on the above qualitative trend, theoretical calculations in the high-field, strictly 2D limit [9], and numerical calculations for 20 nm wells at magnetic fields of 9T and above. Naïve comparisons with D⁻ are very misleading in some cases due to the unusual behavior of the internal transitions of X⁻ which result from the translational symmetry.

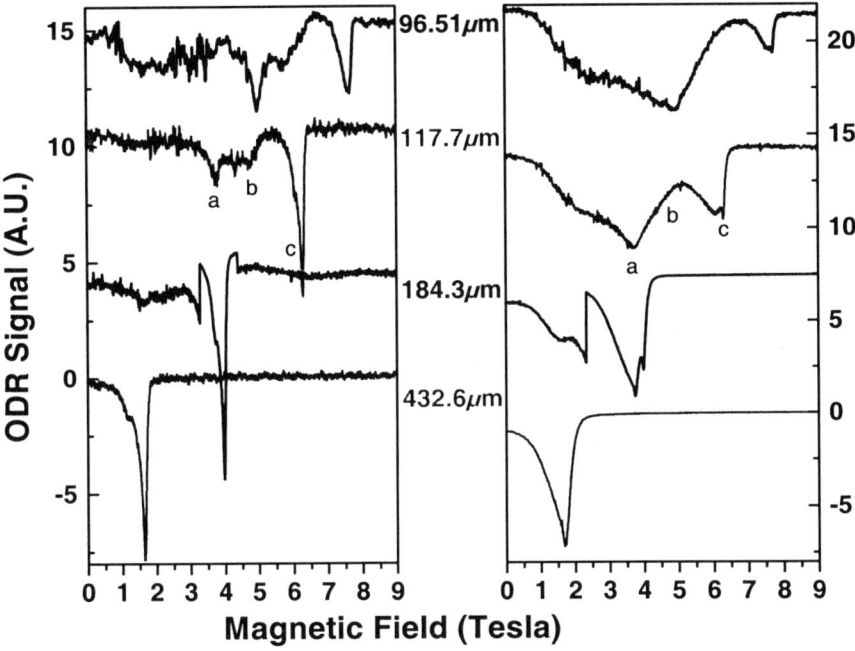

Figure 8. Compilation of ODR field-scans for samples 3 (left panel) and 4 (right panel) at several laser wavelengths. The features labeled a, b and c for the 117.7 :m laser line are the internal transitions of X⁻ discussed in the text.

Figure 9 summarizes the energy positions of the various resonant lines vs. magnetic field for sample 3. The pair of lines with the same slope indicated by the solid upright triangles corresponds to features a and b of Fig. 8. These lines are assigned to singlet continuum transitions analogous to transitions 2 and 3 of Fig. 1. Although they overlap in energy at high fields, the $1s \rightarrow 2p_+$ IETs of X for a 20 nm well begin at an energy (43 cm⁻¹) well-above either of these features at zero field, and have a smaller slope. Thus the possibility that one of the two lines is an IET of X can be excluded. The calculated positions of these transitions for a 20 nm well with the appropriate barrier height are shown by the open circles with x's. The agreement is excellent. The line that lies just below e-CR in magnetic field (just above in energy) is assigned to the dominant triplet (T⁻) transition, from the localized ground state ($|0,0,M_z;T>$) to the $X_{00} + e_1$ continuum in Fig. 1. The peak of this band corresponds to the edge of the continuum, so the observed peak is shifted to higher energies from e-CR by the (small) triplet binding energy with a tail to higher frequencies (lower fields). All the oscillator strength for the X⁻-triplet internal transitions in the Φ^+-polarization is shifted from the bound-to-bound transitions that dominate D⁻ to bound-to-continuum states lying below the first excited LL. For X⁻ the bound-to-bound transitions would correspond to $|0,0,M_z;T>$ to $|1,0,M_z+1;T>$ (arrow 1 in Fig. 1), which have vanishing oscillator strength in 2D in a magnetic field. Numerical calculations predict a peak in absorption about 5.5 cm⁻¹ above e-CR. The experimental peaks lie approximately 3-4 cm⁻¹ above e-CR, in fair agreement.

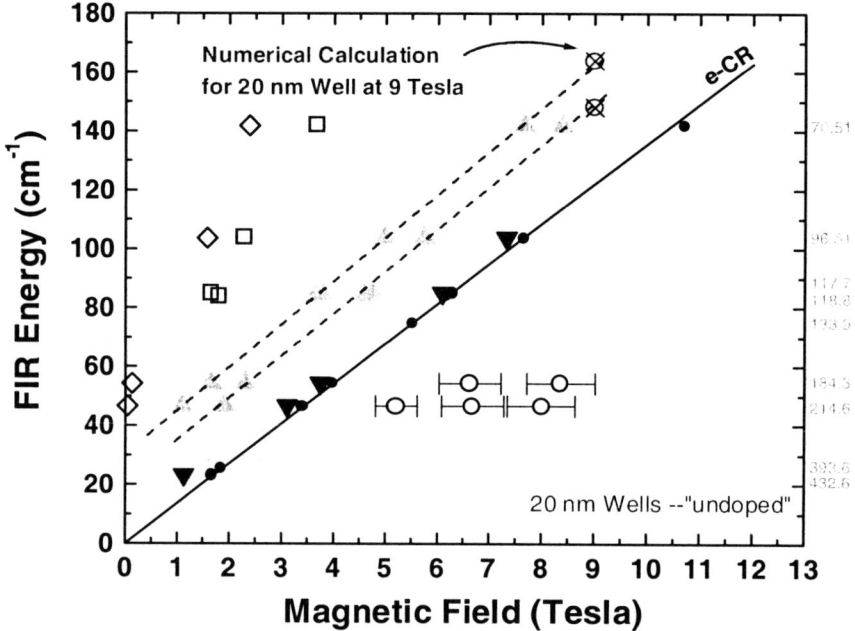

Figure 9. Summary plot of ODR data for sample 3 showing the bands attributed to the X¹-singlet (upright solid triangles) and -triplet (inverted solid triangles) localized-ground-state-to-continuum transitions predicted by theory. The results of a numerical calculation for the singlet transitions are shown by the open circles with x's at 9 T. The small solid circles represent e-CR data points. The open symbols represent weak features discussed in the text. Lines are guides to the eye.

The broad and weak features at high fields and low photon energies (open circles in Fig. 9) are not fully understood. They are observed only in sample 3. It is possible that these broad lines are weak 1s - 2p IETs of X, which dominate at the high fields where they are observed. Such internal transitions may become more effective at high fields in modulating the PL. At low fields the free exciton is weaker, and the dynamics of the processes seem to favor internal transitions of X⁻ in modulating the PL.

The absence of these transitions in sample 4 is understandable based on the dominance of X⁻ and the very weak PL corresponding to neutral excitons.

Finally, the weak features with large slopes in the plot of Fig. 9 appear to be higher excited-state transitions of X⁻ (to continua associated with the $N = 2$ LL).

6. Summary and conclusions

The high sensitivity and general utility of the ODR technique are clear from the present results. The simultaneous detection of e-CR, $m_J = \pm3/2$ h-CRs, and several internal excitonic transitions has allowed a clear observation of the consequences of the symmetry for this electron-hole system. The predicted relationship between the energy differences between the np_+ and np_- internal transitions and the energy differences between e-CR and the $m_J = \pm3/2$ h-CRs has been verified. Calculations of the h-CR energies vs. B are generally in good agreement with the measurements.

Results of ODR measurements on samples having excess electrons in the wells and exhibiting both X and X⁻ PL lines have shown internal transitions of X⁻ from localized, discrete ground states to continua of final states. In the case of the triplet-related transitions, this leads to the peak in the triplet resonance appearing at higher frequencies than e-CR in contrast to analogous results for negative donor ions (D⁻). This is the first experimental evidence for internal transitions of negatively charged excitons (X⁻) in semiconductor quantum wells, and the predicted consequences of the hidden (translational) symmetry of the Hamiltonian [9] are clearly demonstrated.

Experimental observation of IETs for both neutral and charged excitons provide an opportunity to test detailed theoretical models of excitons in confining structures, and to explore the effects localization and magnetic-field-induced "metal-insulator transitions", and exciton-exciton interactions as well as interactions with free carriers.

Acknowledgements. The work at SUNY at Buffalo was supported in part by the National Science Foundation under grant DMR 9722625. We are grateful to G. S. Herold for important contributions to the understanding of the IETs of neutral excitons and to R. J. Heron for a careful reading of the manuscript.

References

1. Kheng, K., Cox, R. T., Daubigne, Y. M. *et al.*, (1993) *Phys. Rev. Lett.*, **71**, 1752.
2. Finkelstein, G., Shtriktman, H. and Bar-Josef, I. (1995) *Phys. Rev. Lett.*, **74**, 976.
3. Shields, A. J., Pepper, M., Ritchie, D. A. *et al.*, (1995) *Phys. Rev. B*, **51**, 18049.
4. Gekhtman, D., Cohen, E., Ron, A. *et al.*, (1996) *Phys. Rev. B*, **54**, 10320.
5. O. V. Volkov, V. E. Zhitomirskii, I. V. Kukushkin *et al.* (1997) *JETP Lett.*, **66**, 766.
6. Kioseoglou, K., Cheong, H. D., Ryu, S. R., Petrou, A., McCombe, B. D. and Schaff, W. (1999*) Phys. Rev B* , submitted.
7. See *e.g.*, Jiang, Z. X., McCombe, B. D., Zhu, Jia-Lin, *et al.* (1997) *Phys. Rev. B* **56**, R1692, and references therein.
8. Dzyubenko, A. B. (1997) *JETP Lett*, **66**, 617.
9. Dzyubenko, A. B. and Sivachenko, A. Yu. (1999) *Phys. Rev. Letters*, submitted.
10. Palacios, J. J., Yoshioka, D. and MacDonald, A. H. (1996) *Phys. Rev. B*, **54**, 2296.

11. Whittaker, D. M. and Shields, A. J. (1997) *Phys. Rev. B*, **56**, 15185.
12. Dzyubenko, A. B. and Sivachenko, A. Yu. (1993) *Phys. Rev. B*, **48**, 14690.
13. Dzyubenko, A. B., to be published.
14. Kono, J., Lee, S. T., Salib, M. S. *et al.*, (1995) *Phys. Rev. B*, **52**, R8654.
15. Salib, M. S., Nickel, H. A., Herold, G. S. *et al.*, (1996) *Phys. Rev. Lett.*, **77**, 1135.
16. Černe, J., Kono, J., Sherwin, M. S. *et al.*, (1996) *Phys. Rev. Lett.*, **77**, 1131.
17. Wright, M. G., Ahmed, N., Koohian, A. *et al.*, (1990) *Semicond. Sci. Techn.*, **5**, 438.
18. Gubarev, S. J., Dremin, A. A., Kukushkin, I. V., *et al.* (1991) *JETP Lett.*, **54**, 355.
19. Warburton, R. J., Michels, J. G., Nicholas, R. J. *et al.* (1992) *Phys. Rev. B*, **46**, 13394.
20. Herold, G. S., Nickel, H. A., Tischler, J. G., *et al.* (1997) *Physica E*, **2**, 39.
21. Kioseoglou, G., Cheong, H. D.,Nickel, H. A., Yeo, T. M., Petrou, A., McCombe, B. D. and Schaff, W. (1999) *Phys. Rev. B*, submitted.
22. Bajaj, K. K., private communication.

TAILORING OF SPIN-DEPENDENT EXCITONIC INTERACTION IN QUANTUM WELLS BY AN ELECTRIC FIELD

G. AICHMAYR[a] , L.VIÑA[a], E.E. MENDEZ[b]
[a] *Dept. Física de Materiales. Universidad Autónoma. E-28049 Madrid. Spain*
[b]*Dept. of Physics and Astronomy, SUNY at Stony Brook, N.Y. 11794-3800, USA*

Abstract. We have studied the dynamics of a spin-polarised excitonic gas under the influence of an external electric field using time-resolved photoluminescence spectroscopy. This extends our previous investigations of spin-dependent excitonic interactions in semiconductor quantum wells and allows us to assess the relative importance of exchange and vertex corrections in the inter-excitonic interaction and to verify the predictions of a recently developed model.

1. Introduction

The spin dynamics of low-dimensional semiconductor heterostructures has been studied intensively both experimentally [1-6] and theoretically [7-9] in the last decade. One of the most intriguing findings, first observed by Damen et al. in GaAs quantum wells (QW's) using time-resolved photoluminescence (TRPL) [1], is the appearance of an energy splitting between excitons with spin +1 and -1 in the absence of any magnetic field. Using circularly polarised (σ^+) light to excite the samples, close to resonant formation of heavy-hole excitons, an energy splitting between the two components, σ^+ and σ^-, of the heavy-hole (*hh*) exciton luminescence has been reported. The σ^+ component is always at higher energies than that of σ^- helicity. The breaking of the spin degeneracy at high excitonic densities was confirmed by femtosecond pump and probe measurements in GaAs QW's under an external magnetic field [2], and later on by TRPL [3,4,10-12]. Theoretical models have been proposed [8,13] to explain only the exciton spin relaxation without taking into account the exciton-exciton interaction. The models give alternative explanations to spin relaxation processes in terms of intraexcitonic exchange [14-17], Dyakonov-Perel [18] and Elliot-Yafet [19,20] mechanisms. However, these free-exciton models fail to describe the spin level splitting. Only recently, a model has been developed to explain the spin splitting as arising from inter-excitonic exchange interaction [9].

Closely related to this behaviour of spin-polarised excitons is the existence of a

M.L. Sadowski et al. (eds.), Optical Properties of Semiconductor Nanostructures, 117–132.
© 2000 *Kluwer Academic Publishers. Printed in the Netherlands.*

blue shift of excitonic transitions, observed in pump and probe experiments in GaAs QW's under high excitation [21,22]. The shift in two-dimensional (2D) systems has been attributed to the repulsive interaction among excitons due to the Pauli exclusion principle [23,24]. The long-range Coulomb correlation interaction is reduced in 2D systems, giving rise to the blue shift of the excitons with increasing density. Some other manifestations of spin-dependent exciton-exciton interactions have been also shown recently [12,25]. They include a spin-dependent optical dephasing time and a linewidth difference between the two photoluminescence components σ^+ and σ^- (the luminescence component co-polarised with the laser excitation is narrower than the counter-polarised one). All these can be interpreted as a result of interexcitonic exchange.

In this work, we will review the current understanding of the splitting in a polarised 2D exciton gas and report new results of the electric field dependence of the exciton interaction obtained by TRPL. The rest of the manuscript is organised as follows: Section 2 gives the experimental details. The splitting between polarised excitons is shown in Section 3. Section 4 presents the tuning of the splitting by an external electric field. Finally, we summarise in Section 5.

2. Experimental details

The experiments were performed in a temperature variable cryostat, exciting the samples with light pulses. These were obtained from either a mode-locked Nd-YAG laser, which synchronously pumped a cavity-matched dye laser, or a Ti:Za mode-locked laser pumped by an Ar$^+$-ion laser. The pulse widths were 5 ps and 1.2 ps for the dye and the Ti:Za systems, respectively. The PL was time-resolved in a standard up-conversion spectrometer. A double grating monochromator was used to disperse the up-converted signal. The exciting light was circularly polarised by means of a $\lambda/4$ plate, and the PL was analysed into its σ^+ and σ^- components.

For the study of the dependence of the exciton interaction on initial carrier concentration we concentrate on the results of a GaAs/AlAs multiquantum well, consisting of 50 periods of nominally 77 Å-wide GaAs wells and 72 Å-wide AlAs barriers. The sample presented a small Stokes shift of ~2.5 meV, which allowed us to perform quasi-resonant excitation at the free heavy-hole exciton, observed in pseudo-absorption experiments (PL excitation), detecting at the weakly-bound exciton seen in PL. The influence of the exciton localisation has been reported in the literature [6]. The electric field dependence of the splitting has been investigated using coupled double quantum wells (cDQW). This sample consisted of 10 periods of two 50 Å wide GaAs wells separated by a 20Å $Al_{0.3}Ga_{0.7}As$ barrier. Each cDQW period is separated by a 200Å $Al_{0.3}Ga_{0.7}As$ layer. The structure is grown on a 1 μm thick n$^+$-doped GaAs:Si layer on a [100] GaAs substrate. The cDQW stack is capped by a 500 nm p$^+$-doped GaAs:Be layer. An undoped $Al_{0.3}Ga_{0.7}As$ buffer layer of 100 nm and 80 nm is placed between the n$^+$- and p$^+$-layers and the 10-period cDQW structure, respectively.

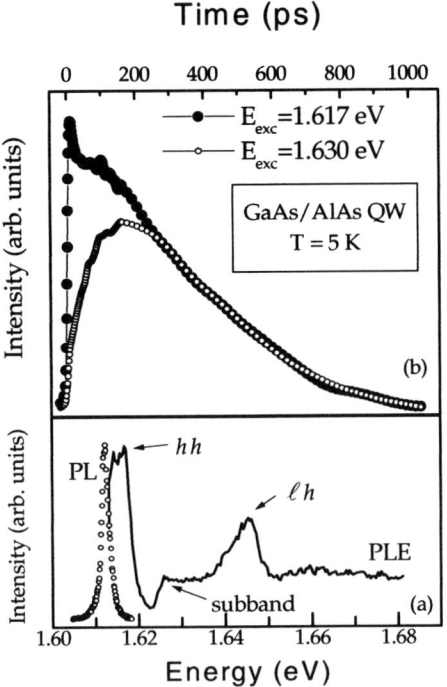

Figure1 a) Photoluminescence (dots) and excitation spectra of an intrinsic 77Å GaAs QW. b) Time decay of the PL exciting at two different energies and detecting at the PL peak.

3. Carrier density dependence of the spin splitting

Figure 1a shows the cw-PL (dots) and the photoluminescence excitation (PLE) spectra (line) of the intrinsic GaAs/AlAs QW. The spectra were recorded at 5K under very low excitation density (5 mW cm^{-2}). The first peak in both spectra is due to the heavy-hole (*hh*) exciton. The Stoke shift between the *hh* exciton peaks in PL and PLE, amounting to 2.5 meV, indicates that the PL originates from localised excitons. However, when the temperature is increased (not shown), the Stokes shift decreases and vanishes at 40K. The linear temperature dependence of the decay time (see below), which is characteristic of free excitons [26], also indicates that localisation effects are not very important. The step at 1.626 eV in the PLE spectra corresponds to the onset of the *hh-e* intersubband continuum, and the peak at 1.646 eV originates from the light-hole (*ℓh*) exciton.

Figure 1b illustrates the dynamics of the light emission. It depicts the time evolutions of the total PL excited with σ^{+} polarised light at two different energies, 1.617 eV –(i), filled circles– and 1.630 eV –(ii), open circles–. Both sets of time profiles were recorded detecting the maximum of the PL band (1.612 eV), with an excitation

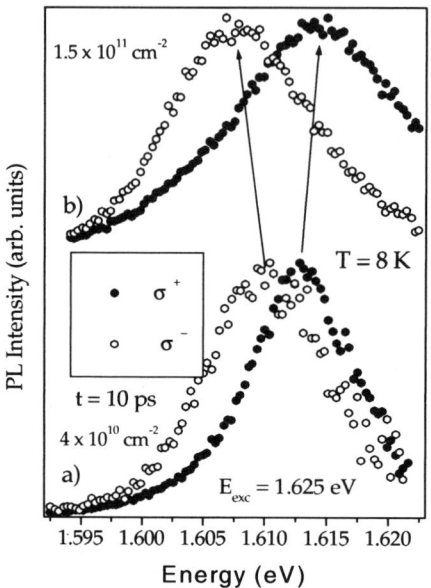

Figure 2 a) RPL spectra of the intrinsic QW taken 10 ps after excitation for two different carrier densities: a) 4×10^{10} cm^{-2} and b) 1.5×10^{11} cm^{-2}. The excitation pulses are σ^+ polarised and the emission is resolved into σ^+ (solid points) and σ^- (solid points) polarisations.

power creating an initial excitonic density of 4×10^{10} cm^{-2}. The large difference in the rise times of the two curves in Fig. 1b reflects the fact that (ii) corresponds to the excitation of electron-hole pairs in the subband continuum ("nonresonant excitation"), and excess energy has to be lost before the exciton can recombine. On the other hand, in the case of curve (i), the $1s$-hh excitons have been created resonantly with $K=0$, and they can couple directly to light. The fast initial decay observed in trace (i) is related to the filling of dark excitonic states due to hole spin flip.

Time-resolved PL spectra at 8 K taken 10 ps after the excitation with σ^+ pulses for two different densities, 4×10^{10} cm^{-2} and 1.5×10^{11} cm^{-2}, excited with 1.625 eV, are shown in Fig. 2. The solid points show the polarised (spin +1) emission while the open circles correspond to the unpolarised (spin -1) PL. With this quasi-resonant excitation, an energy splitting of $\delta = 2.5$ meV is clearly seen between the two peaks in Fig. 2a. When the excitation density is increased, both a broadening of the lines and a strong enhancement of the splitting is observed. Under these conditions, the splitting is mostly due to the red shift of the σ^- polarised emission and exhibits marked time- and excitation-energy dependences. It has been contradictorily reported that the splitting is either due to the energy shift of the luminescence component with the same helicity of the laser pump, the other component being only slightly red shifted [4], or due to the component of the opposite helicity [27]. Theory predicts that the absolute positions of the σ^+ and σ^- emission components depend on the quasi-3D vs. quasi-2D character of the semiconductor system, [9] and we will show in Sect. 4 that those positions can be varied by an external electric field applied to the heterostructures.

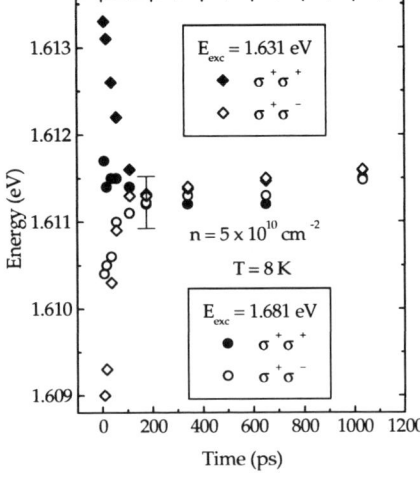

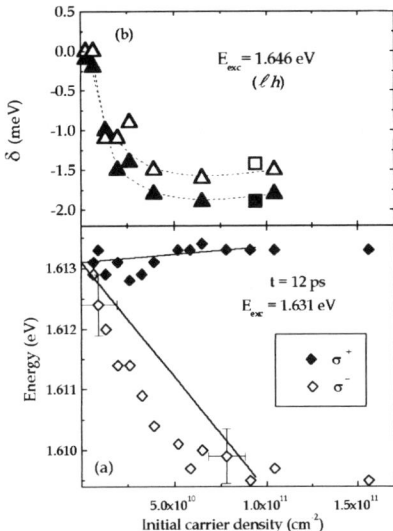

Figure 3. Time dependence of the emission energies of the intrinsic QW for two different excitation energies and polarisations of the emission.

Figure 4. a) Carrier density dependence of the emission energies of the intrinsic QW at 12 ps after excitation for both circular polarisations, excited below the light-hole exciton. b) Same as in exciting at the light-hole exciton. Solid (open) symbols correspond to σ^+ (σ^-) emission. The solid lines are the results of the calculations given by Eq. (1).

The time dependence of the polarised (solid points) and unpolarised (open points) PL is shown in Fig. 3 for two different excitation energies and an initial carrier density of 5×10^{10} cm^{-2}. At short times, the splitting amounts to 4 meV, with an excitation at 1.631 eV, below the light-hole (ℓh) exciton (diamonds). However, if the excitation is moved to energies above the ℓh it becomes only 1.2 meV (1.681 eV, circles). We have found that the splitting is strongly correlated with the degree of polarisation ($\wp$) of the luminescence, which for a given helicity of the exciting light, i.e. σ^+, is defined as the fractional difference of the PL intensities of the two circular polarisations, σ^+ and σ^-, $\wp = (\Gamma^+ - \Gamma^-)/(\Gamma^+ + \Gamma^-)$. Time-resolved measurements on the same sample have shown that $\wp$ at t = 10 ps amounts to 80% and 20% at 1.631 eV and 1.681 eV, respectively. [6] The behaviour of the peak positions of the PL with time, seen in the figure, is common for all excitation energies: the polarised (unpolarised) emission shifts towards lower (higher) energies with increasing time until both emission bands merge at ~150 ps.

Figure 4a depicts the dependence of the energy positions of the PL on the initial carrier density (open and solid points). Under the conditions presented in this figure,

12 ps after excitation at 1.631eV, the σ^+ emission remains practically constant, while the σ^- red shifts with increasing carrier density up to 9 x 10^{10} cm^{-2}. The lines correspond to a model that takes into account interexcitonic exchange interaction and screening [9], which gives the changes in the energies of the interacting ±1 excitons as a function of the populations as:

$$\Delta E^{\pm} = 2\,(n^{\pm} + f\,n^{\mp})\,(I_{VC} - I_{SE}) - \frac{0.82\pi\,e^2\,na}{\varepsilon} \qquad (1)$$

where n is the total density of excitons, $n^{\pm}$ are those of ±1 excitons, ε the dielectric constant and a the three-dimensional Bohr radius. I_{SE} describes a "self-energy" (SE) correction, also known in the literature as "exchange" correction, which weakens electron-electron and hole-hole repulsion. I_{VC} is a "vertex" correction (VC) that, due to the Pauli exclusion, reduces the inter-excitonic electron-hole attraction. The term involving f is a small coupling between ±1 excitons, essentially due to valence band mixing. The last term in Eq. (1) is a screening correction using the random phase approximation.

The energy splitting, $\delta = \Delta E^+ - \Delta E^-$, neglecting the second term of Eq.(1) is given by:

$$\delta \propto n\,\wp\,(I_{VC} - I_{SE}) \qquad (2)$$

The two terms I_{VC} and I_{SE} are part of the inter-excitonic exchange, I_x, or the unscreened exchange interaction between fermions of the same type. It is important to distinguish between inter-excitonic exchange and intra-excitonic exchange. The former is a many exciton intraband (conduction-conduction or valence-valence) exchange and the latter is a single exciton or interband (conduction-valence) effect [14-16]. The intra-excitonic exchange does not break the symmetry between spin +1 and spin -1 exciton and has a very weak influence in the exciton energy levels. I_{VC} and I_{SE} can be obtained by perturbation theory [9]:

$$I_{VC} = \pi\,a^2\left|E_0^{2D}\right| \quad , \quad I_{SE} = \pi\left|E_0^{2D}\right|\frac{315\,\pi^2}{2^{12}} \qquad (3)$$

where $\left|E_0^{2D}\right|$ is the 2D Rydberg, $E_0^{2D} = -\dfrac{2e^2}{a} = -\dfrac{2h^2}{\mu a^2} = 4E_0^{3D}$, and μ the reduced exciton mass. Substituting the numerical values of I_{VC} and I_{SE}, with $\varepsilon=13$ for GaAs and neglecting the small f term, Eq. (1) can be rewritten as:

$$\Delta E^{\pm}\,(eV) = 2.214\mathrm{x}10^{-16}\;a(\mathring{A})\times\left[1.515n^{\pm}\,(cm^{-2}) - 0.4\,\ln n\,(cm^{-2})\right] \qquad (4)$$

For the lines in Fig.4a, the energy of a single exciton has been taken from the experimental energy of the +1 exciton at the lowest carrier density; a three-dimensional Bohr radius, a, of 150Å and an initial degree of polarisation of 80% have been used. In spite of the strong approximations used in the theory, which are not borne out by the experiments, such as neglecting the presence of dark ±2 states, and assuming that the excitons are all at $\mathbf{K}=0$, the agreement with the experiments is satisfactory.

According to Equation 4, the inter-excitonic exchange interaction produces a blue shift of both levels. The flatness of the calculated position for the polarised σ^+ PL arises from an almost perfect cancellation of the "self-energy", the vertex correction and the screening in Eq. (4) for ΔE^+. On the other hand, the theory predicts a less pronounced red shift of the unpolarised σ^- PL with increasing carrier density than that observed experimentally, and does not reproduce the saturation of the shift. These discrepancies between the experiments and the calculations can originate from different sources: i) the theory depends on excitonic density, while the experiments are plotted against carrier density, which is not necessarily the same as the excitonic one, especially at high densities when the number of created excitons saturates; ii) the theory assumes free excitons with zero kinetic energy, while the excitons are actually weakly bound, [6] and have some excess energy; iii) the theory considers a strictly two-dimensional system while the actual one has a width in the growth direction.

The interdependence between the magnitude, and also the sign, of the splitting and $\wp$ is further demonstrated in Fig. 4b, which depicts the shift of the PL peaks versus initial carrier density, exciting at the ℓh exciton with t = 12 ps. In this case, the unpolarised emission (open triangles) lies at slightly higher energies than the polarised one (closed triangles), in concordance with the small value of $\wp = -10\%$ [6]. The adequacy of the theory to explain the experiments is also confirmed by the results obtained with Eq. (4) for $\wp = -10\%$, depicted in Fig. 4b as solid (σ^+) and open (σ^-) squares for an excitonic density of 9.5×10^{10} cm^{-2}. The calculated shift of the +1 excitons and the splitting amount to 2 meV and -0.4 meV, respectively, in very good agreement with the experimental values.

The splitting is strongly correlated with the degree of polarisation of the exciton gas, as can be observed in Fig. 5, which depicts the time evolution of the splitting (solid diamonds) and $\wp$ (grey dots) for a carrier density of 9.5×10^{10} cm^{-2}. Both quantities show a single exponential decay with a time constant of 40 ps. This correlation has also been confirmed by experiments where the degree of polarisation of the exciting light has been varied from circular to linear. [27]

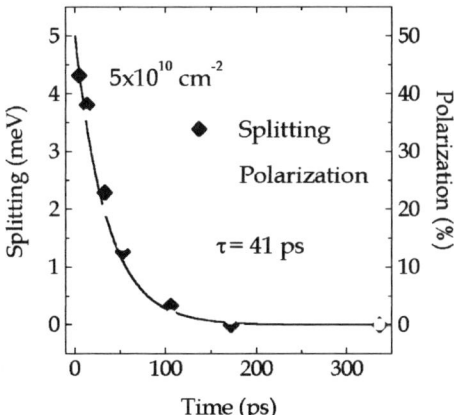

Figure 5. Time dependence of the splitting (diamonds), between the two circularly polarised components of the PL, and of the degree of polarisation (dots) for the intrinsic QW. The line depicts a single exponential decay with a time constant of 41 ps.

The experiments corroborate the theoretical predictions concerning the relative strength of the electron-hole vertex (I_{VC}) and of the electron(hole)-electron(hole) (I_{SE}) "self-energy" corrections: the splitting grows with increasing initial carrier density corresponding to $I_{VC}/I_{SE}=6.28/4.76>1$. (see Eqs. (A4) and (A12) in [9]). The possible reversal of this inequality by an external perturbation could have important consequences in the state of polarisation of the excitonic gas. The variation of the total energy with the degree of polarisation $\wp$ can be written as:

$$\frac{\partial E_T}{\partial \wp} = \frac{2\,e^2\,n^2}{\varepsilon\,a}\left(I_{VC}-I_{SE}\right)(1-f)\,\wp \tag{5}$$

Taking into account that $f \ll 1$ if $I_{VC}>I_{SE}$, Eq. (5) predicts that the system prefers to have zero polarisation because there it attains an energy minimum. However, if an external perturbation, such as an electric field, could cause that $I_{SE}>I_{VC}$ then the excitonic gas would prefer to be polarised. This will be considered in the following Section.

4. Tuning of the spin splitting by an external electric field

Up to now, we have reported only on positive splittings. Thus the majority +1 excitons always have higher energies than the minority -1 excitons and the interaction is dominated by the VC. Fernández Rossier et al. expanded their model from Ref. [9] to a system where the electrons and holes forming an exciton can be separated locally. [28] In their model, it is assumed that the electrons and holes are confined to separate planes at a distance d. The overlap between holes and electrons is tuned by changing d, therefore modifying the inter-excitonic scattering mechanisms: I_{VC} (I_{SE}) tends to

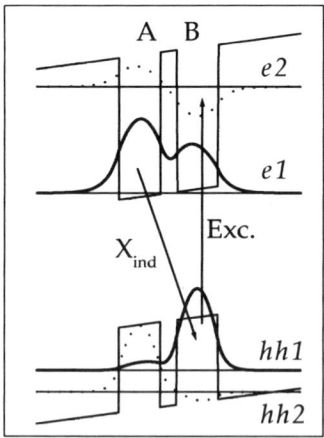

Figure 6. Band profile for the cDQW at a field strength of 14 kV/cm. The curves depict the envelope functions for the electron and heavy-hole levels (the light-hole are not shown for simplicity). The arrows indicate the excitation (Exc.) and detection (X_{ind}) used in the TRPL experiments.

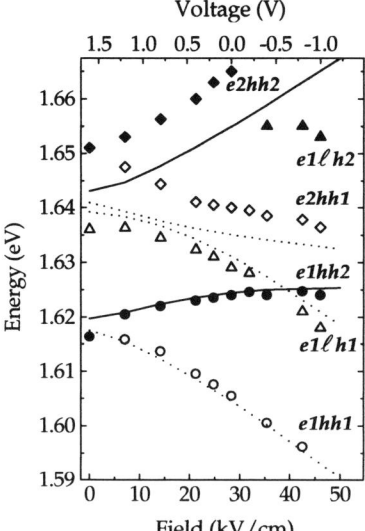

Figure 7. Electric field dependence of the excitonic transitions in the cDQW heterostructure. The lines are the results of solving the Schrödinger equation in the envelope function approximation.

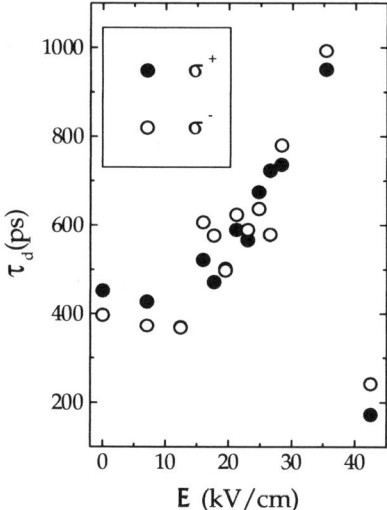

Figure 8. Electric field dependence of the PL decay time in the cDQW heterostructure. Solid (open) points correspond to +1 (-1) excitons.

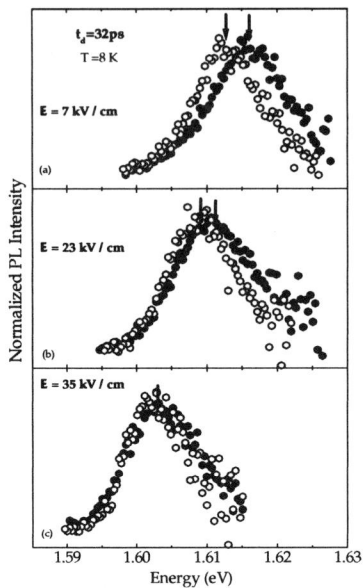

Figure 9. TRPL spectra for the cDQW at different field strengths, taken 32 ps after excitation. The arrows mark the positions of the PL maxima.

decrease (slightly increase) with increasing d. According to Eq. (2), this should lead to a reduction of δ (at unchanged n and $\wp$) and, eventually, it could become negative at a certain $d = d_{cr}$. A $\delta < 0$ means that the minority excitons have higher energies than the majority ones, thus promoting a repolarisation of the exciton gas or at least a stabilisation of the polarisation as the system can win energy going to the majority states. The exciton gas would be in a ferromagnetic phase. Hence, by tuning d and n, one could obtain a quantum phase transition between a ferromagnetic phase (dominated by self-energy) and a paramagnetic phase (dominated by the vertex corrections). It has been argued that a ferromagnetic phase could make the Bose-Einstein condensation of the exciton gas experimentally more accessible [29]. In practice, the local separation of electrons and holes in 2D systems can be achieved by the use of a type-II QW or by applying an electric field to a type-I QW, where the parameter d is proportional to the field $\mathcal{E}$

     We have used a double quantum well imbedded in a p-i-n junction to investigate the influence of an external electric field on the strength of the splitting. A schematic representation of the potential profiles of the cDQW structure at a field of 14 kV/cm is shown in Fig 6. Two states, one symmetric ($e1$) and one antisymmetric ($e2$), are formed in the CB. The coupling of the heavy- (hh) and light-hole (ℓh) valence band states of each well establishes the four states, $hh1$, $\ell h1$, $hh2$ and $\ell h2$ (the ℓh levels are not shown for simplicity). As it is well documented in the literature [30-32], for increasing $\mathcal{E}$ the ground state exciton X_{e1hh1} becomes spatially indirect due to the localisation of electrons and holes in wells A and B, respectively. One should also mention that in this case, where the excitons become indirect in real space with drastically reduced electron-hole overlap, the spin flip processes could be dominated by those of electrons forming excitons [33].

The electric field dependence of the different excitonic transitions, obtained from PLE measurements, is shown in Fig. 7. The lower lying indirect transitions X_{e1hh1} and X_{e1lh1} red shift with increasing field, because of the potential tilting and the Stark effect. For the identification of the excitonic transitions, we have numerically solved the Schrödinger equation using the envelope function approximation (see lines in Fig. 7). This procedure also allows us to deduce the relationship between the voltage applied to the whole structure and the electric field in the intrinsic region. For the TRPL experiments, the excitation energy was always tuned to the X_{e2hh1} direct exciton and the light pulses were circularly polarised in order to create a +1 exciton population. The detection was at the lower excitonic transition, X_{e1hh1}. An excitation power of 15 mW, which was used in most of our experiments, leads to an approximate exciton density of $n \sim 5 \times 10^{10}$ cm^{-2}. The PL rise time was found to be almost independent of $\mathcal{E}$ and it amounts to 110 ± 10 ps for the X^+_{e1hh1} and to 140 ± 10 ps for the minority X^-_{e1hh1}. The PL decay time, τ_d, shown in Figure 8, nearly independent of the spin polarisation, increases linearly with increasing field from 400 ps at $\mathcal{E} = 10$ kV/cm to 1 ns at $\mathcal{E} = 35$ kV/cm, due to the reduction of wave function overlap [34]. For higher fields the decay time decreases again due to tunnelling of the carriers out of the wells [34,35], at $\mathcal{E} = 40$ kV/cm it drops to $\tau_d = 250$ ps and the photocurrent also increases exponentially.

Figure 9 depicts the TRPL spectra taken at 32 ps after excitation with 15 mW for different field strengths. At flat band (Fig. 9a) and low fields (Fig. 9b) we observe a splitting between +1 and -1 excitons. The splitting at zero field amounts to approximately half of the binding energy of the exciton. Increasing the field while maintaining the same initial exciton density, n, causes the splitting to decrease and finally vanish. Figure 10 compiles the dependence of the initial splitting δ_0 on the electric field and demonstrates that it decreases linearly with increasing $\mathcal{E}$. We have defined δ_0, for a given field, as the mean value of three measurements taken at delays $t_d = 7$, 15 and 32 ps. The maximum splitting of 4 meV is reached at zero field. At a field of $\mathcal{E} \sim 35$ kV/cm the splitting becomes zero, this is in agreement with the predictions of the model developed in [28]. To ensure that the decrease of the splitting is actually due to a change in exciton-exciton interactions by the local separation of the electrons and holes, the field dependence of $\wp$, which is the only additional parameter in Eq. 2, has to be checked. Exciting at X_{e2hh1}, an initial ($t \sim 0$), field independent, X_{e1hh1}-polarisation of $\wp_0 = 0.5 \pm 0.05$ is obtained. Since the spin splitting, δ, is proportional to the total exciton population n and to the polarisation, and $\wp$ does not change significantly with $\mathcal{E}$, any change in δ, at a given excitation power, can be unambiguously related to a variation in the exciton-exciton interaction (I_{VC} and/or I_{SE}). The dynamics of the splitting is depicted in Fig. 11 for three different electric field strengths. At a low field of $\mathcal{E} = 7$ kV/cm, the splitting between the two exciton populations has an initial value of $\delta_0 \sim 3.5$ meV. The time dependence of the ± 1 excitonic energies (Fig. 11a) reveals that the majority +1 excitons red shift on approaching the equilibrium population where $n^+ = n^-$.

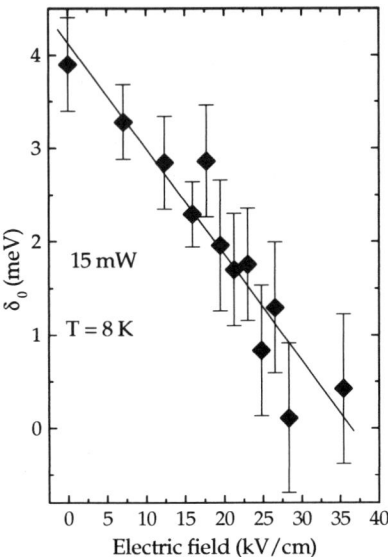

Figure 10. Electric field dependence of the splitting in the cDQW at an excitation power of 15 mW, the magnitude of δ is a mean value of those at 7, 15 and 32 ps.

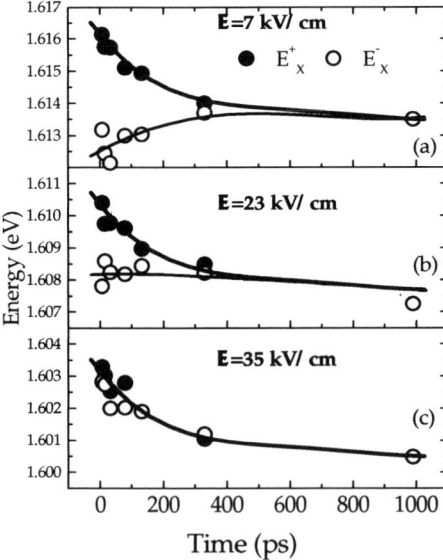

Figure 11. Time dependence of the emission energies in the cDQW at different field strengths. Solid (open) points correspond to +1 (-1) excitons.

The minority -1 excitons, in contrast, blue shift with time approaching the +1 exciton energies, and when the splitting vanishes the system also becomes unpolarised. Raising the field to $\mathcal{E}$=23 kV/cm (Fig. 11b), the initial splitting becomes ~2 meV, and its dynamics changes appreciably. The position of the +1 excitons as a function of time shows practically the same behaviour as in the previous case, except that the whole energy scale is red-shifted by ~ 6 meV due to the tilting of the QW-potential and the Stark effect. The most remarkable difference is seen in the dependence of the -1 exciton's energy, which now remains constant within experimental error. When the field is further increased to $\mathcal{E}$=35 kV/cm, the +1 exciton's energy maintains the same time-dependence. Despite the lack of dependence of the initial PL-polarisation on $\mathcal{E}$, the splitting has now vanished and the minority -1 excitons have the same energy positions as the +1 excitons.

The power dependence of the ±1 exciton's energies at short delays, and hence at constant polarisation, allows a deeper insight into the excitonic interaction. The energy positions of the two spin polarisations as a function of excitation power are displayed in Fig. 12. At a low field of $\mathcal{E}$= 7kV/cm (Fig. 12a) the energy of the +1 excitons does not change within experimental error; however, the -1 excitons red shift, thus δ grows with increasing power. The behaviour is similar to that described in Sect. 3 and is in agreement with the calculations. Already to explain those measurements it was necessary to include corrections in the calculations due to screening of the VC and SE effects. This screening at zero field does not depend on the individual exciton densities $n^\pm$ but only on the total population, n, and therefore obtains the same shift of the energy positions of both exciton components, leaving the splitting unchanged. However, it modifies significantly the individual ±1 excitonic energies: $\Delta E^\pm \propto n^\pm$ (I_{VC}-I_{SE}), is cancelled and overcompensated by the screening for the +1 and -1 components, respectively. At an intermediate field of $\mathcal{E}$= 23 kV/cm (Fig. 12b) δ still increases with power, but its values are considerably reduced as compared to the low field case. At a field $\mathcal{E}$=35 kV/cm the splitting vanishes, as shown in Fig. 12c, indicating that the two corrections become equal and cancel mutually.

The absolute energy positions of the two spin components as a function of power show a strong field dependence as clearly seen in Fig. 12. In Fig. 12b the absolute energies of both components increase on increasing the power. Pictorially, we can describe the effect as a counter clockwise rotation of the power-dependence behaviour of the excitonic energies around their position at vanishing excitation power. With increasing $\mathcal{E}$, the energies of the -1 excitons are rising faster than that of the +1 excitons, resulting in a smaller splitting. At $\mathcal{E}$=35 kV/cm, where δ=0 meV, the -1 exciton energies have moved up into the +1 energies.

The model proposed by Fernández-Rossier et al. [28] accounts well for the field dependence of the splitting δ between the two spin components, but does not deal with the field dependence of the absolute energies of the excitons as measured in our work. These positions depend on the screening corrections to the excitonic interactions (while the splitting does not). In the model, the field dependence of the screening effects is not taken into account, and it is implicitly assumed that both interaction terms, I_{VC} and I_{EC}, are affected in the same way by the screening.

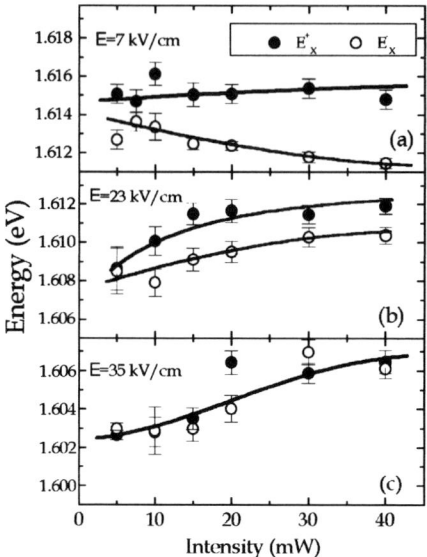

Figure 12. Power dependence of the emission energies in the cDQW at different field strengths. Solid (open) points correspond to +1 (-1) excitons.

Since at zero and low fields the screening is responsible for the independence of the +1 energies on power, and our results clearly show an increase of those energies with power when the electric field strength is augmented, we argue that these screening corrections are also subjected to changes under an electric field. According to our findings this would mean that on spatially separating the electrons and holes (i.e. increasing d), the screening of the VC becomes less efficient than that of the SE, leading to an overall blue-shift of the exciton energies.

Figure 13 shows the phase diagram, obtained by Fernández-Rossier and Tejedor [28], for the sign of the splitting, δ, in terms of the density of an exciton condensate and the separation of the electrons and holes forming the excitons, d. For large d and/or n the kinetic energy dominates the electron-hole attraction and the electron-hole pairs are not bound (unshaded, not hatched region). The thick line (d_{cr}) separates the region of positive splitting, and therefore unpolarised exciton gas (to the right of the line), from that of a polarised exciton gas (negative splitting). For $d<d_{cr}$, I_{VC} always dominates over I_{SE} and the unpolarised state δ>0 has the minimum energy. When d is increased, the excitons are less bound, I_{VC} decreases while I_{SE} increases, allowing the appearance of a polarised phase with δ<0. The maximum field that can be applied to our structure, 35 kV/cm, yields a zero splitting between +1 and -1 excitons. Unfortunately, any further increase in $\mathcal{E}$ leads to a strong ionisation of the excitons hindering the investigation of the possible existence of a negative splitting and of a ferromagnetic phase in the exciton gas.

5. Summary

We have shown the existence of a splitting between the two circularly polarised components of a dense, 2D, exciton gas in the absence of any external magnetic field. The dynamics of the splitting is strongly correlated with that of the degree of polarisation of the light emitted by the gas. We have also demonstrated, using an external electric field and a cDQW heterostructure to spatially separate the electrons and holes forming the excitons in different wells, that the splitting decreases with increasing separation, due to the reduction of the vertex correction to the inter-excitonic interaction with increasing electron-hole separation. The splitting vanishes for electric fields of ~ 35 kV/cm. The existence of a ferromagnetic phase, predicted by theory, could not be demonstrated due to leaking of the carriers out of the quantum wells at high field strengths.

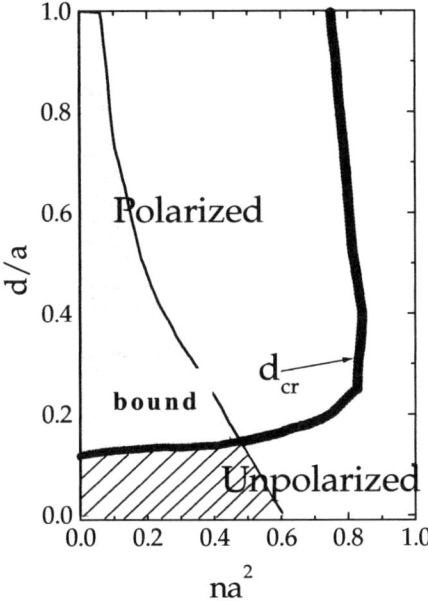

Figure 13. Phase diagram of the sign of the splitting in an excitonic condensate as a function of electron-hole distance (*d*) and exciton density (*n*), both defined as a function of the three-dimensional Bohr radius (*a*). (after Ref. [28]).

Acknowledgements. This work would not have been possible without the collaboration with many people, especially with: L. Muñoz, E. Pérez, J. Fernández-Rossier and C. Tejedor. Some samples mentioned in this manuscript were kindly provided by K. Ploog. This research has been partially supported by the Fundación Ramón Areces, the Fullbright Comission, and the Spanish DGICYT under contract PB96-0085, CAM 7N/0026/1998 and ARO.

132

References

1   T. C. Damen, L. Vina, J. E. Cunningham, and J. Shah, Phys. Rev. Lett. **67**, 3432 (1991).
2   J. B. Stark, W. H. Knox, and D. S. Chemla, Phys. Rev. B **46**, 7919 (1992).
3   B. Dareys, X. Marie, T. Amand, J. Barrau, Y. Shekun, I. Razdobreev, and R. Planel, Superlattices and Microstructures **13**, 353 (1993).
4   T. Amand, X. Marie, B. Baylac, B. Dareys, J. Barrau, M. Brosseau, R. Planel, and D. J. Dunstan, Phys. Lett. A **193**, 105 (1994).
5   A. Vinattieri, J. Shah, T. C. Damen, K. W. Goossen, L. N. Pfeiffer, M. Z. Maialle, and L. J. Sham, Appl. Phys. Lett. **63**, 3164 (1993).
6   L. Muñoz, E. Pérez, L. Viña, and K. Ploog, Phys. Rev. B **51**, 4247 (1995).
7   G. Bastard and R. Ferreira, Surface Sci. **267**, 335 (1992).
8   M. Z. Maialle, E. A. de Andrada e Silva, and L. J. Sham, Phys. Rev. B **47**, 15776 (1993).
9   J. Fernandez-Rossier, C. Tejedor, L. Muñoz, and L. Viña, Phys. Rev. B **54**, 11582 (1996).
10   L. Muñoz, E. Pérez, L. Viña, J. Fernández-Rossier, C. Tejedor, and K. Ploog, Solid State Electron. **40**, 755 (1996).
11   L. Viña, L. Muñoz, E. Pérez, J. Fernandez-Rossier, C. Tejedor, and K. Ploog, Phys. Rev. B **54**, 8317 (1996).
12   P. L. Jeune, X. Marie, T. Amand, F. Romstad, F. Perez, J. Barrau, and M. Brousseau, Phys. Rev. B **58**, 4853 (1998).
13   G. L. Bir, A. G. Aronov, and G. E. Pikus, Sov. Phys. JETP **42**, 705 (1976).
14   M. M. Denisov and V. P. Makarov, Phys. Stat. Sol. (b) **56**, 9 (1973).
15   R. Knox, *Solid State Physics, Supl. 5* (Academic Press, New York, 1963).
16   G. E. Pikus and G. L. Bir, Sov. Phys. JETP **33**, 108 (1971).
17   V. A. Kiselev and A. G. Zhilich, Sov. Phys. Solid State **13**, 2008 (1972).
18   M. I. D'yakonov and V. I. Perel', Sov. Phys. JETP **33**, 1053 (1971).
19   R. J. Elliot, Phys. Rev. **96**, 266 (1954).
20   Y. Yafet, *Solid State Physics, Vol 14* (Academic Press, New York, 1963).
21   N. Peyghambarian, H. M. Gibbs, J. L. Jewell, A. Antonetti, A. Migus, D. Hulin, and W. T. Masselink, Phys. Rev. Lett. **53**, 2433 (1984).
22   D. Hulin, A. Mysyrowicz, A. Antonetti, A. Migus, W. T. Masselink, H. Morkoc, H. M. Gibbs, and N. Peyghambarian, Phy. Rev. B **33**, 4389 (1986).
23   S. Schmitt-Rink, D. S. Chemla, and D. A. B. Miller, Phys. Rev. B **32**, 6601 (1985).
24   H. Haug and S. Schmitt-Rink, Prog. Quantum Electron. **9**, 3 (1984).
25   T. Amand, X. Marie, D. Robart, P. L. Jeune, M. Brousseau, and J. Barrau, Solid State Commun. **100**, 445 (1996).
26   J. Feldmann, G. Peter, E. O. Gobel, P. Dawson, K. Moore, C. T. Foxon, and R. J. Elliott, Phys. Rev. Lett. **59**, 2337 (1987).
27   D. Robart, T. Amand, X. Marie, B. Baylac, J. Barrau, M. Brousseau, and R. Planel, Ann. Phys. (Paris) Suppl. **20**, C2 91 (1995).
28   J. Fernandez-Rossier and C. Tejedor, Phys. Rev. Lett. **78**, 4809 (1997).
29   C. Tejedor, in *24th Intern. Conf. on the Physics of Semiconductors*, edited by M. Heiblum (World Scientific, Jerusalem, 1998).
30   Y. J. Chen, E. S. Koteles, B. S. Elman, and C. A. Armiento, Phys. Rev. B **36**, 4562 (1987).
31   E. E. Mendez, F. Agulló-Rueda, and J. M. Hong, Phys. Rev. Lett. **60**, 2426 (1988).
32   P. Voisin, J. Bleuse, C. Bouchet, S. Gaillard, C. Alibert, and A. Regreny, Phys. Rev. Lett. **61**, 1639 (1988).
33   E. A. de Andrade e Silva and G. C. L. Rocca, Phys. Rev. B **56**, 9259 (1997).
34   K. Köhler, H.-J. Polland, L. Schultheis, and C. W. Tu, Phys. Rev. B **38**, 5496 (1988).
35   A. Alexandrou, J. A. Kash, E. E. Mendez, M. Zachau, J. M. Hong, T. Fukuzawa, and Y. Hase, Phys. Rev. B **42**, 9225 (1990).

EXCITON CONDENSATION

P. B. LITTLEWOOD[1,2] AND PAUL EASTHAM[1]
[1] *TCM Group, Cavendish Laboratory, Madingley Road, Cambridge CB3 0HE, UK*
[2] *Bell Laboratories, Lucent Technologies, Murray Hill NJ 07974, USA*

Abstract. Recent progress in understanding theoretical models of condensed excitonic systems is reviewed. We exhibit the close parallels between the mean-field theory of the interacting electron-hole condensate, and a model of localised excitons interacting via the photon field in a microcavity. Beyond the mathematical similarity of these two problems, this allows us to comment on the qualitative and quantitative distinctions between a conventional semiconductor laser, and a state with substantial "excitonic" coherence. More generally, coupling of an excitonic condensate to radiative modes of an optical cavity offer an opportunity to manipulate the condensate and its optical response.

1. Introduction

Suggestions for Bose condensation of excitons, and the excitonic insulator state were made long ago, and this problem provides a continuing challenge to experiment. [1, 2, 3] Excitons are of course composite particles, and the internal fermionic structure will significantly affect the properties once there is substantial overlap between them. At low densities, when the internal structure of an exciton is rather inert and unimportant, the centre-of-mass motion contributes the most to the entropy. The condensation should occur in much the same way as in a collection of weakly interacting bosons. As the density increases, the binding energy of a single electron-hole pair will eventually become small, changing its character from single exciton binding to collective electron-hole pairing due to both the screening and the band (phase space) filling effects. Therefore, at high densities, as the temperature is raised from zero, free fermions due to broken pairs dominate the entropy.

M.L. Sadowski et al. (eds.), Optical Properties of Semiconductor Nanostructures, 133–141.

At these high densities the Bose-Einstein condensation transition coincides with the pair formation. The situation is then rather analogous to that in a weak-coupling superconductor.

While a naive picture of exciton condensation might imagine this as a property of noninteracting bosonic particles, starting from the fermionic picture of interacting electrons and holes makes it clear that the underlying physics is driven by the Coulomb interaction. The mean field theory of the excitonic insulator introduced by Keldysh is a very useful starting point for the discussion of the theoretical aspects of the problem; it provides a smooth interpolation between the low density limit of weakly interacting bosons, and the high density limit which is very analogous to BCS pairing in a superconductor.

Recently, there has been some considerable effort on another problem, which seems superficially related, that of "polariton" condensation in microcavities.[4] A polariton is well-known as a coupled mode of excitons and photons; since the mass of this mode can be very light, a straightforward application of weak coupling theory of Bose-Einstein condensation would give a large transition temperature. The naive idea is immediately suspect on many grounds. One is that the polariton modes are simply transverse eigenmodes in linear response theory, and do not allow for real excitation of excitons. If the modes become significantly occupied, the dispersion will be changed, and at high density the fermionic character of the constituents will be revealed.[5] Furthermore, since the evidence for condensation is coherence, and this evidenced by coherent light emission, how is one to distinguish exciton condensation from lasing?

In this paper, we will discuss the theoretical relationships between these problems. We briefly review the mean field theory of the excitonic insulator. Next, to compare to this problem, we then introduce a simple model for the microcavity problem, namely an array of non-interacting localised two level systems coupled to a single photon mode in a cavity.[6] This model describes localised excitons (on traps for example), but explicitly does not include exciton-exciton interactions, except for those mediated by the photon field. It turns out that the (nearly exact) solution of our model of polariton condensation bears close similarity to the *mean field* solution of Keldysh for the excitonic insulator problem. This is not an accident, because in both cases the same underlying symmetry is broken by the condensate.

2. BCS mean-field theory for the exciton condensate

We shall consider an isotropic, parabolic, direct band gap model for a semiconductor, and assume that electrons and holes have been excited and reached thermal equilibrium in their respective bands. This problem has been extensively reviewed elsewhere, and we will be as brief as possible in covering the basics of the theory. [1, 3, 7]

In second-quantised notation, the Hamiltonian consists solely of the kinetic energy plus the Coulomb interaction, viz.

$$H = H_o + H_{coul},$$ (1)

where

$$H_o = \sum_{\vec{k}} \left[\epsilon_{ck} a_{c,k}^\dagger a_{c,k} + \epsilon_{vk} a_{v,k}^\dagger a_{v,k} \right],$$ (2)

and

$$H_{coul} = \frac{1}{2} \sum_q \left[V_q^{ee} \rho_q^e \rho_{-q}^e + V_q^{hh} \rho_q^h \rho_{-q}^h - 2 V_q^{eh} \rho_q^e \rho_{-q}^h \right].$$ (3)

$a_{c,k}^\dagger$ and $a_{v,k}^\dagger$ are creation operators for electrons in the conduction and valence bands. The density operators are $\rho_q^e = \sum_k a_{c,k+q}^\dagger a_{c,k}$, $\rho_q^h = \sum_k a_{v,k} a_{v,k+q}^\dagger$. For convenience, we shall consider equal-mass, parabolic bands, so that $\epsilon_{vk} = -E_g - \epsilon_{ck}$. Note that recombination of electrons and holes is not allowed, and the electron and hole densities are separately conserved.

The mean-field theory starts from the following ansatz for the ground-state wavefunction of the interacting electrons and holes [1]. Evidently, it is in the form of a BCS wavefunction for a superconductor [8]:

$$|\Psi_0> = \prod_{\vec{k}} [u_{\vec{k}} + v_{\vec{k}} a_{e,k}^\dagger a_{v,k}] |vac>,$$ (4)

with $|u_k|^2 + |v_k|^2 = 1$. The vacuum state $|vac>$ corresponds to one in which the valence band is completely filled and the conduction band completely empty.

To proceed, we introduce the chemical potential μ for the introduction of electron-hole pairs, and minimise the free-energy f:

$$f = <h_{eh}> - \mu <n>,$$ (5)

with respect to the variational parameters v_k. Setting $\partial f / \partial v_{\vec{k}} = 0$ and considering only s-wave pairing in which case all quantities are functions of k, the magnitude of $\vec{k}$, we find in close analogy to the BCS algebra a set of self-consistent equations [8]:

$$\xi_k = \epsilon_k - \mu - 2 \sum_{k'} V_{k-k'}^{ee} n_{k'} = \epsilon_k - \mu - \sum_{\vec{k'}} V_{k-k'}^{ee} (1 - \xi_{k'}/E_{k'}),$$ (6)

$$\Delta_k = 2 \sum_{k'} V_{k-k'}^{eh} <a_{e,k}^\dagger a_{v,k}> = \sum_{k'} V_{k-k'}^{eh} \Delta_{k'}/E_{k'},$$ (7)

$$E_k^2 = \xi_k^2 + \Delta_k^2.$$ (8)

ξ_k is the renormalised single-particle energy (per pair) measured from the chemical potential.($\epsilon_k = \frac{k^2}{2m_e} + \frac{k^2}{2m_h}$.) Δ_k is the gap-function and is also the order-parameter. E_k can be identified as the pair-breaking excitation spectrum: it is the energy cost of taking one-pair out of the condensate and placing them in plane-wave states of momentum $\vec{k}$ [8].

In terms of these quantities, the variational parameters in the wavefunction are:

$$v_k^2 = \frac{1}{2}\left(1 - \xi_k/E_k\right), \tag{9}$$

$$u_k^2 = \frac{1}{2}\left(1 + \xi_k/E_k\right), \tag{10}$$

$$u_k v_k = \frac{\Delta_k}{2E_k}. \tag{11}$$

Within the present approximations, the spin degrees-of-freedom play no role other than changing the density of electron-hole pairs by a factor of two. The natural unit of length is the Bohr radius (m is the reduced mass) $a_0 = \hbar^2\varepsilon/me^2$, and of energy the Rydberg $E_0 = e^2/2\varepsilon a_0$. The single dimensionless parameter of density is r_s: $\frac{4\pi}{3}(r_s a_0)^3 = 1/n$ in 3D, and $\pi(r_s a_0)^2 = 1/n$ in 2D.

In the high density limit ($r_s \ll 1$), the self-consistent equations approach the Hartree-Fock solution of a two-component fermion system, subject to pairing interactions. The kinetic energy dominates the total energy of the system. In this limit, pairing is a collective phenomenon and mainly affects the properties of the system in the vicinity of the Fermi surface. Here μ lies very close to ϵ_{k_F} and ξ_k changes it sign from $\xi_k < 0$ to $\xi_k > 0$ as k goes from $k < k_F$ to $k > k_F$. This means that $|u_k|^2$ becomes nearly a step function at k_F, just as in a Fermi liquid. Also in this limit, Fermi surface nesting between that of the electrons and that of the holes is crucial to the stability of the paired phase.

In the low density limit ($r_s \gg 1$), the interaction energy is dominant, and the chemical potential lies below the bottom of the band. The equation for the order parameter Δ_k becomes the Wannier equation for a single, isolated, exciton. To see this, we let $\phi_k = \Delta_k/E_k$ and rewrite Eq. 7 as:

$$E_k \phi_k = \sum_{\vec{k}'} V^{eh}_{\vec{k}-\vec{k}'} \phi_{k'}. \tag{12}$$

Remembering that in the low density limit $E_k = \epsilon_k - \mu$, we find that Eq. 12 is precisely the exciton Wannier equation written in momentum space, in which μ plays the role of the eigenvalue and ϕ_k is the eigenfunction. Therefore, $\mu = -|E_{ex}|$ and $\phi_k \propto \phi_{1s}(k)$ where E_{ex} is the exciton binding energy and $\phi_{1s}(k)$ is the exciton $1s$-state wavefunction.

Of course, in both regimes there is a special feature which is the existence of the off-diagonal order parameter $< a^\dagger_{e,k} a_{v,k} >$, that is the signature of coherence in the system.

3. Two-level systems in a microcavity

We now turn to a different model, that for excitons coupled to photons in a microcavity. We shall treat the excitons here as N two-level systems with excitation energy E_i, which could represent physically the exciton bound to a localised trap, so E_i may be taken from a distribution of inhomogeneously broadened levels. In this model, the excitons cannot ionise, and there are no free fermion states. We assume a microcavity with a single relevant optical mode at an energy ω_c described by a field ψ, to which all the two-level systems have the same (dipole) coupling. This is physically reasonable provided the wavelength of the cavity mode is much larger than the typical spacing. Details of the solutions will be presented elsewhere [6] and here we shall just exhibit the formal similarity of this model to that of the previous section.

The Hamiltonian is then

$$H = \frac{1}{2} \sum_i E_i(b^\dagger_i b_i - a^\dagger_i a_i) + \omega_c \psi^\dagger \psi + \frac{g}{\sqrt{N}} \sum_i (b^\dagger_i a_i \psi + \psi^\dagger a^\dagger_i b_i) \quad (13)$$

This is the widely studied Dicke model[9]. We have used a fermionic representation for the two level systems, but adopt the restriction $\left\langle b^\dagger_i b_i + a_i\dagger a_i \right\rangle = 1$ so that the system is undoped.

This Hamiltonian conserves the total number of photons and electronic excitations, governed by the operator $L = \psi^\dagger \psi + \frac{1}{2}\sum(b^\dagger b - a^\dagger a)$. When $\rho_{exciton}$ is small, the two-level system can be approximately represented by a bosonic Harmonic oscillator, and the Hamiltonian is just that of two interacting oscillators. This is the familiar context in which one derives the polariton.

Following the same line of attack as before, we now investigate the variational wavefunction

$$|\Psi_0> = e^{\lambda\psi^\dagger} \prod_i [u_i + v_i b^\dagger_i a_i]|vac> \quad (14)$$

and then minimise $\langle(H - \mu L)\rangle$, where the vacuum state has all the lower states occupied, i.e. $|vac> = \prod_i a^\dagger_i|0>$. After rescaling $\lambda \to \lambda\sqrt{N}$, we obtain the following

$$\tilde{\omega}_c \lambda + \frac{g}{N} \sum_i u_i v_i = 0, \quad (15)$$

$$2\tilde{\epsilon}_i u_i v_i - g\lambda(v^2_i - u^2_i) = 0, \quad (16)$$

where $\tilde{\omega}_c = \omega_c - \mu$, $\tilde{\epsilon}_i = (E_i - \mu)/2$. The total excitation density is

$$\rho = \frac{1}{N} \langle L \rangle = \rho_{photon} + \rho_{exciton} = \lambda^2 + \frac{1}{2N} \sum (v^2 - u^2). \qquad (17)$$

The new feature of this mean field approach is that the order parameter has two components, with λ measuring the amplitude of the coherent light field, and $u_i v_i$ the coherent component of the exciton field. They are related by Eq. 15, and the condensate will have proportionately more or less photon/exciton character depending on the dimensionless detuning $\tilde{\omega}_c/g$. Using Eq. 15 to eliminate λ Eq. 16 becomes

$$\Delta - \frac{g^2}{N\tilde{\omega}_c} \sum \frac{\Delta}{2E} = 0, \qquad (18)$$

where we have used the transformations Eqs. 9, 10 &11 to bring Eq. 18 into the same form as the gap equation Eq. 7.

In the case where $\mu = 0$, Eq. 18 corresponds to the solution of Hepp and Lieb[10], since rederived by many authors. The calculation (with the detailed results to be given elsewhere [6]) extends it to finite, fixed excitation densities.

4. Discussion

In the derivation we have presented above the parallels between the two problems are clear. One (important) technical point is that while the mean field theory for the excitonic insulator is an *approximate* solution of Eq. 1, the mean field solution for our polariton problem Eq. 13 is essentially exact (for $N \to \infty$). This may be shown by constructing the mean field theory as a $1/N$ expansion[6, 11]; physically the reason is that the photon mode couples democratically to all of the two-level systems, providing a nearly infinite-range coupling, insomuch as the cavity contains many localised exciton states.

Furthermore, it is clear that the two models can be combined; if one adds to Eq. 1 the dipole interaction with the photon field, the algebra can be carried through smoothly and the gap equation will have a kernel

$$V_{eff} = V^{eh}_{k-k'} + g^2/\tilde{\omega}_c \qquad (19)$$

so the effect of the photon field will be to enhance the condensation. One also notices that the coherent optical field and the condensate amplitude are coupled, and they have the same phase - while the overall phase of the condensate is free, the relative phases of the two components are fixed. Consequently a signature of the excitonic insulator state is coherent radiation (assuming that dipole recombination is allowed), but of course this does not allow a fundamental distinction between an excitonic condensate (whether bosonic ($r_s \gg 1$) or fermionic ($r_s \ll 1$) in nature) and a semiconductor laser.

There are three characteristics of a Bose-Einstein condensate: macroscopic occupation of a single eigenstate (usually the zero-momentum state in a translationally invariant system); phase coherence; and possibly superfluidity. The relationship between these phenomena is worth a comment.

The simplest picture of Bose-Einstein condensation arises is usually couched as macroscopic occupation of the lowest energy state of an ideal gas. But this begs the question as to why the condensed particles don't occupy states so nearby in energy that there is negligible difference in the thermodynamic limit. In fact, the stability of the condensate *requires* interactions between the particles, and the interactions stabilise macroscopic phase coherence.[12, 13] Phase coherence is generally now regarded to be the most fundamental property of a Bose-Einstein condensate. The mean field solutions of both models above have phase coherence - more explicitly long range order in the off-diagonal correlator $< a^\dagger b >$. But notice that there is nothing particularly special about such an order parameter - it is generated automatically by any linear mixing of two degenerate bands. In bilayer electron hole systems, it appears automatically if there is an amplitude for tunnelling from one layer to another.

Superfluidity requires further that there exists a Goldstone mode corresponding to the broken gauge symmetry; this will never appear in an infinite range model, because there is no concept of space - we cannot impose a spatial modulation in the phase of the order parameter without incorporating other spatial modes. In our microcavity model, other modes are non-existent - in practice separated by a gap from the ground state - so the system cannot support superflow. Without the dipole interaction however, the excitonic insulator (in a large box) should indeed be superfluid, and further calculations beyond the mean-field approximation do indeed yield and acoustic mode whose stiffness is the superfluid order parameter.[15, 16] In the electron-hole system this symmetry is generally broken; it is a consequence of the separate conservation of particle and hole occupation by the Hamiltonian of Eq. 1. Terms that break this symmetry (e.g. electron-hole recombination) fix the order parameter phase, and introduce a gap into the spectrum of collective modes.[3, 14]

A more practical issue is the existence of a gap in the pair excitation spectrum. The mean field theories all have such a gap, which is in the limit of small r_s just the order parameter 2Δ, and in the low density limit evolves to become the exciton binding energy. The two-level model of Section 3 always shows a (renormalised) Rabi splitting; here of course there is no BCS-like limit because the electron and hole occupations are bounded. In both models, the key is that the fermionic quasiparticle occupations are strongly renormalised by the order-parameter field.[17] In contrast, the usual theory (and practice) for a semiconductor laser is that the fermion distribution is assumed to be unchanged, so there is no gap at the Fermi surface. This relies on there being rapid dephasing processes (included, for example, in the Bloch equation picture by a rapid T_2 for the relaxation of the polarisation and short lifetimes for the single particle states).[18] Such phase-breaking processes are familiar in the theory of superconductivity, particularly in the context of magnetic impurities. Here these processes depress the superconducting transition temperature (eventually to zero), but in an intermediate range close to T_c introduce a regime of *gapless* super-conductivity. Of course the specific processes that produce phase breaking in the excitonic insulator problem are different, but the principle remains. It is possible to have a phase which is gapless for single pair excitations, but still has an s-wave order parameter.

In a purely electronic model, it is likely that phase breaking processes (should they be important) will rapidly destroy the condensate, so that the regime of parameters where there is a gapless but condensed phase, is small. But when there is a subsidiary order parameter (by coupling to the photon field), the gapless regime is likely to be large. In the extreme limit of strong dephasing, the order parameter will be almost completely photon-like, and there will be little change in the excitation spectrum for the renormalised fermions from the uncondensed state. This is the conventional semiconductor laser. Consequently, the practical issue for the optically-coupled excitonic insulator is the relative amplitude of the two components of the order parameter. In the theory given here, it is of order unity, and there is a large gap in the spectrum. In a conventional semiconductor laser, the ratio is at least 10^5 (in favour of photons), and there is no discernible gap in the quasiparticle spectrum (Since almost all measurements are near room temperature, which is much larger than the exciton binding energy in GaAs, this should not come as a surprise).

We conclude that while the symmetry of the order parameter is a cleanly defined issue, the question of the composition of the order parameter (fermionic and laser-like, or bosonic and exciton-like) is a matter of degree. This latter distinction is best made by studying the absorption/gain spectrum (gapped or not?) or by non-linear probes that give a direct estimate of the coherent polarisation [19].

References

1. L. V. Keldysh and Y. V. Kopaev, Fiz. Tverd. Tela **6**, 2791 (1964) [Sov. Phys. Solid State **6**, 2219 (1965)]; A. N. Kozlov and L. A. Maksimov, Zh. Eksp. Teor. Fiz. **48**,

1184 (1965) [Sov. Phys. JETP **21**, 790 (1965)]; L. V. Keldysh and A. N. Kozlov, Zh. Eksp. Teor. Fiz. **54**, 978 (1968) [Sov. Phys. JETP **27**, 521 (1968)].

2. *Bose-Einstein Condensation*, Eds. A. Griffin, D. W. Snoke, S. Stringari, (Cambridge Univ. Press, New York, 1994).

3. P.B.Littlewood and X.J.Zhu, Physica Scripta, (1997).

4. S.Pau et al., Phys.Rev.A **54** R1789 (1996). P.Senellart and J.Bloch, Phys.Rev.Lett. **82**, 1233 (1999), and references therein

5. M.Kira et al., Phys.Rev.lett. **79**, 5170 (1997). F.Tassone and Y.Yamamoto, Phys.Rev.B **59**, 10830 (1999)

6. P.R.Eastham, and P.B.Littlewood, to be published.

7. C. Comte and P. Nozieres, J. Phys. (Paris), **43**, 1069 (1982); P. Nozieres and C. Comte, *ibid.*, 1083 (1982); P. Nozieres, Physica **117B/118B**, 16 (1983).

8. J. R. Schrieffer, *Theory of Superconductivity*, (Addison-Wesley, New York, 1964); J. M. Blatt, *Theory of Superconductivity*, (Academic Press, New York, 1974).

9. R.H.Dicke, Phys.Rev. **93**, 99 (1954).

10. K.Hepp and E.Lieb, Ann.Phys. (NY) **76**, 360 (1973).

11. V.N.Popov and V.S.Yarunin, *Collective effects in Quantum Statistics of radiation and matter*, Mathematical Physics Studies 9, (Kluwer 1988).

12. P.W.Anderson, rev.Mod.Phys. **38**, 298 (1966)

13. P.Nozieres, in [2], p. 15.

14. R. R. Guseinov and L. V. Keldysh, Zh. Eksp. Teor. Fiz. **63**, 2255 (1972) [Sov. Phys. JETP **36**, 1193 (1973)].

15. R. Côté and A. Griffin, Phys. Rev. B **37**, 4539 (1988).

16. G. E. W. Bauer, Phys. Rev. Lett. **64**, 60 1989; G. E. W. Bauer, Physica Scripta T45, 154 (1992).

17. V. M. Galitskii, S. P. Goreslavskii, and V. F. Elesin, Zh. Eksp. Teor. Fiz. **57**, 207 (1969) [Sov. Phys. JETP, **30**, 117 (1970)].

18. G. P. Agrawal and N. K Dutta, "Semiconductor Lasers", (Van Nostrand, New York, 1993).

19. Th. Östreich, K.Schönhammer, and L. J. Sham, Phys.rev.B **58**, 12920 (1998). F.Jahnke et al., Phys.Rev.lett. **77** 5257 (1996).

COHERENT RESPONSE TO OPTICAL PULSES IN QUANTUM WELLS

J. FERNÁNDEZ-ROSSIER, D. PORRAS, C. TEJEDOR
Departamento de Física Teórica de la Materia Condensada,
Universidad Autónoma de Madrid,
Cantoblanco, 28049, Madrid, Spain.

AND

R. MERLIN
Department of Physics,
University of Michigan,
Ann Arbor, MI 48109-1120, USA

Abstract. We present a theoretical analysis of the coherent response of quantum wells to one or two short optical pulses. A bosonic description is used for the two physically different cases of undoped and doped systems. Coherent-control and optical beat experiments in intrinsic systems are studied in terms of the induced electrical polarisation associated with non-interacting excitons. In the case of a doped quantum well, we use a simple bosonisation procedure to show that the non-linear response presents exactly the same Fermi edge singularity as the well established one in linear response.

1. Introduction

Ultra-fast spectroscopy of semiconductor quantum wells (QW) is a field of enormous current interest due to the possibility of studying experimentally the dynamics of photoexcited carriers. QWs are excited by one or two low intensity laser (sub)picosecond pulses spectrally peaked around the lowest energy resonance. In the absence of free carriers, this resonance corresponds to an excitonic transition [1, 2, 3, 4, 5, 6, 7, 8, 9]. For modulation doped quantum wells, the transition threshold is the energy gap plus the Fermi energy of the free carriers. In such a case, ultra-fast spectroscopy has been

M.L. Sadowski et al. (eds.), Optical Properties of Semiconductor Nanostructures, 143–157.

performed for the coherent non-linear response within a four wave mixing (FWM) framework [10]. The common characteristic of all these experiments is that they can be described in terms of the low energy excitations of the particular system treated as a set of bosons. The aim of this work is to present such a theoretical description paying attention to a few specific cases, each of them corresponding to a different set of bosons.

Only *cw* spectroscopies have been used to study the Fermi edge singularity (FES) in doped QWs mainly by optical absorption and emission [11]. Ultra-fast experiments have not been able to detect any feature related to this singularity mainly because they have not been performed in suitable conditions. In conditions discussed below, FWM experiments in doped QWs could be an excellent tool for analysing possible singularities. In particular, we are going to use a simple bosonisation technique to show that the coherent non-linear response presents exactly the same FES as the one appearing in linear response.

In spite of the high interest of the analysis for the FES in FWM, we devote the major part of this work to the case of intrinsic QWs where a much larger body of experimental data is available. In this case, we present a theoretical analysis of the transient linear coherent response (TLCS) of excitons. A macroscopic description of these experiments is possible in terms of classical electrodynamics: an electric dipole linearly proportional to the exciting electric field of the laser is induced in the QW, considered as a dielectric [12]. The induced macroscopic dipole decays in a dephasing time T_2 characterising the loss of coherence. We discuss here a microscopic theory focussed on the description of excitonic coherence and on the effect of a weak disorder potential. The induced electric dipole is understood in terms of "coherent excitons". In the linear regime, excitonic coherence is identified by means of interference experiments. There are two possible kinds of interference:

a) Classical interference of the electromagnetic fields emitted by the induced electric dipole related with excitons

b) Quantum interference of the exciton wave function.

One of our main interests is to elucidate which kind of interference is the one measured experimentally.

The decay of the macroscopic dipole observed in TLCS experiments [1, 2, 3, 4, 5, 6, 7, 8, 9] is mainly due to disorder induced broadening of the exciton lines [12, 13], something which is usually neglected in the Semiconductor Bloch Equations approach [14]. Another frequently used model is the few level description [13] which can be certainly used for strongly localised excitons [15]. However, few level models are not suitable for weakly localised excitons such as those present in most experiments [1, 2, 3, 4, 5, 6, 7, 8, 9].

Our microscopic description of weakly localised coherent excitons is done in terms of non-interacting bosons in a weak disorder potential, by solving exactly the many-exciton Hamiltonian [16]. The collective wave function of excitons resonantly created by a laser pulse turns out to be a Glauber-like coherent state, even in the presence of disorder. The exact solution allows the description of excitonic coherence, which is in agreement with the main experimental findings [1, 2, 3, 4, 6, 7, 8, 9].

2. The bosonic model for excitons in QWs

The experiments analysed here are performed at temperatures below a few Kelvin, with very low exciton density (below $10^{10} cm^{-2}$). Excitons are created under resonant conditions in order to minimise the density of unbound electron-hole pairs and to have a negligible non-linear response. Therefore, we neglect the two sources of nonlinearity, i. e. exciton-exciton interaction and band filling effects in the light matter coupling [17]. On the other hand, it is well known that the deviation of the excitonic operators from a bosonic behaviour is proportional to the average exciton density [18]. This deviation is very small (below 0.1%) in all the experiments we analyse. Therefore, our model is that of non-interacting bosonic excitons coupled to the electromagnetic field. The bosonic model is obtained from the standard fermionic Hamiltonian [19] in the low excitation limit, by means of a controlled truncation of the Hilbert space [20]. The Hamiltonian is

$$\hat{H} = \hat{H}_0 + \hat{U} + \hat{H}_{rad} + \hat{H}_{int}. \tag{1}$$

The exciton kinetic energy term is

$$\hat{H}_0 = \sum_{\mathbf{k}_{||},\alpha,M} \hbar\omega_{\mathbf{k}_{||},\alpha} A^{\dagger}_{\mathbf{k}_{||},\alpha,M} A_{\mathbf{k}_{||},\alpha,M} \tag{2}$$

where $A^{\dagger}_{\mathbf{k}_{||},\alpha,M}$ creates an exciton with in-plane centre of mass momentum $\mathbf{k}_{||}$, third component of the angular momentum M and valence band index α. We consider only optically active excitons $M = \pm 1$ and heavy hole $(\alpha = HX)$ or light hole$(\alpha = LX)$ excitons in the lowest energy quantum well state. The single exciton energy is $\hbar\omega_{\mathbf{k},\alpha} = E^g_\alpha - \epsilon_\alpha + \hbar^2 k^2/2m_\alpha$, where E^g_α is the semiconductor gap and ϵ_α is the exciton binding energy.

The elastic scattering of excitons due to the QW disorder potential [16]

$$\hat{U} = \sum_{\alpha,\mathbf{q}_{||},\mathbf{p}_{||}} V_\alpha(\mathbf{q}_{||} - \mathbf{p}_{||}) \left(A^{\dagger}_{\mathbf{q}_{||},\alpha} A_{\mathbf{p}_{||},\alpha} + h.c. \right) \tag{3}$$

is assumed to be weak enough to leave unchanged the relative coordinate part of the electron-hole exciton wave function. This is equivalent to two different statements: *i)* the localisation length associated with the disorder potential is much bigger than the free exciton Bohr radius; *ii)* the disorder potential is weaker than the exciton binding energy. Since the typical excitonic linewidth is both bigger than the disorder potential, and less than 1 meV in good samples, statement *ii)* is fulfilled in the experiments we are interested in [1, 2, 3, 4, 5, 6, 7, 8, 9]. We must stress that LX and HX are not mixed by the elastic disorder potential (3).

The energy of the free electromagnetic field with three dimensional photon wave vector $\mathbf{k} = (\mathbf{k}_{||}, k_z)$ and polarisation index λ is

$$\widehat{H}_{rad} = \sum_{\mathbf{K},\lambda} \hbar c |\mathbf{k}| b^\dagger_{\mathbf{k},\lambda} b_{\mathbf{k},\lambda} \qquad (4)$$

where $b_{\mathbf{k},\lambda}$ is the photon operator.

We perform a quantum mechanical treatment of the emitted light [21] due to the very low occupation of the emitted photonic modes. The quantum character of the emitted light is manifest in the fact, shown below, that the light emitted by coherent excitons is coherent in the sense defined by Glauber [21, 22]. However, as in all available theoretical treatments [12, 16, 17], the exciting laser is considered as a classical object. Thus, within the rotating wave approximation for the long wavelength limit, the laser-exciton interaction is

$$\widehat{H}_{int} = -V \sum_{\mathbf{k}_{||}, \mathbf{k_z}, \lambda} \mathbf{E}_{\mathbf{k}_{||}, k_z, \lambda}(t) \cdot \left(\widehat{\mathbf{P}}^{(+)}_{\mathbf{k}_{||}} + h.c. \right) \qquad (5)$$

where V is the volume of the QW. $\mathbf{E}_{\mathbf{k}_{||}, k_z, \lambda}(t)$ is the classical electric field of the exciting laser while the density of electric dipole is

$$\widehat{\mathbf{P}}^{(+)}_{\mathbf{k}_{||}} = \frac{1}{\sqrt{V}} \sum_{\alpha, M} G_{\alpha, M} A^\dagger_{\mathbf{k}_{||}, \alpha, M} \mathbf{u}_M \qquad (6)$$

where $G_{\alpha, M}$ is the electric dipole matrix element [13] and $\mathbf{u}_{\pm 1}$ the unitary vectors associated with left and right circular polarisations. It must be pointed out that $\widehat{H}_{int}$ scales with $\sqrt{V}$, an important fact in discussing the linear approximation schemes frequently used in this problem.

We first solve the model (1) with no disorder. Later on we include $U \neq 0$ in order to discuss the effect of disorder in the linear spectroscopic properties of the QW. We neglect any quantum fluctuation of the electromagnetic field ($\widehat{H}_{rad} = 0$) and we assume that $E(t)$ is a known function.

When $\widehat{U} = \widehat{H}_{rad} = 0$, we have excitons in a perfect crystal driven by an external function. Performing the sum over k_z in (5)

$$\widehat{H}' = \widehat{H}_0 + \widehat{H}_{int} = \widehat{H}_0 + \sqrt{V} \sum_{\mathbf{k}_{||}, \alpha, M} E_{\mathbf{k}_{||} M}(t) G_{\alpha, M} (A^\dagger_{\mathbf{k}_{||}, \alpha, M} + A_{\mathbf{k}_{||}, \alpha, M}) \quad (7)$$

The Hamiltonian (7) describes excitons as a set of independent harmonic oscillators linearly driven by an external classical field. The state of the perfect QW assuming that in $t = -\infty$ all the excitonic modes are empty is [23]:

$$|\Xi(t)\rangle \equiv e^{i\widehat{H'}t}|0(t = -\infty)\rangle =$$

$$= e^{-iH_0 t/\hbar} \prod_{\mathbf{k}_{||},\alpha,M} e^{-|K_{\mathbf{k}_{||},\alpha,M}|^2/2} e^{iK_{\mathbf{k}_{||},\alpha,M}(t)A^\dagger_{\mathbf{k}_{||},\alpha,M}}|0\rangle \qquad (8)$$

where

$$K_{\mathbf{k}_{||},\alpha,M}(t) = \frac{\sqrt{V}G_{\alpha,M}}{2\hbar} \int_{-\infty}^{t} E_{\mathbf{k}_{||},M}(s)e^{i\omega_{\mathbf{k}_{||},\alpha}s} ds \qquad (9)$$

The important features of this exact state are:

i) (8) is a product of states of different excitonic modes. There is no quantum correlation between excitonic modes with different quantum numbers. This is because the initial state was the product of the vacuum of the excitonic modes and because the Hamiltonian (7) is a sum of the different excitonic modes without any intermode coupling.

ii) The factors of the product state (8) are Glauber coherent states. Therefore, under all the approximations stated above, *the state of the QW interacting with a classical electric field is a multimode coherent state.* "Coherent" has a precise meaning: the excitonic coherence induced by the laser is mathematically equivalent to the Glauber coherence of the photons in a coherent electromagnetic field [24].

iii) In the coherent regime, each excitonic mode, $(\mathbf{k}, \alpha, \mathbf{M})$, is fully characterised by a single complex function $K_{\mathbf{k},\alpha,M}(t)$ determined by the applied external field (see (9)).

The inclusion of disorder is not going to change these three conclusions.

We are interested in measurable quantities, namely, the exciton density and the induced electric dipole. The average occupation of the excitonic mode $\mathbf{k}_{||}, \alpha, M$ in the state (8) can be straightforwardly calculated

$$\langle \Xi(t)|A^\dagger_{\mathbf{k}_{||},\alpha,M} A_{\mathbf{k}_{||},\alpha,M}|\Xi(t)\rangle = |K_{\mathbf{k}_{||},\alpha,M}(t)|^2 \qquad (10)$$

while the total exciton density will be given by $\sum_{\mathbf{k}_{||},\alpha,M} |K_{\mathbf{k}_{||},\alpha,M}(t)|^2$. In coherent control (CC) experiments, the exciton density is measured as the average of the system response to many pulses. Hence, the well defined quantum average (10) is measured.

Another measurable quantity is the $\mathbf{k}_{||}$ component of the laser-induced excitonic electric dipole density given by:

$$\langle\Xi(t)|\hat{P}_{\mathbf{k}_{||}}^{(+)}|\Xi(t)\rangle = \frac{1}{\sqrt{V}}\sum_{\alpha}G_{\alpha,M}ie^{i\omega_{k,\alpha}t}K_{\mathbf{k}_{||},\alpha,M}(t). \qquad (11)$$

This is linearly proportional to $E_{M,\mathbf{k}_{||}}(t)$. Due to the in-plane translational invariance, $\langle\hat{P}_{M,\mathbf{k}_{||}}^{+}\rangle$ is only nonzero if the electric field has a nonzero $E_{\mathbf{k}_{||}}$ component. The linearity between the induced electric dipole and the laser electric field is an *exact* property of both (1) and (7), i. e. no linear response approximation is made. Within a linear approximation one obtains the same the electric dipole, but a different result for the average exciton density.

Let us now analyse the effect of disorder. Most of the features of the perfect case remain unaltered. Without disorder, the exciton basis is that of plane waves. When disorder is included, there is a new localised basis $\psi_{\alpha,M,\xi}(\mathbf{r}_{||})$ with ξ standing for the quantum number associated with the disorder potential. It is convenient to argue in terms of the projection of such a basis onto the plane-wave one

$$c_{\alpha,M,\xi}(\mathbf{k}_{||}) \equiv \int dV e^{i\mathbf{k}_{||}\cdot\mathbf{r}_{||}}\psi_{\alpha,M,\xi}(\mathbf{r}_{||}) = \langle\mathbf{k}_{||}|\alpha,\mathbf{M},\xi\rangle. \qquad (12)$$

The analysis performed previously for the perfect crystal can be repeated using the localised basis, but now the coupling between light and excitons, which depends on the coefficients $c_{\alpha,M,\xi}(\mathbf{k}_{||})$, is independent of the volume of the system. Once again, one obtains a bunch of coherent excitons described by a Glauber type state. The average number of excitons is

$$\langle\tilde{\Xi}(t)|N_{\xi,\alpha,M}|\tilde{\Xi}(t)\rangle = \sum_{\alpha,M}\int d\omega_{\xi}\frac{d\xi}{d\omega_{\xi}}\frac{G_{\alpha,M}^{2}}{\hbar^{2}}|c_{\omega_{\xi},\alpha,M}(\mathbf{k_L})|^{2}|\tilde{E}(\omega_{\xi\alpha},t)|^{2} \qquad (13)$$

where $d\xi/d\omega_{\xi}$ is the density of localised excitons. Information about the integrand in (13) could be also obtained from continuous wave spectroscopy experiments.

Although it is not discussed here, our formalism allows[25] the study of coherent secondary emission. There, disorder produces coherent elastic scattering of excitons so that the emitted light is in a different direction from that of the excitation.

3. Coherent control of the exciton density

In this section we give a microscopic description of experiments on the CC of the exciton density in GaAs QWs [2, 4, 5, 7]. In these experiments a QW, at temperatures below 10 K, was excited by a sequence of two identical *phased locked* pulses of 150 fs, tuned to the HX frequency ω_H. In references [2, 5], the exciton density was measured by means of the reflectivity change (ΔR), as measured by a third pulse, as a function of the delay between the pulses τ. In [4, 7], the exciton density was inferred from the time integrated luminescence. In all the cases, the exciton density after the excitation with two pulses, $\langle n_2 \rangle$, turns out to be:

$$n_2 = 2n_1 \left[1 + A(\tau)cos(\omega_H \tau)\right] \tag{14}$$

where n_1 is the density of excitons created with a single pulse. The amplitude of the oscillations, $A(\tau)$, is smaller than one and it tends to zero for $\tau > T_2$.

We consider a circularly polarised plane wave,

$$E_M(t) = \delta_{\mathbf{k},\mathbf{k_L}} \left(E_0 sin[\Omega t]e^{-(t/\Delta t)^2} + sin[\Omega(t-\tau)]e^{-([t-\tau]/\Delta t)^2}\right). \tag{15}$$

where Δt is the pulse duration, and Ω the position of the spectral maximum of the laser.

For simplicity, we begin our analysis with the simplest case in which only one heavy excitonic mode, $A_{\mathbf{k_L}}$, of frequency ω_H, $M = +1$ and coupling constant G, is excited by the laser pulses (15). This would be the case of a perfect crystal whose HX-LX splitting is much bigger than the spectral width $(\Delta t)^{-1}$. Following (10), the average density of excitons in a QW excited by (15) is:

$$\langle n(t,\omega) \rangle = |\frac{G}{2\hbar} \int_{-\infty}^{t} E(s)e^{i\omega s}ds|^2. \tag{16}$$

For a single pulse the density of photoexcited excitons would be:

$$\langle n_1(\omega) \rangle = \left(\frac{GE_0 \Delta t \sqrt{\pi}}{2\hbar}(e^{-[(\omega-\Omega)\Delta t]^2/4} - e^{-[(\omega+\Omega)\Delta t]^2/4})\right)^2 \approx$$

$$\frac{G^2 E_0^2 \Delta t^2 \pi}{4\hbar^2} e^{-[(\omega-\Omega)\Delta t]^2/2} \tag{17}$$

which is independent of τ. In figure 1 we plot $n(t)/(n_1)$ for four slightly different values of the delay, τ, all of them lying in an interval of two fs. One can see that, as an the effect of the second pulse, the density of excitons can take any value between zero (destructive interference) and an increase by a factor of 4.

In this case of a single mode, the density of excitons created by the two pulses, $n_2 \equiv n(t \gg \tau)$ is

$$\langle n_2(\tau) \rangle = |\frac{G}{2\hbar} \int_{-\infty}^{+\infty} E(s)e^{i\omega s}ds|^2 = 2n_1 \left(1 + cos(\omega\tau)\right) \tag{18}$$

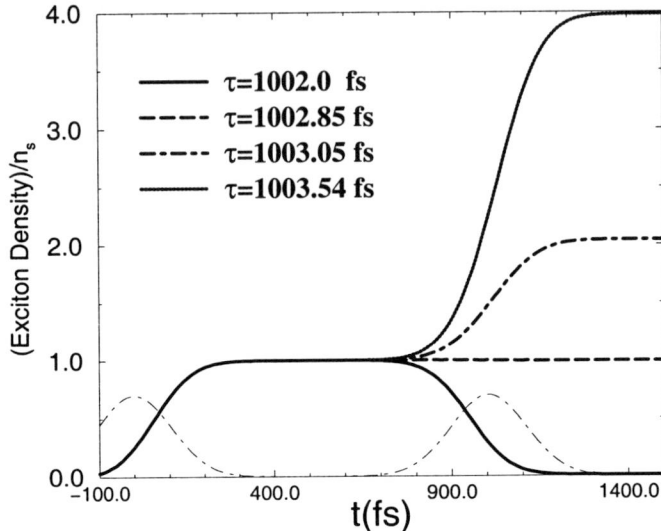

Figure 1. Density of excitons created by two laser pulses, as a function of time, in units of the density of excitons created by a single pulse. The density after the second pulse depends dramatically on the delay between the pulses, which are also displayed (dashed line).

Therefore, for a single excitonic mode and neglecting any dephasing mechanism, we obtain the experimental CC result (14) with $A(\tau) = 1$. In this simplified case the interference between the first and the second pulse would take place for arbitrarily long τ.

In a realistic case, in which disorder is present, several excitonic modes are excited by the laser. If only $M = +1$ HX are excited, (13) can be used to show[25] that:

$$n(t) = \sum_{\xi} |c_{\xi,\alpha,M}(\mathbf{k_L})|^2 \langle n(\omega_\xi, t)\rangle. \tag{19}$$

For $t >> \Delta t + \tau$ we have:

$$n_2(\tau) = \int \frac{d\xi}{d\omega_\xi} d\omega_\xi |c_{\xi,\alpha,M}(\mathbf{k_L})|^2 2\langle n_1(\omega_\xi)\rangle (1 + cos(\omega_\xi \tau)) \tag{20}$$

with $\langle n_1(\omega)\rangle$ given by (17). Equation (20) implies that when disorder is taken into account, the density of excitons photoexcited by the phase locked pulses is a sum over single modes of slightly different energy. As a consequence of disorder, many modes of slightly different energies are in a coherent state driven by the laser pulses. In the sum, each mode is weighted

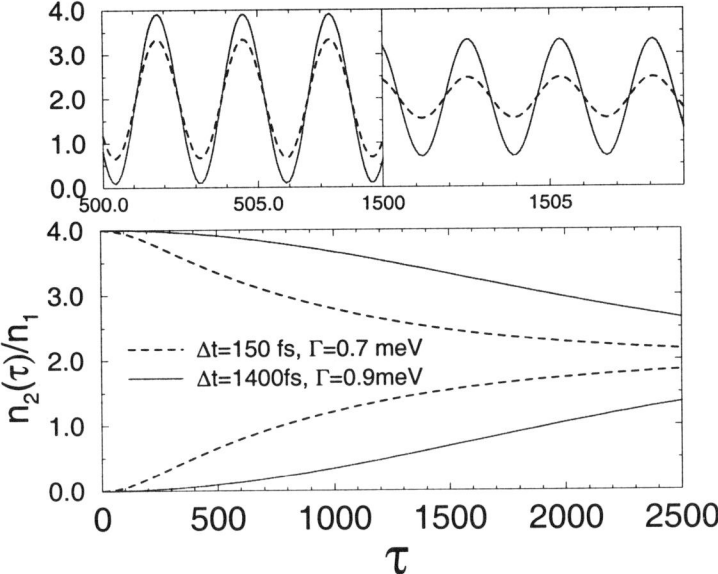

Figure 2. (a) Oscillations of the exciton density created by two pulses. (b) Envelope of the exciton density. Both magnitudes are given as a function of the delay τ in fs.

by the square of the overlap between the quantum state of the mode ξ and the plane wave state $\mathbf{k_L}$, $\mathbf{k_L}$, $|c_{\xi,\alpha,M}(\mathbf{k_L})|^2$. For a particular disorder configuration we assume[25]

$$\frac{d\xi}{d\omega_\xi}|c_{\xi,\alpha,M}(\mathbf{k_L})|^2 = \frac{1}{\pi}\frac{\Gamma_\alpha}{(\omega_\xi - \omega_{0\alpha})^2 + \Gamma_\alpha^2} \tag{21}$$

where Γ_α is the disorder induced broadening of the α exciton. For $\Omega = \omega_{0\alpha}$ we obtain [26]

$$A(\tau) = e^{-\Gamma_\alpha\tau}erfc[\frac{\Delta t\Gamma_\alpha}{\sqrt{2}} - \frac{\tau}{\sqrt{2}\Delta t}] + e^{\Gamma_\alpha\tau}erfc[\frac{\Delta t\Gamma_\alpha}{\sqrt{2}} + \frac{\tau}{\sqrt{2}\Delta t}] \tag{22}$$

The density of excitons created by the field (15), for the distribution (22) is plotted in figure 2 for two different values of Γ_{HX}. In contrast with the single mode case, the amplitude of the oscillations $A(\tau)$ decreases as τ increases, in agreement with the experimental observation (14).

A few interesting facts must be pointed out:

i) The amplitude of the CC interferences, $A(\tau)$, decays due to the broadening induced by the disorder potential without any inelastic dephasing. The decay takes place even if all the excitonic modes remain in a coherent quantum state.

ii) $A(\tau)$ is not always of the form $exp(-\tau/T_2)$. Instead, the shape of the decay depends on the particular disorder potential.

iii) The decay rate is a function of Δt and the broadening Γ_α. When $\Delta t << \Gamma_\alpha^{-1}$, the decay time is given by Γ_α^{-1}.

We should mention that our description in terms of a coherent state of non-interacting excitons is able to properly describe an interesting set of experiments in which the exciton spin is coherently controlled [4]. Excitation is performed by two oppositely polarised mode-locked laser pulses, delayed by τ, in resonance with the HX. The experiments were done in two configurations: i) exciting with linearly polarised (x, y) pulses and detecting the degree of circular polarisation of the emitted light and ii) exciting with circularly polarised $(+, -)$ light and detecting the degree of linear polarisation. In both cases the resulting degree of polarisation was an oscillating function of τ, which is adequately described by our bosonic scheme [25].

4. Optical beats produced by a short pulse

In the previous section we have just considered the case in which only one excitonic energy is involved. Now we take a step forward and include a second excitonic mode corresponding to the LX. The exciting pulse is very short in time so that its spectral width is broad enough to excite both HX and LX (with a well defined M). We can use an effective Hamiltonian

$$H = \hbar\omega_H N_H + \hbar\omega_L N_L + E(t)(G_H P_H + G_L P_L) \qquad (23)$$

where $N_X \equiv A_X^\dagger A_X$ and $P_X \equiv A_X^\dagger + A_X$

The wave function, as calculated from equation (8) for this two mode case, is a *product* of HX and LX coherent states:

$$|\Xi_{lh}\rangle \propto e^{-i\omega_H t N_H} e^{iK_H(t)A_H^\dagger} e^{-i\omega_L t N_L} e^{iK_L(t)A_L^\dagger} |0\rangle. \qquad (24)$$

From this wave function, one obtains that the induced electric dipole is a sum of the LX and HX dipoles:

$$\langle \Xi(t)|P_M|\Xi(t)\rangle = 2(G_H)^2 cos(\omega_H t) + 2(G_L)^2 cos(\omega_L t) \qquad (25)$$

Since the state (24) is certainly not a sum of LX and HX states, it is not possible to state that a certain exciton is a quantum superposition of HX and LX states. This is not only true for the particular wave function (24) but also applies to the general state (8). Therefore, quantum interference phenomena, like quantum beats, are *ruled out* for non interacting excitons. *The interference phenomena are due to polarisation interferences.* This negative result is very important. It could be argued that the wave function

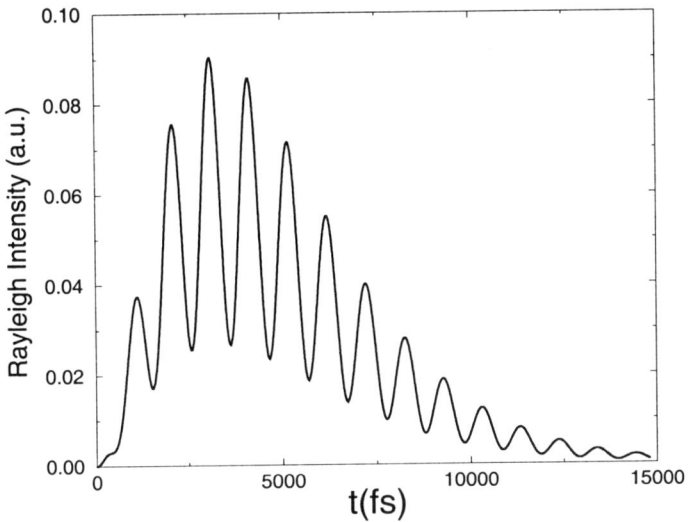

Figure 3. RSE intensity, in arbitrary units, emitted by a QW having $\omega_L = 1549 \pm 0.2$ meV and $\omega_H = 1545 \pm 0.2$ meV when excited by a laser pulse of frequency 1538 meV with $\Delta t = 180$ fs.

(24) is a linear combination of Fock states with a well-defined number of states. One could write the state (24) as:

$$|\Psi\rangle = \sum_{N_H, N_L} \frac{C(N_H, N_L)}{\sqrt{N_H!}\sqrt{N_L!}} (A_H^\dagger)^{N_H} (A_L^\dagger)^{N_L} |0\rangle. \tag{26}$$

This state is certainly a sum of states but there is no correlation between the number of HX and that of LX.

$$\langle\Psi|(N_H)(N_L)|\Psi\rangle - \langle\Psi|N_H|\Psi\rangle\langle\Psi|N_L|\Psi\rangle = 0 \tag{27}$$

while this magnitude should be finite in order to have Quantum Beats. Photoexcitation by a laser can not prepare the QW in a Fock state

$$\frac{1}{\sqrt{N!}} (c_H A_H^\dagger + c_L A_L^\dagger)^N |0\rangle \tag{28}$$

with a well-defined number of excitons.

The experimental method of studying optical beats due to the simultaneous excitation of HX and LX is usually more complicated than the analysis performed up to here. In general, resonant secondary emission, or

Rayleigh scattering, is measured. Photons are emitted in a direction different from that of the exciting laser. Disorder in the QW provokes the elastic scattering required for such a process. The analysis must be performed once again in terms of the localised basis $\psi_{\alpha,M,\xi}(\mathbf{r}_{||})$. Using similar approximations [25] to the ones used in the previous section, one gets the following results as observed in figure 3:

  i) The intensity of emitted photons initially increases as t^2 and later on decreases exponentially.

  ii) Beats of the induced polarisation produce an emitted intensity with classical beats having a period $(\omega_H - \omega_L)^{-1}$.

FWM experiments [28] also show beats but with different physical origin. FWM is a non-linear technique able to detect many excitons effects [10]. Coupling between different excitonic modes is responsible for the FWM signal. Many-exciton coupling may produce states which are quantum superpositions of different excitonic modes which implies non zero *quantum beat* correlations. In any case, the understanding of the QW exciton quantum beats phenomena in terms of single exciton physics is out of the question: only exciton-exciton interaction could produce an excitonic state being a sum instead of a product.

5. Fermi edge singularities in FWM

In this section we analyse the non-linear response of an n-doped QW within the experimental framework of FWM. Our goal is to study the possible singularities in the non-linear response when the excitation pulse embraces a narrow range around the frequency of the transition from the valence band to the Fermi energy at the conduction band. Our interest is to compare this FES with the well-known singularity appearing in linear response as measured in optical absorption. We are going to make use of two main ingredients: first, polarisation is, once again, the physical magnitude controlling the non-linear response, and second, a bosonisation of the low energy excitations around the Fermi level [29] is able to describe both the linear and non-linear optical responses.

In FWM experiments, the system is excited by two phase locked pulses having momenta $\mathbf{k_1}$ and $\mathbf{k_2}$ with $|\mathbf{k_1}| = |\mathbf{k_2}|$ and a relative delay τ. In the case $\mathbf{k_1} = \mathbf{k_2} = 2\mathbf{k_2} - \mathbf{k_1}$, one recovers the experimental situation of CC. The detection of a signal with momentum $2\mathbf{k_2} - \mathbf{k_1}$ can be either time-integrated or time-resolved [13]. A theoretical analysis can be easily performed if dephasing is ignored. The excitation of the system produces an electrical polarisation $P(t)$ in the sample. By extending up to third order the usual perturbation scheme developed for obtaining the Kubo formula, one gets the FWM signal in the $2\mathbf{k_2} - \mathbf{k_1}$ direction

$$F(t) = -i\mu^4 \int_{-\infty}^{t} dt_1 dt_2 dt_3 E_1^*(t_3) E_2(t_1) E_2(t_2)$$

$$\times \left[\langle P(t)\mathcal{T}\{P(t_1)P(t_2)P(t_3)\}\rangle + \langle \mathcal{T}^{-1}\{P(t_2)P(t_1)\}P(t)P(t_3)\rangle \right]$$

$$+h.c. \qquad (29)$$

where $E_1(t)$ and $E_2(t)$ are the amplitudes of the incident pulses in the directions $\mathbf{k_1}$ and $\mathbf{k_2}$ respectively, $\mathcal{T}$ and $\mathcal{T}^{-1}$ stand for the time ordering and anti-ordering operators and the brackets $\langle\rangle$ mean thermal averages. Equation (29) has been obtained within a rotating wave approximation in which the width Δt of each pulse is much larger than the period $\propto \omega_0^{-1}$ of the exciting light.

As mentioned above, we are interested in a possible FES in the FWM signal. Therefore, the frequency of the exciting light ω_0 must be as close as possible to $E_g + E_f$ where E_g is the energy gap of the QW and E_f is the Fermi energy of the electron gas at the conduction band. The excitation covers a frequency range $\Delta\omega$ determined by the the pulse width Δt. If $\Delta\omega$ is only a few meV, the excitations it provokes are of low energy and can be described as bosons given by a Tomonaga-Luttinger approach to the Hamiltonian [29]. The FWM signal $F(t)$ can be straightforwardly obtained when the inter-pulse delay τ is much larger than the inverse of the energy cut-off Δ for the excitations of the bosonised model. As in the case of the linear response, the singularity is determined by the response at very long times $t \gg \tau$, which is given by $F(t) = \chi^{(3)}(t)E_1^* E_2^2$ with

$$\chi^{(3)}(t) \propto -i\mu^4 e^{i2(E_g+E_f)\tau} (i\tau\Delta)^{2\alpha} \frac{sinh^{2\alpha}(\pi k_B T\tau)}{(\pi k_B T\tau)^{2\alpha}}$$

$$\times e^{i(E_g+E_f)t} (it\Delta)^{\alpha} \frac{sinh^{\alpha}(\pi k_B Tt)}{(\pi k_B Tt)^{\alpha}} + h.c. \qquad (30)$$

where T is the temperature. As in the case of the linear response, $\alpha = -(1 - V_h\overline{\rho})^2$ with V_h being the potential felt by the electrons due to the hole and $\overline{\rho}$ the mean electron density at the conduction band. Equation (30) shows an extremely important result: apart from the first row, which is characteristic of FWM, the factor in the second row shows exactly the same time dependence as the linear response [11]. In other words, when $-1 < \alpha < 0$, *FWM presents exactly the same FES as optical absorption.* That means that in time-integrated FWM, the signal $F(\omega)$ must have, at low temperature, a singular behaviour $(\omega - E_g - E_f)^{-\alpha-1}$[11].

6. Conclusions

We have used a bosonic picture of short pulsed excitation of QW to get the following conclusions:

i) For intrinsic QWs, we put on an equal footing the concepts of photonic and excitonic coherence;

ii) A description in terms of non-interacting excitons is simple and exact for understanding several interesting ultra-fast spectroscopy experiments;

iii) In the case of a doped QW, a simple bosonisation procedure shows how the non-linear response presents exactly the same FES as the well-known singularity appearing in linear response.

In the case of intrinsic QWs, we focus on CC experiments [2, 4, 6] in which excitons in a single QW are resonantly excited with two low intensity laser pulses, in the linear regime, delayed by a time τ. When $\tau < T_2$ the "coherently controlled" QW property is an oscillatory function of $\omega_H \tau$ superimposed on a decay associated with dephasing. Some examples are the intensity of the terahertz emission, the density of excitons [2, 4], the optical orientation [4].

For doped QWs the non-linear response measured by FWM can described in terms of bosonic charge excitations around the Fermi level. We show how the FES could be studied in these FWM experiments.

Acknowledgements. Work supported in part by MEC of Spain under contract PB96-0085, Fundacion Ramón Areces and CAM under contract 07N/0026/1998.

References

1. H. Wang, J. Shah, T.C.Damen and L.N. Pfeiffer, Phys. Rev. Lett., **74**, 3065 (1995).
2. A.P. Heberle, J.J. Baumberg and K. Köhler, Phys. Rev. Lett., **75**, 2598 (1995); J. J. Baumberg, A.P. Heberle and K. Köhler, Phys. Stat. Sol (b), **204**, 9 (1997).
3. S. Haacke, R.A. Taylor, R. Zimmermann, I. Bar-Joseph and B. Deveaud, Phys. Rev. Lett., **78**, 2228 (1997).
4. X. Marie, P. LeJeune, T. Amand, M. Brousseau, J. Barrau, M. Paillard and R. PLanel, Phys. Rev. Lett., **79**, 3222, (1997); P. Le Jeune, X. Marie, T. Amand, M. Brousseau and J. Barrau, Phys. Stat. Sol. **164**, 527 (1997).
5. J.J. Baumberg, A.P. Heberle, A.V. Kavokin, M.R. Vladimirova, and K. Kohler, Phys. Rev. Lett., **80**, 3567 (1998).
6. D. Birkedal and J. Shah, Phys. Rev. Lett., **81**, 2372 (1998).
7. M. Woerner and J. Shah, Phys. Rev. Lett., **81**, 4208 (1998).
8. S. P. Kennedy, N. Garro, M.J. Snelling, R.T. Phillips and K.H. Ploog, in Proceedings of ICPS 24, ed. by D. Gershoni, (World Scientific, Singapore, 1998), to be published.
9. W. Langbein, J.M. Hvam and R. Zimmermann, Phys. Rev. Lett., **82**, 1040 (1999).
10. D. S. Kim, J. Shah, J. E. Cunningham, T. C. Damen, S. Schmitt-Rink and W. Schafer, Phys. Rev. Lett., **68**, 2838 (1992).
11. See, for instance, K. Ohtaka and Y. Tanabe, Rev. Mod. Phys., **62**, 929 (1990).
12. H. Stolz, *Time-Resolved light scattering from excitons*,(Springer-Verlag, Berlin, 1994).

13. J. Shah, *Ultrafast Spectroscopy of Semiconductors and Semiconductors Nanostructures* (Springer, Berlin, 1996).

14. H. Haug and S.W. Koch, *Quantum Theory of Optical and Electronic Properties of Semiconductors* (World Scientific, London, 1993).

15. N. H. Bonadeo, J. Erland, D. Gammon, D. Park, D.S. Katzer and D.G. Steel, Science, **282** 1473 (1998).

16. R. Zimmermann, Il Nuov. Cim., **17**D, 1801 (1995).

17. D. S. Chemla, *Nonlinear optics in quantum confined structures*, Physics Today **June** , 46 (1993).

18. L. V. Keldysh and A.N. Kozlov, Zh. Eksp. Teor. Fiz., **54**, 978 (1968) [Sov. Phys. JETP, **27**, 521 (1968)].

19. E. Hanamura and H. Haug, Physics Reports, **33C**,209 (1977).

20. Th. Oestreich, K. Schöenhammer and L. J. Sham, cond-mat 9807135.

21. J. Fernández-Rossier, C. Tejedor and R. Merlin, Solid State Comm., **108**, 473 (1998).

22. R.J. Glauber, Phys. Rev., **131**, 2766 (1963).

23. A. Galindo and P. Pascual, *Quantum Mechanics*, (Springer Verlag, Berlin, 1991).

24. C. Cohen-Tannoudji, J. Dupont-Roc and G. Grynberg, *Atom-Photon interactions* (John Wiley and Sons, New York, 1992).

25. J. Fernandez-Rossier, C. Tejedor and R. Merlin, to be published.

26. If the probability density is approximated by a gaussian, $|c_{\xi,\alpha,M}(\mathbf{k_L})|^2 d\xi/d\omega_\xi = e^{-[(\omega_{\xi,\alpha}-\omega_\alpha)/\Gamma_\alpha]^2}/\Gamma_\alpha^2$, one obtains $A^G(\tau) = e^{-(\tau/T_2)^2}$ where $T_2 = \sqrt{2\Delta t^2 + 4/\Gamma^2}$.

27. M. Guriolo, F. Bogani, S. Ceccherini and M. Colocci, Phys. Rev. Lett., **78**, 3205 (1997).

28. E.O. Göbel, K. Leo, T.C. Damen, J.Shah, S. Schmitt-Rink, W. Schäfer, J.F. Müller and K. Köhler, Phys. Rev. Lett., **64**, 1801 (1990).

29. K. D. Schotte and U. Schotte, Phys. Rev., **182**, 479 (1969).

HIGH DENSITY TWO-DIMENSIONAL ELECTRON-HOLE PLASMAS IN QUANTISING MAGNETIC FIELDS

S. SCHOSER, C. KUTTER, AND M. POTEMSKI
Grenoble High Magnetic Field Laboratory, MPI/FKF and CNRS
BP 166, 38042 Grenoble Cedex 9, France

B. ETIENNE
Laboratoire des Microstructures et de Microélectronique, CNRS
92 220 Bagneux, France

Abstract. Luminescence spectra of highly excited undoped GaAs/GaAlAs quantum wells subjected to quantising magnetic field are analysed and compared with conventional photoluminescence excitation (quasi-absorption) measurements. Transitions observed in emission show very little change in the energy position when the excitation intensity, i.e. the quasi-equilibrium population of electron-hole pairs, is varied over a wide range. These transition are, however, very different from magneto-exciton transitions seen in absorption-type spectra. The results are in contradiction with a standard model of two interpenetrating Fermi fluids, usually applied to describe the luminescence spectra of highly excited semiconductors. They indicate the importance of individual electron-hole correlations (excitonic effects) in the case of systems with quasi-discrete one-particle energy levels.

1. Introduction

Relatively simple experiments, such as measurements of interband luminescence spectra of undoped semiconductors, create an interesting possibility of investigating the properties of electron-hole pairs which occupy the conduction and valence bands under quasi-equilibrium condition, and whose density can be tuned with the intensity of laser excitation. Interband optical experiments performed at low excitation intensities (diluted electron-hole systems) depict the properties of excitons, i.e., individual electron-hole pairs involving "hydrogen-like" Coulomb biding. High density systems, which are expected under intense optical excitation, imply a more complex interaction scheme: not only electron-hole but also electron-electron and hole-hole terms should be taken into account. A well-established phenomenon of the formation of an electron-hole liquid condensate in indirect semiconductors (such as germanium or silicon) [1] shows the relevant role of many body effects in the description of luminescence spectra of high-density electron-hole systems. The formation of an electron-hole condensate is

159

understood as a consequence of the minimum in the mean energy of the electron-hole system, calculated as a function of the density [2]. This minimum results from the competition between the energy increase due to kinetic terms and energy lowering due to exchange and correlation terms. The measured spectra of the electron-hole liquid are well reproduced by the theoretical model of interpenetrating electron and hole Fermi fluids [1,2]. This model implies the rigid (no changes in the dispersion relation of electron and holes) collapse of the bands ("band gap renormalisation") due to exchange and correlation effects but neglects individual electron-hole correlations [2]. Investigations of direct gap semiconductors do not show the formation of an electron-hole condensate. Direct gap semiconductors are considered as systems for which the density of electron-hole pairs can be rather arbitrarily tuned with the excitation power [3]. Most often, the experiment [4,5] and theory [6,7] meet at the line-shape analysis. This many-parameter procedure is not always appreciated, nevertheless, the model of two-interpenetrating Fermi fluids with a rigid band shift seems to be quite satisfactory. The density dependence of the bandgap renormalisation has been determined in various bulk and 2D semiconductor structures [4,5].

In this paper we discuss the luminescence spectra of highly excited quantum well structures under conditions of high magnetic fields applied perpendicular to the 2D layers [8,9,10]. We show that under these conditions, the rigid band-shift model fails to explain the experimental results. We think that the individual electron-hole correlation is an important ingredient in understanding the experimental data and speculate that this is due to the discrete nature of the electronic states involved.

2. Experimental

Here, we discuss the data obtained for relatively narrow GaAs/GaAlAs quantum wells (well widths of 4 and 6 nm). Other systems show similar experimental results [10]. Single quantum well structures were used in our experiments. Special attention was paid to probing the homogenous carrier distribution in the lateral direction. The samples were excited with 20 ns pulses of a frequency doubled Nd-YAG laser. The luminescence signal was detected with a 0.4 nm-spectral and 1 ns-time resolution. Experiments were performed at pumped-helium bath temperatures and in magnetic fields up to 30T.

Representative magneto-luminescence spectra of highly excited quantum wells are shown in Figures 1 and 2. Figure 1 shows a typical magnetic field evolution of the luminescence spectra measured at an excitation level corresponding to an actual density of electron hole pairs of $\sim 4.1 \times 10^{12}$ cm^{-2}. As seen in this figure, the broad luminescence spectrum observed at B=0 splits into several peaks when sufficiently high magnetic fields are applied. Figure 2 show the data at fixed magnetic field of B=19 T but at different excitation powers, in the range of corresponding densities of electron-hole pairs from 10^{11} cm^{-2} to 4.1×10^{12} cm^{-2}. In Figure 3 we compare high-excitation luminescence spectra with a standard photo-luminescence-excitation spectrum, all measured at B=19 T.

Figures 2 and 3 depict the characteristic features of magneto-luminescence spectra of highly excited quantum wells. The spectra measured as a function of (high-power) laser excitation do show the effects of progressive population of higher and higher energy levels (band-filling effects), but the observed, well-resolved luminescence peaks hardly change their energy positions. On the other hand, the peak energies observed in emission spectra are very different from those measured in absorption spectra. These observations are in clear conflict with the standard image of smooth bandgap-shrinkage with density of electron-hole pairs.

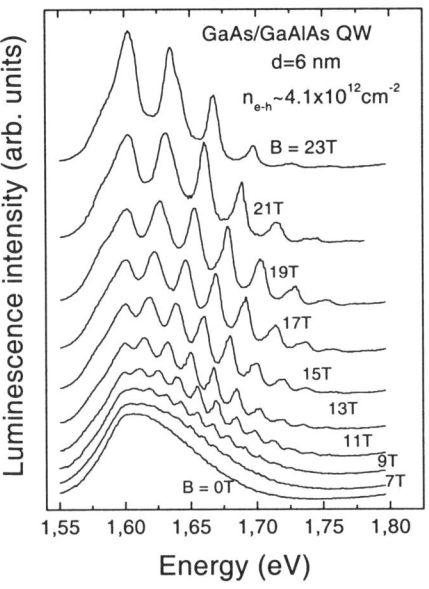

Figure 1. Luminescence spectra measured at different magnetic fields (B = 0 - 23T) and fixed (high) excitation intensity generating the quasi equilibrium population of electron-hole pairs at the estimated density of $4.1 \times 10^{12} \text{cm}^{-2}$.

3. Discussion

Studies of conventional (linear) magneto-absorption spectra of undoped quantum wells are a mature subject [11,12,13]. These spectra can be understood in terms of the energy levels of a two-dimensional hydrogen atom in a magnetic field [12,13]. In the low field limit, the observed magneto-absorption resonances are identified as 1s, 2s, 3s, ... hydrogen-like excitonic states. In the high field limit, these states can be recognised as magnetoexcitonic states, correspondingly associated with subsequent pairs (0,0), (1,1), (2,2), ..., of conduction and valence band Landau levels with the same indexes. At sufficiently high magnetic fields, the energy position of an Ms magnetoexciton follows the field dependence characteristic for the interband transition between the valence and conduction band Landau levels with the indexes M-1, i.e., it depends on magnetic field as (M-1/2) $\hbar\omega_r$ ($\hbar\omega_r$ is the reduced cyclotron energy). Coulomb binding is important for each magnetoexcitonic state, and it increases with the magnetic field strength. However, higher energy states are less bound and the binding energy of each state changes only as

162

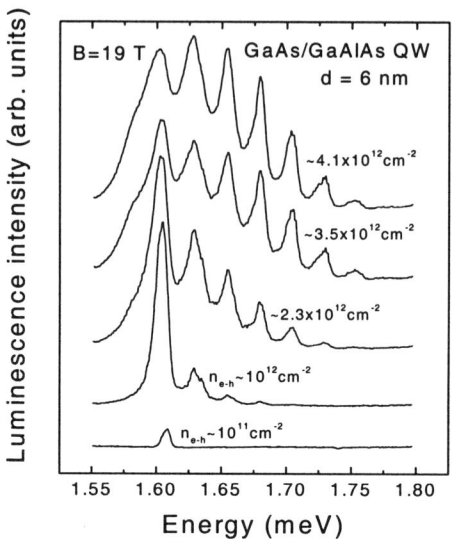

Figure 2. Luminescence spectra measured at fixed magnetic field of B=19T and different (high) intensities of laser excitations generating a quasi equilibrium band population in the density range of electron-hole pairs from 10^{11} to $4.1 \times 10^{12} cm^{-2}$.

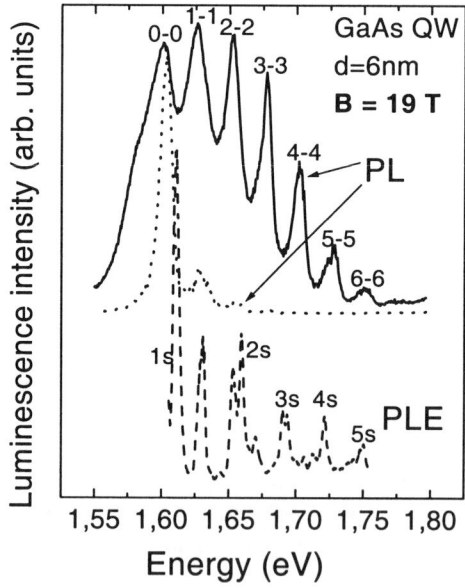

Figure 3. Comparison of luminescence spectra under high excitation powers (two upper traces which correspond to different intensities of laser excitation generating the density of electron hole pairs of n_{e-h}= $4.1 \times 10^{12} cm^{-2}$ and $1 \times 10^{12} cm^{-2}$, correspondingly for upper and lower trace) and the conventional photoluminescence excitation spectrum (lowest trace), all measured for the same sample at B=19T.

the square root of the magnetic field. In consequence, the linear term (M-1/2) $\hbar\omega_r$) always becomes dominating in the high field range. Mixing of heavy and light hole valence states implies the usual complications, nevertheless, as can be seen in Figure 3, the most pronounced magneto-absorption transitions are those related to s-states of the heavy- hole exciton (with the exception of the 1s light hole resonance visible in Figure 3 at 1.63 eV). The Landau level fanchart of the observed heavy-hole magnetoexcitonic states can be very well described within the simplified "hydrogen-like model" [12,13]. Such a model applied to the experimental data allows us to determine three fitting parameters involved, i.e., the two-dimensional energy gap, exciton binding energy and the reduced cyclotron energy (reduced mass). The fitting procedure is particularly accurate in the case of narrow quantum wells studied here. Solid lines in Figure 4 show the calculated field-dependence of magneto-excitonic transitions for a 6nm-wide quantum well. These calculations reproduce the experimental points perfectly.

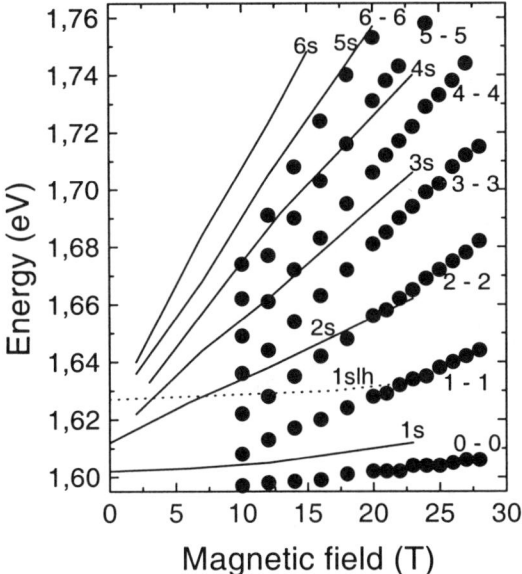

Figure 4. Landau level fanchart of transitions observed in high-excitation-power magneto-luminescence (solid circles). The solid lines represent the theoretical fit, perfectly reflecting the experimental data of magnetic field dependence of the main transitions observed in conventional photo-luminescence spectra. The data are shown for a 6nm-wide GaAs/GaAlAs quantum well structure.

We now turn to the analysis of transitions observed in high-excitation-power magneto-luminescence spectra. The energy positions of peaks observed in these spectra have been investigated as a function of the magnetic field and excitation power. As mentioned above, these peaks show very little change as a function of the excitation intensity but clearly develop as a function of the magnetic field (see Figures 1 and 4). The high power data shown in Figure 4 have been obtained at fixed excitation power, generating an electron-hole density of ~4.1x10^{12}cm^{-2}. As shown in this figure, the energy positions of transitions observed in high-power luminescence spectra are very different from those seen in (linear) absorption (photo-luminescence-excitation). However, investigating the field behaviour one can conclude that the 1s absorption transition shows a dependence similar to the 0-0 emission transition, while the 2s absorption transition behaves like the 1-1 emission transition. In the same way, higher energy absorption transitions 3s, 4s, 5s, 6s, can be paired with, correspondingly, 2-2, 3-3, 4-4, and 5-5 emission transitions. We deduce that the latter transitions are associated with successive pairs of conduction and valence Landau levels, similarly to the corresponding magneto-excitonic states. As a matter of fact, a plot of energy positions of luminescence peaks versus the magnetic field represents a quite regular Landau level fanchart. The resulting effective cyclotron energy is usually somewhat smaller than the reduced cyclotron energy deduced from the analysis of the absorption spectra.

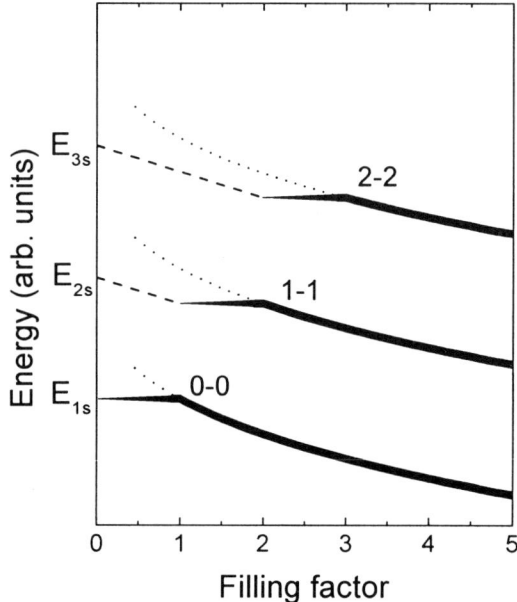

Figure 5. A schematic scenario of the changes of the emission line positions as a function of the Landau level filling factor, i.e., as a function of the density of electron-hole pairs (see text for the details).

The puzzling experimental observation reported here is the remarkably small energy shift of magneto-luminescence peaks investigated as a function of the excitation intensity, i.e., as a function of the electron-hole density. We believe that in our experiments the quasi population of electron-hole pairs can be changed in the density range from $\sim 10^{11} cm^{-2}$ up to $\sim 5 \times 10^{12} cm^{-2}$. In this density range, one would expect energy shifts up to 50meV, assuming this shift to be equal to the expected value of the bandgap renormalisation concluded from the studies at zero magnetic field [6]. The observed red shifts of magnetoluminescence peaks are at most a few meV. However, our data are surprising only if we assume that that the behaviour of the electron-hole plasma in a magnetic fields follows the model applied in the absence of the magnetic field, except a simple effect of the modification of the density of states into discrete Landau levels. As pointed out in Refs. 14 and 15, the application of magnetic fields leads in fact to more profound changes in the description of the emission spectra associated with a dense two-dimensional electron-hole system. Although these theories [14,15] are not as yet sufficiently elaborated to be directly compared with the experimental data, they indicate a possible reason of surprisingly stable energy structure of the magneto-luminescence transitions in respect to changes of the electron-hole density. Strong theoretical statements are drawn under simplified approximations of zero carrier temperature and in the high field limit. Following Ref. 14, we present in Figure 5 a tentative scenario of the evolution of the magneto-luminescence transitions as a function of the Landau level filling factor, i.e., as a function of the density of electron-hole pairs. In the low-density limit (low excitation intensity) the single emission peak is identified with the 1s magneto-exciton transition. At low excitation intensities, the higher energy 2s, 3s, ... magnetoexcitons states are not observed in emission but can be seen in absorption. The increase of the excitation intensity leads to an increase of a number of 1s magneto-excitons, i.e., the electron-hole pairs associated with the lowest pair (0,0) of conduction and valence band Landau levels. When the density of electron hole pairs equals the degeneracy of the Landau level (filling factor one), the recombination process of an electron from a completely full conduction band Landau level with a hole from a completely empty (completely full of holes) valence band Landau level can be viewed as due to a "deexciton" transition [14]. The Coulomb interaction (binding energy) of a single exciton in the initial state (before recombination) is the same as the Coulomb interaction for the "deexciton" in the final state (after recombination).

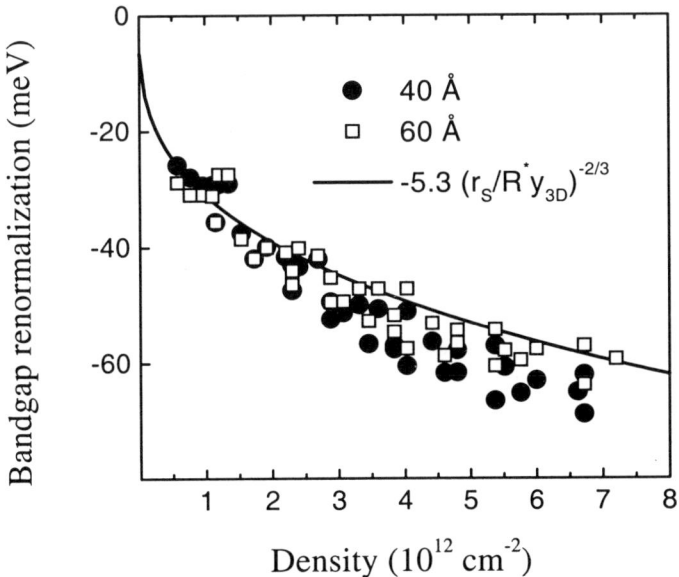

Figure 6. Zero field bandgap renormalisation as extrapolated from the data measured at high magnetic fields. Open squares and solid circles represent the experimental data. The solid line is a theoretically postulated dependence for strictly two-dimensional systems [6,7].

This implies that both exciton and deexciton transitions have the same energy or in other words that the energy of the emission transition in the low-density limit (filling factor zero) is the same as in the case of filling factor one. We further expect that this transition energy is constant in the whole range of filling factors between zero and one (subtle effects which are possible in this intermediate range have been discussed in Ref. 16). The increase of the number of electron hole pairs associated with (0,0) Landau levels leads to the change in the position of the excited 2s, 3s, 4s, ..., magnetoexcitonic resonances. This change could be possibly observed in pump-probe experiments [17] but not in simple emission experiments reported here. The emission associated with the second pair (1,1) of Landau levels becomes visible at filling factors slightly bigger then one (1^+). At $v=1^+$, the lowest 0th conduction band Landau level is fully populated with electrons, whereas the upper 0th valence band Landau level is completely populated with holes. In this case, the recombination process associated with electron-hole pairs from (1,1) conduction and valence band Landau levels can be viewed as a usual excitonic transition, except for the effect of the "renormalisation" due to presence of electrons and holes in the 0th Landau levels. Similarly as for a 0-0 transition in the case of $0<v<1$, the 1-1 transition does not change its energy position when the filling factor is tuned in the range between $v=1$ and 2. An analogous behaviour can be expected for recombination processes associated with (2,2), (3,3), ..., pairs of Landau levels in the ranges of filling factors $2<v<3$, $3<v<4$, ..., correspondingly. This behaviour can be summarised in a different way when we introduce the quasi Fermi energies for electron and holes. Loosely speaking, we expect no change in the energy positions for the emission transitions associated with the states from the vicinity of the quasi Fermi energies. The density evolution of the emission peaks associated with deeper, fully occupied levels is hardly reproduced in calculations [14,15]. One may only expect that once a given pair of Landau levels becomes fully populated, further increase in the density of electron hole pairs (population of higher energy levels) leads to a renormalisation (red shift) of the emission transition associated with these fully populated levels. In Figure 5 we assume that this red shift follows the bandgap renormalisation inferred from the model of the rigid band collapse developed on the basis of zero field studies.

The expected scenario of the density evolution of the emission peaks presented in Figure 5 still remains in conflict with the experimental data. In contrast to theoretical expectations, the experiment indicates that even when levels become fully populated, the corresponding transitions do not change their energy positions under further increase of the electron hole density. This disagreement might be due to the fact that theory is not sufficiently developed or that the experiment is not sufficiently accurate. Neglecting this disagreement, we further assume that both theory and experiment give a correct answer in case of integer filling factors and only for the highest energy transitions. In the following, we determine, from the experiment at given magnetic field B, the energy positions $E_0(B)$, $E_1(B)$, $E_2(B)$, ..., etc. of the subsequent transitions observed in the high-excitation-power magneto-luminescence spectra. We assume that these energies are characteristic for the recombination of electron-hole pairs associated with (0,0) Landau levels at filling factor one, from (1,1) Landau levels at filling factor two, from (2,2) Landau levels at filling factor three, ..., etc., respectively. We determine the effective reduced cyclotron frequency $\hbar\omega_r$ and the 2D energy gap E_g at zero magnetic field from the analysis of the "magneto-absorption" data and calculate the energy difference $\Delta_N(B) = E_N(B) - (N+1/2)\ \hbar\omega_r-E_g$. This procedure of compiling the data obtained in a magnetic field can be viewed as an approach to extrapolating the zero-field bandgap renormalisation Δ_N corresponding to the density of electron hole pairs $n_{e-h} = (N+1)2eB/h$ (2eB/h is the degeneracy of Landau level and $(N+1)2eB/h$ is the electron-hole density at filling factor $\nu=N+1$). Figure 6 shows the plot of Δ_N versus $(N+1)2eB/h$. The data are obtained from measurements at different magnetic fields and for two different quantum wells with different quantum well thickness (and therefore with different E_g and $\hbar\omega_r$ parameters). These data reproduce very well the effect of bandgap renormalisation expected for the recombination processes associated with electron-hole plasmas at zero magnetic fields [6].

4. Conclusions

The conventional model of the rigid band gap collapse, developed on the basis of the zero-field studies of emission spectra associated with electron-hole plasmas, can not be easily extended to the case of dense two-dimensional electron-hole systems in magnetic fields. The experiment shows that energies of magneto-luminescence transitions associated with dense 2D electron-hole systems change very little as a function of the density of electron-hole pairs. Theory accounts for stable, density independent transition energies but only in the vicinity of quasi Fermi levels. The origin of this effect is a two-particle electron-hole correlation which become important in the case of discrete energy levels even at high carrier densities. The theory is not sufficiently developed or the experiment is not sufficiently accurate to achieve an agreement in the case of recombination processes associated with states below the Fermi levels. The compilation of the magnetic field data presented in Figure 6 indicates that the experiment and theory of emission processes associated with dense two-dimensional electron-hole systems subjected to magnetic fields are in good agreement as far as transitions from the vicinities of quasi Fermi levels are considered. We think that a similar limitation of the model of the rigid bandgap collapse can be apparent in case of other systems, such as for example quantum wires [18,19], which show quasi-discrete energy levels even in the absence of magnetic fields.

References

1. Hensel, J.C., Phillips T.G., and Thomas G.A. (1977) The Electron-Hole Liquid in Semiconductors: Experimental Aspects, in H. Ehrenreich, F. Seitz, D. Turnbull (eds), *Solid State Physics*, **32,** Academic Press, New York, pp. 87-314.
2. Rice, T.M. (1977) The Electron-Hole Liquid in Semiconductors: Theoretical Aspects, in H. Ehrenreich, F. Seitz, D. Turnbull (eds), *Solid State Physics*, **32,** Academic Press, New York, pp.1-86.
3. Nozierés, P. (1983) Phase Transitions in electron-hole liquids, in M. Averous (ed*), Proceedings of the 16th International Conference on the Physics of Semiconductors*, North-Holland, Amsterdam, pp.16-22.

168

4. Tränkle, G., Lach., E., Forchel, A., Scholz, F., Ell, C., Haug, H., Weimann G., Griffiths, G, Kroemer, H., and Subbanna, S. (1987), General relation between band-gap renormalization and carrier density in two-dimensional electron-hole plasmas, *Physical Review B* **36**, 6712-6714.

5. Cappizzi M., Modesti, S., Frova, A., Staehli, J.L., Guzzi, M., and Logan R.A. (1984) .*Physical Review B* **29**, 2028.

6. Kleinmann, D.A., (1986) Binding energy of electron-hole liquid in quantum wells, *Physical Review B* **33**, 2540-2549.

7. Schmitt-Rink, S., Ell, C., and Haug, H. (1986) Many-body effects in the absorption, gain, and luminescence spectra of semiconductor quantum-well structures, *Physical Review B* **33**, 1183-1189.

8. Potemski, M., Maan, J.C, Ploog, K., and Weimann, G. (1988) Electron hole plasma in quantum wells in high magnetic fields, in W. Zawadzki (ed), *Proceedings of the 19th International Conference on the Physics of Semiconductors,* Institute of Physics of Polish Academy of Sciences, Warsaw, pp. 119-123.

9. Potemski, M., Maan, J.C., Ploog, K., Weimann, G. (1990) Properties of a dense quasi-two-dimensinal electron-hole gas at high magnetic fields, *Solid State Communication* **75**, 185-188.

10. Butov, L.V., Kulakovskii, V.D., Bauer, G.E.W., and Grutzmaher, D. (1992) Excitons in dense two-dimensional electron-hole magnetoplasmas, *Physical Review B* **46**, 12765-12768.

11. Lerner, I.V. and Lozovik Yu.E. (1980) Mott exciton in quasi two-dimensional semiconductors in high magnetic fields, *Zhurnal-Eksperimental'noi i Teoreticheskoi Fiziki [Soviet Physics - JETP]* **78**, 1167-1175

12. MacDonald, A.H. and Ritchie, D.S. (1986) Hydrogenic energy levelsin two-dimensions at arbitrary magnetic fields, *Physical Review B* **33**, 8336- 8344.

13. Potemski, M., Vina, L., Bauer, G.E.W., Maan, J.C., Ploog, K., Weimann, G. (1991) Magneto-excitons in narrow GaAs/GaAlAs quantum wells, *Physical Review B* **43**, 14707-14710.

14. Bychkov, Yu.A., Rashba, E.I. (1991) Excitons and deexcitons in a neutral two-dimensional magnetoplasma with a strong population inversion, *Physical Review B* **44**, 6212-6219.

15. Bauer, G.E.W., (1992) Precursor of the excitonic insulator in excited quantum wells, *Physica-Scripta* **45**, 154-157

16. Paquet, D., Rice, T.M., Ueda, K. (1985) Two dimensional electron-hole fluid in a strong perpendicular magnetic field: exciton Bose condensate or maximum density two-dimensional droplet, *Physical Review B* **32**, 5208-5221.

17. Schmitt-Rink, S., Stark, J.B., Knox, W.H., Chemla, D.C., Shafer, W. (1991) Optical properties of quasi-zero-dimensional magneto-excitons, *Applied Physics A* **53**, 491-502.

18. Wegscheider, W., Pfeiffer, L.N., Dignam, M.M., Pinczuk, A., West, K.W., McCall, S.L., Hull, R. (1993) Lasing from excitons in quantum wires, *Physical Review Letters* **71**, 4071-4074

19. Ambigapathy, R., Joseph, I.B., Brasil, M.J., Oberli, D., Haacke, S., Kapon, E., Deveaud, B. (1996) The recombination of a dense electron-hole plasma in a quantum wire: evidence for excitonic effects, in S. Scheffler and R. Zimmermann (eds) *Proceedings of the 23th International Conference on the Physics of Semiconductors* (World Scientific, Singapore), pp1193-1196.

RECOMBINATION DYNAMICS OF NEGATIVELY CHARGED EXCITONS IN GATED GaAs QUANTUM WELLS

V. CIULIN, S. HAACKE, J.-D. GANIERE, B. DEVEAUD
Physics Department, Institute of Micro- and Optoelectronics, Swiss Federal Institut of Technology Lausanne, CH-1015 Lausanne EPFL, Switzerland

G. FINKELSTEIN, V. UMANSKY, I. BAR-JOSEPH
Department of Condensed Matter Physics, The Weizmann Institute of Science, Rehovot 76100, Israel

Abstract. We study the dependence of the photoluminescence decay of the negatively charged exciton on the electron concentration for different excitation energies. The radiative lifetime of the trion, directly measured when exciting the trion resonantly, is found to be constant for the studied concentration range from about 3×10^{10} cm^{-2} upwards.

Excitons (X) dominate the low-temperature optical spectra of undoped semiconductor quantum wells (QWs). The introduction of low-density intrinsic excess electrons results in the appearance of an energetically lower spectral line assigned to the negatively charged exciton (X$^-$), which is formed of two electrons and a hole [1]. With higher electron concentration, the X$^-$ line has been observed to increase in intensity and the X line to decrease rapidly. As the exciton disappears, the X$^-$ transition evolves into an emission characteristic of the recombination of an electron gas with photoexcited holes [2].

The recombination dynamics of excitons in undoped QWs has been extensively investigated. It is found to depend on the excitonic radiative lifetime and on non-radiative processes, such as the scattering time of excitons to non-radiative states with larger k vectors and on the spin relaxation time. These factors lead to a complex picture, where a lifetime in the tens of picoseconds range is found to be the signature of radiative free excitons. This fast radiative lifetime can be determined by the use of resonant excitation of X [3].

For low electron concentration (about 3×10^{10} cm^{-2}) and at low temperature, the recombination of X$^-$ is primarily governed by radiative recombination. Under these conditions, using resonant excitation of X$^-$, we have directly measured the radiative lifetime of X$^-$ and found 60 ps in GaAs QWs [4]. In the following, we present the dependence of the trion radiative lifetime on the electron concentration in a given sample. It is found to be constant for concentrations higher than 3×10^{10} cm^{-2}. We stress the fact that the X$^-$ radiative lifetime can be measured only using resonant excitation on

M.L. Sadowski et al. (eds.), Optical Properties of Semiconductor Nanostructures, 169–172.
© *2000 Kluwer Academic Publishers. Printed in the Netherlands.*

the trion; resonant excitation being necessary to isolate the evolution of the trion population from the excitonic population decay.

The studied sample consists of a 20 nm QW separated from the Si delta-doped region by an undoped 50 nm AlGaAs spacer layer. The electron density is varied by applying a voltage to the gate with respect to the contacts, which are annealed into the two-dimensional electron gas (2DEG). The electron density after illumination is 2×10^{11} cm^2 and is estimated to reach the lowest value of 3×10^{10} cm^2. The high mobility of this structure, 4×10^6 cm^{-2}/Vs, indicates a very low impurity concentration and assures us that the low energy spectral line is due to X$^-$.

The sample is excited by short light pulses (3 ps) from a mode-locked Ti-sapphire laser with an average power density of 100 mW/cm^2 ($\sim 10^9$ X$^-$ / cm^{-2}). The time-resolved photoluminescence (PL) is recorded in linear cross polarisation by a 2D streak camera using a 0.27 m spectrometer with better than 1 meV spectral resolution. The temperature was kept at 2 K.

Fig. 1 dispays several time-integrated PL spectra at different gate voltages following a non-resonant excitation. The PL spectra remain the same up to a gate voltage of 0.17 V, where the electron concentration starts to change. One can identify X and X$^-$, which are equally strong in PL in the first spectra. As the electron concentration is increased, the exciton is quenched. At a gate voltage of 0.2 V, the PL can still be attributed to X$^-$ but the exciton has disappeared. For higher gate voltages, the PL spectra are more characteristic of the 2DEG as the line begins to broaden due to the increasing Fermi energy.

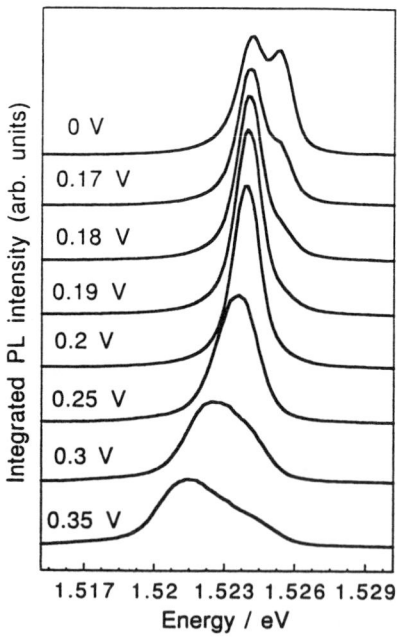

Figure 1: Time-integrated PL spectra at different gate voltages upon a non-resonant laser excitation at 1.61 eV.

For non-resonant excitation below the barrier, the X time-resolved PL rises faster than X⁻ PL and their decay times are similar [4, 5]. Both the rise and decay times of the X⁻ PL, shown in Fig. 2, are extracted from a simple fit to the experimental curves and decrease with increasing electron concentration. An important contribution to non-resonant dynamics is the cooling of the photoexcited carriers to form excitons. This effect can be observed to become faster with an increasing 2DEG concentration as the PL risetimes of both X and X⁻ decrease simultaneously (Fig. 2). Indeed, the scattering of excitons with cold electrons has been found to be a very efficient process and to vary linearly with the electron concentration [6]. As for the measured X⁻ PL decay, it does not represent the true lifetime of X⁻. For the lowest electron concentration, the decay time of the trion is strongly influenced by the dynamics between X⁻ and X and in particular by the trion formation time which is about 100 ps at 2 K [4]. The decrease in the X⁻ trion decay time can tentatively be explained by a decrease in the X⁻ formation time due to an enhanced interaction with the cold electrons.

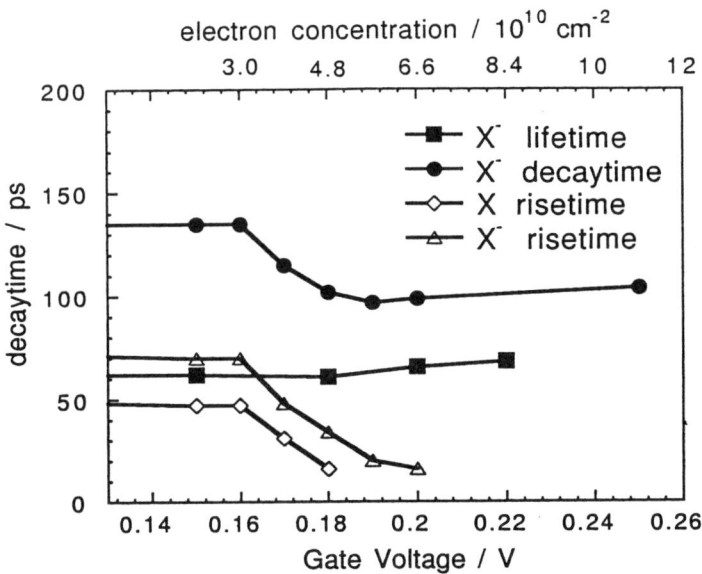

Figure 2: Decay time of the charged exciton PL as a function of gate voltage at 2K for non-resonant excitation (circles) and resonant excitation with the charged exciton (squares). Non-resonant excitation risetime of the X (open diamonds) and X⁻ PL (open triangles). The electron concentration indicated has been extracted from the Fermi filling following [9] and is only indicative.

Resonant excitation on the trion yields a trion PL risetime within the laser resolution followed by a single exponential decay. This decay, as trion dissociation is negligeable at 2 K, represents the X⁻ radiative lifetime [4]. Here the radiative lifetime is found to be 60 ps and is essentially independent of the 2DEG concentration, as seen in Fig. 2 [7]. The same dependence on concentration has been observed for acceptor bound excitons (A_0X) [8]. In the latter, it can be explained by the fact that the radiative lifetime is a measure of a well-defined transition, which is not influenced by the number of acceptors in the QW. This argument cannot be simply applied to X⁻ for two reasons.

First, the available electronic ground states are expected to depend on the 2DEG concentration. Second, as the number of electrons in the QW increases, the interaction between them increases. It has been suggested that X^- then contains more than two electrons [10]. These two effects would tend to decrease the X^- lifetime when increasing the concentration. The negligible dependence of the X^- radiative lifetime on the electron concentration is therefore surprising. Quantitative theoretical calculations are needed in order to estimate the importance of these effects on the trion radiative lifetime.

In conclusion, we have studied the trion lifetime as a function of the 2DEG density. We find that only resonant excitation yields a short lifetime which we interpret as the radiative lifetime of X^-. This radiative lifetime essentially does not depend on the 2DEG density and does not reflect either the expected change in the electron ground state or the existence of multiply charged excitons.

Acknowledgements. We wish to thank Dr. P. Kossacki for helpful discussions. This work was supported by the Swiss Fond National pour la Recherche Scientifique under contract No. 2100-049538.96/1.

References

1. K. Kheng, Y. Merle d'Aubigné, F. Bassani, K. Saminadayar, and S. Tatarenko, Phys. Rev. Lett. **71**, 1752, (1993); G. Finkelstein, H. Shtrikman, I. Bar-Joseph Phys. Rev. Lett. **74**, 976, (1995).
2. A. J. Shields, M. Pepper, D. A. Ritchie, M. Y. Simmons, and G. A. C. Jones, Phys. Rev. B **51**, 18049, (1995)
3. B. Deveaud, F. Clerot, N. Roy, K. Satzke, B. Sermage, and D. S. Katzer, Phys. Rev. Lett. **67**, 2355, (1995); A. Vinattieri, J. Shah, T. C. Damen, D. S. Kim, L. N. Pfeiffer, M. Z. Maialle, and L.J. Sham, Phys. Rev. B **50**, 10868, (1994).
4. G. Finkelstein, V. Umansky, I. Bar-Joseph, V. Ciulin, S. Haacke, J. -D. Ganière, and B. Deveaud Phys. Rev. B **58**, 12637, (1998).
5. A. Ron, H. W. Yoon, M. D. Sturge, A. Manassen, E. Cohen, and L. N. Pfeiffer, Solid State Commun. **97**, 741, (1996).
6. M. Koch, R. Hellmann, G. Bastian, J. Feldmann, and E. O. Goebel, Phys. Rev. B **51**, 13887, (1995).
7. The error on the extracted lifetime is estimated at about 10 ps. Each point shown for the radiative lifetime of X^- is the average of several individual measurements.
8. C. I. Harris, B. Monemar, H. Kalt, P. O . Holtz, M. Sundaram, J. L. Merz, and A. C. Gossard, Phys, Rev. B **50**, 18367 (1994).
9. A. J. Shields, C. L. Foden, M. Pepper, D. A. Ritchie, M. P. Grimshaw, and G. A. C. Jones, Superlatt. Microstruct. **15**, 355, (1994).
10. G. Finkelstein, H. Shtrikman, and I. Bar-Joseph, Physica B **249-251**, 575, (1998).

EXCITONS IN THE TWO-DIMENSIONAL HOLE GAS AT THE Al₀.₅Ga₀.₅As/GaAs INTERFACE

EXCITONS IN THE TWO-DIMENSIONAL HOLE GAS AT THE
$Al_{0.5}Ga_{0.5}As$/GaAs INTERFACE

M. CIORGA, K. RYCZKO, M. KUBISA L. BRYJA, J. MISIEWICZ
Institute of Physics, Wrocław University of Technology
Wybrzeże Wyspiańskiego 27, 50-370 Wrocław, Poland
O. P. HANSEN
Niels Bohr Institute, University of Copenhagen
Universitetsparken 5, DK-2100 Copenhagen, Denmark

Radiative recombination processes associated with the $Al_xGa_{1-x}As$/GaAs heterojunction interface have been intensively studied recently. Yuan *et al.* [1] first observed the PL line on the low energy side of the GaAs exciton, the so-called the H-band, from a single heterojunction. They attributed the H-band emission to two-dimensional (2D) carriers that tunnel from the triangular well at the interface and recombine with free carriers (3D) in the flat band region. Most later papers concerned n-type samples with 2D electrons recombining with free holes [2-4]. Some papers [2,3,5] agreed with Yuan's explanation of the H-band peak, but other authors concluded that recombining electrons and holes are bound by Coulomb interaction [4, 6]. In this paper we present results of experimental and theoretical studies which shed new light on the origin of the H-band emission phenomenon.

The studied sample was obtained by molecular-beam epitaxy growth on a [001] semi-insulating substrate, using the following sequence of layers: 1 μm GaAs, 7 nm undoped $Al_{0.5}Ga_{0.5}As$ spacer, 50 nm Be-doped 1×10^{18} cm⁻³ $Al_{0.5}Ga_{0.5}As$ and 5 nm Be-doped 2×10^{18} cm⁻³ GaAs cap. All measurements were carried out at liquid-helium temperatures down to 2 K in magnetic fields up to 5 T, applied both perpendicular ($B \perp z$) and parallel ($B \| z$) to the growth axis. In the former case both σ^+ and σ^- polarisations of emitted light were analysed, while in the latter case σ and π light polarisations were studied. As the excitation source we used an Ar⁺-ion laser (PL) and a tuneable dye laser with Styryl 9M dye (PL and PLE) pumped by the argon laser. The power density of the excitation beam on the sample was approximately 50 mW/cm². Luminescence was analysed by a 2-m double grating monochromator with a photomultiplier as a detector.

The PL spectrum of the sample is shown in the inset of Fig. 1. The emission lines with energies close to 1.515 eV were identified as free excitons (X), neutral-donor-bound excitons (D⁰X) and neutral-acceptor-bound excitons (A⁰X) of GaAs. Similarly, the donor - carbon acceptor (D-C_{As}) and the free electron - acceptor (e-C_{As}) lines with energies close to 1.491 eV originate from the GaAs layer. The peak H at the energy of 1.5033 eV was assigned to the H-band emission. We found that its position depends on temperature and excitation intensity, but not so significantly as reported by other

173

M.L. Sadowski et al. (eds.), Optical Properties of Semiconductor Nanostructures, 173–176.
© 2000 *Kluwer Academic Publishers. Printed in the Netherlands.*

authors [1,3,4]. Fig. 1 also shows the PLE spectra: one measured with the detector-hold position on the low energy slope of the H-band and the second obtained for detector-hold position at 1.4905 eV – close to the D-C$_{As}$ line. In the former case we observe a single strong peak at the energy 1.5153 eV of the bulk GaAs exciton, which supports the interpretation that electrons and holes involved in the H-band emission arise mainly from free GaAs excitons. In the PLE spectrum obtained in the latter case, apart from the weak free exciton X line at 1.5153 eV, we observe two stronger peaks with energies of 1.5096 eV (A) and of 1.5123 eV (B). They are located below the lowest GaAs exciton line A^{0}X (1.5127 eV) and above the H-band position. We attribute the lines A and B to the excitation of interface excitons, first proposed by Balslev [7].

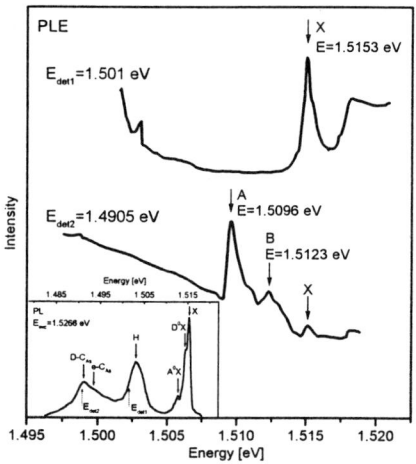

Figure 1. Low temperature PL spectrum from studied sample (inset) and PLE spectra obtained with the detection energy slightly below the energy of the H-band emission (E$_{det1}$) and close to the D-C$_{As}$ line (E$_{det2}$). For details, see text.

Our interpretation of PLE phenomena in the region of H-band emission is based on theoretical results presented in [8]. We performed detailed calculations of the potential distribution and hole energy levels in a p-doped Ga$_{1-x}$Al$_x$As/GaAs heterojunction, taking into account exchange-correlation effects. We also showed that holes from excited 2D levels can bind free electrons and create interface excitons. These excitons however have a quasistationary character: the interface electric field repels electrons from the junction and destroys the electron-hole coupling. For this reason, they can be observed in a PLE (quasi-absorption) experiment, but they do not contribute to PL emission. Calculated interface exciton energies allowed us to attribute line A of Fig. 2 to the HH2 excitonic transition and line B to the LH1 (or HH3) excitonic transition. An excitonic state with the hole from the ground HH1 level cannot be formed because of the phase space filling effect.

On the basis of our theoretical results we also proposed an improved interpretation of the H-band emission [8]. As demonstrated in [4,7,9] bulk excitons, which are photocreated in the flat-band region of the GaAs layer, drift toward the interface and recombine, giving rise to line H. However, they recombine as free particles, because the excitonic binding is destroyed by the interface electric field. In p-doped heterostructures, free electrons recombine with 2D holes from the ground HH1 level, since the occupation of excited hole levels is substantially smaller even under conditions of light excitation. Comparing the measured H-band position with the calculated potential distribution and the HH1 level energy, we conclude that the recombination occurs at the distance of about 6 nm from the junction. On the other hand, the valence band electrons excited by light from 2D hole levels (with the exception of HH1 electrons) form short living interface exciton states. Electrons in these states are situated about 35 nm from the interface and after a few picoseconds escape towards the flat band region. They are trapped by ionised donors and recombine, contributing to the D-C_{As} emission line of PL spectrum. Because of the interface electric field, these electrons cannot contribute to the H-band emission, which takes place closer to the interface. Consequently, no interface exciton lines are observed in the PLE spectrum with the detector-hold position at the H-band.

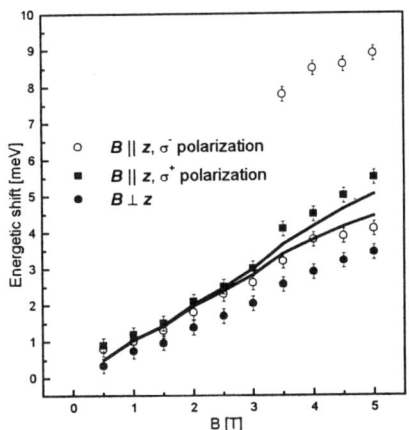

Figure 2. Energetic shift of the H-band line in a magnetic field. Squares and circles represent experimental points. Solid lines represent the shift of the H-band observed in the **B**⊥**z** added to the shift of the spin-split lowest hole Landau level calculated by Ekenberg and Altarelli in [10].

In a magnetic field perpendicular to the interface the H-band splits and shows a pronounced shift to higher energies. The energies of the peak positions as a function of magnetic field are shown in Fig. 2. In σ⁻ light polarisation, at a field of 4 T a new line appears, located 5 meV above the main H-band. Fig. 2 also presents the results of measurements in the **B**⊥**z** configuration (closed dots). In the latter case no splitting is observed in any of the emitted light polarisation (σ and π). The energetic shift of the H-band in the **B**⊥**z** configuration (3.5 meV in B = 5 T) is comparable to the shift of a free GaAs electron in the first Landau level. This confirms our interpretation of the H-band line as originating from the recombination of free GaAs electrons with 2D holes confined at the interface. We attribute the lower energy line observed in the σ⁻ polarisation and the line observed in the σ⁺ polarisation to transitions to the spin split

176

lowest Landau level of the HH1 hole subband. Ekenberg and Altarelli [10] calculated the magnetic field dependence of 2D hole levels in a p-doped $Ga_{1-x}Al_xAs/GaAs$ heterostructure. We used their results and added the shift of the H-band observed in the $B \perp z$ configuration to calculated shifts of the spin-split lowest hole Landau level (solid lines shown in Fig. 2). A good agreement with the measured positions of two lower energy components of the H-band confirms our interpretation. The origin of the higher energy line observed in the σ^- polarisation is not clear. A similar line was reported by Ossau *et al.* [5] and attributed to the recombination of a free electron with a 2D hole from the LH1 subband. Our calculations [8] show however that the energetic distance between the ground HH1 subband and excited hole subbands at $B = 0$ (≥ 18 meV) is too large to confirm this interpretation.

In that paper we discussed the observation of excitons in a two-dimensional hole gas in a p-type $Al_{0.5}Ga_{0.5}As/GaAs$ interface. We showed that the H-band observed in low temperature PL spectra is the effect of recombination of free GaAs electrons with 2D holes confined on the HH1 level of a triangular quantum well. Recombining carriers arise from bulk excitons photocreated in the flat-band region of GaAs, which drift towards the interface, where they are destroyed by the electric field. However, the holes confined on excited 2D levels can bind free electrons and form so-called interface excitons. Because of the quasi-stationary character of such excitons we observe them only in the PLE experiment, with a detection energy slightly below the energy of the $D-C_{As}$ emission line.

We thank C. B. Soerensen for making the sample.

References

1. Yuan, Y.R., Pudensi, M.A.A., Vawter, G.A., Merz, J.L. (1985), New photoluminescence effects of carrier confinement at an AlGaAs/GaAs heterojunction interface, *Journal of Applied Physics* **58**, 397-403.

2. Kukushkin, I.V., v. Klitzing, K., Ploog K., Optical spectroscopy of two-dimensional electrons in $GaAs-Al_xGa_{1-x}As_x$ single heterojunctions, *Physical Review B* **37**, 8509-8512.

3. Zhao, Q.X., Bergman, J.P., Holtz, P.O., Monemar, B., Hallin, C., Sundaram, M., Merz, J.L., Gossard, A.C., (1990), Radiative recombination in doped AlGaAs/GaAs heterostructures, *Semicond. Science and Technology* **5**, 884-889;

4. Reynolds, D.C., Look, D.C., Jogai, B., Yu, P.W., Evans, K., Stutz, C.E., Radomsky, L (1994), Radiative recombination at the $Al_xGa_{1-x}As$-GaAs heterostructures interface by two-dimensional excitons, *Physical Review B* **50**, 7461-7466.

5. Ossau, W., Bangert, E., Weimann, G., (1987), Radiative recombination of a 3D-electron with a 2D-hole in p-type GaAs/(GaAl)As heterojunctions, *Solid State Commun.* **64**, 711-715.

6. Gilland, G.D., Wolford, D.J., Kuech, T.F., Bradley, J.A., (1994), Luminescence kinetics of intrinsic excitonic states quantum-mechanically bound near high-quality (n-type GaAs)/(p-type $Al_xGa_{1-x}As$) heterointerfaces, *Physical Review B* **49**, 8113-8125.

7. Balslev, I. (1987), Recombination via two-dimensional excitons in GaAs-(AlGa)As heterojunctions, *Semiconductors Science and Technology* **2**, 437-441

8. Ciorga, M., Ryczko, K., Kubisa, M., Bryja, L., Misiewicz, J., Hansen, O.P., Observation of quasistationary excitons in p-doped $Ga_{1-x}Al_xAs/GaAs$ single heterojunctions, submitted for publishing

9. Shen, J. X., Oka, Y, Ossau, W., Landwehr, G., Friedland, K.J., Hey. R., Ploog, K., Weimann, G., (1998), Vertical transport of photo-excited carriers for excitonic recombinations in modulation doped $GaAs/Ga_{1-x}Al_xAs$ heterojunctions, *Solid State Communications* **106**, 495-499.

10. Ekenberg, U., Altarelli, M., (1985), Subbands and Landau levels in the two-dimensional hole gas at the $GaAs-Al_xGa_{1-x}As$ interface, *Physical Review B* **32**, 3712-3722

PHOTOREFLECTANCE STUDY OF GaAs/AlAs SINGLE QUANTUM WELL STRUCTURES

J. KAVALIAUSKAS[1], G. KRIVAITĖ[1], B. ČECHAVIČIUS[1],
V. I. KADUSHKIN[2], E. L. SHANGINA[2], S. N. LARIN[2], AND
YU. A. KOTOVA[2]
[1]*Semiconductor Physics Institute, A.Goštauto 11, 2600 Vilnius, Lithuania*
[2]*Ryazan State Pedagogical University, Svobody 46, 390000 Ryazan, Russia*

Abstract. The photoreflectance spectra of type-I GaAs/AlAs single quantum well structures near the Γ-X crossover are studied. The optical data indicate that excitation at an energy below or above the indirect band-gap of AlAs barriers changes in a different manner the potential profile of the heterostructure and the reflectance spectra. A red shift and significant broadening of exciton resonances in the GaAs QW were related to the Stark effect and the Γ-X electron scattering. It was found that initial exciton characteristics can be restored by additional excitation above the AlAs barrier band-gap.

1. Introduction

GaAs/AlAs quantum well (QW) structures and superlattices have attracted considerable attention for their potential use in the novel optoelectronic devices. Either type-I or type-II band alignment in GaAs/AlAs QW systems can be achieved by varying the GaAs well width or by application of an external electric field [1] or hydrostatic pressure [2]. Usually, built-in electric fields exist naturally in close-surface GaAs/AlAs heterostructures and can be changed via a surface photovoltage and/or a photovoltage created across the GaAs buffer layer [3]. This suggests that illumination of the QW structure may be used as an alternative way in order to vary the Γ(GaAs)–X(AlAs) subband alignment and, in turn, the optical properties. The photo-induced modifications of the energy band structure and potential profile in GaAs/AlAs QW structures may be observed by means of modulation spectroscopy [4].

In this work, we report a detailed photoreflectance (PR) study of GaAs/AlAs type-I single QW structures just below ($\leq$ 80 meV) the Γ–X subband crossover. In addition to PR, contactless electroreflectance (CER) and wavelength-modulated reflectance (WMR) measurements were also performed.

177

M.L. Sadowski et al. (eds.), Optical Properties of Semiconductor Nanostructures, 177–182.
© 2000 *Kluwer Academic Publishers. Printed in the Netherlands.*

2. Experimental

The samples used in this study were grown by molecular-beam epitaxy on a semi-insulating GaAs (001) substrate. The structure consists (Fig. 1) of an 0.4 μm unintentionally doped GaAs buffer layer followed by a nominally undoped 45-Å/200-Å GaAs/AlAs single QW and a 100-Å GaAs cap layer.

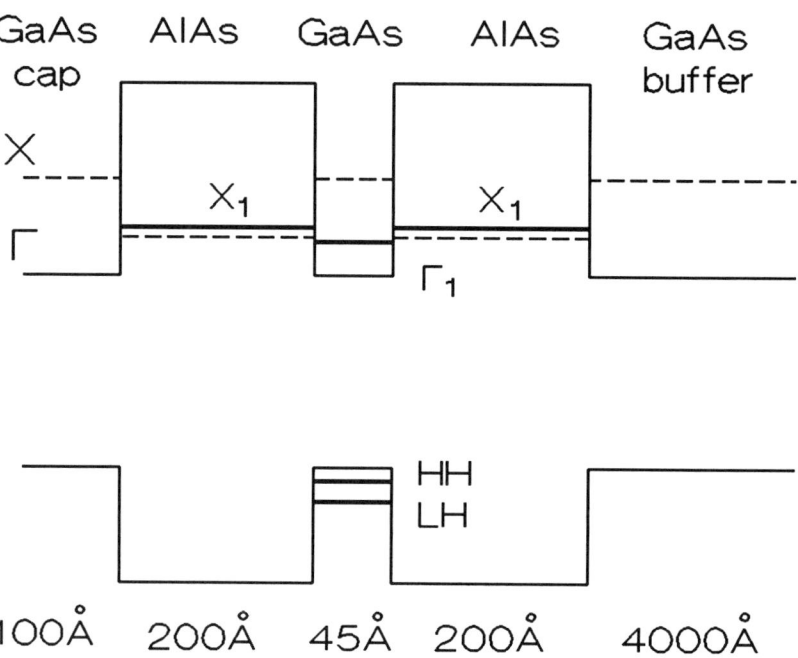

Figure 1. A schematic band diagram of the GaAs/AlAs single QW structure

The PR measurements (T=300 K) were performed using the three-beam (modulation, bias, and probe light) photoexcitation technique. The He-Ne (6328 Å) and/or Ar^+ (4765 Å) laser beams (≤4 mW cm^{-2}) were used alternatively as modulation (ac) and optical bias (dc) pump sources. Modulation frequencies from 20 to 700 Hz were used. Note that the excitation energy E_{ex} is lower (E_{ex} =1.96 eV, He-Ne beam) or higher (E_{ex} =2.6 eV, Ar^+ beam) than the indirect band-gap E_g^X =2.17 eV of the AlAs barrier layers. Moreover, the penetration depths of the He-Ne (≈260 nm) and Ar^+ (≈80 nm) beams are quite different. In the CER studies, a capacitor-like arrangement was utilised. In measuring the WMR spectra, the wavelength of the incident probe light was modulated.

Figure 2 displays PR spectra of the GaAs/AlAs single QW structure modulated at 20 Hz by He-Ne (full curves) and Ar^+ (broken curves) laser beams. CER and WMR spectra of the same sample are presented in Fig. 3a and Fig. 3b, respectively. All the spectra were measured with and without additional dc He-Ne laser illumination. A sharp feature E_g(GaAs) at 1.424 eV corresponds to the bulk GaAs band gap and the peak at ~1.39 eV denoted as BSR is due to back-surface reflection. The PR features just above E_g(GaAs) that we relate to the optical transitions in the buffer layer (Fig. 1) depend considerably both on the modulation frequency and optical bias as well as on the wavelength of the pumping beam. It was revealed that a complex line shape (Fig. 2a, large positive peak at ~1.46 eV) in the buffer-related PR spectra is composed actually of the slow and fast signal components. A slow PR component is associated with the photomodulation process in the front-side of the QW structure studied and its lineshape is almost independent of the pump beam wavelength. In contrast, the fast PR component originates from the photomodulation in the buffer layer.

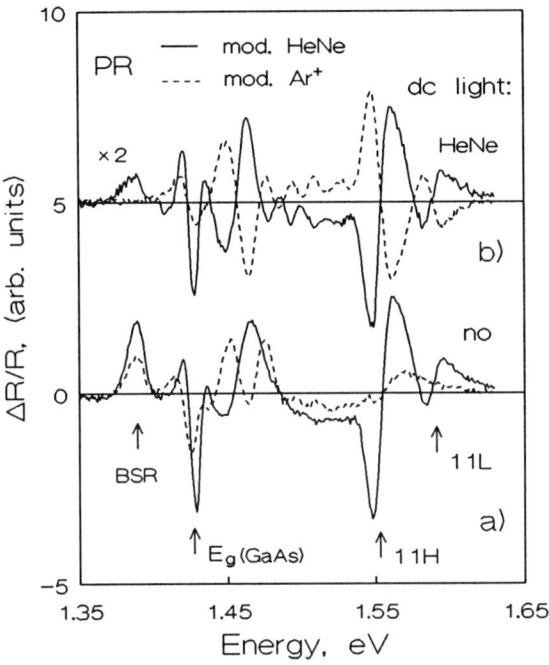

Figure 2. PR spectra of the GaAs/AlAs QW structure modulated at 20 Hz by He-Ne (full curves) and Ar$^+$ (broken curves) laser beams, measured without (a) and with (b) additional dc He-Ne laser illumination.

The high-energy features denoted as 11H (1.553 eV) and 11L (1.584 eV) are due to the ground-state heavy- and light-hole, respectively, interband excitonic transitions in the GaAs single QW (Fig. 1). The lineshapes for 11H and 11L are different in the PR spectra (Fig. 2a) as compared to the WMR ones (Fig. 3b). The lineshape analysis of the PR spectra shows that illumination with the He-Ne laser induces significant exciton line broadening, especially for the 11H transitions, accompanied by a small red shift. These results were also confirmed by investigations of the WMR spectra (Fig. 3b). Under dc illumination with the He-Ne laser beam, we observed a red shift (≤1 meV) and a broadening (≤3 meV) of the excitonic resonances. Such behaviour of excitonic lines cannot be explained by the quantum-confined Stark

180

effect (QCSE) associated with the surface photovoltage effect. It may be a signature of some processes increasing the in-well electric field under He-Ne laser illumination. It is important to note that the above described effect is less evident for excitation with the Ar$^+$ pump beam. As a result, the QW-related signal, well defined in the He-Ne laser modulated PR spectra, is strongly reduced in the spectra taken with Ar$^+$ beam excitation (Fig. 2a). Therefore, the results of photomodulation at low (E_{ex} = 1.96 eV) and high (E_{ex} = 2.6 eV) photon energy are quite different. In addition, it is rather surprising that GaAs QW-related features do not appear in the CER spectrum (Fig. 3a, full curve). This may be due to the negligible QCSE in the narrow QW, as studied here, and/or too weak built-in electric field modulation across the QW. On the other hand, an apparent QW signal with oscillatory features near E_g(GaAs) is induced in the CER when the sample is illuminated by He-Ne laser beam (Fig. 3a, broken curve).

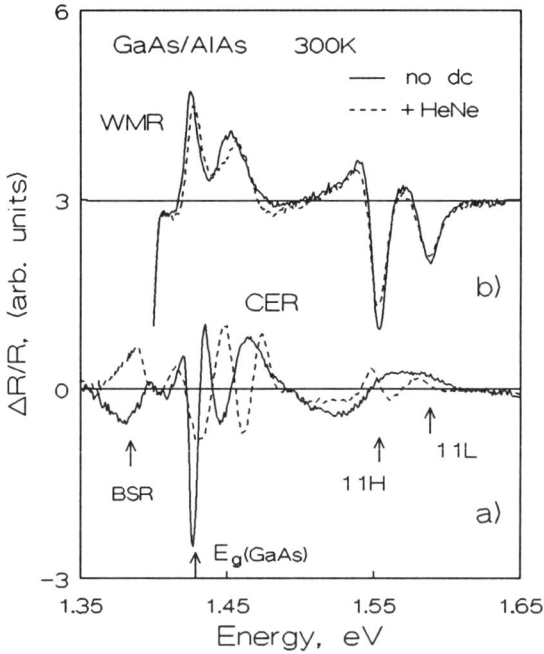

Figure 3. CER (a) and WMR (b) spectra of the GaAs/AlAs QW structure at 300K measured without (full curves) and with (broken curves) additional dc He-Ne laser illumination.

Two distinct effects are also observed in the PR spectra, when the He-Ne bias light is added (Fig. 2b). First, the slow signal component in the PR spectrum modulated by the He-Ne beam is eliminated (probably due to saturation effects) and the PR signal is dominated only by the fast component. Second, both buffer-related oscillations and QW-related features in the Ar$^+$ beam modulated spectra (dashed lines) are reversed with respect to those in the He-Ne beam modulated ones (solid lines). A 180° phase change of the QW signal in the PR spectra indicates that the Ar$^+$ beam excitation partly restores the exciton lineshape, its energy and broadening, modified by the He-Ne laser beam.

3. Discussion

The experimental data described above indicate that the photoexcitation in front of (Ar$^+$ pump beam) and beyond (He-Ne pump beam) the AlAs barrier/GaAs buffer interface affects differently the potential profile near the QW. We interpret these results as follows. When the QW structure is illuminated by the He-Ne laser beam with a photon energy of $E_g(GaAs)<E_{ex}<E_g^X(AlAs)$, the dominant photoexcitation will be in the buffer. The internal electric field in the buffer layer separates the photocarriers sweeping the electrons towards the heterojunction (assuming downward band bending) and the holes – towards the substrate. The photoexcited electrons accumulate at the lower AlAs barrier/GaAs buffer interface and can form two-dimensional electron gas, while the holes may be trapped by negatively charged background acceptors. Thus, spatial charge separation and carrier localisation effects should decrease the band bending in the buffer layer and raise the potential at the barrier/buffer interface. As a consequence, the photoinduced electric field adds to the built-in electric field in the QW region and subtracts from that in the buffer layer. This behaviour is similar to that observed by Shields et al. [3] for remotely doped GaAs/AlGaAs QW's. Alternatively, it may be that built-in electric field across the QW is a superposition of the surface and the buffer electric fields having opposite directions. In such a case the resulting field near the QW should increase under illumination due to reduction of field component coming from the buffer.

Thus, a red shift of excitonic features under illumination with the He-Ne laser may be related to the QCSE, which arises due to the photocreated electric field near the GaAs QW. A linewidth broadening of excitonic features could also be a consequence of the QCSE. However, in contrast to the spectroscopic data discussed above, the QCSE predicts rather small linewidth broadening as compared to the Stark shift. Therefore, the pronounced broadening of excitonic lines, which cannot be attributed to the QCSE alone, may be related to the light-induced increase of the Γ–X transfer of photoexcited electrons. The room temperature Γ–X transfer time is expected to be of the order of sub-ps. The escaping time for the holes out of the QW will be much longer. Therefore, the excess holes created in the QW under illumination can be assumed to be an additional broadening factor of excitonic resonances.

When the excitation photon energy $E_{ex}>E_g^X(AlAs)$, the illumination has no significant influence on exciton transitions regardless of e-h pairs created in the GaAs buffer and AlAs barrier layers. This may be because of a competition between two opposite photomodulation effects, originating from the surface and buffer layer, which cancel each other near the QW under certain photoexcitation conditions. Such a balance is probably destroyed by adding the dc He-Ne laser beam to the modulating Ar$^+$ beam. This follows from a change in the phase of the QW-related PR features, shown in Fig. 2b. This means that the electric field induced by photoexcitation in the buffer layer is partly cancelled (screened) by photoexcited e-h pairs in the QW/barrier regions. Note that a similar effect can also be achieved by an external electric field, as follows from the CER spectrum (Fig. 3a, broken curve). Concluding, these results demonstrate that the electric field strength near the GaAs QW and, consequently, the energy and broadening of exciton resonances in the structures, can be controlled by illumination.

References

1. Meynadier, M. -H., Nahory, R.E., Worlock, J. M., Tamargo, M. C., de Miguel, J. L., and Sturge, M. D. (1988) Indirect-direct anticrossing in GaAs-AlAs superlattices induced by an electric field: evidence of Γ-X mixing, *Phys. Rev. Lett.* **60**, 1338-1341.
2. Nunnenkamp, J., Reimann, Kuhl, K., J., and Ploog, K. (1991) Pressure-induced Γ-X electron-transfer rates in a $(GaAs)_{15}/(AlAs)_5$ superlattice, *Phys. Rev.* B **44**, 8129-8137.
3. Shields, A. J., Osborne, J. L., Simmons, M. Y., Ritchie, D. A. and Pepper, M. (1996) Comparison of optical and transport measurements of electron densities in GaAs/Al_xGa_{1-x}As quantum wells, *Semicond. Sci. Technol.* **11**, 890-896.
4. Pollak, F.H. (1991) Modulation spectroscopy of semiconductor microstructures, *Superlattices Microstruct.* **10**, 333 -346.

LANDAU LEVEL MIXING IN ASYMMETRIC QUANTUM WELLS

R. WYSOCKI, W. BARDYSZEWSKI

Institute of Theoretical Physics, Warsaw University,
ul. Hoza 69, 00-681 Warsaw, Poland

AND

S. SCHOSER, M. POTEMSKI

Grenoble High Magnetic Field Laboratory,
MPI/FKF and CNRS, BP 166 Grenoble Cedex 9, France

1. Introduction

Lowering of the symmetry of electronic states in quantum wells with respect to bulk semiconductors has a profound effect on the selection rules governing the optical transitions between the valence and the conduction subbands. These transitions involve excitonic states of electron-hole pairs and thus provide useful information about the properties of excitons. The situation is particularly interesting when a magnetic field is applied along the growth axis of a quantum well system. The interband magnetoabsorption spectrum is then composed of a series of excitonic lines associated with different Landau levels. The symmetry of those states can be resolved by analysing the absorption or photoluminescence excitation spectra taken at circular polarisation of incident light parallel to the quantum well symmetry axis ($\hat{z}$ axis).

A model quantum well is typically assumed to be ideally symmetric with respect to a mirror reflection in the xy plane with a rectangular potential profile. This symmetry is however easily perturbed by an external or built-in electric field perpendicular to the interface plane. The k·p coupling between the light and heavy hole valence subbands in quantum well systems made of zinc-blende structure semiconductors may then lead to a substantial modification of the selection rules for the magnetoexcitonic absorption. In this paper we present a theoretical and experimental study of the effect of asymmetry of the quantum well potential on the magne-

M.L. Sadowski et al. (eds.), Optical Properties of Semiconductor Nanostructures, 183–186.
© *2000 Kluwer Academic Publishers. Printed in the Netherlands.*

toexcitonic spectra. According to our multiband model of magnetoexcitons the additional lines in the magnetoabsorption spectra may be explained by the presence of a residual electric field across the quantum well.

2. Exciton wave function

Since the coupling between the conduction and valence subbands is usually negligible, the electron wave function has a well defined spin $s_c = \pm\frac{1}{2}$. We limit our discussion to the lowest conduction level, for which it is assumed that the envelope function has no nodes along the $\hat{z}$ axis. In particular, the electron envelope function is even under a mirror reflection with respect to the symmetry plane in the special case of a symmetric potential profile.

The hole wave function is given by a linear combination of the valence band Bloch functions at the Γ_8 point, labelled by their respective projections of spin $\frac{3}{2}$ onto the $\hat{z}$ axis, multiplied by envelope functions. Consequently the excitonic state is represented by a vector

$$
\Phi(\mathbf{r}) = \begin{pmatrix} f_{0,\,\frac{3}{2}}(\mathbf{r}, z_e, z_h) \\ f_{0,-\frac{3}{2}}(\mathbf{r}, z_e, z_h) \\ \vdots \\ f_{0,-\frac{1}{2}}(\mathbf{r}, z_e, z_h) \\ f_{0,\,\frac{1}{2}}(\mathbf{r}, z_e, z_h) \end{pmatrix}
$$

where f_{n,s_v} denote the envelope functions with the $\hat{z}$ component of spin $s_v = \pm\frac{1}{2}$ for light hole levels and $s_v = \pm\frac{3}{2}$ for heavy hole levels, while n is equal to the valence subband number in the quantum well. The arguments of these functions refer to the relative motion of the electron and hole in the xy plane described by the two-dimensional vector $\mathbf{r}$ and the $\hat{z}$ components of the positions of the electron and hole respectively.

In the cylindrical approximation (i.e. neglecting the warping of the valence band) the effective Hamiltonian for the exciton is invariant under the rotation along the $\hat{z}$ axis, so that each eigenstate of the exciton has a well defined $\hat{z}$ projection of the total angular momentum J_z. Each component of the exciton wave function Φ is therefore characterised by the $\hat{z}$ projection of the orbital angular momentum equal to $M_z = J_z - s_c + s_v$. Since in general all the components of a given exciton eigenstate Φ are different from zero, the quantum number M_z characterises this state only approximately. It is a common practice however to refer to the exciton states as being of s,p,d... type according to the value of $|M_z|$ of the dominant component of the vector Φ. Thus the strongest absorption lines are associated with excitons of type s.

Apart from axial symmetry we consider also a mirror symmetry with respect to the xy plane. This symmetry is exact in symmetric quantum wells and can be also used as a starting point for the analysis of weakly asymmetric potential profiles. As it has been pointed out in ref. [1] the parity under reflection with respect to the xy plane (that is under the transformation $z \rightarrow -z$) is a good quantum number in symmetric quantum wells. The components ϕ_v of a state Φ of fixed parity $\mathcal{P}$ are even or odd with respect to the symmetry plane of the quantum well according to the formula $p_v = \mathcal{P}(-1)^{s_v+1/2}$. The parity of the excitonic states giving the dominant contribution in the absorption can be determined by setting $p_v = 1$ and $s_v = s_c - J_z$ where $J_z = 1$ for the σ^+ and $J_z = -1$ for the σ^- polarisation of the incident light. For example, using the σ^+ polarisation we can generate even, s-type, light hole excitons with $s_v = -\frac{1}{2}$ denoted by lhs^+ and odd, s-type, heavy-hole excitons with $s_v = -\frac{3}{2}$ denoted by hhs^-.

The quantum well exciton problem can be formulated within the so called mini- **k·p** model in which the valence band effective Hamiltonian is obtained from the Luttinger-Kohn Hamiltonian by replacing its z-dependent components with the appropriate matrix elements taken between envelope functions [2]. The conduction subband is taken to be spherically symmetric and parabolic. In this approach the Coulomb interaction between the electron and the hole is explicitly isotropic, so there is no direct Coulomb mixing between the light and heavy hole excitons due to their different spins. The Luttinger-Kohn Hamiltonian on the other hand couples excitons containing different types of holes *via* **k·p** terms originating from the d-type components of the Hamiltonian [3]. Only excitons of the same parity with $\Delta M_z = \pm 2$ are coupled in the symmetric quantum wells. This coupling is particularly spectacular in narrow quantum wells leading to a strong anticrossing behaviour of the lowest hhd^- and lhs^- excitonic lines as discussed in [4]. The asymmetry of the quantum well potential results in mixing between the states of different parity with $\Delta M_z = \pm 1$. One would expect therefore that the spectral lines associated with the lhp^+ and hhp^- excitons should also appear in the polarisation σ^+.

3. Results and conclusions

The magnetic field dependence of the excitonic transitions for the polarisation of the incident light σ^+ obtained from the photoluminescence excitation spectra of a single 8nm wide GaAs quantum well lattice matched to the GaAlAs barrier taken at the temperature $T = 4K$ is presented in fig. 1a. Clearly visible are the lowest hh and lh magnetoexcitonic lines. The next line from the bottom begins to appear at $B \approx 13T$ and is identified as the transition to the lowest hhp^- exciton activated by the external electric

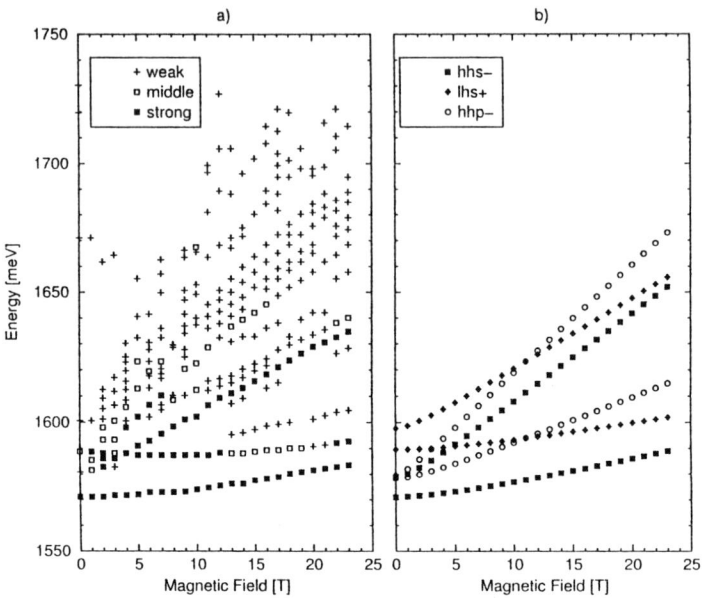

Figure 1. Magnetic field dependence of magnetoexcitons for the σ^+ polarisation for an 80 Å wide GaAs/GaAlAs quantum well: a) experimental data, b) theoretical model with no **k·p** coupling

field. Since the magnitude of the electric field could not be determined by an independent measurement it was treated as a fitting parameter in our model calculations. In fig. 1b we present the theoretical magnetoexciton energies obtained for the estimated value of the electric field $F = 25\text{kVcm}^{-1}$. For the purpose of better identification we have retained only the lowest energy transitions and neglected the effect of the **k·p** coupling. The energy level identified as $hh0^-$ in this figure clearly coincides with the observed weak transition line in fig. 1a. Our more detailed calculations involving the numerical diagonalisation of the exciton Hamiltonian incorporating two topmost light and heavy hole levels also give a proper estimation of the relative intensity of the "forbidden" and allowed transitions. Similar results are also obtained for the polarisation σ^+ of the incident light with the same value of the electric field.

References

1. Uenoyama, T. and Sham, L.J. (1990-I) Carrier relaxation and luminescence polarization in quantum wells, *Phys. Rev. B*, **42**, pp. 7114–7123
2. Bardyszewski, W. and Yevick, D. (1994-II) Electroabsorption and electrorefraction effects in quantum-well modulators, *Phys. Rev. B*, **49**, pp. 5368–5378
3. Baldereschi, A. and Lipari, N.O. (1971) Energy levels of direct excitons in semiconductors with degenerate bands, *Phys. Rev. B*, **3**, pp. 439–450
4. Potemski, M., Viña, L., Bauer, G.E.W., Maan, J.C., Ploog, K., and Weimann, G. (1991-II) Magnetoexcitons in narrow GaAs/Ga$_{1-x}$Al$_x$As quantum wells, *Phys. Rev. B*, **43**, pp. 14 707–14 710

MICROCAVITIES WITH 3D OPTICAL CONFINEMENT

A. FORCHEL[A], J. P. REITHMAIER[A], M. BAYER[A], T. L. REINECKE[B],
V. D. KULAKOVSKII[C]
[A] *Lehrstuhl für Technische Physik, Universität Würzburg, 97070
Würzburg*
[B] *Naval Research Laboratory, 20375 Washington DC, USA*
[C] *Institute of Solid State Physics, Russian Academy of Sciences,
Chernogolovka*

Abstract. We have fabricated microcavities with three-dimensional optical confinement
by the combination of epitaxially grown vertical cavity surface emitting cavity
structures and lithographic patterning. The emission spectra of semiconductor pillars
fabricated by this technique show a number of optical modes, which can be tuned by
changing the lateral dimensions of the structures. By using angle-resolved
photoluminescence spectroscopy, the dispersion of the optical modes in these photonic
dots has been investigated. In contrast to microcavities with confinement in one
direction only, microcavities with three-dimensional optical confinement show no
dispersion of the mode energies in angle resolved photoluminescence experiments.
Studies of 1D photonic wires show suppression of dispersion perpendicular to the wires
whereas dispersion is observed along the 1D structures. By investigations of the far
field intensity distribution of different modes, information on the internal
electromagnetic field distributions is obtained. The energies of the optical modes in the
photonic dots as well as the intensity variations in angle resolved experiments have been
modelled by numerical approaches, which agree quantitatively with the experimental
results.

1. Introduction

Dimensionally controlled electronic and photonic properties constitute a very active
area of solid state physics research at present. For electronic properties a wealth of new
information has become available with the advent of self assembled quantum dots. For
photonic systems, E. Yablonovitch and S. John proposed independently about a decade
ago that a strong modulation of the reflective index in three dimensions could be used to
form an optical band gap [1,2]. This band gap will suppress the propagation of light. In
addition to the interest in these structures based on a physics point of view, applications
of photonic band gap systems became immediately apparent. Potential applications
include the guiding of light by photonic band gap based mirror structures, dispersion
compensation in photonic transmission systems, highly efficient filters etc. [3] Parallel

M.L. Sadowski et al. (eds.), Optical Properties of Semiconductor Nanostructures, 187–200.
© 2000 *Kluwer Academic Publishers. Printed in the Netherlands.*

to the proposal of the photonic band gap concept Björk and Yamamoto presented model calculations and experimental results on semiconductor based microcavities. [4] They pointed out for example that a system which allowed to suppress the spontaneous emission outside a well-defined optical mode could be used in order to obtain thresholdless laser emission.

We have prepared and investigated microcavities with three-dimensional confinement in the InGaAs/AlGaAs material system. The structures investigated are based on a GaAs cavity sandwiched between two GaAs/AlAs Bragg reflectors. At the centre of the cavity an InGaAs quantum well is inserted. The emission of this quantum well is used to probe the optical mode in the structures. In order to obtain lateral optical confinement, electron beam lithographically defined pillars are used. [5] The emission spectra of the structures allow to investigate in detail three-dimensional optical confinement effects when the lateral dimensions or shapes are varied. Due to the three-dimensional optical confinement we observed the formation of a set of discrete optical modes in these spectra. The emission energy of the modes can be tuned over an energy range of more than 10 meV by changing the width of the structures.

To a certain degree the present structures can be seen as the photonic analogues to electronic quantum dots in which the energy levels are defined by the dimensions of the quantum dot structure. The density of states of confined photons as well as of confined electrons for three - dimensional confinement is given by delta - functions. In the electronic system, the size quantisation requires quantum dot dimensions of the order of the de Broglie wavelength whereas in the photonic case the characteristic length scale is given by the optical wavelength.

A very interesting difference between electronic and photonic dot systems arises from the fact that, in the photonic case, the photon observed is the particle which underwent size confinement. In electronic quantum dots, on the other hand, photons are merely used in order to detect transitions between electronic states. For photonic dots this results in the very interesting possibility to investigate dispersion effects in systems with optical confinement by angle resolved spectroscopy. By changing the angle of detection in the experiments the wave vector components parallel to the epitaxial layer are gradually varied. If the optical system permits propagation in the epitaxial layer plane, this will result in an increase of the transition energy as the detection angle is tilted off the surface normal. By comparing angle resolved studies on microcavity layers without lateral optical confinement to studies on microcavities with three-dimensional and 2D optical confinement we obtain experimental evidence for the suppression of dispersion along optically confined directions. In addition, the relation between the far field intensity patterns and the internal fields by a simple Fourier transform permits to obtain detailed insight into the size controlled field distributions of the photonic modes by angle resolved far field spectroscopy.

The paper is organised as follows: After a short discussion of the technology used to fabricate microcavities with three-dimensional confinement, section III addresses the size dependence of the spectra in individual microcavities. The final chapter focuses on angle-resolved studies in microcavities with 3D optical confinement (photonic dots), photonic wires and microcavities without lateral optical confinement.

2.Technology and experiment

The microcavity layer structures used to fabricate systems with three-dimensional optical confinement were grown by solid source molecular beam epitaxy. On a GaAs substrate a Bragg mirror stack consisting of about 20 $\lambda/4$ AlAs/GaAs layers is grown followed by the cavity section. The cavity is made of GaAs and has a thickness of 251 nm corresponding to one wavelength of the active $In_{0.14}Ga_{0.86}As$ single quantum well included at its centre. On top of the cavity the top Bragg reflector consisting of 15 to 20 quarter wave stacks of GaAs/AlAs layers is deposited. As can be shown by reflectivity experiments, the Bragg reflector structures result in the formation of a wide stop band with a reflectivity of about 1 centred at about 1.4 eV. The insertion of the λ cavity leads to a sharp resonance close to the centre of the stop band. The exciton transition in the InGaAs quantum well occurs about 10 to 20 meV at higher energy than the resonance of the cavity. This implies that the present structures are in the so-called weak coupling regime of photonic and excitonic transitions. We will in the following therefore assume that interaction effects between excitons and photons can be neglected. In this case the excitons may essentially be seen as a light source illuminating the cavity in the energy range of the resonance.

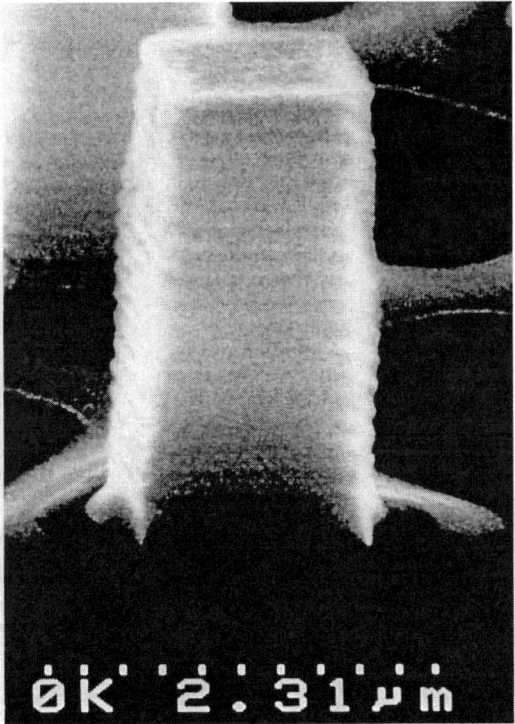

Figure 1: Scanning electron micrograph of a square photonic dot with a lateral size of 1,2μm and a height of 2.5μm.

In order to obtain lateral optical confinement, we have prepared pillar structures from the epitaxial layers. By electron beam lithography etch masks with lateral dimensions between approximately 6 μm x 6μm and 1μm x 1μm were defined. By using a reactive ion etching process with electron cyclotron resonance enhancement with Ar/Cl$_2$ the mask patterns were transferred into the semiconductor structures. Figure 1 shows a scanning electron micrograph of a photonic dot structure realised by this technique. The pillar has a width of 1.2 μm and a height of about 2.5 μm. The pillar is well defined by etched surfaces, which are very close to vertical. Indications for the alternating AlAs and GaAs layers can be seen from the corrugations at the edges of the pillar. As can be seen from Fig. 1 only the upper Bragg reflector is etched completely while the lower Bragg reflector is only partially affected by the patterning process. Optical confinement in the lateral directions is therefore obtained by the change of the refractive index between the semiconductor material forming the upper part of the pillar (approximately 3.2) and the surrounding air.

For the optical experiments, samples with arrays of microcavities with three-dimensional optical confinement as well as individual structures with varying lateral dimensions were studied by angle integrated and angle resolved luminescence spectroscopy. The samples were placed in an optical cryostat, which permitted variable temperature studies (1 K $\leq$ T $\leq$ 200 K). An Ar$^+$ laser (514.5 nm) was used for optical excitation. For angle integrated experiments, the light was collected along the surface normal of the structures within an acceptance angle of about 20 degrees. For angle resolved experiments, the detection was limited by an aperture placed at about 20 cm distance to the samples providing angle resolution of 1 degree. The emitting light was dispersed by a monochromator (f = 0.6 m) and detected by a charge coupled device camera.

3. Investigation of the size dependence of the confined optical mode energies

Figure 2 shows a luminescence spectrum of a two-dimensional reference sample in comparison with spectra taken on three-dimensionally confining microcavities of different lateral sizes [5]. The emission spectrum of the two-dimensional mesa reference shows a single line at about 1.4015 eV. By comparison with reflectivity measurements, this line is associated with the strong decrease of the reflectivity in the resonance of the vertical cavity. The cavity resonance is seen in the present photoluminescence spectrum due to the high reflectivity of the cavity in the energy range of the exciton. This suppresses the exciton emission located about 15 meV above the resonance and results in spontaneous emission from the low energy tail of the exciton via the resonance. Because of the high quality factor of the two-dimensional cavity of about 4000, this intensity enhancement is very strong and leads to the dominant line in the emission spectra of the microcavities.

The luminescence spectra of the laterally patterned microresonators show striking differences compared to the two-dimensional reference. As shown in Fig. 2 for a microresonator width of 4.7 μm, we observe a splitting of the optical resonance into several peaks. These peaks are located at higher energies than the mode of the vertical cavity. By reducing the microresonator width, the energy splitting of the modes is

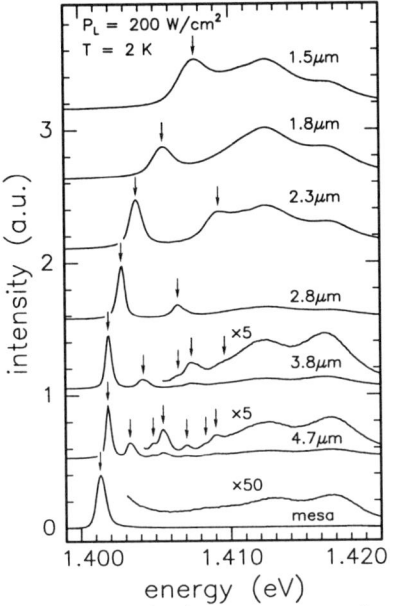

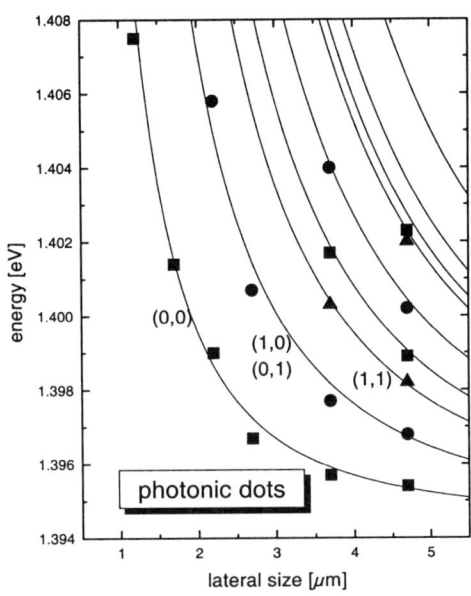

Figure 2: Photoluminescence spectra of square photonic dots of varying lateral sizes in comparison to the spectrum of an unpatterned planar cavity.

Figure 3: Energies of the low lying confined optical modes in photonic dots plotted versus the lateral photonic dot size. Symbols give the experimental data, lines the results of calculations using the boundary element method.

significantly increased (cf. spectra of 3.8 and 2.8 μm resonators). In parallel the low energy peaks show a well-pronounced width dependent shift to higher energies at smaller sizes.

With decreasing lateral size we observe a broadening of the optical modes. This broadening may be partially due to increased damping by excitonic absorption for small cavities. In addition the quality of the cavity decreases somewhat due to an increasing influence of the lateral patterning in small pillars. In the spectra shown in Fig. 2 this reduction of the quality of the cavity at small sizes results in a relative increase of the exciton emission lines in the spectra. In spite of the reduced quality factor of the cavities the optical modes even for the smallest pillars investigated are well separated and have linewidths of the order of a few meV, i.e. the optical modes are well suited for detailed spectroscopic studies.

Figure 3 displays the measured energies of the resonances (filled symbols) for different lateral dimensions of the resonators in comparison with the results of model calculations (solid lines). In order to describe quantitatively the three-dimensional confinement effects and, in particular, the dependence on the lateral size, we have made detailed numerical calculations of the optical mode energies of these cavities. In a first step we solve Maxwell's equations for a three-dimensional cavity with a constant vertical extension as a function of the lateral width L using an effective refractive index of n = 3.22. The vertical extension is characterised by the resonance wavelength λ_0 of the unpatterned microcavity structure, which is obtained, along with the effective refractive index, from a transfer matrix calculation. The present model includes only weak coupling effects between the cavity resonance and the electronic transition.

By using the "boundary element method", we calculate a series of confined photon modes with energies which depend on the lateral size L [5]. These modes have mostly transverse electromagnetic character with only small components of the fields in the z direction. In order to account for the relatively small effects of the infinite lateral extent of the mirrors below the active region in the experimental structures (Fig. 1), we obtain the energies of the confined photon modes from an average of those calculated for a completely isolated, etched cavity and that of the unpatterned microresonators weighted by the integral of the square of the electric field in the etched and unetched regions. By using a transfer matrix calculation and the known depths of the etching, we obtain values of the integrals of about 75 % and 25 % for the etched and unetched regions of the microresonators, respectively.

The results of these numerical calculations are given by the solid lines in Fig. 3 [5]. The confined optical modes in the three-dimensional confined microresonators are characterised by M_{m_z,m_x,m_y} where m_z is the vertical quantum number and m_x and m_y are the lateral quantum numbers. M_x and m_y indicate the number of nodes of the electric field in a given direction. The calculation describes the experimentally observed energies of all modes with very good accuracy. As illustrated in Fig. 3 by the experimental and theoretical results, in going from a one-dimensionally confining vertical resonator structure to a three-dimensionally confining microresonator structure with a lateral width of about 1.5 μm, the energy of the fundamental mode (M_{000}) in the present structure is increased by about 10 me V. Similar values are obtained for the separation in energy of the next higher modes. The present three-dimensional microcavities therefore permit to realise a size controlled discrete set of optical eigenstates.

The size dependence of the energies of the resonances can be understood qualitatively by considering a three-dimensional cavity defined in the vertical direction by the Bragg mirror stacks and the lateral directions by discontinuities of the refractive index at the semiconductor surfaces of the etched structures. As can be shown by calculations to lowest order in (λ_0/L) the electric field vanishes at the lateral surfaces. The energy of the three-dimensionally confined optical modes can then be written as

$$E_{ph} = \frac{\hbar c}{n}\sqrt{k_0^2 + k_x^2 + k_y^2} \tag{1}$$

where $k_0 = 2\pi n/\lambda_0$ denotes the wave vector of the vertical cavity, n the refractive index, λ_0 the vacuum wavelength due to the vertical cavity, and k_x and k_y are related to the lateral directions. The wave vectors of the lateral modes are discretised according to

$$k_{x,y} = (m_{x,y} + 1)\frac{\pi}{L}. \tag{2}$$

correspond to the number of nodes of the modes in x and y direction and L is the lateral width of the square shaped structures.

The photonic eigenstates observed in our structures have important similarities to the electronic eigenstates of semiconductor quantum dots. For example, due to the lateral width dependence of the wave vector components k_x, k_y, the energies of the confined optical modes increase approximately proportional to the square of the inverse lateral width. A similar relation describes the size dependence of the electronic states in quantum dots. The qualitative picture of the refractive index confinement discussed above is similar to the simplest model of electronic quantisation in a quantum dot, in which infinite potentials are assumed in all directions. Because of the much smaller extent of the envelope wave functions of electrons and holes compared to the extent of the optical modes, lateral structure sizes on the order of 10 nm are required in order to observe lateral carrier quantisation effects of similar magnitude as those obtained here for the confined optical modes in microresonators of about 1μm size. Different to the case of carrier confinement, however, there is also a continuum of photon states down to zero energy. Therefore the confined photonic states are strictly speaking strong resonance states with a pronounced size dependence.

4. Investigation of dispersive effects in microcavities with different degree of optical confinement

Angle resolved photoluminescence studies provide a particularly interesting access to the properties of microcavities with optical confinement in more than one direction. Due to the vertical cavity, a single optical mode with respect to k_z is selected in planar structures. Depending on the direction of observation, however, the parallel wave vector components k_x, k_y of the optical modes are expected to vary continuously. In contrast, in photonic dots the additional lateral confinement results in a discretisation of the parallel component of the wave vector. The allowed $k_{||}$ values are controlled by the width of the structures, the refractive index and the order of the mode. In angle resolved experiments, we therefore expect strong differences between a planar microcavity and a photonic dot. For a planar microcavity, the optical mode should display a continuous shift to higher energies when the angle of observation is increased from the surface normal. In photonic dots the energies of the modes should be independent of the detection angle. Dispersion effects in planar microcavities had been observed earlier by other groups. [6]

Figure 4 defines the angles relevant for the angle resolved photoluminescence studies. Here Θ denotes the angle by which the direction of detection is tilted from the growth direction (z-axis). The angle ϕ denotes the direction of the projection of the observation direction into the xy - plane.

Figure 5 shows a set of photoluminescence spectra taken at different angles Θ on a planar microcavity. When the observation direction is tilted off the surface normal, we observe a continuos shift of the optical mode to higher energies. This behaviour is due to the influence of increasing parallel wave vector components on the energy of the optical modes. By using simple arithmetic the energy of the optical mode can be written as a function of the wave vector as

$$E = \sqrt{\hbar^2 c^2 k^2 \sin^2 \Theta + E_0^2}$$

Figure 6 shows angle-resolved studies of microcavities with three-dimensional confinement [7]. In this example, photonic dots with a lateral extension of 6 μm x 6 μm

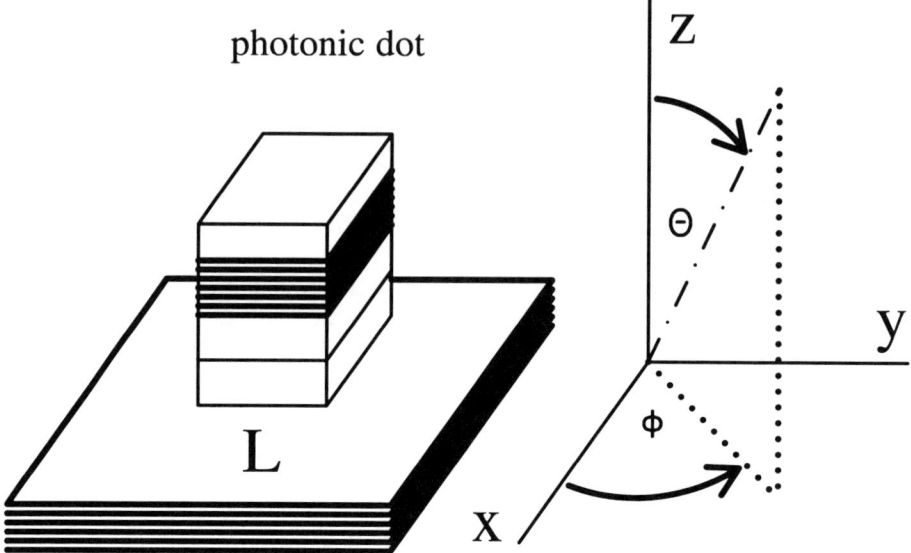

Figure 4: Definition of the polar and the azimuthal angles by which the direction of detection is characterised in the angle-resolved studies.

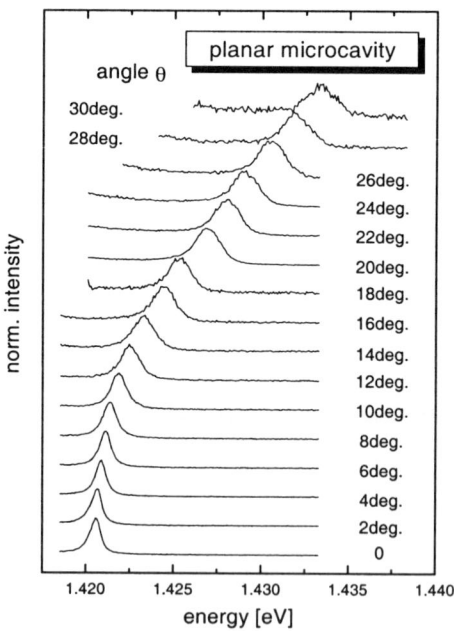

Figure 5: Angle-resolved photoluminescence spectra of a planar microcavity. The polar angle was varied between 0 and 32 deg.

have been investigated. The angle of observation is varied between zero and sixteen degrees of the surface normal in a plane aligned parallel to one of the edges of the photonic dot (parallel to x axis in figure 4). As can be seen from figure 6, the optical modes of the photonic dots show a strikingly different dependence on the angle of observation compared to planar microcavities. The energy of the various optical modes is independent of the angle of observation. This is a clear confirmation of the suppression of dispersive effect in microcavities with three-dimensional optical confinement. As the energies of the optical modes are controlled by the vertical and lateral cavity dimensions

$$E = \sqrt{E_x^2 + E_y^2 + E_0^2}$$

there is no degree of freedom from which a continuous dependence on wave vector could arise.

The intensity of the different modes of the photonic dots shows a characteristic angle dependence. The ground mode has maximum emission intensity for observation along the surface normal and vanishes when the angle of observation is increased above about ten degrees. The next higher lying mode has zero emission intensity for observation along the surface normal and is maximum at an angle of around seven degrees. For higher lying modes, more complex emission intensity patterns are observed, which include maxima of the emission intensity which shift to higher and higher angles.

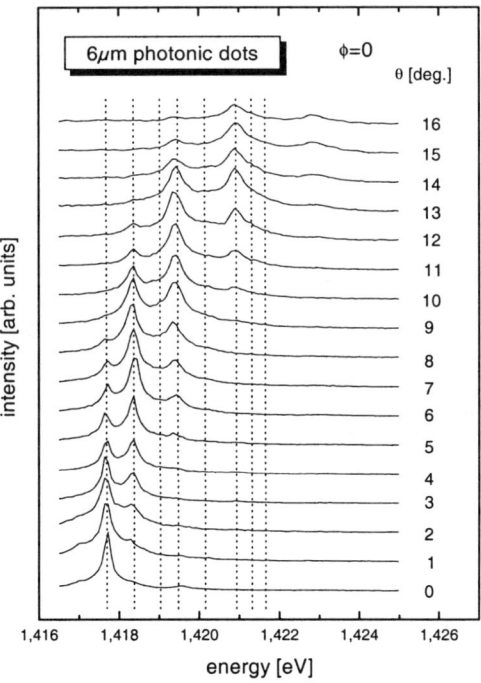

Figure 6: Angle resolved spectra of a 6μm photonic dot. The polar angle was varied between 0 and 32 deg. in a plane along on of the sides of the photonic dot.

The angle variation observed for different modes in the present far field experiments may be used to obtain information on the internal field distribution of the optical modes in the photonic dots. The far field emission of a photonic dot corresponds to the internal electromagnetic field distribution diffracted by a square aperture with the cross section of the size of the dot. In the case of Fraunhofer diffraction, where the signal is detected far from the aperture given by the photonic dots in the present experiments, the interference pattern is described by the Fourier - transform of the aperture function. The intensity distribution at the detector is then given by

$$I_{m_x m_y}(\Theta, \Phi) = \left| \int \varepsilon_{m_x m_y}(x, y) \exp\left(i\left(k_x x + k_y y\right)\right) dx dy \right|^2$$

where $\varepsilon_{m_x, m_y}(x, y)$ is the internal electromagnetic field distribution and $k_x = k$ $\sin\Theta\cos\Phi$ and $k_y = k \sin\Theta\sin\Phi$.

The electromagnetic fields in the photonic dots are well confined for the present dot sizes, which correspond to about 10 wavelengths in the semiconductor or more. In this case the field amplitudes vanish at the boundaries of the dots. Therefore Maxwell's equations can be separated in a first approximation in the x and y co-ordinates. Then the fields can be approximated by those of a cavity with perfectly reflecting side walls. The field components in x and y direction can be described in this case by cos and sin functions depending on the parity of the mode in the different directions and the order of the mode. Figure 7 shows the calculated internal field distribution for the ground mode (0,0) and a higher lying mode (1,1). As expected from symmetry considerations, the ground mode has approximately a Gaussian shape with the maximum of the field strength at the centre of the cross section of the photonic dots. In contrast, the (1,1) mode has a node at the centre of the x and also along the y direction.

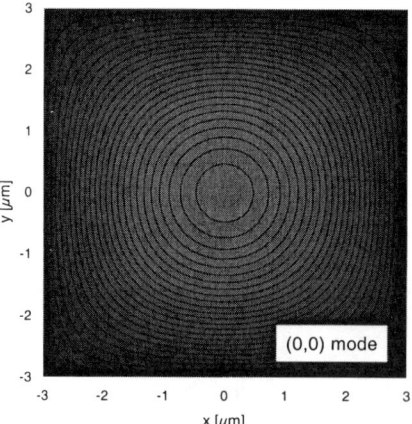 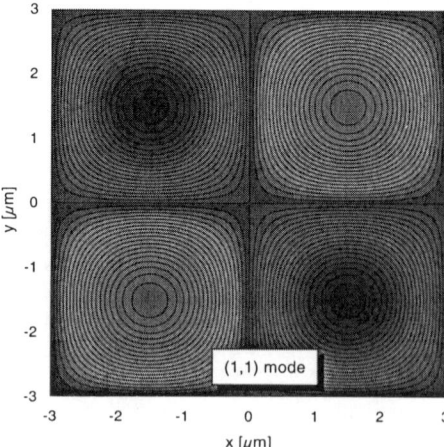

Figure 7: Contour plot of the calculated electric field distributions of the (0,0) mode and the (1,1) mode in a 6μm photonic dot.

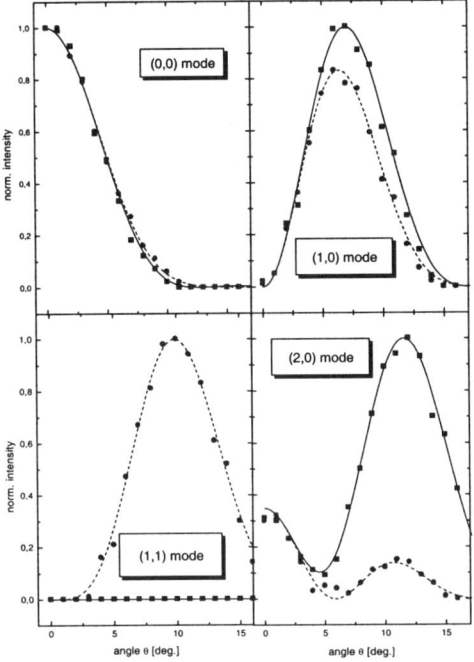

Figure 8: Dependence of the far field emission intensity from the four lowest optical modes (0,0), (1,0), (1,1) and (2,0) on the detection angle Θ. The azimuthal angle Φ was fixed either at 0 deg. (squares) or at 45 deg. (circles). The lines give the results of calculations.

Figure 9: Scanning electron micrograph of a photonic wire with a lateral width of about 1 μm.

For the far field intensity pattern, the different parities have striking consequences. Due to the Gaussian shape of the electromagnetic field of the ground state, the far field intensity is maximum for perpendicular observation and decays strongly when the direction of observation is tilted with respect to the surface normal. For the (1,1) mode, in contrast, the odd parity results in a complete suppression of emission intensity for any tilt angle Θ of the surface normal as long as the observation plane is parallel to one of the edges of the photonic dots. In figure 6 this absence of emission due to the (1,1) mode is indicated by the dashed line at 1.419 eV. Emission from this mode can only be observed for angles Φ unequal to zero, 90, 180,..

Figure 8 displays the angle dependence of the far field emission intensity of the four lowest modes in 6 μm wide photonic dots, versus the polar angle Θ for different azimuthal angles Φ (squares $\Phi = 0°$, circles $\Phi = 45°$). [7] For the (0,0) mode, we observe for both values of Φ a maximum for detection along surface normal and a complete decay of the intensity for angles smaller than ten degrees. The small difference for the two values of Φ is due to the square shape cross section of the photonic dots. Much stronger differences for the two values of Φ arise for higher order modes. These are particularly pronounced in the case of the (1,1) mode. As discussed above, here we observe no emission intensity for any angle Θ aligned parallel to an edge of the photonic dots (squares) whereas emission intensity with a maximum at around ten degrees is observed when $\Phi = 45°$. The dashed and solid lines in figure 8 have been calculated numerically by using the boundary element method in order to derive the internal fields of the structures. [7]. As can be seen from the figure, experiment and calculation for the far field emission intensities, are in agreement for the Θ and Φ dependence including even subtle features of the emission patterns.

We would like to point out once again to a particularly beautiful feature of the angle resolved investigations of optical modes. The far field emission intensity traces shown in figure 8 are directly related to the internal field distribution. In other words, by studying the far field dependence for various angles Θ and Φ, we obtain experimental data which, by use of a reverse Fourier transform, gives direct information on the internal field distributions in the photonic dots.

Figure 9 shows an SEM micrograph of a microcavity with two degrees of optical confinement, a photonic wire. [8] In this case we expect to observe dispersion for angle resolved studies in which the angle of detection is varied in a plane given by the axis of the wire and the surface normal. In contrast, if the angle of detection is varied perpendicular to the axis of the wire the system should display dot like behaviour.

Figure 10 shows angle resolved spectra for the two cases taken from photonic wires with a width of 5.25 μm. In the left part of the figure, spectra for different values of Θ are shown when the variation is carried out perpendicular to the wire axis. Very similar to the experimental results on photonic dots (compare figure 6) we observe different optical modes, which occur at characteristic energies independent of the angle of observation. A change of the angle of observation results in a change of the dominant optical mode. Dispersive effects, however, are not observed. This is in strong contrast to the variation of the spectra observed when Θ is varied in the plane along the wires. Here the emission of the optical mode is seen to shift continuously to higher energies very similar to the case of the microcavity without lateral confinement (compare figure 5), i.e. along the wire we have strong dispersive effects in the photonic wires.

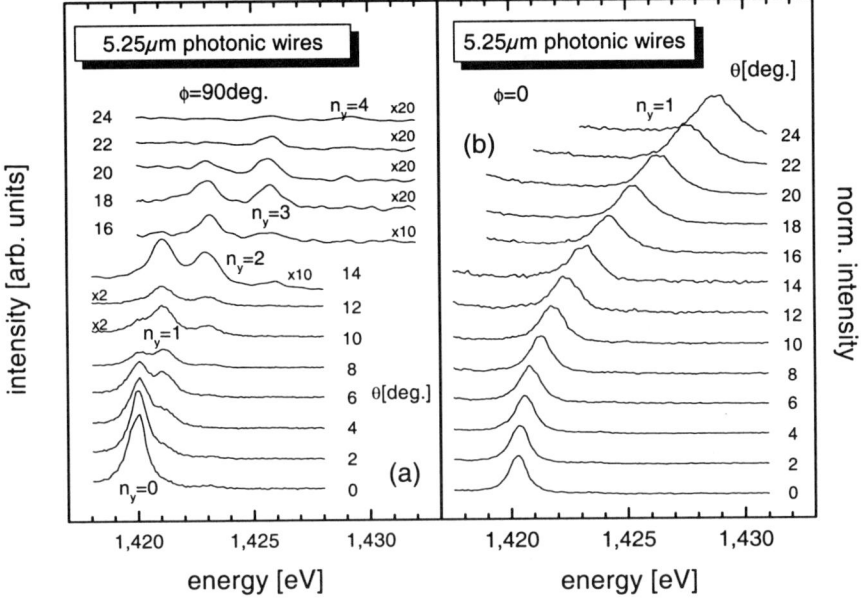

*Figure 10:*Angle-resolved photoluminescence spectra of a photonic wire with a width of 5.25μm. In the left panel the detection was done normal to the wire axis, in the right panel it was done along the wire axis.

5. Conclusions

By lateral patterning of epitaxial Bragg reflector structures microcavities with 3 and 2 D optical confinement have been obtained. The energies of the optical modes in the structures can be tuned over an energy range of about 10 meV by varying the lateral dimensions of the structures. By using angle-resolved studies the suppression of dispersive effects in the photonic dots can be demonstrated. In addition the intensity variation obtained in angle resolved experiments provides experimental access to the internal field distributions in the dots.

The present experiments clearly indicate the potential of structures with 3D optical confinement. In addition to possible applications in low threshold lasers a large area of interesting basic physics studies becomes available concerning a controlled light matter interaction in the structures. By varying the photonic properties significant changes in the electronic properties may be obtained. These include e.g. a significant increase of the spontaneous recombination rate [9] in appropriately designed microcavities or the observation of indication of a coherent exciton phase in the structures [10].

Acknowledgements. We would like to acknowledge the expert assistance of F. Schäfer, R. Werner, P. Knipp, T. Gutbrod, A Tartakovskii, N.A. Gippius and S.G. Tikhodeev for their contributions to the technology, experiments and theory presented here. We are grateful for the financial support of the activities in Würzburg by the State of Bavaria, for the support of the US work by the U.S. Office of Naval Research, and for the support of the work in Russia by INTAS.

References

1.   E. Yablonovitch, Phys. Rev. Lett. **58**, 2059 (1987).
2.   S. John, Phys. Rev. Lett. **58**, 2486 (1987).
3.   see, for example, J.D. Joannopoulos, P.R. Villeneuve, S. Fan, Photonic Crystals, Molding the Flow of Light, (Princeton University Press, NJ, 1995).
4.   G. Björk et al., Phys. Rev. A **44**, 669 (1991).
5.   J.P. Reithmaier et al., Phys. Rev. Lett. **78**, 378 (1997).
6.   R. Houdre et al., Phys. Rev. Lett. **73**, 2043 (1994)
7.   T. Gutbrod et al., Phys. Rev. B **59**, 2223 (1999).
8.   A. Kuther et al., Phys. Rev. B **58**, 15744 (1998).
9.   B. Ohnesorge et al., Phys. Rev. B **56**, 4367 (1997), J.M. Gérard et al., Phys. Rev. Lett. **81**, 1110 (1998).
10. Le Si Dang et al., Phys. Rev. Lett. **81**, 3920 (1998).

SPONTANEOUS EMISSION CONTROL IN In$_x$Ga$_{1-x}$As/GaAs PLANAR MICROCAVITIES WITH DBR REFLECTORS

T.OCHALSKI, J.MUSZALSKI, M.ZBROSZCZYK, J.M.KUBICA, K.REGIŃSKI, J.KĄTCKI, M.BUGAJSKI
Institute of Electron Technology, Al.Lotników 32/46, 02 668 Warszawa, Poland

Abstract: Spontaneous emission control has been achieved in In$_x$Ga$_{1-x}$As/GaAs planar microcavities with DBR reflectors. The room temperature emission in λ-sized cavities is enhanced compared to its free space value whilst in $\lambda/2$-sized cavities suppression of spontaneous emission is observed. The characteristics of spontaneous emission in microcavities depend on the wavelength difference between the emitter and the cavity resonance. It has been shown that ideal tuning of the cavity can be achieved by adjusting the sample temperature. In general, the observed trends are in agreement with theoretical predictions. These changes in the spontaneous emission process directly effect VCSEL properties. An increased coupling efficiency of spontaneous emission into the lasing mode is observed in VCSELs with λ-sized cavities.

1. Introduction

Vertical cavity surface emitting lasers (VCSELs) have been attracting much attention in recent years. Owing to their characteristics such as very low threshold current, parallel optical beams, array capabilities, high speed, etc., VCSELs are considered to be very promising devices [1], [2]. The laser cavity of a VCSEL is usually constructed perpendicular to the substrate plane by stacking multilayer films including an active region, spacer and two dielectric mirrors. Such a structure forms a one dimensional Fabry – Perot cavity resonator. A dielectric mirror can be formed with a periodic stack of quarter wavelength thick layers of alternating high and low refractive index materials. Such a mirror is referred to as a distributed Bragg reflector (DBR). The dielectric layers can be semiconductor layers deposited via molecular beam epitaxy (MBE) growth. The active region usually consists of a spacer layer of a thickness equal to an integer multiple of half the wavelength, and of one or several quantum wells (QWs). The quantum wells are typically situated at the antinodes of the standing wave pattern. Some of the attracting features of VCSELs are connected with the possibility of controlling the spontaneous emission (SE) in the microcavity structure, and in particular with enhanced coupling of the spontaneous emission into the lasing mode [3].

M.L. Sadowski et al. (eds.), Optical Properties of Semiconductor Nanostructures, 201–210.
© *2000 Kluwer Academic Publishers. Printed in the Netherlands.*

2. Distributed Bragg reflectors and microcavities

The cavity, which has an optical length of either one wavelength (λ-type cavity) or one-half wavelength ($\lambda/2$-type cavity), is sandwiched between two dielectric mirrors. At the center of the cavity there is a thin planar quantum well. Excitonic transitions in the quantum well are used as a source of spontaneous emission. The cavity is made of GaAs, the quantum well uses the ternary alloy InGaAs. The Bragg mirrors on both sides of the cavity are composed of an equal number of pairs of quarter-wavelength AlAs/GaAs layers. What determines the spontaneous emission rate in such microcavities is the amplitude of the electric field of the standing-wave of the cavity mode at the location of quantum well. In a λ-cavity the quantum well is located in the antinode of the cavity, whereas in a $\lambda/2$-cavity the exciton dipole sees the node of the cavity as shown in Fig.1. These two distinctively different situations should be reflected in the spontaneous emission properties of the discussed microcavities.

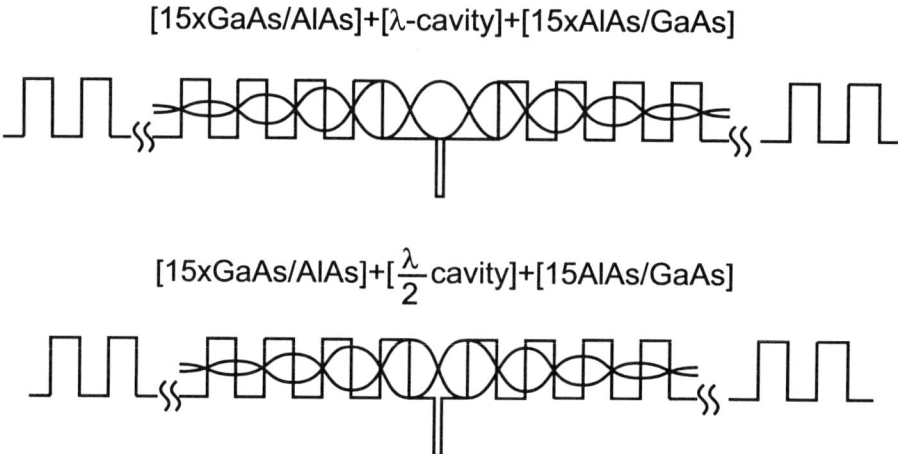

$$[15xGaAs/AlAs]+[\lambda\text{-cavity}]+[15xAlAs/GaAs]$$

$$[15xGaAs/AlAs]+[\tfrac{\lambda}{2}\text{cavity}]+[15AlAs/GaAs]$$

Figure 1. Schematic illustration of the band diagram of λ-sized and $\lambda/2$-sized microcavities. Note the shape of the standing wave pattern in both cases.

The optimisation of the microcavity requires proper tuning of the wavelength of radiation emitted from the active region, the peak reflectivity of the DBRs, and the cavity resonance. This is the reason why the structure performance is very sensitive to variations in the thicknesses of the layers and their composition. The wavelength of radiation from the QW depends on both the composition and thickness. The spectral shape of reflectivity of DBRs in the case of GaAs/AlAs reflectors depends on the layer thickness in the mirrors. Similarly, the position of the cavity resonance depends on the thickness of the spacer layers between the mirrors and the QWs region and the phase of the reflection from the mirrors. Thus, the optimum growth of the structure requires a simultaneous alignment of all three features.

From theoretical considerations it follows that even a 10% error in layer thickness does not significantly degrade the peak reflectivity, but in this case the stop band is shifted in wavelength [4]. Also of interest is the effect of having one layer in the DBR thinner than a quarter wavelength and one thicker. The result of the offsetting thickness variations is to decrease slightly the peak reflectivity of the mirror. Thus, the goal in growing the mirrors for a VCSEL is to obtain layers each approximately a quarter wavelength thick each and to get the reflection band centred at the right wavelength. If the layers differ slightly from a quarter wavelength, it is not so important as long as the reflection band is situated in the right place. The position of the cavity resonance will ultimately determine the lasing wavelength. From theoretical considerations it follows that in a properly fabricated laser structure the wavelength of the reflectivity peak may be shifted by +/- 2%, or each layer thickness may vary by +/- 2%. This means that the accuracy of the control of gallium and aluminium fluxes should be of the order of 2%.

The final issue to consider is the effect that a change in layer thickness and composition has on the wavelength emitted from the active region. For an InGaAs quantum well there are two effects to account for: a change in band gap with composition and a change in electron and hole confinement energies with well width. From simple calculations it follows that even an error of 5% in the Ga flux causes negligible change in the transition energy since the increased confinement energy is compensated by the decreased band gap. The same calculations for a 5% error in the In flux show a wavelength shift one order of magnitude greater. Whereas a small change in the Ga flux will cause no noticeable shift in the transition energy, the In flux should be extremely accurate. In reality, it is very difficult to obtain stabilisation of the flux better than 2% without a real – time control of the growth rate in the MBE system.

The required accuracy can hardly be achieved without additional internal control in the MBE system. Thus, besides the careful calibration of the growth rate and composition, some non-standard methods of internal control should be applied. The real time control of growing layers has been achieved in our case by applying pyrometric interferometry. This method allows a continuous monitoring and readjustment of growth rate to maintain a given thickness of the layers [5], [6]. The advantage of the method is that it does not require any additional equipment, nor does it necessitate any modifications to the MBE machine. It simply makes use of the already installed pyrometer.

We have used a standard IRCON pyrometer to sample the temperature in the centre of the substrate. This particular model is especially designed to measure the GaAs surface temperature by monitoring radiation emitted in a narrow range of wavelengths $(0.940\pm0.03\mu m)$ which are shorter then the band edge of GaAs (but longer then $Al_xGa_{1-x}As$, x>0.25) at the temperatures which are of interest for MBE (400-750°C). For these wavelengths the absorption coefficient is of the order of $10^4 cm^{-1}$ so the $1\mu m$ thick GaAs layer can be still considered as opaque, i.e. the pyrometer registers the radiation emitted by the surface of the structure but not the radiation emitted by the substrate heater.

The apparent temperature oscillations due to interference effects caused by refractive index steps at heterointerfaces are shown for the case of the growth of a microcavity structure in Fig.2. The structure consisted of two Bragg reflectors (15 pairs of quarter wave AlAs and GaAs layers, λ=1000nm) separated by a λ-size cavity with $In_{0.2}Ga_{0.8}As$ QW in the centre. In this case each layer of the Bragg reflector contributes new interfaces to the multiple internal reflections of radiation. In effect, the amplitude of temperature oscillations was increased up to ~15°C (cf. Fig.2.). The sudden change in the temperature readout at the AlAs/GaAs interfaces is caused by a change of thermal radiation background when the Ga and Al shutters are opened and closed, respectively. Analysing the data taken during the whole growth process one can notice that although the readout temperature changes drastically with time, the average temperature stays constant. This proves that substrate temperature control based on the thermocouple readout gives satisfactory results, although the real temperature value must be established in a different way. One can also notice that the shape of the oscillations stabilises after the growth of approximately 8 pairs of AlAs/GaAs layers composing the Bragg mirror. This proves that the pyrometer measures the radiation emitted from the top ~1µm of the structure as it can be expected from a simple analysis of the absorption of thermal radiation in the structure.

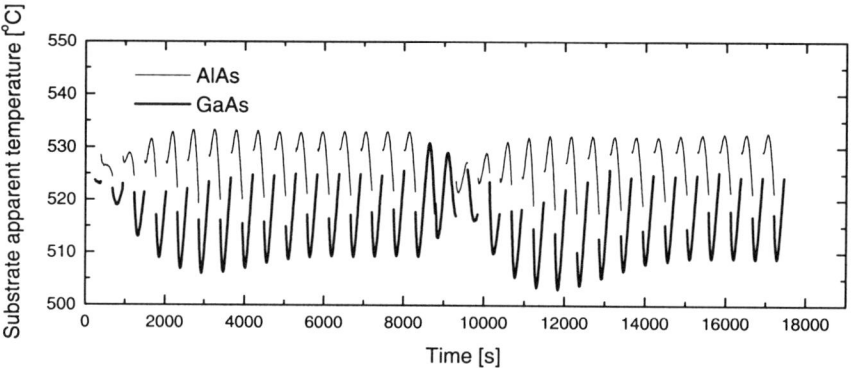

Figure 2. Pyrometer readout during the MBE growth of $In_{0.2}Ga_{0.8}As$/GaAs VCSEL structure.

The apparent temperature oscillations can be used to determine the thickness of the layers, which otherwise would be difficult to assess. Since the actual phase of the interference signal depends on the thickness of the grown layer, they can be used to calibrate the growth rate. The growth rate (G) can be therefore related to the oscillation period (T) by the relation:

$$G = \left(\frac{1}{T}\right)\left(\frac{\lambda}{2n}\right)\cos(\alpha) \qquad (1)$$

where λ is the pyrometer operating wavelength (0.940µm), n is the refractive index of the layer at growth temperature and α is the angle of incidence. Calibrating the refractive index the pyrometric oscillation can be later used for the growth rate measurements over the thick layers i.e., the verification of the group III flux stability. An accurate gallium flux control is also essential to maintain λ_{QW} within ±2nm to assure matching with VCSEL cavity resonance. To achieve low threshold currents in VCSELs it is necessary to maintain the difference between these two below ~5nm. Such tight requirements are difficult to fulfil unless an *in-situ* control technique is applied.

Fig.3 shows a transmission electron microscope (TEM) picture of λ-sized, and λ/2-sized In$_{0.2}$Ga$_{0.8}$As/GaAs planar microcavities grown with pyrometric interferometry. The reflectivity spectrum of the λ-sized microcavity is shown in Fig.4. The width of the high reflectivity wavelength band at normal incidence is roughly given by [7]:

$$\Delta\lambda_{stopband} = \frac{2\lambda_B \Delta n}{\pi n_{eff}} \qquad (2)$$

where Δn is the refractive index difference between the dielectric layers, n_{eff} is the effective refractive index of the mirror (arithmetic mean of the refractive indices in the stack) and λ_B is the Bragg wavelength. For an AlAs/GaAs mirror, the stopband predicted by eq. (1) for $\lambda_B = 1000$ nm is 98.4 nm, which is in good agreement with 103.4 nm observed experimentally. The stopband of the reflector is flat and peak reflectivity as expected theoretically for this type of reflector, i.e. 0.979. Note also the accuracy of the resonance frequency; 1000.2 nm vs. 1000 nm, which was attempted. The resonance width is given by

$$\Delta\lambda = \frac{\lambda_B^2 (1-R)}{2\pi L\sqrt{R}} \qquad (3)$$

which leads to 0.97 nm for λ-sized cavity. The experimental value is almost twice that much, i.e. 1.8 nm. The broadening of the resonance is mainly caused by the reflectivity measurements conditions. The probing beam divergence is the source of inhomogeneous broadening of the reflectance dip. Estimations show that to produce the observed broadening a beam divergence of about 2.5° is enough. Calculated from reflectivity spectra, the cavity finesse is 277.7 for the λ-sized cavity and 833.3 for the λ/2-sized cavity ($\Delta\lambda$=1.2 nm).

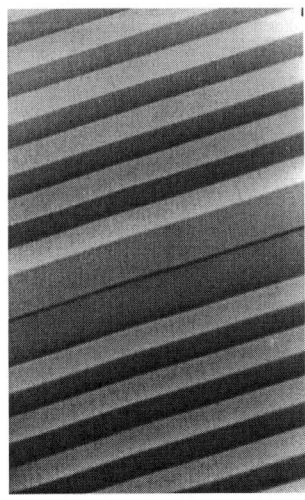

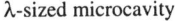

λ-sized microcavity

λ/2-sized microcavity

Figure 3. TEM pictures of λ-sized, and λ/2-sized In$_{0.2}$Ga$_{0.8}$As/GaAs planar microcavities

3. Spontaneous emission control in planar microcavities

Let us consider an elementary excitation (exciton) in a solid coupled to the quantized radiation field. The coupling can be described by a perturbation term in the exciton-field Hamiltonian,

$$V = e\mathbf{dE} \qquad (4)$$

The exciton is in an initial (excited) state and drops into the final (ground) state emitting a photon of energy $\hbar\omega$. The rate of spontaneous emission γ_{sp} is given by Fermi's golden rule:

$$\gamma_{sp} = \frac{2\pi}{\hbar}\left|\langle i|e\mathbf{dE}|f\rangle\right|^2 \rho(\omega) \qquad (5)$$

where -e**d** is the (vector) dipole moment, **E** is the electric field at the location of the exciton dipole and $\rho(\omega)$ is the density of optical modes per unit energy at the angular frequency ω. It is clear from eq.(5) that the spontaneous emission rate can be altered either by modifying the mode density, or by modifying the electric field at the location of the exciton dipole. Both the mode density and the electric field can be modified in a microcavity whose size is properly adjusted. To observe cavity related effects, the field does not have to be confined in all three dimensions. One dimensional microcavities already give sizeable effects, although their strength scales with the degree of confinement. Inside the cavity, the electromagnetic field forms a standing wave which meets the resonance condition, namely the round trip phase shift equal to an integer multiple of 2π, as in Fabry-Perot resonators made of metallic mirrors.

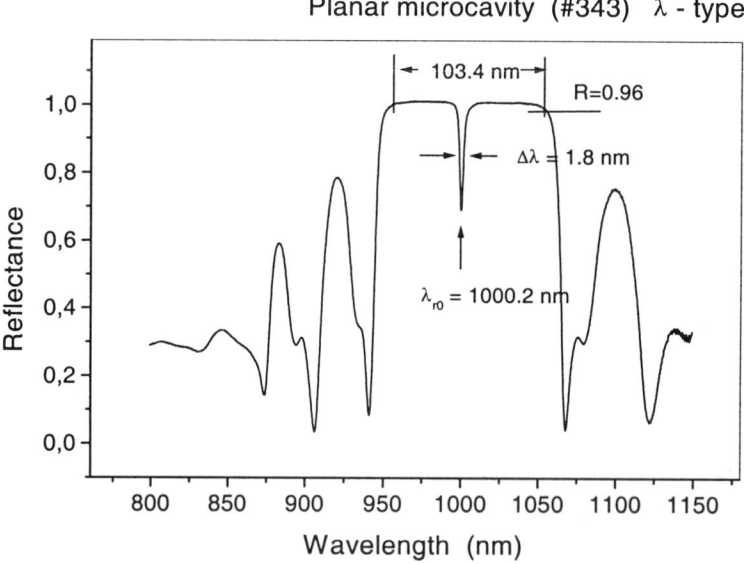

Figure 4. Reflectance spectrum of a λ-sized $In_{0.2}Ga_{0.8}As$/GaAs planar microcavity, resonant at 1000 nm

Reflectance, photoluminescence perpendicular to the Bragg reflector (PL), and photoluminescence from the edge of the structure (PL in plane) of a λ-sized $In_{0.2}Ga_{0.8}As/GaAs$ planar microcavity, resonant at 1000 nm are shown in Fig.5. The photoluminescence signal from the edge of the structure can be regarded as a reference spontaneous emission (SE) unaffected by the cavity. In the direction perpendicular to the cavity plane the PL signal is concentrated in a narrow line forced by cavity resonance and its intensity increases roughly by a factor of 10 (the integrated intensity of the PL line increases 1.9 times). The characteristics of spontaneous emission in microcavities depend on the wavelength difference between the emitter and the cavity resonance. By lowering the sample temperature it is possible to shift the QW PL line to higher energies, while the position of the cavity resonance remains practically unaffected. In the case of the discussed λ-sized cavity, temperature tuning produced a further increase of the PL signal by another factor of 10, which leads to a total enhancement of the PL signal by the microcavity of about 100. This is to be compared with a calculated cavity enhancement factor equal to 190.

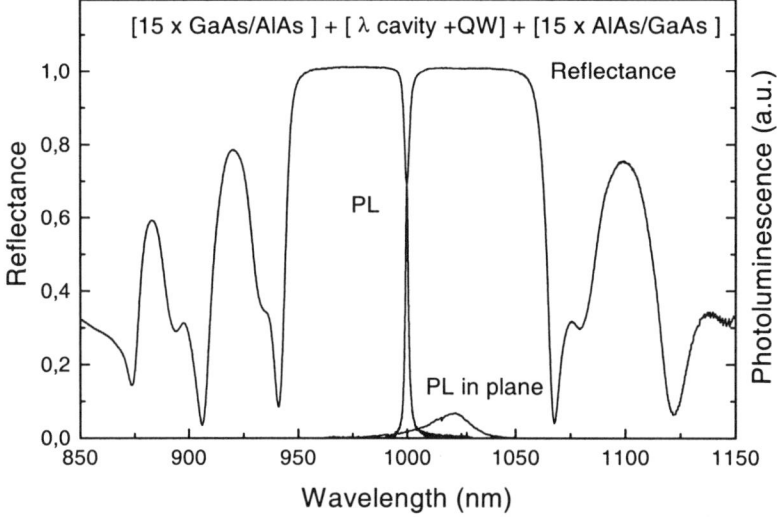

Figure 5. Reflectance, photoluminescence perpendicular to the Bragg reflector (PL), and photoluminescence from the edge of the structure (PL in plane) of λ-sized $In_{0.2}Ga_{0.8}As/GaAs$ planar microcavity, resonant at 1000 nm

Another type of cavity is the $\lambda/2$-sized microcavity, in which the QW positioned in the centre of the spacer is located at the node of standing wave pattern of the cavity mode. Reflectance, photoluminescence perpendicular to the Bragg reflector (PL), and photoluminescence from the edge of the structure (PL in plane) of a $\lambda/2$-sized $In_{0.2}Ga_{0.8}As/$ GaAs planar microcavity, resonant at 1000 nm are shown in Fig.6. As it might be expected this cavity effectively quenches the PL signal (the ratio of the integrated intensity of perpendicular PL to in-plane PL is 1/12). The fact that there is

208

still some PL emitted from the cavity is due to two factors. The first is that we collect the PL signal from a small solid angle around the direction perpendicular to the cavity, the second is that the QW might be slightly off centre with respect to the intended position and consequently the exciton dipole interacts with a nonzero field amplitude. Nevertheless the PL quenching by a $\lambda/2$-sized cavity is beyond any doubt. This is an important result because it once again proves that spontaneous emission is not an inherent property of the emitter but is indeed a stimulated emission, stimulated by vacuum field fluctuations of the electromagnetic field in the cavity.

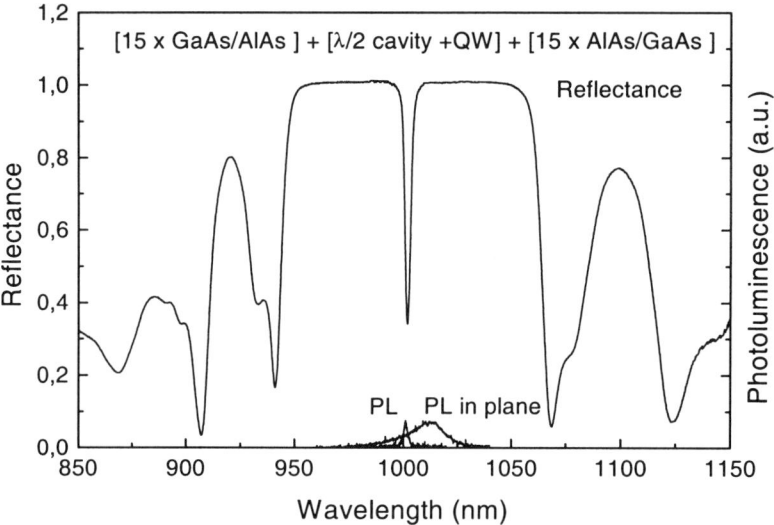

Figure 6. Reflectance, photoluminescence perpendicular to the Bragg reflector (PL), and photoluminescence from the edge of the structure (PL in plane) of $\lambda/2$-sized $In_{0.2}Ga_{0.8}As$/GaAs planar microcavity, resonant at 1000 nm

4. Spontaneous emission in optically excited VCSEL

Complete VCSEL structures with doped Bragg reflectors were also grown and optically tested. The sequence of the layers in VCSEL was the following: [15xGaAs/AlAs: Be]+[λ-cavity+QW $In_{0.2}Ga_{0.8}As$]+[24xAlAs/GaAs: Si]. The structure was grown on a p-type substrate and was designed for emission through the substrate. The reflectivity spectrum of the VCSEL structure is shown in Fig.7 together with the calculated mode spectrum of the laser. Although weak, cavity modified emission was observed at 1000 nm, we failed to make this structure lase because the QW exciton energy was 25 nm off the cavity resonance towards longer wavelengths. Lowering the sample temperature, similarly to the previous case, we were able to increase the PL signal but that was still not enough to reach the lasing threshold. We have estimated the increased coupling of the spontaneous emission into the lasing mode (β factor) to be of the order of 10^{-3} which is to be compared with a number of the order of 10^{-5} for conventional edge emitting lasers.

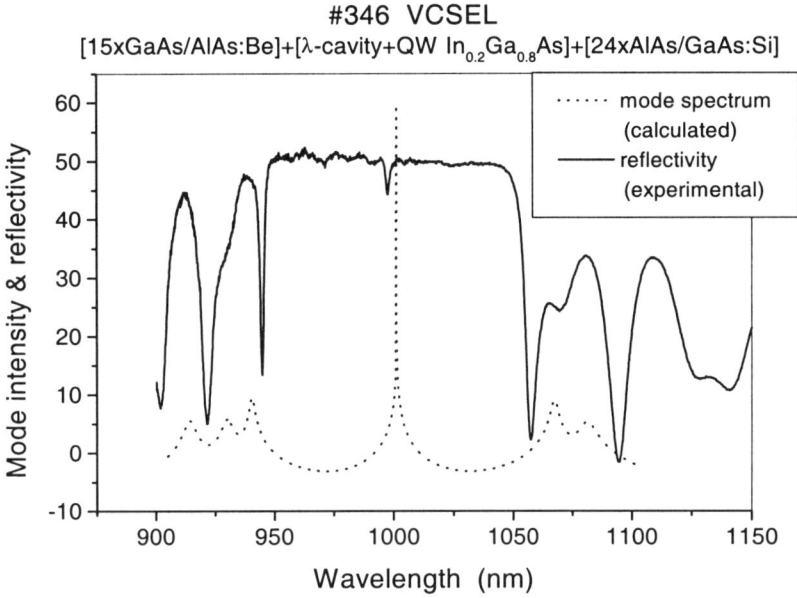

Figure 7. Reflectance spectrum of $In_{0.2}Ga_{0.8}As/GaAs$ VCSEL structure. The calculated mode spectrum of the laser, above threshold, is also shown.

5. Conclusion

Spontaneous emission control has been achieved in $In_xGa_{1-x}As/GaAs$ planar microcavities with DBR reflectors. The room temperature emission in λ-sized cavities is enhanced compared to its free space value whilst in $\lambda/2$-sized cavities suppression of spontaneous emission is observed. The characteristics of spontaneous emission in microcavities depend on the wavelength difference between the emitter and the cavity resonance. It has been shown that ideal tuning of the cavity can be achieved by adjusting the sample temperature. In general, the observed trends are in agreement with theoretical predictions. These changes to the spontaneous emission process directly effect VCSEL properties. An increased coupling efficiency of spontaneous emission into the lasing mode is observed in VCSELs with λ-sized cavities.

We have also demonstrated that for reproducible growth of microcavities and in particular vertical cavity surface emitting lasers (VCSELs) by MBE, the growth rate of the individual layers has to be controlled with an accuracy better than 2%. To achieve this level of process control, a real time monitoring of the growth is required. In this work we also report the apparent substrate temperature oscillations observed by infrared pyrometry during the MBE growth of multilayer laser structures and demonstrate their usefulness in process control. The phase information and the period of oscillations provide information on the actual growth rate, whereas the mean value of the pyrometer readout correlates with the true substrate temperature. Interference pyrometry offers a convenient alternative to standard laser reflectometry when it is necessary to control the growth rate of thick layers with high precision. Using the above described method we have grown a number of microcavities and VCSEL structures with precisely tailored Bragg reflector characteristics and cavity resonance tuned to the centre of DBR stopband and quantum well emission wavelength.

Acknowledgements. This work has been supported by the State Committee for Scientific Research (Poland) under Contract No. PBZ–28.11/P7.

References

1. Chang-Hasnain, C.J. (1995) Vertical-Cavity Surface-Emitting Lasers, in G.P.Agrawal (ed.), *Semiconductor Lasers; Past, Present, and Future,* AIP Press, Woodbury, pp.145-180.
2. Coldren, L.A., Hegblom, E.R., Akulova, Y.A., Ko, J., Strzelecka, E.M., Hu, S.Y. (1998) VCSELs in 98: What we have and what we can expect, in K.D.Choquette and R.A.Morgan (eds.), *Vertical-Cavity Surface-Emitting Lasers II, Proceedings of SPIE,* **3286,** pp. 2-16.
3. Baba, T., Hamano, T., Koyama, F., Iga, K. (1991) Spontaneous Emission Factor of a Microcavity DBR Surface Emitting Laser, *IEEE J.Quantum Electron.* **27,** 1347-1358.
4. Geels, R. S. (1991) Vertical-Cavity Surface-Emitting Lasers: Design, Fabrication and Characterization. PhD thesis, University of California, Santa Barbara.
5. Muszalski, J., Ochalski, T., Bugajski, M. (1999) Apparent Substrate Temperature Oscillations during MBE Growth of Laser Heterostructures, *Acta Physica Polonica,* to be published.
6. Regiński, K., Muszalski, J., Bugajski, M., Ochalski, T., Kubica, J.M., Zbroszczyk, M., Kątcki, J. Ratajczak, J. (1999) MBE growth of planar microcavities with distributed Bragg reflectors, *Thin Solid Films,* to be published.
7. Bjork, G., Yamamoto, Y. (1995) Spontaneous Emission in Dielectric Planar Microcavities, in H. Yokoyama and K. Ujihara (eds.), *Spontaneous Emission and Laser Oscillation in Microcavities,* CRC Press, Boca Raton, pp. 189-235.

FERROMAGNETIC SEMICONDUCTORS AND THEIR NANOSTRUCTURES: NEW OPPORTUNITIES AND CHALLENGES

J. K. FURDYNA, P. SCHIFFER, Y. SASAKI, S. J. POTASHNIK, AND X. Y. LIU
Department of Physics, University of Notre Dame, Notre Dame, IN 46556, USA

1. Introduction

Research on magnetic semiconductors has for the last two decades been largely focused on II-VI semiconductor alloys containing Mn. These materials (e.g. HgMnTe, ZnMnSe) have served to firmly establish the potential role of electron spin in semiconductor physics. Giant Faraday rotation in CdMnTe, enormously large negative magnetoresistance in p-type HgMnTe [1] and - in the case of multilayers - the achievement of spatial segregation of electrons with opposite spin in so-called "spin superlattices" [2,3] are examples of the many new and exciting effects which are made possible by these materials. Although the phenomena just illustrated are of a magnitude that readily suggests a number of spin-based device applications (such as optical isolators [4]), the realisation of such devices has been very limited because all the effects mentioned above are determined by the magnetisation of the magnetic sublattice, which rapidly decreases as one approaches room temperature. Increasing the concentration of magnetic ions to enhance these effects in II-VI-Mn alloys is ineffective (and in fact counterproductive), because the dominant Mn-Mn interaction in these materials is antiferromagnetic, limiting the magnetisation - and thus also the spin-dependent effects just described.

In the last several years this situation has, however, changed in an important way due to the emergence of semiconductors with *ferromagnetic* coupling between magnetic ions. The resulting increase in magnetisation, and the "survival" of this magnetisation to significantly higher temperatures, thus hold out new prospects for device applications. These new developments have occurred on several fronts. First, in the "traditional" II-VI-based magnetic semiconductor alloys with Mn, it was recently demonstrated that heavy p-type doping can render these materials ferromagnetic via the RKKY interaction, automatically leading to a very significant enhancement of spin-based effects [5,6]. Furthermore, ferromagnetic alloys of III-V semiconductors with Mn (most notably GaMnAs and InMnAs) have also recently emerged on the scene, holding enormous possibilities for both the basic and the applied sectors [7,8]. It is noteworthy that specimens of GaMnAs have now been achieved with Curie temperatures in the neighbourhood of 100 K [9]. Their ferromagnetism, relatively high critical temperature, and the possibility of integration with "standard" III-V-based electronic and

M.L. Sadowski et al. (eds.), Optical Properties of Semiconductor Nanostructures, 211–224.
© *2000 Kluwer Academic Publishers. Printed in the Netherlands.*

optoelectronic structures make this development especially significant.

In this paper we will review these new advances in the area of ferromagnetic semiconductor alloys and their multilayers. We will also attempt to identify those major issues that still stand in the way of optimisation of these materials, with a view of making them attractive for realistic device applications.

2. Ferromagnetism in II-VI-based magnetic alloys

As has been noted, the $II_{1-x}Mn_xVI$ magnetic alloys (often referred to as "diluted magnetic semiconductors" or "semimagnetic semiconductors") display a number of spectacular effects, which have made them the focus of intense research activity. Giant Faraday rotation [10], dramatic negative magnetoresistance in p-type $Hg_{1-x}Mn_xTe$ [11] and, in the context of epitaxially-grown heterostructures, band offset tuning in DMS/non-DMS multilayers [12] are but a few examples of these very interesting effects, all arising from the marriage of semiconductor physics and magnetism.

The attractive magneto-transport and magneto-optical properties of magnetic semiconductors all have their origin in the large Zeeman splitting of electronic levels, which is in turn determined by the magnetisation M of the magnetic sublattice (comprised, in the specific case discussed here, of Mn^{++} ions) [1]. Although these level splittings - and thus the effects arising from them - are typically quite large at cryogenic temperatures, they diminish rapidly as the temperature increases, to the point where they may not be sufficiently large at room temperature for practical use in magneto-optical and/or spin transport devices. This is because, as noted earlier, the size of all these effects (Zeeman shift of the band edges, Faraday rotation, etc.) is ultimately determined by the value of the magnetisation M. In Mn-based II-VI semiconductor systems the magnetisation is limited by the fact that, *as a rule*, Mn-Mn interactions are antiferromagnetic. Many authors have considered the possibility of enhancing the magnetisation by reducing the number of nearest neighbours in an ordered fashion (by using planar, or "digital" Mn inserts, or by fabricating entirely new crystal structures such as stannites [13]) to reduce the effect of this interaction.

But clearly one would make much greater strides toward increasing magnetisation - and thus toward enhancing the spin-based effects referred to above - by fabricating materials where interactions between magnetic ions are *ferromagnetic*. Indeed, recent advances in materials design have now firmly established the existence of ferromagnetic inter-ion interactions in Mn-based magnetic semiconductors, and we outline these below as a promising step toward significant enhancement of magneto-optical and magneto-transport spin-based effects. We begin with the discovery of ferromagnetism in the "traditional" $II_{1-x}Mn_xVI$ systems.

Recently it has been predicted theoretically [5] and confirmed experimentally [6] that the two-dimensional hole gas in modulation-doped CdMnTe quantum wells between CdMgZnTe barriers produces, via the RKKY interaction, a ferromagnetic interaction between Mn spins. This RKKY interaction due to the presence of holes overcompensates the "traditional" antiferromagnetic coupling between Mn^{++} ions mediated by superexchange, and thus results in a $II_{1-x}Mn_xVI$ ferromagnet. This seminal work demonstrates that the well-established techniques of modulation doping can be put to use to engineer and tailor *magnetic* interactions in magnetic semiconductor heterostructures. As a consequence, in the experimental work cited above a colossal enhancement of Zeeman splitting (as observed by inter-band magnetooptical transitions) has been achieved [14]. It should be noted, finally, that the process of "ferromagnetising" the $II_{1-x}Mn_xVI$ alloys has now been extended to heavily p-type-doped ZnMnTe both in bulk and in epilayer form [15], thus firmly establishing the universality and reproducibility of this important process.

3. III-V-based ferromagnetic semiconductors

As has already been said, II-VI-based magnetic semiconductors (and particularly those involving Mn) have been explored the most extensively, and this has paved the way for our understanding of the role of electronic spin in semiconductor physics. However, recently very significant strides have also been made in the area of III-V-based magnetic semiconductors (and, again, the III-V alloys involving Mn are leading the way). This work, which was begun with the pioneering success in growing $In_{1-x}Mn_xAs$ films by MBE [7], and which has more recently been extended to $Ga_{1-x}Mn_xAs$ [8], is important for several reasons. First, it holds out the promise of integrating magnetic semiconductors with III-V-based electronic and opto-electronic circuits in a natural way; and second, the work on the $III_{1-x}Mn_xV$ alloys already carried out clearly establishes that they are ferromagnetic. We now know that, although the Mn ions enter the III-V lattice by substituting for the group-III element, they retain their strong preference to the divalent state. The Mn^{++} ions in the III-V host thus act *both* as the localised moments *and* as acceptors. The $III_{1-x}Mn_xV$ alloys are thus automatically p-type, and it is assumed that the holes provide the necessary mechanism for mediating the ferromagnetic RKKY coupling between the Mn^{++} spins similar to that already discussed in the preceding section in connection with p-type $Cd_{1-x}Mn_xTe$ and $Zn_{1-x}Mn_xTe$. One should note that the credit for this achievement goes in large measure to the availability of non-equilibrium epitaxial techniques such as MBE, since the low equilibrium solubility of Mn in the III-V matrix puts severe limits of Mn incorporation in bulk crystals (below 1 at. %). Successful growth of $III_{1-x}Mn_xV$ alloys with x>0.01 has thus (at least so far) only been achieved in thin film form. This linkage of III-V ferromagnetic systems and epitaxy makes integrating such ferromagnetic films into existing III-V-based electronic and optoelectronic heterostructures all the more likely.

3.1. $Ga_{1-x}Mn_xAs$: A REPRESENTATIVE III-V FERROMAGNET

We begin our discussion of III-V-based ferromagnetic semiconductors with the case of $Ga_{1-x}Mn_xAs$, a material which has been studied the most extensively, and which illustrates most of the properties common to (or expected in) other III-V systems containing Mn.

Fabrication: Most commonly, GaMnAs is grown on (001) GaMnAs substrates. After depositing a GaAs buffer layer at some "standard" temperature (*ca.* 700°C), the substrate is cooled to between 200 and 300°C. A buffer layer is typically deposited at this low temperature, and at some point the Mn shutter is opened to commence the

GaMnAs growth [8,16,17]. Growth at such a low temperature is made necessary by the fact that at higher temperatures there is a tendency to form ferromagnetic MnAs precipitates (in NiAs structure) within the GaMnAs alloy, as has been convincingly demonstrated by annealing experiments of Van Esch et al. [16]. However, as will be seen later, the low growth temperature also results in the formation of As antisites that tend to reduce the quality of the electrical properties, by counteracting the effectiveness of the Mn^{++} acceptors through compensation. Optimisation of Mn incorporation and p-type doping can be achieved by a judicious balance of temperature and As overpressure, as has been demonstrated by systematic studies of Shimizu et al. [17]. With such precautions, zinc blende $Ga_{1-x}Mn_xAs$ epilayers of excellent quality can be grown without formation of the unwanted MnAs phase. The high quality of the growth is supported by clear RHEED oscillations that, under carefully optimised conditions, persist indefinitely [18], signalling excellent two-dimensional deposition. Because the lattice parameter of GaMnAs is very close to GaAs, pseudomorphic strained layers of more that 1μm have been readily obtained in this growth mode.

Structure and Mn concentration: The lattice parameter of the zinc blende $Ga_{1-x}Mn_xAs$ increases with the Mn content, because the tetrahedral radius of Mn is larger than that of Ga [19]. By determining the Mn content of the alloys via chemical analysis, a dependence of the lattice parameter on x has been obtained [8,16]. Using this calibration, one can now rely on the value of the lattice parameter measured by x-ray diffraction as a means of determining the Mn content of $Ga_{1-x}Mn_xAs$ using the formula

$$x = [a(x) - a_{LT-GaAs}]/[a_{MnAs} - a_{LT-GaAs}], \tag{1}$$

where a(x) is the measured lattice constant, a_{LT}-$_{GaAs}$ is the lattice constant of LT-GaAs (see below), and a_{MnAs} is the lattice parameter of the hypothetical zinc blende form of MnAs (also see below).

Two comments should be made at this point. First, a linear extrapolation of a(x) observed on $Ga_{1-x}Mn_xAs$ of various Mn concentration to x = 1.00 should yield the hypothetical value of the lattice parameter of zinc blende MnAs. This value is variously quoted as 5.98Å [8], 6.014Å [16], and (as extrapolated from $In_{1-x}Mn_xAs$ data) 6.01Å [7]. If one were to use the bond length expected from the tetrahedral radii of Mn and As (1.326 and 1.225Å, respectively [19]), one would obtain a value for a_{MnAs} of 5.901. This is significantly less than the extrapolated values of a $_{MnAs}$ quoted above. It is possible that the discrepancy originates from the fact that the GaMnAs specimens used in the extrapolation are strained. But also one must recognise that, by making such a Vegard's law extrapolation from data obtained on crystals with small concentrations of Mn, one is subject to fairly significant error. In our view the precise value of the lattice parameter of zinc blende MnAs given in the literature requires further verification.

The second aspect involving crystal structure pertains to the consequences of the low-temperature (LT) growth of GaMnAs. GaAs itself, when grown at low temperatures (the so-called LT-GaAs), has a slightly higher lattice parameter ($a_{LT-GaAs}$ = 5.6572Å) [16] than pure GaAs (5.6533Å), due to the presence of As precipitates [20] that distort the lattice. We note in this connection the systematic annealing studies described in Ref. [16], which show that the lattice parameter of $Ga_{1-x}Mn_xAs$ gradually shrinks as the substitutional Mn in the system is precipitated out in the form of MnAs clusters as a result of annealing. But this Mn remains in the system, and would therefore contribute to chemical analysis, without contributing to Vegard's-law behaviour of the alloy. We believe therefore that further work is needed (including careful EXAFS and TEM studies) focused on a more precise characterisation of the crystallography of $III_{1-x}Mn_xV$ alloys.

Magnetic properties of GaMnAs: The most striking property of GaMnAs is of course its ferromagnetic behaviour at low temperatures. Figure 1 shows the magnetisation curve, with a characteristic hysteresis loop, observed at three temperatures on one of the $Ga_{1-x}Mn_xAs$ layers ($x \approx 0.06$) grown at Notre Dame. At this time the highest value reported for the Curie temperature is about 110K, observed by Matsukara *et al.* [9]. However, more typical values currently obtainable by various groups still remain below 100K.

The ferromagnetism of GaMnAs (and of its sister alloy InMnAs) originates from two ingredients: the obvious presence of magnetic ions, in the form of Mn^{++}; and the high concentration of holes in these systems. It is an elegant aspect of $III_{1-x}Mn_xV$ alloys that the Mn^{++} ion, which has a magnetic moment and at the same time acts as an acceptor, satisfies both requirements simultaneously. It is widely accepted that (as was already mentioned above) in entering the III-V lattice the Mn ion retains its preference for the divalent state.

There are two models that are currently proposed for ferromagnetism arising in this situation. One scenario rests on the assumption that the large number of holes produced via Mn-doping provides a means for RKKY exchange between the Mn ions [8]. While the hole concentration is very large for a semiconductor (10^{18} to 10^{20} cm^{-3}), it is sufficiently low for the RKKY wavelength to be longer than the distance between nearest Mn neighbours (i.e., $2k_Fr_{ij} \ll 1$), where k_F is the Fermi wave vector of the free hole gas and r_{ij} is the distance between nearest-neighbour Mn ions. This situation tends to align all Mn spins along the same direction.

The alternative model [16,21] is based on the assumption that the Mn^{++} ions form Mn^{++}-hole complexes. Briefly, in this model the ferromagnetic alignment of Mn spins arises as follows. There is a strong antiferromagnetic coupling between the spin S of the $2D^5$ core of Mn and the spin σ of the hole surrounding it. The holes in turn interact with other holes ferromagnetically, thus automatically aligning the Mn moments to which they are coupled. For further details regarding this model the reader is referred to Refs. [16,21]. Although at the present time the RKKY picture is gaining wider acceptance, it should be said that neither model has been definitively proven or disproven. For example, core-level photoemission study of $Ga_{1-x}Mn_xAs$ by Okabayashi *et al.* [22] was unable to distinguish conclusively between the presence of neutral acceptors formed by a Mn^{++} ion at a Ga site plus a bound hole, or of ionised acceptors consisting of Mn^{++} ions and free holes. In this paper we have chosen to use the RKKY process as a working hypothesis, with the understanding that the identification of the mechanisms responsible for ferromagnetism in $III_{1-x}Mn_xV$ alloys still awaits definitive resolution.

We now return to magnetic properties as observed experimentally. The most direct and unambiguous fingerprint of ferromagnetism in $III_{1-x}Mn_xV$ alloys is the observation of hysteresis loops such as that shown in Fig. 1. Such behaviour is observed in the range of concentrations from *ca.* $x \approx 0.02$ to $x \approx 0.07$ [23]. The disappearance of ferromagnetic order below 2% Mn is not surprising because of the separation between nearest Mn neighbours (as well as the automatically lower density of free holes that is correlated with the Mn concentration). At the other end of Mn concentrations (x > 0.08) the quality of the crystal deteriorates because of formation of MnAs inclusions, and possibly also of an excess of compensating centres that reduce the hole concentration required for the RKKY mediation process. Thus it appears that the optimal Mn concentration at the present state of the art is x ≈ 0.05, and samples in that range have indeed shown the highest values of the Curie temperature [9].

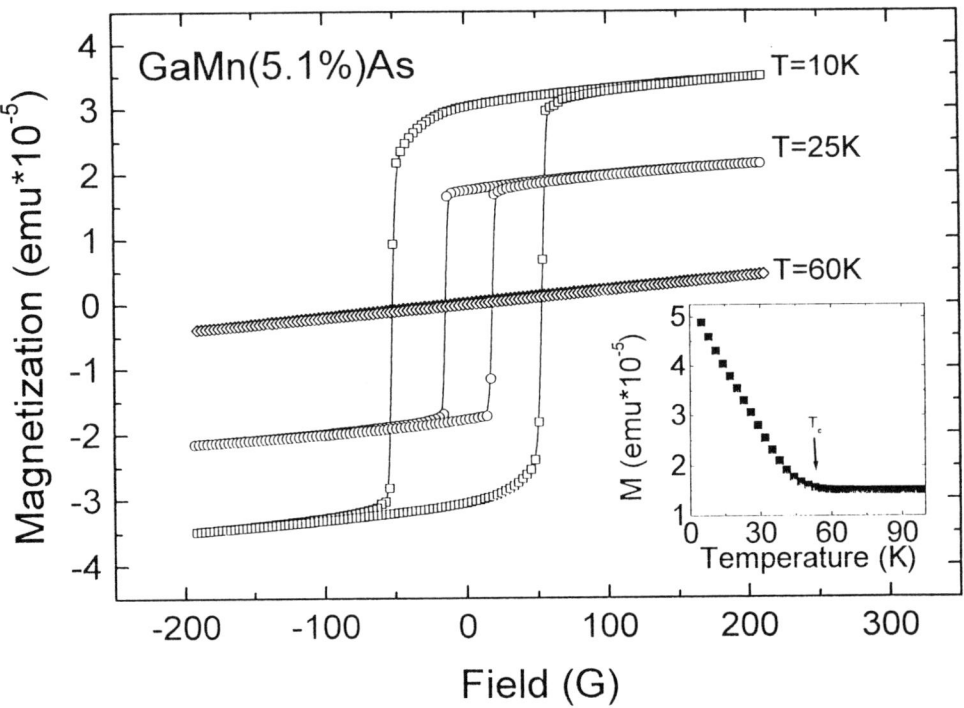

Figure 1: Magnetic hysteresis data observed on a representative $Ga_{1-x}Mn_xAs$ layer (x ≈ 0.051) at three temperatures, illustrating the ferromagnetism of the material. The applied field is in the plane of the layer. In the inset we show saturation magnetisation as a function of temperature, T_c indicating the estimated Curie temperature.

Quantitative magnetisation measurements on samples which are believed to be free of MnAs inclusions indicate the value of saturation magnetisation M_s close to the calculated value $N_{Mn}g\mu_\beta S$, where N_{Mn} is the Mn concentration, g = 2, μ_β is the Bohr magneton, and S is the spin of Mn^{++} (S = 5/2 for Mn^{++}). Using the nominal value of N_{Mn}, Ohno *et al.* [8] obtain S = 2.2, indicating that practically all Mn contributes to the magnetisation.

The p-type doping by Mn is not as efficient, since the effectiveness of Mn in contributing a hole is counteracted by compensation [9]. Typically the hole concentration is below 20% of the Mn concentration in a given alloy. As already noted, it has been shown by Shimizu *et al.* [17] that the number of holes for a given Mn concentration x can be maximised during growth by a careful balance between the substrate temperature and As overpressure, so as to reduce the number of As antisites. It was then demonstrated that, for a given x, there is an almost direct proportionality between the hole concentration p and the corresponding value of the Curie temperature T_c [17]. It therefore appears that there is still considerable room for improvement in this regard, i.e., that one should be able to push the value of T_c to still higher temperatures by devising means of increasing the hole concentration originating from Mn, or by some other means of external doping.

To complete the sketch of our present understanding of magnetic behaviour, we finally note that the direction of the easy axis of magnetisation *is a function of strain.* GaMnAs grown on GaAs is under biaxial compressive strain in the basal plane, owing to the larger value of the GaMnAs lattice parameter (e.g., 5.672Å for 5% Mn) relative to LT-GaAs (5.657Å). It is found that this compressive strain renders the easy axis of magnetisation to be in the plane of the layers [8]. In contrast, if the biaxial strain is tensile in the layer plane (as, e.g., for GaMnAs grown on GaInAs buffers with sufficient In content to exceed the lattice parameter of GaMnAs), the easy axis was found to be normal to the layers [24]. One thus has the possibility of "engineering" the orientation of the axis of magnetisation by strain that can be built into the epitaxial material through judicious choice of lattice parameters of the constituent layers.

Transport and magneto transport: It is certainly clear that electrical properties - and in particular the population of holes - have a direct and decisive effect on the magnetic behaviour of the system. But the converse is also true: transport properties of the alloys appear to be strongly linked to its magnetic character. We will provide a brief overview of electrical properties of GaMnAs below.

The ferromagnetic $Ga_{1-x}Mn_xAs$ layers can be divided into three concentration regions [23]: samples in the region $0.02 \leq x \leq 0.03$ are ferromagnetic but *insulating* (i.e., their resistivity increases dramatically with decreasing temperature); those in the range $0.03 \leq x \leq 0.06$ are *metallic* (i.e., their resistivity below T_c is approximately independent of temperature); and those samples with Mn concentration in the range above $x \approx 0.06$ again show *insulating* behaviour, possibly because of the increased number of deep donors, that may be linked (directly or indirectly) to the high Mn content. It is a striking feature of these alloys that the metallic samples consistently have the highest values of the Curie temperature.

As the temperature is lowered from room temperature down, GaMnAs manifests a resistivity peak at the Curie temperature, which is assumed to arise from enhanced spin scattering as one passes through the critical temperature. This is especially evident in metallic samples (because of the relatively slow variation of resistivity with temperature), but is actually present in the rapidly varying resistivity of the insulating samples as well, in the form of a clearly distinguishable "knee".

A very dramatic feature of the insulating samples is their enormous negative magnetoresistance occurring at low temperatures [23]. Both the rapid rise of the resistivity with decreasing temperature, and its dramatic drop with magnetic field at cryogenic temperatures can be understood in terms of magnetic polarons, as follows. Magnetic polarons, which form continuously as the temperature decreases, are extremely efficient scatterers, thus accounting for the rise of resistivity. Application of a magnetic field destroys the polarons, restoring the resistivity to what it would have been in their absence [25].

We finally note that the Hall resistivity is anomalous in the materials under discussion, and serves as an extremely useful tool for detecting ferromagnetism and measuring its relative temperature dependence. Briefly, the Hall resistivity ρ_{Hall} in a ferromagnetic conductor can be written in the form [26]

$$\rho_{Hall} = R_0 B + R_s M \tag{2}$$

where R_0 is the ordinary Hall coefficient, B is the applied field, M is the magnetisation, and R_s is the so-called anomalous Hall coefficient [26], which is itself proportional to the resistivity ρ of the material. The term involving M generally dominates Eq. (2), so that ρ_{Hall} is directly proportional to M, and is often used as a means of measuring its behaviour. Because of this domination by $R_s M$, however, the Hall resistivity is relatively insensitive to the hole concentration, and that important parameter is therefore difficult to measure precisely. One possibility of determining the free hole concentration p is to use very high magnetic fields, such that the first term in Eq. (2) becomes the leading term.

Exchange integral: An aspect of magnetic semiconductors which makes them extremely interesting is the exchange interaction between the electronic states and the magnetisation of the medium. This interaction is responsible for the enormous Zeeman splittings encountered in II-VI-based magnetic semiconductors [1], which lead to the many transport and magnetooptical effects that continue to attract attention of the scientific community. For completeness, therefore, we note that similar dependence of Zeeman splittings of electronic states on magnetisation is expected in ferromagnetic semiconductors [27]. However, the key parameter governing these splittings, - i.e., the exchange integral - is in the case of $III_{1-x}Mn_x V$ materials still a matter of investigation, and the values reported by various authors vary rather widely. For example, from transport measurements Matsukura *et al.* report a value of 3.3 eV for the p-d exchange integral $N_0\beta$, which determines the splitting of the heavy-hole valence band [9], while Szczytko *et al.* [28] report 2.5 eV from optical data. And Okabayashi *et al.* [22] find $N_0\beta = -1.2$ eV from photoemission studies. From this diversity of results one must conclude that it is premature to attempt pinpointing the exact values of exchange constants at this time. However, it is important to note that all values reported are essentially comparable to those established for the II-VI-based magnetic semiconductors. Thus - since the Zeeman splitting involves a product of the exchange integrals and of macroscopic magnetisation M - one should expect to observe similarly spectacular magnetooptical effects in ferromagnetic semiconductors that are known to occur in $II_{1-x}Mn_x VI$ alloys, with the difference that in the latter case such effects typically require fields of several Tesla, while in ferromagnetic samples they are expected to occur in fields of the order of only tens of Gauss.

In connection with Zeeman splitting of electronic states - and especially in considering optical band-to-band transitions - one should finally note that care must be exercised in making comparisons between the well-established behaviour of the II-VI-based magnetic systems and that expected in the ferromagnetic $III_{1-x}Mn_x Vs$. Indeed, there is a good analogy between the $II_{1-x}Mn_xVI$ behaviour and that to be expected for the *insulating* $III_{1-x}Mn_xVs$. However, optical transitions in *metallic* $III_{1-x}Mn_xVs$ need to take into account the *degeneracy* of the hole gas and the resulting Moss-Burstein effect, which can have dramatic consequences in the magnetic field dependence of optical transitions. For a discussion of this aspect the reader is referred to the elegant exposition of the subject by Szczytko *et al.* [29].

Finally, we wish to point to yet another difference between phenomena based on Zeeman splitting in the now familiar $II_{1-x}Mn_xVI$ systems and in the new ferromagnetic alloys. If we ignore magnetisation fluctuations and magnetic polaron effects, then the Zeeman splitting in the (paramagnetic) $II_{1-x}Mn_xVs$ is induced by magnetic field, and there is no essential difference between such splitting on the local and on the macroscopic scale. This aspect is fundamentally different in ferromagnetic alloys, since *on the local scale* the magnetisation has its saturation value within each ferromagnetic domain even in the absence of an external field, and thus a corresponding Zeeman splitting of electronic levels will also exist on the local scale. The role of the field is only to align the domains, and to achieve a net *macroscopic* magnetisation. The consequences of this fundamental difference between the (predominantly paramagnetic) $II_{1-x}Mn_xVs$ and the ferromagnetic semiconductor alloys has not as yet been adequately explored in the literature.

3.2 InAs/GaSb HETEROSTRUCTURES CONTAINING Mn: A SPECIAL NEW OPPORTUNITY

In the preceding discussion we have singled out $Ga_{1-x}Mn_xAs$ for the sake of specificity, and also because this material holds out the promise of integration into GaAs-based electronic and opto-electronic structures. Such integration is attractive from the point of view of achieving devices based either on spin-polarised transport or on the kind of band-offset tuning [12] that has played such an important role in the research on $II_{1-x}Mn_xVI$ heterostructures. But there are other devices, not involving GaAs, that should be mentioned here because of their potentially even greater sensitivity to magnetic band-offset tuning. Consider, for example, the well-known lattice-matched type-II heterostructure GaSb/InAs. This system is unique because at the interface of the two materials the top of the valence band of GaSb lies *above* the bottom of the InAs conduction band [30]. Such alignment makes this system a semimetal, i.e., electrons from the GaSb valence band automatically spill over to the InAs conduction band, so that the InAs region near the interface is populated with free electrons, and the adjacent GaSb by free holes.

Superlattices and other quantum well systems have additional degrees of freedom, that can render the system either semimetallic or semiconducting, depending on the dimensions of the constituent layers. Consider, for example, the GaMnSb/InAs/GaMnSb single quantum well [31] shown in Fig. 3. Even though the bottom of the InAs quantum well is below the top of the valence band in the barrier, quantum confinement produced by the dimensions of the well can push the ground state of the well above the GaMnSb valence band, rendering the arrangement insulating, as shown in Fig. 3a for B = 0. Now, if the barrier is made of GaMnSb, the very large spin splitting of the valence band in this material can change this situation dramatically. For one spin polarisation (spin up in Fig. 3b) the valence band of the barrier is pushed further down, retaining the insulating character, and there is no qualitative change for that spin orientation. But for the spin-down states in the figure the valence band is pushed *above* the ground state of the well, creating holes in the interface region of the GaMnSb barrier, and electrons in the well. The holes are completely spin-polarised, while the electrons *may* have both spin orientations, but here it is expected that the spin polarisation of the well will manifest interesting spin lifetime characteristics. Thus applying and removing a (small) magnetic field can render the structure conducting and insulating, respectively, with the conduction corresponding to a definite spin polarisation, holding out prospects of "spintronic" applications.

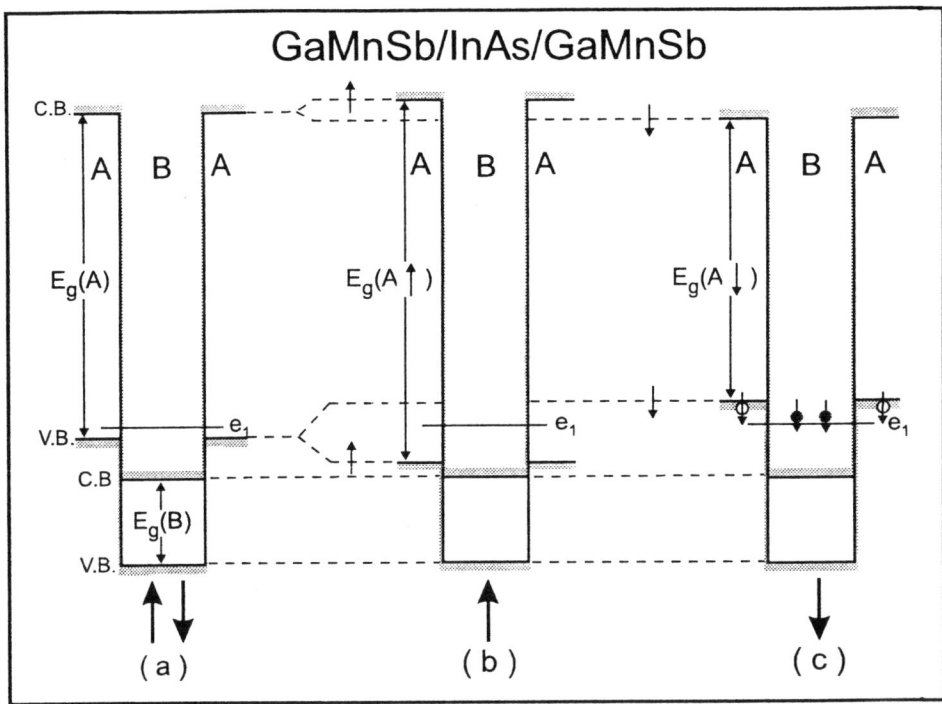

Figure 2: Band alignment for a thin InAs layer (indicated by A) sandwiched between GaMnSb (indicated by B). In the figure (a) corresponds to zero magnetic field; and (b) and (c) correspond to spin-up and spin-down states in a finite field. Note that the heterostructure becomes conducting in the spin-down situation. The energy gap for each material is indicated as E_g in the three cases.

The above structure is intended to serve only as an example of the sort of possibilities provided by InAs/GaSb heterostructures containing Mn - of which there is a wide variety - in transport, tunnelling, and in mid-infrared optical applications. For example, in Fig. 4 we show a simple InAs/GaSb superlattice, with ground-state subbands for the electrons (e_1) and for heavy holes (hh_1) indicated. The unique properties of these subbands have already been exploited in a host of optoelectronic devices, such as mid-infrared lasers, second-harmonic generators, and optical modulators [32]. If now we introduce Mn into the system, it is readily seen that application of a magnetic field will - by generating a large splitting of the band edges - move the energy separation of the subbands, introducing the feature of tunability into such devices. The fact that the energy separation between the subbands is small (it can even be zero) makes the InAs/GaSb:Mn system especially sensitive (and therefore attractive) for such magnetic tuning.

Two further comments should be made regarding such GaSb/InAs heterostructures containing manganese. First, if Mn is introduced into the InAs layers, one should note that in GaSb/InMnAs multilayers the holes introduced via Mn, and required for the ferromagnetism of InMnAs, are expected to immediately move to the GaSb layers of the heterostructure. If this occurs, the RKKY channel will be eliminated, and InMnAs will cease to be ferromagnetic. We have therefore focused on the (yet to be grown) GaMnSb as the ferromagnetic layer in the examples chosen in Fig. 3.

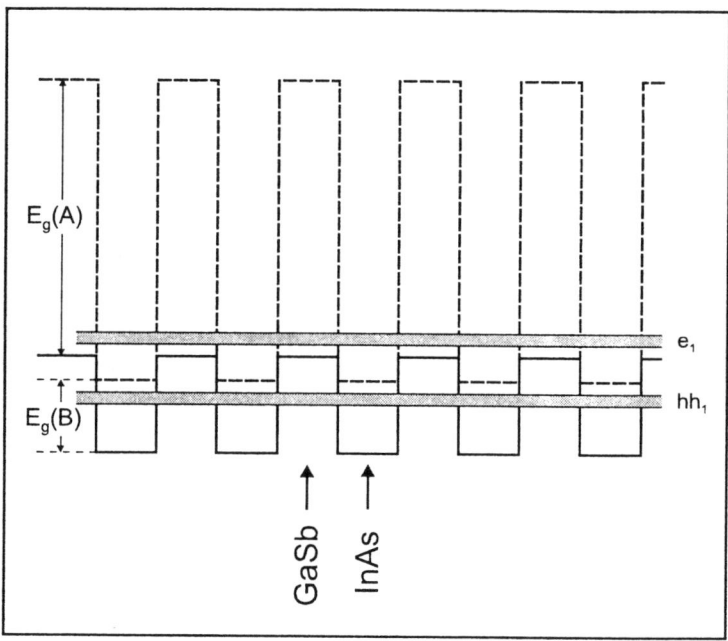

Figure 3: Band alignment for a GaSb/InAs superlattice. Solid and dashed lines represent, respectively, the valence and conduction band edges; e_1 denotes the first conduction subband of the superlattice; and hh_1 denotes the first heavy hole subband. Note that if Mn is added to GaSb and/or InAs, the separation between the hh_1 and e_1 can be tuned by an applied magnetic field, thus resulting in tunable optical properties of the system as a whole.

Second, in discussing tuning of band edges, we have ignored the degeneracy of the populated hole band. Our considerations do apply readily to relatively dilute Mn concentrations (e.g., 2 at. % Mn), where the $III_{1-x}Mn_xV$ material is "insulating" [23]. However, metallic $III_{1-x}Mn_xV$ systems are more complex, and require further discussion along the lines developed in Ref. [29], that are beyond the scope of this article.

Finally, we note that the above lattice matched system has already been investigated, in the form of InMnAs/GaSb, in yet another context by the very interesting experiments of Koshihara et al. carried out on GaSb/InMnAs junctions [33]. In the as-grown structures the InMnAs in this case proved to be paramagnetic (probably because of the drift of the holes to GaSb, as noted in the preceding paragraph). However, these workers have succeeded in rendering the InMnAs layer ferromagnetic by *illuminating* the structure with light, and thus producing a sufficient hole concentration in InMnAs. Although the detailed band structure of this system at the GaSb/InMnAs interface is still a matter of debate, the very elegant experiment of Koshihara et al. must be credited with two seminal achievements. First, it demonstrated that ferromagnetism can be switched on and off by illumination. And second, it showed the ferromagnetism induced in this way to be very long-lived (very much like persistent photoconductivity). Both these properties have potential device implications, adding to the already growing number of possibilities that ferromagnetic $III_{1-x}Mn_xV$ alloys appear to be holding in store.

4. Concluding remarks

The rich spectrum of physical phenomena which occur in magnetic semiconductors arises because of the interplay of two disciplines - semiconductor physics and magnetism - thus resulting in a host of new (and often spectacular) effects. With the advance of epitaxial fabrication - which enables the growth of multilayer systems involving magnetic semiconductors - one can incorporate this spectrum of new effects into the equally rich context of quantum wells, superlattices, and also structures of lower dimension, such as quantum dots and wires.

In this paper we have attempted to review the status of magnetic semiconductor films, and have pointed to the interesting transport and magneto-optical behaviour of these materials, that may in the future result in attractive semiconductor devices based on electron spin. We also noted the challenges that still need to be overcome to further increase the desired effects, with special attention to strategies for increasing the Curie temperature of $III_{1-x}Mn_xV$ alloys. In addition to possible device options (that still need to be identified in greater detail), the materials discussed - and their heterostructures - provide an unexpectedly rich spectrum of fundamental physical phenomena: re-entrant insulator-metal-insulator transition, anomalous Hall effect, spin-scattering-based giant magnetoresistance, the effect of degeneracy on the Zeeman splitting that has not been encountered before, and a host of others. All these features, basic and applied, hold promise of considerable scientific excitement, and possibly important device pay-off as well.

Acknowledgements The authors would like to thank T. Dietl, J. Kossut, J. Sadowski, M. Sawicki, and A. Twardowski for valuable discussions; and P. M. Reimer and J. R. Buschert for structural studies on our GaMnAs specimens. This work was supported in part by U.S. National Science Foundation Grants DMR97-05064, DMR97-01548 and DMR98-03218; and by U.S. Army Research Office Grant DAAG 55-98-1-0032.

References

1. See, e.g., *Diluted Magnetic Semiconductors*, edited by J. K. Furdyna and J. Kossut as Vol. 25 of the series *Semiconductors and Semimetals*, (Academic, Boston, 1988); T. Dietl, in Handbook on Semiconductors, edited by T. S. Moss (North Holland, Amsterdam, 1994), Vol. 3b, p. 1251, J. Kossut and W. Dobrowolski, in Handbook of Magnetic Materials, vol. 7, edited by K.H.J. Buschow (North Holland, Amsterdam, 1993), p. 231; J. Kossut and W. Dobrowolski, in Narrow-gap II-VI Compounds for Optoelectronic and Electromagnetic Applications, edited by P. Capper (Chapman and Hall, London, 1997), p. 401..

2. W. C. Chou, A. Petrou, J. Warnock, and B. T. Jonker, Phys. Rev. Lett. **67**, 3820 (1991).

3. N. Dai, H. Luo, F. Zhang, N. Samarth, M. Dobrowolska, and J. K. Furdyna, Phys. Rev. Letters **67**, 3824 (1991).

4. K. Onodera, T. Matsumoto, and M. Kimura, Electron. Lett. **30**, 1954 (1994).

5. T. Dietl, A. Haury, and Y. Merle d'Aubigné, Phys. Rev. B **55**, R3347 (1997).

6. A. Haury, A. Wasiela, A. Arnoult, J. Cibert, S. Tatarenko, T. Dietl, and Y. Merle d'Aubigné, Phys. Rev. Lett. **79**, 511 (1997).

7. H. Munekata, H. Ohno, S. von Molnar, A. Segmuller, and L.L. Chang, Phys. Rev. Lett. **63**, 1849 (1989); H. Ohno, H. Munekata, T. Penny, S. von Molnar, and L.L. Chang, Phys. Rev. Lett. **68**, 2664 (1992).

8. H. Ohno, A. Shen, F. Matsukura, A. Oiwa, A. Endo, S. Katsumoto, and Y. Iye, Appl. Phys. Lett. **69**, 363 (1996).

9. F. Matsukara, H. Ohno, A. Shen, and Y. Sugawara, Phys. Rev. B **57**, R2037 (1998).

10. See, e.g., J. A. Gaj, R. R. Galazka, and M. Nawrocki, Solid State Commun. **25**, 193 (1978); also Eunsoon Oh, A. K. Ramdas, and J. K. Furdyna, Journal of Luminescence **52**, 183 (1992).

11. T. Wojtowicz and A. Mycielski, Physica B **117 & 118**, 476 (1983).

12. See, e.g., M. Dobrowolska, H. Luo, and J. K. Furdyna, Acta Phys. Polonica A **87**, 95 (1995); also J. K. Furdyna, Solid-State Electronics **37**, 1065 (1994).

13. Y. Shapira, E. J. McNiff, Jr., N. F. Oliveira, Jr., E. D. Honig, K. Dwight, and A. Wold, Phys. Rev. B **37**, 411 (1988).

14. For very recent developments in magneto-optics of this system, see P. Kossacki, *et al.*, this workshop.

15. D. Ferrand *et al.*, Extended Abstracts of the 28th Int. School on the Physics of Semiconducting Compounds, Jaszowiec, Poland, June 1999 (communicated to the author by M. Sawicki).

16. A. Van Esch, L. Van Bockstal, J. De Boeck, G. Verbank, A. S. van Steenbergen, P. J. Wellman, B. Grieteus, R. Bogaerts, F. Herlach, and G. Borghs, Phys. Rev. B **56**, 13103 (1997).

17. H. Shimizu, T. Hayashi, T. Nishinaga, and M. Tanaka, Appl. Phys. Lett. **74**, 398 (1999).

18. J. Sadowski, private communication.

19. J. K. Furdyna and J. Kossut, Superlatt. Microstruct. **2**, 89 (1986).

20. See, e.g., M. R. Melloch, D. D. Nolte, J.C.P. Chang, D. B. Janes, and E. S. Harmon, Critical Reviews in Solid State and Materials Sciences **21**, 189 (1996).

21. L. Van Bockstal, A. Van Esch, R. Bogaerts, F. Herlach, A. van Steenbergen, J. De Boeck, and G. Borghs, Physica B **246-247**, 258 (1998).

22. J. Okabayashi, A. Kimura, O. Rader, T. Miyozawa, A. Fujimori, T. Hayashi, and M. Tanaka, Phys. Rev. B **58**, R4211 (1998).

23. A. Oiwa, S. Katsumoto, A. Endo, M. Hirasawa, Y. Iye, F. Matsukara, A. Shen, Y. Sugawara, and H. Ohno, Physica B **249-251**, 775 (1998).

24. H. Ohno, F. Matsukura, A. Shen, Y. Sugawara, A. Oiwa, A. Endo, S. Katsumoto, and Y. Iye, in Proc. 23rd Int. Conf. On the Physics of Semiconductors, edited by M. Scheffler and R. Zimmerman. (World Scientific, Singapore, 1996).

25. The authors acknowledge illuminating discussions with M. Sawicki concerning this model.

26. C. M. Hurd, in *The Hall Effect and its Applications*, edited by C. L. Chien and C. W. Westgate (Plenum, New York, 1980), p. 43.
27. See, e.g., P. Kossacki *et al.*, this conference.
28. J. Szczytko, W. Mac, A. Twardowski, P. Becla, and J. Tworzydlo, Solid State Commun. **99**, 927 (1996).
29. J. Szczytko, W. Mac, A. Twardowski, F. Matsukura, and H. Ohno, Phys. Rev. B **59**, 12935 (1999).
30. L. L. Chang and L. Esaki, Surf. Sci. **98**, 70 (1980).
31. In the figure we have used the combination GaMnSb/InAs/GaMnSb required in later discussion, but qualitatively the band alignment is expected to be the same as that for GaSb/InAs/GaSb.
32. For a recent comprehensive review see, e.g., J. R. Meyer, J. I. Malin, I. Vurgaftman, C. A. Hoffman, and L. R. Ram-Mohan in *Antimonide-Related Strained-Layer Heterostructures*, edited by M. O. Manasreh (Gordon and Breach, New York, 1997), p. 235.
33. S. Koshihara, A. Oiwa, M. Hirasawa, S. Katsumoto, Y. Iye, C. Urano, H. Takagi, and H. Munekata, Phys. Rev. Lett. **78**, 4617 (1997).

MAGNETOOPTICAL STUDIES OF MAGNETIC ORDERING IN MODULATION DOPED QUANTUM WELL OF $Cd_{1-x}Mn_xTe$

P. KOSSACKI [1-3], D. FERRAND,[3] A. ARNOULT,[3] J. CIBERT [3],
Y. MERLE D'AUBIGNÉ [3], A.WASIELA[3], S. TATARENKO [3],
J.-L. STAEHLI[1], AND T. DIETL [4]
[1] *Physics Department, Swiss Federal Institute of Technology, Lausanne*
[2] *Institute of Experimental Physics, Warsaw University, Poland*
[3] *Laboratoire de Spectrométrie Physique, CNRS and UJF-Grenoble, France*
[4] *Institute of Physics, PAS, al. Lotników 32/46, PL-02668 Warszawa, Poland*

1. Introduction

The possibility of creating ferromagnetic order within a subsystem of localised spins by either optically oriented [1] or free carriers [2] was shown for diluted magnetic semiconductors (DMS) [3] some time ago. Two recent developments have, however, renewed considerable interest in the mechanisms underlying carrier-mediated ferromagnetic couplings between the spins. First, low-temperature epitaxy of $GaAs_{1-x}Mn_x$ has been found to result in a single-phase material, in which the critical temperature T_C of a ferromagnetic phase transition routinely attains a value of 50K, the highest T_C reported to date being around 110K [4]. Further progress could, therefore, make it possible to combine the complementary properties of semiconductor and ferromagnetic material systems in useful room temperature devices. Secondly, the experimental confirmation [5,6] of the theoretical prediction [7] that the presence of free holes in structures of II-VI DMS can induce a ferromagnetic order appears to pave the way for quantitative control over the microscopic mechanisms that account for the ferromagnetic couplings in semiconductor matrices.

In this review we summarise the results of investigations aimed at determining the nature of the ferromagnetic phase as well as at tracing the evolution of T_C with the Mn content x and the hole concentration p in modulation-doped heterostructures of $Cd_{1-x}Mn_xTe/Cd_{1-y-z}Mg_yZn_zTe$:N. Due to excellent structural characteristics of MBE-grown II-VI layered structures, their properties near the ferromagnetic phase transition could be examined by sensitive and powerful magnetooptical effects. [5,6] In particular, a Moss-Burstein shift between the photoluminescence (PL) line and the onset of its excitation spectra (PLE) made it possible to determine the value of the carrier concentration. The energy distance between the emission lines collected at various complementary circular polarisations gave, in turn, direct information about the magnitude of the band spin-splitting, and thus on magnetisation inside the domains. At

M.L. Sadowski et al. (eds.), Optical Properties of Semiconductor Nanostructures, 225–235.
© *2000 Kluwer Academic Publishers. Printed in the Netherlands.*

the same time the difference between the integrated intensities of the lines shows how the degree of the domain alignment depends on the magnetic field and temperature. The successful growing of structures with different x, and thus with various T_C values, as well as the detection of the PL signal in a dilution refrigerator, made it possible to examine in detail the novel magnetic phase realised in the semiconductor quantum structure.

2. Samples and their characterisation

The study was carried out on structures grown in a molecular beam epitaxy (MBE) chamber equipped with a home-designed electron cyclotron resonance (ECR) plasma cell as a nitrogen source. The growth of samples was preceeded by finding detailed doping characteristics of the barrier material ($Cd_{1-y-z}Mg_yZn_zTe$). It has been shown [8] that by lowering the growth temperature down to $220°$-$240°C$ it becomes possible to reduce the nitrogen induced diffusion of Mg atoms and to obtain hole concentrations up to 5×10^{17} cm^{-3} in $Cd_{1-y-z}Mg_yZn_zTe$ with $z = 0.07$ and y up to 27%. The studied modulation doped structures consisted of a single 80Å quantum well (QW) of $Cd_{1-x}Mn_xTe$ embedded between $Cd_{0.66}Mg_{0.27}Zn_{0.07}Te$ barriers grown pseudomorphically on (100) $Cd_{0.88}Zn_{0.12}Te$ substrates. Such a layout ensured a large confinement energy for the holes in the quantum well, minimising at the same time the effects of lattice mismatch in the thick barriers (the Mg content mainly determines the valence band offset, while the presence of Zn ensures a good lattice match to the substrate). Due to strain and confinement in the QW, the light-hole heavy-hole splitting is of the order of 30 meV, and in the following one may consider only the heavy holes.

The nitrogen doped region in the front barrier was located at a distance $L_S = 100$ Å or 200 Å from the QW. This distance practically determines the equilibrium density of the hole gas in the quantum well. Furthermore, in order to reduce depleting effects, two additional nitrogen doped layers reside at a distance of 1000 Å from the QW on both sides. The nominal hole concentration in the doped structure, evaluated from a self-consistent solution of the Poisson and Schrödinger equations, is 2×10^{11} cm^{-2} (for $L_S = 200$ Å). The order of magnitude of the hole density was confirmed by Hall measurements on similar structures, with the top barrier modified to allow contacting the 2D hole gas [9]. The presence of a delocalised hole gas in the present samples was checked by Photoluminescence (PL) and PL Excitation (PLE).

Typical PL and PLE spectra are presented in Fig 1. At zero field both spin subbands of the heavy hole band are occupied. As a result, PL and PLE spectra are characteristic of a metallic system. In particular, one observes a step-like PLE spectrum with a blue shift with respect to the PL line. In a sufficiently high magnetic field, the spin splitting in the valence band is larger than the Fermi energy, so that the spin-down band is free of holes and an exciton can be formed between a photocreated electron and a photocreated spin-down hole [5]. A sharp exciton line is observed in the σ^- polarisation, which has actually been shown to be a charged exciton [10]. A step-like PLE spectrum is observed in the σ^+ polarisation, as expected for an optical transition involving the spin-up hole subband which is partially occupied. In such spectra, characteristic of a metallic system, the Moss-Burstein shift E_{MB} between the energy of

PLE step and the PL maximum is generally used to evaluate the hole gas concentration.

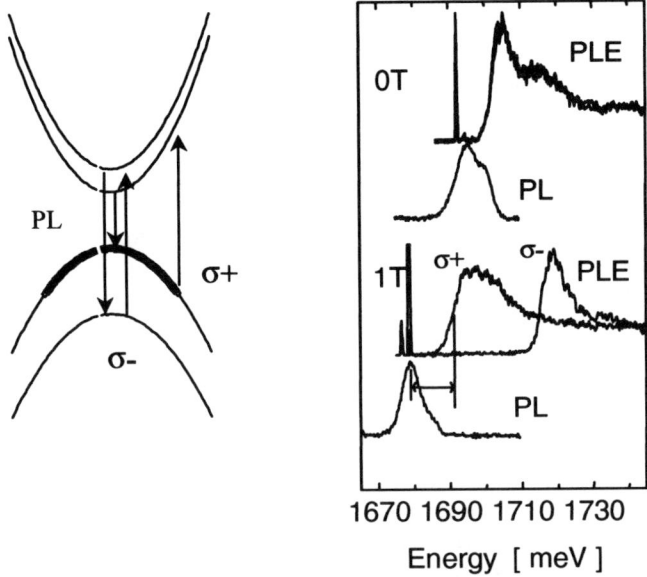

Figure 1. Determination of the hole gas concentration in the metallic region. (a) schematic plot of optical transitions (b) example spectra of PL (dashed line) and PLE (dotted line in the σ⁻ polarisation and solid line in σ⁻⁺) used to measure Moss-Burstein shift,

In the case of full spin polarisation one gets $E_{MB} = 11$ meV for a spacer of $L_S = 20$ nm, which gives (for in-plane effective masses $m_e = 0.1m_o$ and $m_{hh} = 0.25m_o$ [11]) a hole concentration equal to 1.6×10^{11} cm^{-2}. This value is quite close to the one expected from the design of the structure. However, such a determination of the hole density relies on the actual value of the electron mass (and, to a lesser extent since it is larger, of the hole mass), and this determination can be altered by mass renormalisation and excitonic effects.

In order to reduce the carrier concentration in the quantum well, we illuminate the barriers with additional light of controlled power and photon energy higher than the energy gap of the barrier material (a halogen lamp with a blue filter, or Ar-ion laser). The mechanism of this effect was discussed by Shields et al. [12,13] and is based on the diffusion of photo-created electrons from the barrier into the quantum well where they rapidly recombine with the hole. Under continuous illumination the system approaches a stationary state, with a reduced hole gas concentration.

The efficiency of this process was determined experimentally from the value of the Moss-Burstein shift. The accuracy of the hole gas concentration determination was enhanced by applying a small magnetic field (up to 1 T), and measuring only the σ⁺ polarisation which involves the creation of a spin-up hole in the populated subband. Thanks to the giant Zeeman effect, this field remains low enough to make negligible any effects related to the quantisation of the hole gas into Landau levels.

The variation of the hole gas concentration under moderate blue light illumination was employed to study the influence of the carrier density on the ordered magnetic phase. Such an approach allows distinguishing the true concentration phenomena from any small variation of sample parameters.

Exciting above the barrier with high intensity light, or using unfiltered white light in reflectance measurements, efficiently neutralises the 2D hole gas and thus gives a reference in the form of the «emptied» QW. It also allows a precise determination of the quantum well parameters, in particular the concentration of Mn ions in the well material. PL and PLE spectra of these «emptied» QWs are dominated by sharp exciton-like lines and are similar to spectra of undoped samples grown under the same conditions. Under an applied magnetic field, the reflectivity spectra exhibit σ^+ and σ^- heavy-hole exciton lines, symmetrically split by the giant Zeeman effect. The observed linewidths (2meV for a QW with 2.5%, 4meV for 4.2 %) are comparable to those observed for undoped QWs. The positions of the two lines can be fitted by a modified Brillouin function [14]:

$$E_{\pm} = E_0 \mp (N_0\alpha - N_0\beta)xS_0B_{5/2}\left(\frac{5g\mu_B B}{2k_B(T+T_0)}\right)$$

for the σ^+ and σ^- polarisation, respectively, where $N_0\alpha = 0.22$ eV and $N_0\beta = -0.88$ eV represent the spin-carrier exchange in the conduction and valence bands, respectively, and the two adjustable parameters S_O and T_O represent the effect of Mn-Mn antiferromagnetic interaction. The values of these parameters as a function of the Mn concentration in bulk CdMnTe were determined by Gaj et al. [14, 15]. They can also be used for an emptied QW. Special precautions have to be taken due to the possibility of creating charged excitons [10]. If the hole gas is not completely suppressed the σ^- polarisation is dominated by the charged exciton (X+) transition, while the absorption line in σ^+ has a neutral exciton origin (X). Both excitonic species have similar shape, but the X+ energy is diminished by its dissociation energy E_{X+} (about 2.7meV [15, 16]). Therefore, the Zeeman splitting has to be corrected by this value. A typical fit to experimental results is presented in figure 2.

All magnetooptical measurements were performed in the Faraday configuration with the magnetic field applied perpendicular to the sample surface. In the experiment, two types of cryogenic systems were employed for the different temperature ranges required; a pumped liquid ^{4}He cryostat [5] and a dilution refrigerator. A fibre-optic system was used to guide the light to and out of the sample, which was installed on a cold finger in the dilution refrigerator. The temperature was monitored by a calibrated resistor located in close proximity to the sample. A strong temperature dependence of the PL splitting was used to evaluate the actual temperature of the spin-subsystem at a given excitation power. Such an analysis indicated that measurements down to 0.6 K were possible with a reasonable magnitude of the signal-to-noise ratio. For measurements in the temperature range above 1.3 K the samples were mounted strain-free in a superconducting magnet and immersed in liquid helium. PL without additional above-gap illumination and PLE were measured using an Al_2O_3:Ti laser providing less than 2 mW/cm^2. PL under illumination was measured without any further excitation other than the source of light used for hole gas neutralisation.

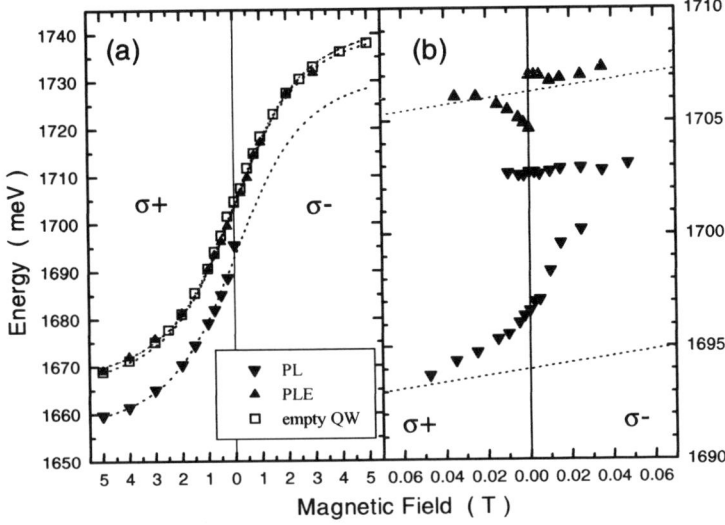

Figure 2. Transition energies at 1.85K (sample M952) (a) at high field (right part for σ⁻ polarisation, left part σ⁺); down triangles = PL; up triangles = PLE; squares = reflectivity with white light; the half- height energy is plotted for the σ⁺ PLE, the line maximum otherwise. Lines are fitted using modified Brillouin functions, with B_O=1500 Oe. (b) at low field; the lines are the same as in (a)

3. High magnetic field

At zero field in the absence of ferromagnetic order both spin subbands of the heavy hole band are occupied. As a result, the PL and PLE spectra are characteristic of a metallic system. Applying a field induces a transfer of holes from the lower spin subband of the valence band to the upper subband. We define a characteristic field for which the splitting in the valence band is equal to the Fermi energy for the holes residing in the one spin subband. The transfer of holes is only partial below this field («low field range»), and total well above it («high field range»). The measurements in the high field range give an opportunity of determining the effective exchange field due to the exchange interaction of the fully polarised hole gas acting on the Mn ion system [17].

In the high field range the PL is fully σ⁺ polarised, i.e. the photoelectrons rapidly relax to the −1/2 electron subband which is at lower energy due to the giant Zeeman splitting. The σ⁺ PLE spectrum, which involves the populated hole band, shows a blue-shifted step. The σ⁻ PLE spectrum on the contrary, which involves the depopulated hole band, exhibits a well-defined excitonic peak. In Fig. 2 we plot the position of the maxima of the σ⁺ PL line and of the σ⁻ PLE line, and the energy of the σ⁺ PLE step (measured at half-height). These data are fitted (dotted lines) using the same modified Brillouin function as for the empty QW, with the same values of the parameters S_0 and T_0, and as adjustable parameters, the exchange field B_0 added to the applied field, and energy shifts $\delta E_\pm$ due to many-body effects:

$$E_\pm = E_0 + \delta E_\pm \mp (N_0\alpha - N_0\beta)xS_0 B_{5/2}\left(\frac{5g\mu_B(B+B_0)}{2k_B(T+T_0)}\right)$$

The values of $\delta E_\pm$ for σ^+ and σ^- polarised PLE and of δE_+ for σ^+ PL, are obtained mainly from data above 1 tesla, where the effect of B_0 is negligible. The value of δE_- for σ^- PL, and the value of the exchange field, B_0, are obtained from data at intermediate fields (0.003 T to 0.2 T). The Zeeman shifts of both the σ^+ and σ^- PL and PLE features are then linear in field with the slope measured for an empty QW (for example 37 meV/T in a QW with 4.2% of Mn at 1.9 K). There, one determines $B_0 = 1500$ Oe ±500. The same value was obtained for samples with different Mn content in the QW. This is in reasonable agreement with the value $B_0 = 1000$ Oe, calculated as in [5] using $g\mu_B B_0 \approx \beta p/(2L_W)$, where $L_W \approx 6$ nm is an effective width of the hole envelope function, and the nominal hole gas concentration $p = 2 \times 10^{11}$ cm^{-2} is the hole concentration estimated for the samples with a spacer of 20 nm. The difference can be explained by uncertainty of the exact value of the effective mass or by introducing many-body effects.

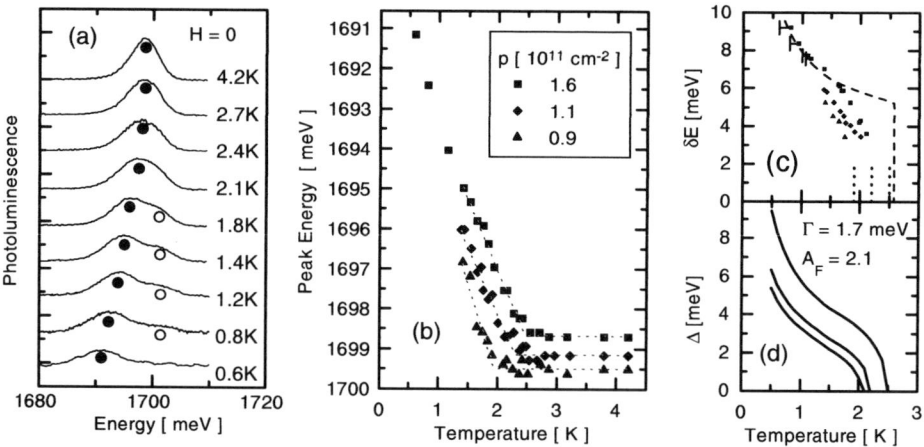

Figure 3. Photoluminescence spectra for modulation-doped p-Cd$_{0.957}$Mn$_{0.043}$Te at selected temperatures, (a). Peak positions of the low- and high-energy lines are marked by full and empty points, respectively. Temperature dependence of the low-energy peak positions and the line splitting are shown for selected values of the hole concentrations in (b) and (c), respectively. Dashed and solid lines in (c) and (d) are calculated neglecting and taking disorder into account, respectively. Vertical lines in (c) show critical temperatures T_c corresponding to slope changes of the points in (b).

4. Phase transition

The most straightforward way of observing the magnetic transition is an experiment at zero field or in a low external magnetic field. The typical findings in the absence of an external magnetic field are summarised for one of the structures in figure 3 [25]. The PL line, which corresponds to the energy gap in the QW region, is seen to show a splitting and a shift of its spectral position below a characteristic temperature T_C. This observation is followed by a critical divergence of the field-induced PL splitting at $T \rightarrow T_C^+$, [5] which follows the critical law $\chi(T) = C/(T - T_C)$. No such effects are detected in undoped structures or under the presence of white-light illumination, which depletes the quantum well of carriers. The critical behaviour found here is in contrast to the gradual changes of $\chi(T)$ associated with the formation of magnetic polarons [18,19], for which the finite volume involved precludes the existence of any second order phase transition [20,18]. These findings constitute the experimental evidence for the presence of a ferromagnetic transition in the modulation-doped p-$Cd_{1-x}Mn_xTe$. As shown in Fig.4, the dependence $T_C(x)$ is well described by the mean-field model [7] with the enhancement factor resulting from the carrier-carrier correlations, [5,7] set at $A_F = 2.1$.

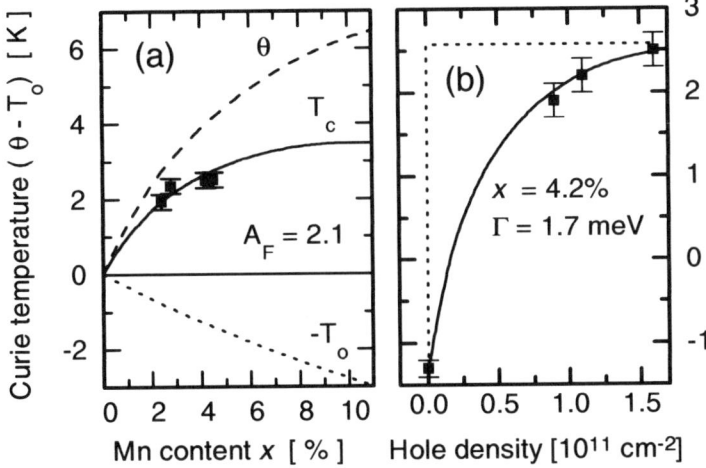

Figure 4. Critical temperature $T_c = \Theta - T_o$ of the ferromagnetic transition as a function of Mn (a) and hole (b) concentrations, respectively. Solid lines are theoretical, taking into account the enhancement factor due to the Coulomb interactions between the holes (a) and also disorder (b). The values of temperature Θ and T_o corresponding to the ferromagnetic and super-exchange antiferromagnetic interactions are shown in (a) by the dashed and dotted line, respectively. The dashed line in (b) is theoretical, neglecting disorder ($\Gamma = 0$).

The applicability of the mean-field approximation (MFA) stems from the long range of the carrier-mediated spin-spin coupling in semiconductors. The corresponding minor importance of the magnetisation fluctuations also explains why the data in Fig.3(b) show no evidence for the red-shift of the gap at $T \rightarrow T_C^+$, in contrast to the case of the ferromagnetic Eu chalcogenides [21], in which the coupling between the Eu spins is driven by a short-range superexchange interaction. The parameter A_F also controls the magnitude of magnetoresistance induced by the spin-splitting [22], $A_F \equiv 1 + g_3 \equiv 1 + F_\sigma/2$. Its value determined here falls in the range of those deduced from the magnetoresistance studies of n-type DMS of various dimensionality [23].

Figures 4(b) and 4(c) show how the energy of the PL peak and its splitting δE vary with the temperature and hole concentration, changed here by white-light illumination. According to the model [7], whose predictions are shown by a dashed line in Fig.3(c), the band spin-splitting Δ, and thus δE, is expected to attain rather abruptly the value corresponding to the full polarisation of the hole liquid. On lowering the temperature in the range $T < T_C$, δE should increase even further with a rate determined by the growth of Mn magnetisation in a temperature-independent molecular field produced by the entirely spin-polarised holes. We see that the experimental values of δE reach the theoretically calculated splitting at the lowest temperatures, a result strongly supporting a general consistency of our interpretation. However, δE decreases with the rising temperature faster than expected theoretically. Furthermore, T_C is predicted to be independent of p in 2D systems [7], a supposition at variance with the results of Fig.3(b). We assign these discrepancies to the effect of static disorder upon the energy dependence of the density-of-states (DOS). To check this conjecture, we insert into the self-consistent procedure [7] the hole polarisation, $p_\downarrow - p_\uparrow$, calculated from the expression,

$$p_{\uparrow\downarrow}(\Delta) = \frac{A_F m_{hh}}{4\pi\hbar^2} \int_{-\infty}^{+\infty} d\varepsilon \, \frac{1 + \mathrm{erf}\left[(\varepsilon \pm \Delta/2)/\gamma\right]}{1 + \exp\left[(\varepsilon - \varepsilon_F)/k_B T\right]}$$

in which the step-like DOS of clean 2D systems with the dispersion $\varepsilon_k = \hbar^2 k^2 / 2m_{hh}$ is modified by a disorder broadening. Its magnitude is characterised by the parameter $\Gamma = \sqrt{2}\gamma$, the FWHM of the Gaussian distribution of the energy states ε_k, which contribute to DOS at energy ε. As shown in Figs.3(d) and 4(b), such a model with $\Gamma = 1.7$ meV qualitatively explains the essential features of our data.

5. Ferromagnetic domains

Turning to the properties of the ferromagnetic phase we note that the easy axis is expected to be oriented along the growth direction as a result of the interaction between the hole spin and the electric field associated with the strain and confinement potential. [7,24]. The measurements at the circular polarisations in the Faraday geometry therefore give information about the relative concentration of domains with the two relevant orientations, as shown schematically in Fig.5(a). The difference in the integrated intensities of the two components contributing to the PL line at a given polarisation, normalised to their values at the initial field $H=0$, is depicted as a function of the magnetic field at 1.53K in Fig.6(c).

A number of conclusions emerge from this plot. First, we see that the domains become aligned in the magnetic field of the order of 100 0e. Second, since we know the magnitude of the spin-splitting [Fig.6(d)], and thus the corresponding magnetisation, we are in a position to evaluate the contribution of the Mn spins to the macroscopic magnetic induction from the system of aligned domains, $B_S = 4\pi M_S = 14 \pm 1$ Gs at 1.5K. Such a small value of B_S makes the dipole interaction weak, and substantiates *a*

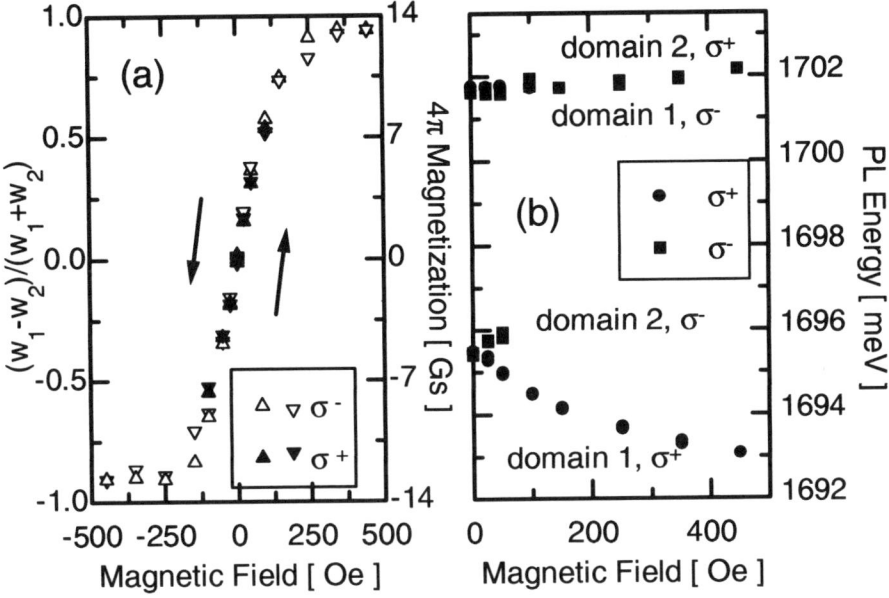

Figure 5 Schematic (a) and experimental results on domain structure studied by polarisation-resolved luminescence spectroscopy of modulation-doped p-$Cd_{0.957}Mn_{0.043}$Te at 1.53 K.

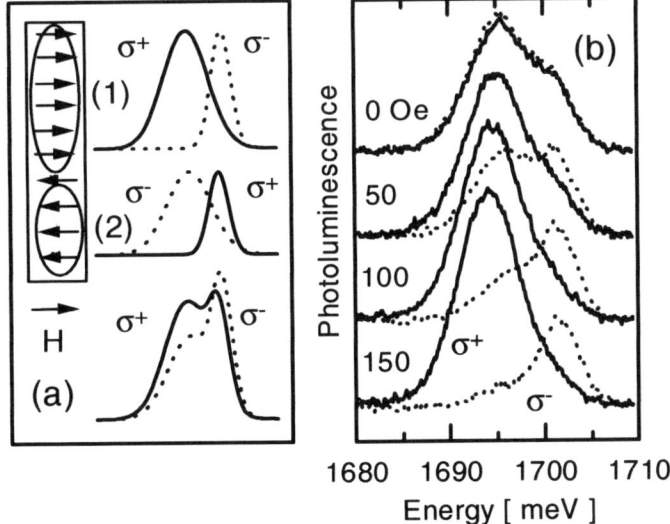

Figure 6. Experimental results on domain structure studied by polarisation-resolved luminescence spectroscopy of modulation-doped p-$Cd_{0.957}Mn_{0.043}$Te at 1.53 K. Peak positions are shown in (a), while (b) presents the difference $(w_1 - w_2)/(w_1+w_2)$, where $w_i = I_i(H)/I_i(0)$, in the normalised integrated intensities I of the two contributions to the line at each polarisation for the field swept up and down.

posteriori our assumption that the spontaneous magnetisation is oriented perpendicularly to a surface. Finally, the data point to a small coercive force, $H_C < 4$ 0e. The virtual absence of hysteresis in the studied ferromagnet is further supported by the observation that the spectra for the two polarisations, that is the concentrations of the two kinds of domains, are identical at $H=0$, even after cooling the sample from above T_C down to 1.5K in the magnetic field of 500 Oe or after 15 min of illumination by circularly polarised light.

In order to explain these unusual domain properties, a previous mean-field theory of ferromagnetic instability in semiconductor structures [7] (see also [26]) has been extended [6] to take into account the possibility of spin density formation. The characteristic temperature $T_{SDW}(q)$ of the spin wave instability corresponding to the wave vector q was found to be given by $T_{SDW}(q) = \Theta F(q)$, where $F(q)$ is the static Lindhard function of the carrier magnetic response function. According to the numerical evaluation, $F(q) = 1$ for $q =< 2k_F$, even if the complex structure of the relevant hole subband is taken into account. This means that the ground state may involve SDW with various q values.

6. Conclusions

The findings presented above demonstrate that indeed p-type doping constitutes a method for substantial enhancement of magnetic, and then magnetooptical effects in DMS, leading to a phase transition to an ordered magnetic system. The determined values of the critical temperature Tc as a function of manganese concentration x confirm that the exchange interaction among the free carriers shifts the magnetic instability toward higher temperatures. At the same time, it is shown how a disorder-induced decrease in the density of states diminishes the magnitude of the spontaneous magnetisation when increasing temperature or decreasing the carrier concentration. Moreover, our results demonstrate that the ordered phase is characterised by a low coercive field, Hc $<$ 4 Oe and a large domain alignment field, Hs $\approx$ 200 Oe. The latter is found to be much greater than the saturation value of the magnetic induction, Bs = $4\pi Ms \approx 15$ Gs at 1.5 K, where Ms is magnetisation of the Mn spins within magnetic domains at H = 0. We suggest that these unusual domain properties result from the proximity to the spin-density-wave instability in this low-dimensional magnetic system.

Acknowledgements. This work has been supported in part by the French – Polish collaboration program Polonium.

References

[1] See H. Krenn *et al.*, Phys. Rev. B **39**, 10918 (1989), and references cited therein.
[2] See P. Lazarczyk, T. Story, M. Arciszewska, and R.R. Gałązka, J. Magn. Magn. Materials **169**, 151 (1997), and reference cited therein.
[3] For a review on DMS see e.g. T. Dietl, in *Handbook on Semiconductors*, edited by T.S. Moss (North-Holland, Amsterdam, 1994) vol.3b, p. 1251; J.K. Furdyna, J. Appl. Phys. 64, R29 (1988).

[4] See H. Ohno, Science **281**, 951 (1998), and references therein.

[5] A. Haury *et al.*, Phys. Rev. Lett. **79**, 511 (1997).

[6] P. Kossacki, D. Ferrand, A. Arnoult, J. Cibert, S. Tatarenko, A. Wasiela, Y. Merle d'Aubigné, J-L. Staehli, J-D. Ganière, W. Bardyszewski, K. Świątek, M. Sawicki, J. Wróbel, and T. Dietl, Proc. of 13th Int. Conf. on the Electronic Properties of 2D Systems, Physica E, submitted.

[7] T. Dietl, A. Haury, and Y. Merle d'Aubigné, Phys. Rev. B **55**, R3347 (1997).

[8] A. Arnoult, D. Ferrand, V. Huard, J. Cibert, C. Grattepain, K. Saminadayar, C. Bourgognon, A. Wasiela, S. Tatarenko., 10th International conference on molecular beam epitaxy, Cannes, France (1998), J. Cryst Growth (in print).

[9] A. Arnoult *et al.*, J. Crystal Growth **184/185**, 445 (1998)

[10] P. Kossacki, J. Cibert, D.Ferrand, Y. Merle d'Aubigné, A. Arnoult, A.Wasiela, S. Tatarenko, and J.A. Gaj, unpublished.

[11] G. Fishman, Phys. Rev. B **52**, 11 132 (1995).

[12] A.J.Shields, J.L.Osborne, M.Y.Simmons, M.Pepper, and D.A.Ritchie, Phys. Rev. B **52**, R5523 (1995).

[13] A.J.Shields, J.L.Osborne, M.Y.Simmons, D.A.Ritchie, and M.Pepper, Semicond.Sci.Technol **11**, 890 (1996).

[14] J. A. Gaj, R. Planel, and G. Fishman, Sol. State Comm. **29**, 435 (1979).

[15] J.A.Gaj, W.Grieshaber, C.Bodin-Deshayes, J.Cibert, G.Feuillet, Y.Merle d'Aubigné, and A.Wasiela, Phys. Rev. B **50**, 5512 (1994).

[16] A.Haury, A.Arnoult, V.A.Chitta, J.Cibert, Y.Merle d'Aubigné, S.Tatarenko, and A.Wasiela, Superlattices and Microstruct. **23**, 1097 (1998).

[17] J.Cibert, P.Kossacki, A.Haury, A.Wasiela, Y.Merle d'Aubigné, T.Dietl, A.Arnoult, S.Tatarenko, J. Crystal Growth **184/185,** 898 (1998).

[18] T. Dietl and J. Spałek, Phys. Rev. Lett. **48**, 355 (1982); Phys. Rev. B **28**, 1548 (1983).

[19] T. Wojtowicz, S. Kolesnik, I. Miotkowski, and J.K. Furdyna, Phys. Rev. Lett. **70**, 2317 (1993).

[20] C. Benoit a la Guillaume, Phys. Stat. Solidi (b) **175**,369 (1993).

[21] P. Wachter, CRC Crit. Rev. Solid State Sci. **3**,189 (1972).

[22] B.L. Altshuler and A.G. Aronov, in *Electron-Electron Interaction in Disordered Systems*, edited by A.L. Efros and M. Pollak (North-Holland, Amsterdam, 1985), p.1; see also, H. Fukuyama, *ibid.*, p.155; P.A. Lee and T.V. Ramakrishman, Rev. Modern Phys. **57**, 287 (1985).

[23] M. Sawicki, T. Dietl, J. Kossut, J. Igalson, T. Wojtowicz, and W. Plesiewicz, Phys. Rev. Lett. **56**, 508 (1986); see also Y. Shapira, N.F. Oliveira, Jr., P. Becla, and T.Q. Vu, Phys. Rev. B **41**, 5931 (1990); J. Jaroszyński, J. Wróbel, M. Sawicki, E. Kamińska, T. Skośkiewicz, G. Karczewski, T. Wojtowicz, A. Piotrowska, J. Kossut, and T. Dietl, Phys. Rev. Lett. **75**, 3170 (1995).

[24] P. Peyla, A. Wasiela, Y. Merle d'Aubigné, D.E. Ashenford, and B. Lunn, Phys. Rev. B **47**, 3783 (1993); B. Kuhn-Heinrich, W. Ossau, E. Bangert, A. Waag, and G. Lanwehr, Solid State Commun. **91**, 413 (1994).

[25] P. Kossacki, J. Cibert, A. Arnoult, S. Tatarenko, D. Ferrand, A. Wasiela, Y. Merle d'Aubigné, K. Świątek, M. Sawicki, J. Wróbel, and T. Dietl *unpublished*

[26] T. Jungwirth et al., Phys. Rev. B **59**, 9818 (1999).

MAGNETOOPTICAL PROPERTIES OF GRADED QUANTUM WELL STRUCTURES MADE OF DILUTED MAGNETIC SEMICONDUCTORS

M. KUTROWSKI[1], T. WOJTOWICZ[1], S. KRET[1], G. KARCZEWSKI[1], J. KOSSUT[1], R. FIEDERLING[2], B. KÖNIG[2], D.R. YAKOVLEV[2], W. OSSAU[2], A. WAAG[2], V.P. KOCHERESHKO[3], F.J. TERAN[4], M. POTEMSKI[4]

[1]Institute of Physics, Polish Academy of Sciences,
Al. Lotników 32/46, 02-668 Warsaw
[2]Physikalisches Institut der Universität Würzburg, Würzburg, Germany
[3]A.F.Ioffe Physiko-Technical Institute, St. Petersburg, Russia
[4]Grenoble High Magnetic Field Laboratory, MPI/FKF and CNRS,
B.P. 166 Grenoble, Cedex 9 France

Abstract. We report on magnetooptical studies of $Cd_{1-x}Mn_xTe$ and $Cd_{1-y}Mg_yTe$ spatially graded quantum well (QW) structures. The photoluminescence excitation and transmission spectra were measured in a magnetic field for trapezoidal, parabolic and half-parabolic quantum wells grown by MBE using the so called "digital" or pulsed flux technique. The composition gradient along the growth direction was checked by cross sectional transmission electron microscopy. The peaks appearing in the spectra were assigned to interband excitonic transitions between different confined levels of electrons and holes. Apart from diagonal transitions (i.e., those preserving the level quantum number) we also observed nondiagonal transitions, usually not dipole-allowed in standard rectangular QWs. This enabled us to determine with high precision the valence band offsets in CdTe/MnTe and CdTe/MgTe systems. The observation of 1s and 2s excitonic transitions demonstrated that the exciton binding energy is particularly large in QWs with a parabolic confining potential.

1. Introduction

Most of the investigations performed on sub–3D systems were performed, up to now, on rectangular quantum wells (RQW) or superlattices (SL). However, modern growth techniques, and especially molecular beam epitaxy (MBE), give the possibility of producing quantum wells with shapes of the confining potential other than a rectangular one. The properties of such graded potential quantum wells are very interesting for several reasons. The pattern of quantised energy levels in such structures depends on the particular shape of the confining potential produced by the band gap grading.

237

M.L. Sadowski et al. (eds.), Optical Properties of Semiconductor Nanostructures, 237–246.
© 2000 *Kluwer Academic Publishers. Printed in the Netherlands.*

Additionally, the distances between energy levels are strongly sensitive to the actual value of the band gap discontinuity between the materials used to produce the graded-gap region. In the middle of the eighties Pötz and Ferry [1] noticed that the sensitivity of optical interband transitions in graded quantum structures to the valence band offset Q_v is much stronger than that in rectangular quantum wells. They showed, in particular, that when the confining potential shape in the growth direction was described by the expression $V(z) \propto |z|^s$ then the smaller the s-value, the greater the sensitivity of quantum confined levels to the band offset. Moreover, they found that the sensitivity of optical spectra with respect to uncertainties in the total width L_z of the potential well is weaker in the case of such potential shapes as parabolic or triangular wells. Therefore, such potentials are very useful in studies aimed at the determination of band offsets.

Among QWs having different possible shapes of the confining potential, the case of parabolic quantum wells (PQW) is particularly interesting due to their unique physical properties and with regard to their potential applications. The parabolic QW is an example of a harmonic oscillator, for which the energy levels can be expressed by a simple analytical formula. The accuracy of this formula is sufficiently high even for finite parabolic QWs if the curvature of the potential is not too large and if it is applied to the first few energy levels in the quantum well [2]. Parabolic shaped quantum wells attracted special attention also because they may lead to a strong enhancement of the exciton binding energy [3], which is the consequence of a new length scale introduced to the excitonic problem by such a potential. Worth noticing is also the behaviour of PQWs in an external electric field. As shown by Chuang et al. [4] all energy levels of parabolic quantum wells shift by an equal amount with increasing electric field (displaced harmonic oscillator). The magnitude of this shift is larger than that of the ground state of a rectangular QW, which is nearly the only level affected by the electric field in the latter case. As a result, the intersubband absorption spectrum is insensitive to the applied electric field. For interband optical transitions, however, the effect of the electric field on absorption spectra should be very strong. A very peculiar property of the parabolic QWs is the possibility of achieving nearly uniform electron spatial density in modulation-doped structures, since the confining potential can be tailored to cancel the Hartree potential of the confined electrons [5].

These and other unique properties of parabolic QWs have been extensively investigated in structures made of III-V compounds since the mid-eighties. In contrast, there were no similar studies made for II-VI compounds although these materials are quite interesting in this context because of their strong excitonic effects. Additionally, magnetic ions such as Mn, which can be easily incorporated into II-VI hosts, make them very sensitive to the presence of an external magnetic field. These so called diluted magnetic semiconductors (DMS) are known for their unusually strong magnetooptic effects such as giant Faraday rotation or giant Zeeman spin splitting, stemming from a strong sp-d coupling of the spin of the conduction band electrons and valence band holes to localised spins of Mn ions. This feature was used for the quantitative characterisation of interface quality in rectangular QW's [6]. Similarly, it is possible to extract precise information concerning the shape of the confining potential from the observed spin splitting of the electrons and holes in the graded QW structures.

2. Experimental results and discussion

The digital method was employed to produce graded potential QW structures in a standard MBE apparatus. The method was originally applied by Gossard *et al.* to grow PQWs of III-V materials [7]. The origin of the digital method is directly connected with the concept of a digital alloy. The digital alloys can be formed by short flux pulses of one component of a ternary compound (e.g., Mn in the case of $Cd_{1-x}Mn_xTe$). In order to obtain the composition grading, variable duration flux pulses are employed. The digital method is particularly appropriate in the case of $Cd_{1-x}Mn_xTe$, because it involves no changes of the Mn-effusion cell temperature during growth. The requirement of continuous change of the Mn molar fraction x and, thus, of the Mn flux, induce in the analogue method a necessity of accurate calibration and precise control of the temperature of the Mn source cell. Particularly in the case of small x very low fluxes should be employed due to a high value of the sticking coefficient of Mn (compared to those of Cd and Te). This is an extremely difficult task and it excludes the analogue method from being useful for the growth of $Cd_{1-x}Mn_xTe$ graded QWs.

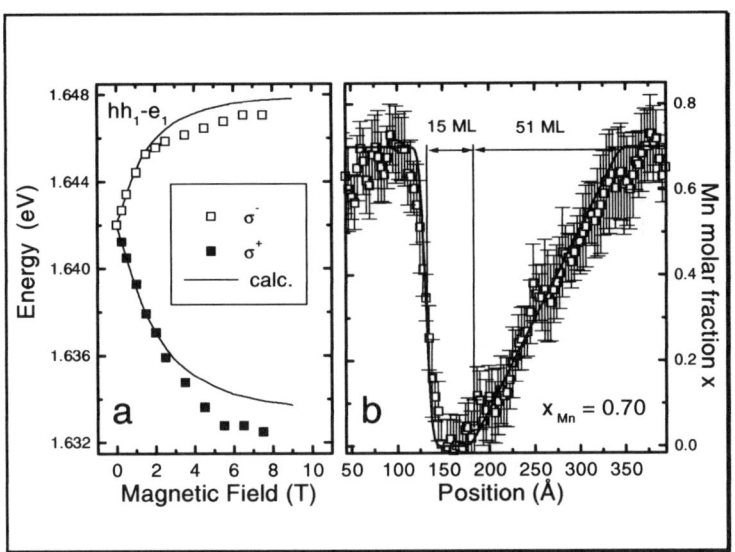

Figure 1. (a) The PLE peak positions for a trapezoidal $CdTe/Cd_{1-x}Mn_xTe$ QW measured in the Faraday configuration for two circular polarisations. The solid lines represent the calculated energy shift for this transition. (b) The comparison of a composition profile derived from HRTEM image analysis (open points with error bars) and the profile for which the calculations of hh_1-e_1 transition were performed.

A high precision of control of the composition, and hence the potential shape profile, achieved in our digitally grown structures was proven by two independent methods. The first proof comes from magnetooptical studies of the spectra of specially designed $CdTe/Cd_{1-x}Mn_xTe$ trapezoidal QW structures. For this purpose we have measured PLE spectra in the Faraday configuration in magnetic fields up to 7.5 T. The well-resolved peaks of excitonic *1s* transitions between the first heavy hole state in the quantum well and the first confined level of electrons (hh_1-e_1), as well as those involving the second heavy holes and second electron state (hh_2-e_2) are observed. We have performed a fitting of the peak positions (Fig.1 a) for the above mentioned transitions, and their Zeeman splitting, while varying the quantum well shape (starting from that assumed in the technological process). The best fitting profile is shown in Fig.1b as a solid line. It was obtained with an interdiffusion length of 3.5 Å, which is a typical value for the QW structures grown at relatively high substrate temperatures. Apart from the excitonic *1s* optical transitions, transitions involving the *2s* and *3s* excited states were also observed. This allowed us to estimate, using the fractional-dimensional approach [8,9], that the hh_1-e_1 exciton binding energy is of the order of 27meV.

The second confirmation is provided by the high-resolution transmission electron microscopy (HRTEM) measurement of the same structure. A numerical analysis of the HRTEM image allows obtaining the relative change of the inter-atomic distance in the growth direction which, in turn, gives information about the relative composition profile. The results are shown in Fig. 1b as open points with error bars. As can be seen the agreement between both kinds of experiment is quite satisfactory.

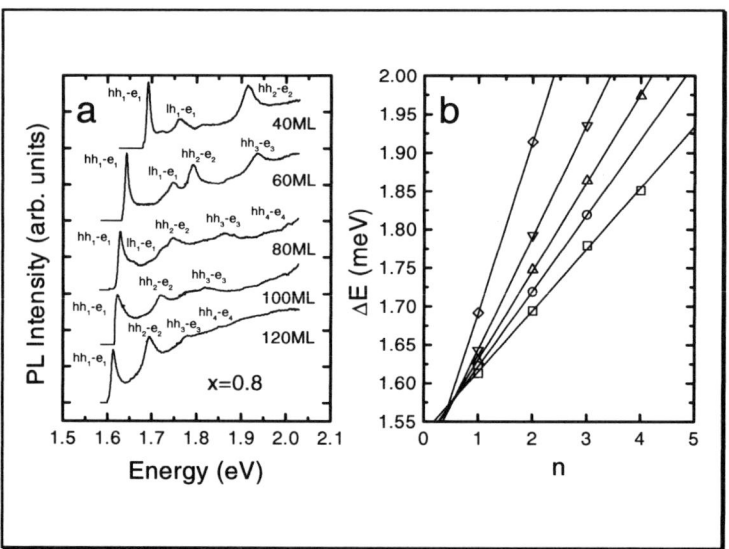

Figure 2. (a) The PLE spectra of $Cd_{1-x}Mn_xTe$ PQWs with different quantum well width and similar Mn content x=0.8. (b) The energy position in heavy hole series for the optical transitions for which spectra are shown on the panel (a). The straight lines represent fits to the experimental points.

The precision of the composition profiling was also confirmed by optical and magnetooptical investigations of parabolic quantum wells. A large set of parabolic and half-parabolic QW structures was produced from $Cd_{1-x}Mn_xTe$ and $Cd_{1-y}Mg_yTe$. A wide range of the curvatures of the confining potential at its centre was obtained either by changing the width (from 40 monolayers (ML) to 120 ML) of the QWs or by varying the composition of the outermost barrier material (from x=0.14 to $x\approx1$ - in the case of Mn, and from y=0.5 to y=0.7 – in the case of Mg). The structures were studied using an optical transmission method (after removal of the nontransparent GaAs substrate) in multiple QW structures and by photoluminescence excitation spectroscopy (PLE).

In Fig2a we collected several PLE spectra of $Cd_{1-x}Mn_xTe$ parabolic quantum wells with similar values of the Mn molar fraction close to $x \approx 0.8$ and different well widths L_z ranging from 40ML to 120ML. Each spectrum reveals a series of peaks. On the basis of their behaviour in an external magnetic field we can assign each of them to either a heavy hole hh_n-e_n or a light hole lh_n-e_n transition (where n is the level index). These are the so-called diagonal transitions which preserve the level index n ($\Delta n=0$). For parabolic quantum wells optical transitions are allowed only for $\Delta n = 0, \pm 2, \pm 4...$ because of the symmetry of the potential. Non-diagonal transitions (with $\Delta n \neq 0$) are allowed in parabolic QWs due to the different extension of the motion of valence band holes and conduction band electrons (their corresponding envelope functions are no longer orthogonal). However, the overlap of the envelope functions with $\Delta n=\pm 2, \pm 4...$ and in turn the oscillator strength is at least one order of magnitude smaller.

In order to describe the spectrum of observed transitions we use the infinite barrier approximation. It was checked by direct comparison with calculations suitable for finite PQWs that such a simple approach is very accurate, particularly in the case of our relatively deep and wide PQWs. The differences between the energy levels calculated for various levels of electrons and heavy holes obtained within the infinite barrier approximation and those resulting from the solution of an exact analytic equation obtained for a finite PQW are much smaller than experimental errors.

Quantised energy levels of a carrier with mass m_i^* belonging to the i-th band ($i=e,lh,hh$) bound in an infinitely high parabolic potential $V_i = (K_i/2)z^2$ are given in the former approximation by:

$$E_{ni} = \left(n-\frac{1}{2}\right)\hbar\sqrt{\frac{K_i}{m_i^*}} \qquad K_i = Q_i K \qquad K = \frac{8\Delta E_g}{L_z^2}$$

where $Q_i = \Delta E_i / \Delta E_g$, with ΔE_i being the discontinuity of the band edges, and ΔE_g being the energy gap discontinuity between the materials at the centre and at a distance $\pm L_z/2$ from the centre of the QW, and $n=1,2,3...$ being the harmonic oscillator quantum number. The energy E_n of the n-th diagonal optical transition in the heavy-hole (hh_n-e_n) series is described by $E_n=E_g+E_{nhh}+E_{ne}-E_{ex}$, where E_g is the energy gap of CdTe and E_{ex} is the exciton binding energy. Consequently the energy positions E_n of peaks belonging to the same series should be linear versus the level index n.

The open points plotted in Fig 2b correspond to the energy peak positions of hh_n-e_n transitions taken for spectra from the left panel of Fig2. As one can notice, these points are collinear and equidistant in energy for each PQW. Moreover, the lines fitted to the experimental points and extrapolated toward $n=0$ intersect with a good accuracy at one point: $n \approx 0.5$ and $E \approx 1.580eV$. This value of E is close to the expected value of $E_{0.5} = E_g-E_{ex}$, where the contributions of E_{nhh} and E_{ne} vanish. These results confirm the applicability of the simple model with infinite barriers to the case of our $Cd_{1-x}Mn_xTe$ PQW structures. PQWs made of a $Cd_{1-y}Mg_yTe$ ternary compound possess the same properties. Therefore we can use this approach to determine the valence band offset in CdTe /Cd$_{1-x}$Mn$_x$Te as well as in CdTe/Cd$_{1-y}$Mg$_y$Te heterostructures.

For this purpose, we have plotted (Fig.3) experimental values of the energy distance between two successive transitions in the hh_n-e_n series, which can be expressed by: $\Delta E=E_{(n+1)hh}+E_{(n+1)e}-E_{nhh}-E_{ne}$, vs. the square root of the total curvature K. These points should follow a straight line. The slope of this line is determined by the value of the valence band offset Q_{hh} ($Q_{hh}+Q_e=1$). The experiments confirm the linear dependence of ΔE vs. $\sqrt{K}$, as clearly visible in Fig.3 for both materials, where the data for all our samples are collected. The curvatures of the PQWs shown in this plot were calculated taking the nominal QW widths and the experimentally determined values of x or y. From a least squares fit of a straight line to the data in Fig. 3a, and b and using the values of the effective masses $m_e = 0.095$ for the electrons and $m_{hh} = 0.6$ for the heavy holes, we obtained the valence band offset $Q_{hh} = 0.44\pm 0.1$ for Cd$_{1-x}$Mn$_x$Te and $Q_{hh} = 0.45\pm 0.1$ for Cd$_{1-y}$Mg$_y$Te. Both of these values are, within experimental error, in agreement with the results of other authors [10,11,12].

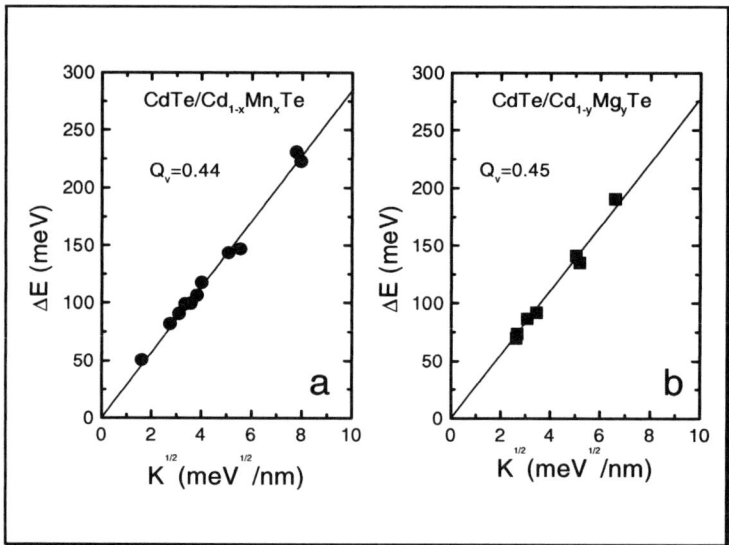

Figure 3. The energy distances ΔE vs. the square root of curvature K of the potential for (a) Cd$_{1-x}$Mn$_x$Te and (b) Cd$_{1-y}$Mg$_y$Te PQWs. The straight lines are least-squares fits to the data. The values of valence band offset Q_{hh}, determined by a fits, are shown in the figure.

Although this simple infinite barrier approximation is so successful in describing the optical properties of PQWs as well as in determining the valence band offset, a more advanced analysis of the experimental data employing a more realistic model of our structures was also performed. We have fitted the observed transitions taking into account the strains (assuming pseudomorphic growth) and the nonparabolicity of the true confining potential (due to the finite time of the shutter response during growth in the MBE machine). We also calculated the energy levels in PQWs in the presence of external magnetic field. As was already mentioned, the magnetic field is a very useful tool in the investigations of quantum well shape for structures with a diluted magnetic semiconductor as a barrier. This method is much more sensitive to small deviations from the assumed shape than even peak positions in the spectra in the absence of the field.

Especially useful for the valence band offset determination are structures for which nondiagonal transitions are observed. Structures with asymmetric quantum well shapes are particularly well suited for such observations. Due to the lack of inversion symmetry of the potential profile the parity forbidden, nondiagonal optical transitions with odd change of level index ($\Delta n = \pm 1, \pm 3, \pm 5 \ldots$) become allowed. The most important are those with $\Delta n = \pm 1$ because their oscillator strength is expected to be larger than that for transitions with $\Delta n = \pm 2$ and their observations should improve the accuracy of the band offset determinations.

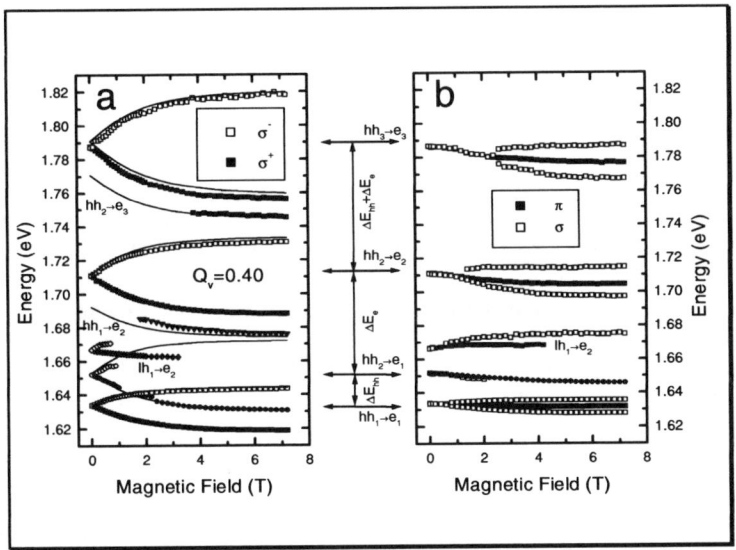

Figure 4. The energy of 1s excitonic features observed in optical magnetotransmission in the structure containing ten 194Å wide $Cd_{0.75}Mn_{0.25}Te$ half-parabolic as functions of the magnetic field at T=1.8K (a) in Faraday and (b) Voigt configuration. The lines in the left panel represent the calculated values of the transition energy involving various spin-split confined states

The half-parabolic quantum well (HPQW) is a special example of such a structure. Having similar properties as the PQWs, it also has the advantages of asymmetric QWs. Figure 4 shows the fan chart of the magnetotransmission spectra of the structure containing ten identical $Cd_{1-x}Mn_xTe$ HPQWs, each 60ML (194Å) wide and with x=0.25 in the barrier, outside of the quantum well. These spectra were collected in two configurations: when the magnetic field was parallel to the wave vector of incident light (Faraday configuration) or perpendicular to this vector (Voigt configuration). In both cases the light was propagating along the growth direction of sample.

The measurements of the transmission in the Faraday and Voigt configurations revealed a set of peaks. In the Faraday configuration (Fig. 4a) they split into σ^+ and σ^-

components (Zeeman splitting) due to the *sp-d* induced hole and conduction band electrons splitting. The value of this splitting strongly depends on the transition energy – it increases with increasing energy. Such behaviour can be easily understood in the case of barriers made of DMSs because the overlap of different electron or hole states in a PQW with magnetic ions (Mn) is different. Thus the splitting of the successive states should increase with increasing level index *n*. This fact helps us to identify the origin of each transition. As we expected, apart from strong diagonal excitonic transitions involving heavy (hh_1-e_1, hh_2-e_2, hh_3-e_3) and light holes (lh_1-e_1) we also observed three nondiagonal transitions (hh_2-e_1, hh_1-e_2, hh_2-e_3). They could be unambiguously identified on the basis of their spin splitting, since the magnetic field dependence of the energy of the transitions involving the same heavy hole states (e.g. hh_2-e_2 and hh_2-e_1) is nearly identical.

The pattern of optical transitions in the Voigt configuration (Fig. 4b) is completely different from that discussed above, where the magnetic field was perpendicular to the sample plane. The Zeeman splitting exhibits a significant anisotropy already reported by other authors [13] for rectangular QW structures [14]. Here we present this anisotropy for transitions involving higher states in the QW. In Fig. 4b the energy positions of different transitions vs. the magnetic field for two linear polarisations: perpendicular (σ) or parallel (π) to applied field are presented. At higher magnetic field the σ-component for the diagonal heavy-hole-originating transition (hh_1-e_1, hh_2-e_2, hh_3-e_3) splits into two branches, while at low fields the corresponding peaks cannot be resolved. The value of the field at which these two branches start to split increases with increasing level index *n*. A broadening of the peaks corresponding to transitions with higher *n* can be a possible explanation. This means that a greater splitting is required to resolve both branches. Below this particular field these two branches are very close. Moreover, they are, in this range of fields, very close to the only π polarised component for which we do not observe the splitting in the whole available range of magnetic fields. For the non-diagonal hh_2-e_1 transition we observe mainly the π polarised component, the only branch of the σ component disappears quickly with the field. The behaviour of spectral features involving the light-hole state is different. We observe only one branch of the σ and π components, and the shapes of their dependences on the field are different from that observed for the heavy holes. Such behaviour was reported in [13], supporting our assignment of this transition in the Faraday configuration to a light-hole.

The lines in Fig. 4a represent a model calculation based on numerical solution of the Luttinger Hamiltonian describing the valence band, and of the conduction band with a quadratic dispersion, with an additional half parabolic potential shape. The calculations were performed with the following values of the Luttinger parameters: γ_1=4.5, γ_2=1.46, and with a constant effective mass of electrons $m_e = 0.09m_0$ [15]. We took into account the fact that, due to finite reaction time of the Mn shutter, Mn was completely absent in those layers that should contain very small average Mn composition. In our calculations the QW potential shape in the presence of a magnetic field was assumed to be determined by the bulk Zeeman splitting [6].

Good agreement between calculations and experimental data was obtained for $Q_{hh} = 0.4$. Similar agreement could be also obtained in the case of remaining HPQW's giving $Q_{hh} = 0.4 \pm 0.05$. The accuracy is mainly limited by the uncertainty in effective mass values. This value confirms $Q_{hh} = 0.44 \pm 0.1$ obtained for $Cd_{1-x}Mn_xTe$ parabolic quantum wells. A model including the x-dependence of the electron effective mass was also considered. It gives, within experimental error, the same value of Q_{hh} [16].

Apart from *1s* excitonic transitions observed in all our structures, in some of them we also observed additional lines, which we attributed to *2s* excitonic transitions. In particular for the HPQW discussed above we observed a *2s*-related kink at zero magnetic field, which evolved into *2s* and *3s* peaks (gaining intensity) with an increase of the magnetic field. They can be distinguished from *1s* excitonic transitions because of their diamagnetic shift. The observed *1s-2s* zero field distance of 18.2 meV gives an exciton binding energy (estimated by the fractional dimension method [8,9]) equal to 21.6 meV in our 60 ML wide HPQW. This confirms the theoretically predicted strong increase of the binding energy of excitons confined in a parabolic potential.

3. Conclusions

In this paper we have discussed the optical and magnetooptical properties of graded QWs. We focussed our attention on QWs with a parabolic and half-parabolic shape of the confining potential. We have shown that apart from other interesting properties these structures are very useful for valence band offset determination. This method was applied to $CdTe/Cd_{1-x}Mn_xTe$ and $CdTe/Cd_{1-y}Mg_yTe$ heterostructures. Moreover we have demonstrated the evolution of transmission spectra in magnetic field in both the Faraday and Voigt configurations. The proof of the strong enhancement of exciton binding energy in a QW with a parabolic confining potential was also given.

Acknowledgements. This work was partially supported by the Committee for Scientific Research (Poland) under Grant No. 2P03B 119 14 and by the Volkswagen Foundation.

References

1. Pötz, W. and Ferry, D. K. (1985), Nonrectangular quantum wells as a basis for studying the band offsets at GaAs-Ga$_{1-x}$Al$_x$As interfaces, *Physical Review B* **32**, 3863-3867.
2. Yuen, W.-P. (1993), Exact analytic analysis of finite parabolic quantum wells with and without a static electric field, *Physical Review B* **48**, 17316-17320.
3. Chu-liang, Y. and Qing, Y. (1988), Sublevels and excitons in GaAs-Al$_x$Ga$_{1-x}$As parabolic quantum well structures, *Physical Review B* **37**, 1364-1367.
4. Chuang, S. L. and Ahn, D. (1989), Optical transitions in a parabolic quantum well with an applied electric field-analytical solutions, *Journal of Applied Physics* **65**, 2822-2826.
5. Karrai, K., Drew, H. D., Lee, M. W., and Shayegan, M. (1989), Magnetoplasma effects in a quasi-three-dimensional electron gas, *Physical Review B* **39**, 1426-1429.
6. Gaj, J. A., Grieshaber, W., Bodin-Deshayes, C., Cibert, J., Feuillet, G., Merle d'Aubigne, Y. and Wasiela, A. (1994), Magneto-optical study of interface mixing in the CdTe-(Cd,Mn)Te system, *Physical Review B* **50**, 5512-5527.
7. Gossard, A. C., Miller, R. C., and Wiegmann, W. (1986), MBE growth and energy levels of quantum wells with special shapes, *Surface Science* **174**, 131-135.
8. Christol, P., Lefebre, P., and Mathieu, H. (1993), Fractional-dimensional calculation of exciton binding energies in semiconductor quantum wells and quantum-well wires, *Journal of Applied Physics* **74**, 5626-5637.

9. Mathieu, H., Lefebvre, P., and Christol, P. (1992), Simple analytical method for calculating exciton binding energies in semiconductor quantum wells, *Physical Review B* **46**, 4092-4101.

10. Kuhn-Heinrich, B., Ossau, W., Litz, T., Waag, A., and Landwehr, G. (1994), Determination of the band offset in semimagnetic CdTe/Cd$_{1-x}$Mn$_x$Te quantum wells: a comparison of two methods, *Journal of Applied Physics* **75**, 8046-8052.

11. Kuhn-Heinrich, B., Ossau, W., Heinke, H., Fischer, F., Litz, T., Waag, A., and Landwehr, G. (1993), Optical investigation of confinement and strain effects in CdTe/(CdMg)Te quantum wells, *Applied Physics Letters* **63**, 2932-2934.

12. Wasiela, A., Peyla, P., d'Aubigne, Y. M., Nicholls, J. E., Ashenford, D. E., and Lunn, B. (1992), Magneto-optical study of CdTe/Cd$_{1-x}$Mn$_x$Te multiple quantum wells with low potential barriers, *Semiconductor Science & Technology* **7**, 571-577.

13. Kuhn-Heinrich, B., Ossau, W., Bangert, E., Waag, A., and Landwehr, G. (1994), Zeeman pattern of semimagnetic (CdMn)Te/(CdMg)Te quantum wells in inplane magnetic fields, *Solid State Communications* **91**, 413-418.

14. Peyla, P., Wasiela, A., d'Aubigne, Y. M., Ashenford, D. E., and Lunn, B. (1993), Anisotropy of the Zeeman effect in CdTe/Cd$_{1-x}$Mn$_x$Te multiple quantum wells, *Physical Review B* **47**, 3783-3789.

15. Le Si, D., Neu, G., and Romestain, R. (1982), Optical detection of cyclotron resonance of electron and holes in CdTe, *Solid State Communications* **44**, 1187-1190.

16. Ossau, W., Fiederling, R., Konig, B., Wojtowicz, T., Kutrowski, M., Karczewski, G., and Kossut, J. (1998), Magnetooptical studies of Cd$_{1-x}$Mn$_x$Te quantum wells with parabolic confining potential, *Physica A* **2**, 209-213.

DYNAMIC SELF-POLARISATION OF MAGNETIC IMPURITY SPINS UNDER OPTICAL PUMPING BY UNPOLARISED LIGHT IN SEMICONDUCTORS

M.I. DYAKONOV[*]
*Laboratoire de Physique Mathématique, Université Montpellier II,
34095 Montpellier, France*

Abstract. The dynamical self-polarisation effect, as predicted originally for the nuclear spin system in a semiconductor, is described in some detail. The possibility of observing this effect for a magnetic impurity spin system is discussed. It is argued that such a phenomenon might be observed in very dilute samples with narrow quantum wells.

1. Introduction

The phenomenon of self-polarisation was predicted in 1972 by V.I. Perel and myself [1,2] (see also [3]) for the nuclear spin system in a semiconductor. The experimental manifestation of this effect should be simple: one illuminates a semiconductor with unpolarised (or linearly polarised) light and looks at the circular polarisation of the luminescence. Normally, such a polarisation should be absent. However, it is predicted that below a certain critical temperature T_c, an appreciable degree of circular polarisation should appear, due to the onset of a spontaneous polarisation of the nuclear spin system. This effect was never observed in a semiconductor (for a recent attempt see [4]), although a very similar phenomenon was seen in an atomic vapour [5].

The purpose of this paper is to point out that the self-polarisation phenomenon should also exist under certain conditions for a spin system of magnetic impurities, such as Mn, in a semiconductor. Since the exchange interaction between carrier spins and the magnetic impurity spins is much stronger than the hyperfine interaction, the critical temperature for the magnetic impurity case could be essentially higher than for the case of a nuclear spin system, making the experimental observation of this effect easier. I will begin by explaining the physics of the self-polarisation phenomenon, and later on I will examine in some detail the conditions needed for its existence in the specific case of a semimagnetic semiconductor.

---

[*] On leave from the A.F. Ioffe Physico-Technical Institute, St. Petersburg, Russia

M.L. Sadowski et al. (eds.), Optical Properties of Semiconductor Nanostructures, 247–253.

2. The self-polarisation phenomenon

Following our original work [1], we will consider the lattice nuclear spin system, interacting with the photocreated conduction electrons in a semiconductor. In all of the general reasoning below the nuclear spin system can be replaced by the magnetic impurity spin system. The self-polarisation phenomenon is a result of the combination of the Overhauser effect [6] and the fact that polarised nuclei (magnetic impurities) produce an effective magnetic field acting on the electron spins. This effective field is due to the hyperfine (exchange) interaction.

If the nuclear spin system interacts with electrons which are artificially maintained in a disordered spin state, a substantial nuclear polarisation must spontaneously arise. Its appearance should have the features of a second-order phase transition. In other words, for $T < T_c$ the disordered state of nuclear spins is unstable, while the stable state corresponds to a high degree of nuclear polarisation. In this state a large nuclear magnetic field exists, acting on the electron spins. The nuclear polarisation, in its turn, is maintained due to the Overhauser effect in this effective magnetic field.

It will be assumed that a small "seeding" external magnetic field is applied, exceeding the local field B_L (B_L is the magnetic field produced at a nuclear site by the neighbouring nuclei and is of the order of several Gauss). The only role of this field is to suppress the nuclear spin-spin interactions, while the equilibrium electron spin orientation in this field is negligible.

2.1. OVERHAUSER EFFECT

Consider, for simplicity, the case of electrons interacting via the hyperfine interaction with one species of lattice nuclei with a spin $I=1/2$. Under stationary conditions the balance equation reads:

$$W_1 N_+ n_- = W_2 N_- n_+ , \qquad (1)$$

where N_+, N_- and n_+, n_- are the populations of the spin up (+) and spin down (-) states for the nuclei and electrons respectively, W_1, W_2 are the corresponding transition probabilities. In thermodynamic equilibrium the populations N_+, N_- and n_+, n_- are related by the respective Boltzmann factor:

$$N_+/N_- = \exp(2\mu_N B/T), \qquad n_+/n_- = \exp(-\mu g B/T) \qquad (2)$$

where μ_N is the nuclear magnetic moment, μ is the Bohr magneton, g is the electron g-factor, B is the magnetic field, and T is the temperature in units of energy. It is assumed that the electrons are non-degenerate.

Hence the following general relation between the transition probabilities should hold:

$$W_1/W_2 = \exp(-\mu g B/T), \qquad (3)$$

where the very small nuclear Zeeman energy is neglected ($\mu_N B \ll T$)

Suppose now that the electron spin populations are maintained equal by some external agent, so that $n_+ = n_-$. Then it follows immediately from Eqs. (1) and (3) that

$$N_+/N_- = W_2/W_1 = \exp(\mu g B/T), \tag{4}$$

or, in other words, the nuclear polarisation $P_N = (N_+ - N_-)/(N_+ + N_-)$ is given by the formula:

$$P_N = \tanh(\mu g B/2T), \tag{5}$$

which means that in these non-equilibrium conditions the nuclear polarisation becomes equal to the equilibrium electron polarisation, but with an opposite sign. This is the essence of the Overhauser effect [6].

Originally, Overhauser proposed to maintain $n_+ = n_-$ by saturation of the electron paramagnetic resonance. Clearly, other possibilities exist. In semiconductors, one can excite electrons by unpolarised or linearly polarised light. In the presence of magnetic field the photocreated electrons will have a tendency toward thermodynamic equilibrium, resulting in an electron spin polarisation following from Eq. (2). The characteristic time for this process is the electron spin relaxation time τ_s. If the recombination time, τ, is much shorter than τ_s, the photocreated electrons will not have enough time to achieve this equilibrium value, the steady state electron spin polarisation will be negligible, and the condition $n_+ = n_-$ will be satisfied with a high degree of accuracy.

For holes in cubic semiconductors the spin relaxation is very fast and normally $\tau_s \ll \tau$. On the other hand, the interaction of the nuclear spins with the holes is not efficient because of the p-character of the valence band Bloch functions. Thus, one can use a p-type semiconductor, and if the number of photocreated electron-hole pairs is not very small, the hyperfine interaction with photocreated electrons will be dominant.

2.2. NUCLEAR FIELD

If the nuclei are polarised, an effective magnetic field (*nuclear field, B_N*) arises which is due to the hyperfine interaction and which acts on the electron spins in the same way as the external field B. The nuclear field may be written as

$$B_N = b_N P_N, \tag{6}$$

where $b_N = A/(2\mu g)$ is the nuclear field which would appear in the case of complete nuclear polarisation, A is the hyperfine constant. The values of b_N may be surprisingly large. For example, in GaAs, if all the lattice nuclei were polarised, this would have an effect on the electron spin equivalent to an external field of ~5T! It should be noted that the concept of a mean nuclear magnetic field is well justified for electrons in semiconductors, either moving freely in the conduction band or localised on a shallow donor. Thus, for a shallow donor in GaAs the Bohr radius is around 10 nm, so that about 10^5 lattice nuclei are simultaneously seen by a localised electron.

2.3. SELF-POLARISATION

If the nuclei are polarised then, for any spin effect, the nuclear field B_N should be added to the external magnetic field, B. In particular, this should be done in Eq. (5) describing the dynamic nuclear polarisation in the Overhauser effect. One may also consider the situation when the external field B is very small so that $\mu g B << T$. Then, using Eqs. (5), (6), we obtain the following self-consistent equation for the nuclear polarisation:

$$P_N = \tanh(\mu g B_N/2T), \qquad (7)$$

which may be rewritten as

$$P_N = \tanh(P_N T_c/T), \qquad (8)$$

with the critical temperature, T_c, defined by

$$T_c = \mu g b_N/2 = A/4. \qquad (9)$$

For $T > T_c$ Eq. (8) has a single solution $P_N = 0$. However, for $T < T_c$ there are also two nonzero solutions (with opposite signs) describing nuclear self-polarisation, and it may be easily shown that these states are stable, while the state with $P_N = 0$ is unstable. The dependence $P_N(T)$ has the character of a second-order phase transition.

The generalisation of Eq. (9) for the case when there are several species of nuclei with different hyperfine constants, A, and spins, I, is given by

$$T_c = (1/3)\Sigma A \, I(I+1), \qquad (10)$$

where the sum is over all the nuclei in the unit cell, and the abundance of different isotopes should be taken into account. Calculations give $T_c = 9K$ for InSb and $T_c = 1.3K$ for GaAs [1,3].

2.4. CONDITIONS FOR SELF-POLARISATION

In practice, the conditions may be different from the ideal situation considered above: the electrons may acquire some polarisation during their lifetime (if τ_s and τ are of comparable magnitude), also, apart from the electrons, there may be other sources of nuclear spin relaxation. Such deviations do not destroy the phenomenon, however they make the critical temperature lower than the value predicted by Eq. (10).

It can be shown that the dependence of T_c on the ratio τ/τ_s is given by an additional factor $\tau_s/(\tau_s+\tau)$ in the right-hand side of Eq. (10). In fact, as it was shown in Ref. [2], the situation is more complicated, since for sufficiently large values of the parameter τ/τ_s the build-up of self-polarisation becomes a phase transition of the first order, which is characterised by a hysteresis in the temperature dependence of P_N. In the case of $I=1/2$ this happens for $\tau/\tau_s>0.5$.

The role of the parameter τ/τ_s is of some practical importance, if one wants to observe the nuclear self-polarisation by optical means. The most favourable situation for self-polarisation is when the disordering of electron spins is complete, which is achieved by pumping by unpolarised (or linearly polarised) light, provided that $\tau_s>>\tau$. In this case, however, the recombinational radiation will be unpolarised, and nuclear self-polarisation, even if it occurs, will not be revealed in the luminescence. For optical detection of self-polarisation it is important that the ratio τ/τ_s should not be too small. Then the appearance of the nuclear field due to self-polarisation will be accompanied by a noticeable electron spin orientation and hence by the appearance of circular polarisation of the luminescence. However, T_c will decrease proportionally to the factor $\tau_s/(\tau_s+\tau)$.

In the ideal situation considered above it was assumed that the hyperfine interaction with electrons is the only process defining the nuclear spin state. Let us denote by T_{1e} the corresponding nuclear spin relaxation time. If other processes of nuclear spin relaxation (characterised by the time T_1) are taken into account, a leakage factor, f $= T_1/(T_1+T_{1e})$, should be introduced in the right-hand side of Eq. (10), resulting in a further decrease of T_c.

Finally, we discuss the role of the external magnetic field. Eq. (8) gives the impression that below T_c self-polarisation appears even in zero magnetic field, and that the direction of the nuclear polarisation is arbitrary. In fact, Eq. (8) is true only if an external magnetic field is present, such that $B>>B_L$ (but $\mu_g B<<T$), where B_L is the local field (of the order of several Gauss) produced at a given nuclear site by the neighbouring nuclei, and the nuclear polarisation should be always directed along B. As discussed in more detail in Ref. [3], the reason is that the average nuclear spin, I_{av}, is always given by the thermodynamic formula

$$I_{av} = (1/3)(I+1)\mu_N B/\theta, \tag{11}$$

where θ is the nuclear spin temperature. The dynamic nuclear polarisation is, in fact, due to the cooling of the nuclear spin system, which makes θ different from the crystal temperature T. This cooling is not effective for $B<<B_L$. A more precise formula, which generalises Eq. (8) for arbitrary small external fields, reads:

$$P_N = B^2(B^2+B_L^2)^{-1}\tanh(P_N T_c /T). \tag{12}$$

This formula says that when B becomes of the order of B_L, or less, the effective critical temperature for self polarisation decreases as $B^2(B^2+B_L^2)^{-1}$.

3. Self-polarisation of the magnetic impurity spin system

The advantage of the magnetic impurity spin system, compared to the nuclear spin system, is that the exchange interaction is much stronger than the hyperfine interaction, which should, in principle, result in much higher critical temperatures. Thus, for the semimagnetic semiconductor $Cd_{1-x}Mn_xTe$ the band spin splittings at saturated Mn spin polarisation are [7]:

$$\Delta E_c = 0.22x \text{ eV} \quad \text{for the conduction band,} \tag{12a}$$
$$\Delta E_v = -0.88x \text{ eV} \quad \text{for the valence band,} \tag{12b}$$

where x is the Mn content. These values correspond to the following spin Hamiltonians for electrons and holes, interacting with the mean exchange field of polarised Mn ions:

$$H_c = -\Delta E_c \boldsymbol{SI}_{av}, \tag{13a}$$
$$H_v = -(1/3)\Delta E_v \boldsymbol{JI}_{av}, \tag{13b}$$

where $\boldsymbol{I}_{av}$ is now the average value of the Mn spin, $\boldsymbol{S}$ is the electron spin, and $\boldsymbol{J}$ is the hole spin ($J=3/2$).

Looking at Eqs. (12), (13), one can see immediately that the electrons are not suitable to produce the self-polarisation effect for Mn spins, since H_c and, as a consequence, the effective magnetic field produced by polarised Mn atoms, have the wrong sign, so that the feedback in the Overhauser effect will be negative. This corresponds to a negative value of A and T_c in Eq. (8), in which case there is no phase transition (at least for positive T !). On the other hand, the interaction with holes has an opposite sign, which is favourable for self-polarisation.

However, for holes in the bulk material it is virtually impossible to have $\tau_s > \tau$, because of the strong spin-orbit interaction in the valence band, which normally makes τ_s of the order of the momentum relaxation time, τ_p. The solution might be found in using holes confined in a narrow enough quantum well. In this case the ground state is formed by the +3/2, -3/2 hole spin states, and the spin relaxation within these states is allowed only for finite values of the momentum k, due to mixing with the higher +1/2 and -1/2 states. Thus, for sufficiently narrow quantum wells and for a sufficiently low hole concentration, τ_s may be made quite long [8].

Another requirement, as discussed above, is that the exchange interaction with holes should be the dominant mechanism of Mn spin relaxation. For high Mn content this is never the case. It is known that the Mn spin-lattice relaxation rate depends strongly on the Mn concentration [9-12], as a result of the presence of antiferromagnetic clusters. Such clusters may increase the spin-lattice relaxation rate by several orders of magnitude, compared to that of isolated Mn impurities, which is characterised by relaxation times in the millisecond range. Thus, the carriers start to contribute to Mn spin relaxation only in sufficiently diluted samples with x of the order of, or less, than 1%, when the spin-lattice relaxation time becomes longer than 10 mK [9, 12]. It should be noted, however, that, as it follows from Eqs. (12), (13), even in these diluted samples one can expect rather high critical temperatures, on the order of 10K. At higher pumping intensities, the effect of carriers is observed even in samples with higher concentrations [13].

Finally, we should discuss the value of the local field, B_L, for the case of the Mn spin system. Similar to the case of the nuclear spin system, this field is determined by the spin-spin interaction of neighbouring Mn atoms. As it can be deduced, for example from Ref. [14], for $x \sim 1\%$ the local field may be estimated as $B_L \sim 100 Gs$. Thus, the external magnetic field, B, should be of the order of, or exceed this value.

4. Conclusions

An elaborated theory of dynamic self-polarisation of the magnetic impurity spin system in a semiconductor still remains to be done. However, the above qualitative discussion leads to the conclusion that this phenomenon hopefully may be observed in very diluted (probably with $x < 1\%$) magnetic semiconductor quantum wells, under pumping by unpolarised (or linearly polarised light). The main conditions are that the well be sufficiently narrow to suppress the hole spin relaxation, and that the concentration of photocreated holes be high enough to dominate the impurity spin relaxation process. Under these conditions one may expect that the critical temperature for the self-polarisation phase transition will be above 2K.

Acknowledgements. I thank Kirill Kavokin, Igor Merkulov, and Denis Scalbert for very useful discussions.

References

1. Dyakonov M.I. and Perel V.I. (1972) *Sov. Phys. JETP Lett.* **16**, 398-400
2. Dyakonov M.I. (1974) *Sov. Phys. JETP*, **38**, 177-181
3. Dyakonov M.I. and Perel V.I. (1984) Theory of Optical Spin Orientation of Electrons and Nuclei in Semiconductors, in: F. Meyer and B. Zakharchenya (eds.), *Optical Orientation*, North Holland, pp. 11-72
4. Farah W., Scalbert D., Dyakonov M.I., and Knap W. (1998) *Phys. Rev. B* **57**, 4713-
5. Klipstein W.M., Lamoreaux S.K., and Forston E.N. (1996) *Phys Rev. Lett.* **76**, 2266-2268.
6. Overhauser A. W. (1958) *Phys. Rev.* **92**, 411-418.
7. Gaj J.A., Planel R., and Fishman G.(1979) *Solid State Commun.* **29**, 435-438.
8. Ferreira R., Bastard G. (1991) *Phys. Rev. B* **43**, 9687-9691.
9. Scalbert D., Cernogora J., and Benoit à la Guillaume C. (1988) *Solid State Commun.* **66**, 571.
10. Strutz T., Witowski A.M., and Wyder P. (1992) *Phys. Rev. Lett.* **68**, 3912.
11. Tyazhlov M.G., Filin A.I., Larionov A.V., Kulakovskii V.D., Yakovlev D.R., Waag A., and Landwehr G. (1997) *JETP*, **85**, 784-796
12. Dietl T., Peyla P., Grieshaber W., and Merle d'Aubigné Y. (1995) *Phys. Rev. Lett.* **74**, 474-477.
13. Kulakovskii V.D., Tyazhlov M.G., Filin A.I., Yakovlev D.R., Waag A., and Landwehr G. (1996) *Phys. Rev. B* **54**, R8333-R8336.
14. Krenn H., Kaltenegger K., Dietl T., Spalek J., and Bauer G. (1989) *Phys. Rev. B* **39**, 10918-10934.

COHERENT SPIN DYNAMICS IN DILUTED-MAGNETIC QUANTUM WELLS

KIRILL KAVOKIN
A.F.Ioffe Physico-Technical Institute, 194021 Politechnicheskaya 26, St-Petersburg, Russia

1. Introduction.

Involving spin degrees of freedom into the operation of semiconductor devices has long been an elusive dream of physicists. During the last years, the list of prospective (while sometimes incredible) applications of spin-related effects has noticeably increased, now spanning a range from fast high-density memories through spin transistors towards quantum computers. In due course, optimum systems are to be found for practical implementation of those ideas which will have proven to be realistic. By now, a few classes of semiconductor or hybrid structures have been outlined, which are most suitable for studying the fundamental physics of coupled spin systems, and as test beds for new ideas. For instance, these are structures based on II-VI ternary and quaternary alloys that incorporate magnetic ions (usually Mn^{2+}), known as diluted-magnetic (semimagnetic) semiconductors (DMSs) [1]. The physics of bulk DMSs, which has been intensively studied for nearly 30 years, is now firmly established. This background knowledge, combined with the advances of optical methods developed for non-magnetic II-VI quantum wells and superlattices, has made DMS-based nanostructures a convenient model for investigating the specific consequences of spin-spin interactions in reduced dimensionalities. The facility of visualising the spin configurations in DMSs is due to the exchange interaction of carriers with localised spins of magnetic ions (usually Mn^{2+}), which is strong enough to produce a considerable splitting of band states when ion spins are polarised. The most widely recognised manifestation of the exchange interaction, the giant Zeeman effect, gives a good illustration of its strength. At helium temperatures, the typical carrier spin splitting in external magnetic fields is of the order of 10 meV per Tesla and can become as large as 50-60 meV when the magnetisation of magnetic ions is saturated. This value is close to typical band offsets in II-VI heterostructures. Thus, the spin state of ions can have a great impact on transport as well as optical properties of the structure. Another important thing is that the orbital angular momentum of the d-shell of the Mn^{2+} ion is zero. For this reason, localised spins are fairly well isolated from the crystal lattice. Even the spin-spin scattering time T_{SS}, responsible for the decay of a non-equilibrium magnetisation in the ensemble of ion spins, is sufficiently long compared with the spin relaxation times of carriers. This gives a possibility of monitoring non-equilibrium states of the magnetic-ion spin system using carriers as a probe and, in principle, to apply the time-dependent magnetic moment of

255

M.L. Sadowski et al. (eds.), Optical Properties of Semiconductor Nanostructures, 255–268.
© 2000 *Kluwer Academic Publishers. Printed in the Netherlands.*

ions as a "spin gate" for the dynamical modulation of electrical or optical response of the structure.

Non-equilibrium magnetisation and modulation of the spin state of magnetic ions can be realised, for example, by traditional EPR techniques, or by optical pumping via the Overhauser effect (see [2]). Here I would like to discuss another possibility which employs the specific features of semiconductor nanostructures. Briefly, the discussed effects consist in a coherent evolution of a great number of localised spins of Mn^{2+} ions due to their exchange interaction with spins of free or weakly-localised *two-dimensional heavy holes*. The pronounced spin anisotropy of holes, resulting in non-conservation of the total angular momentum in the spin system, plays the key role here. Due to this anisotropy, the hole can inject into the magnetic-ion spin system a total spin two orders of magnitude greater than its own spin J=3/2. This massive spin transfer, developing within a picosecond time scale, shows up in various optical experiments.

This paper presents a theorist's view on several spin-dynamical effects that have been already observed experimentally in Mn-based DMS quantum wells, followed by a discussion of possible directions of further experimental efforts. I hope to demonstrate that the physics underlying all these effects is not confined to DMSs, and the experiments touch on some fundamental issues of spin systems and many-body physics, and even on general questions of quantum mechanics. In the following, I will first introduce a simple but instructive model of coherent spin precession under an effective magnetic field. On the base of this model, the key experiments will be described. Then, the main limitations to the precession model will be stated, and advanced theoretical approaches to the problem, as well as ideas of some new experiments, will be discussed.

2. Collective spin precession in an effective field: review of key experiments.

Let us consider, for definiteness' sake, a DMS quantum well (QW) structure with magnetic ions inside the well (Fig.1a). Mn^{2+} ions are isotropic and their spins can be polarised along any direction with an external magnetic field. Let the field be applied along the QW plane. If we then place in the QW a heavy hole with a definite spin projection onto the structure axis (in a real experiment this can be done by shining circularly polarised light on the structure), the exchange interaction between the hole and magnetic ions is switched on. Since there is no matrix element of the hole spin operator $\hat{J}$ between the states of the heavy-hole spin doublet, $J = \pm\dfrac{3}{2}$, the interaction takes the Ising form:

$$\hat{H}_{ex} = a \sum_{i} |\Psi_{hh}(\vec{r}_i)|^2 \hat{S}_{iz} \hat{J}_z \qquad (1)$$

where a is an exchange constant, $\hat{\vec{S}}$ is the spin operator of the i-th ion, $\Psi_{hh}(\vec{r}_i)$ is the heavy-hole wave function at the ion location.

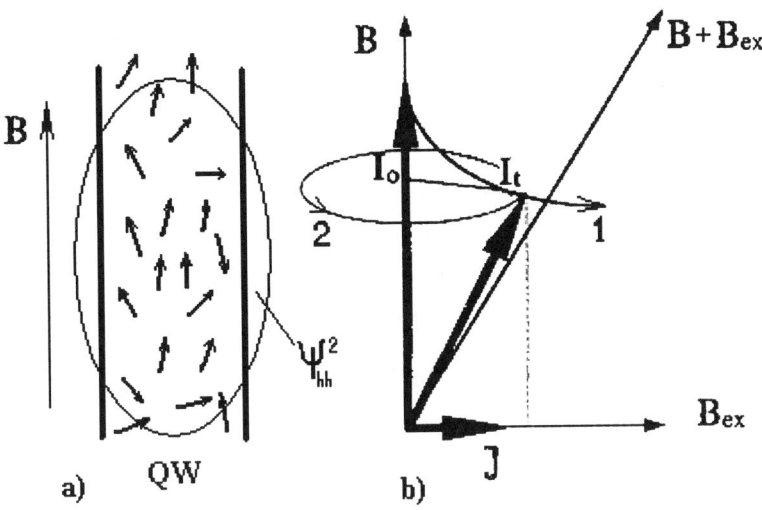

Figure 1. Precession model. a) Sketch of the structure; b) Precession of the magnetic-ion spin: 1 – during the hole lifetime; 2 – after recombination of the hole.

To begin with, we assume that the hole remains in the initial state with a definite z-component of the spin operator. This is not far from reality if we consider short time intervals after the hole creation, so that the hole spin has no time to flip due to some scattering event (note that hole spin flips between +3/2 and –3/2 states are forbidden in the first order of perturbation theory). In this case, the exchange interaction can also be expressed by an *exchange field* of the hole,

$$\vec{B}_{ex} = a \sum_i \left| \Psi_{hh}(\vec{r}_i) \right|^2 J_z \vec{e}_z \,, \tag{2}$$

acting upon magnetic-ion spins. Here $\vec{e}_z$ is a unit vector along Z. On average, the field is in inverse proportion to the hole localisation volume, or in direct proportion to the concentration of holes when they are free. Under this assumption, we come to a very simple picture of the spin-system evolution, shown in Fig.1b: Magnetic-ion spins, initially oriented along the external magnetic field, after creation of the hole are subjected to the total effective field which is just a vector sum of the external field and the exchange field of the hole. As the initial orientation of the Mn mean spin does not coincide with the direction of the effective field, the spins begin to rotate about the effective field. Fig.1b schematically shows the main qualitative consequences of this collective precession (their experimental manifestations are collected in Fig.2):

1) The projection of the total Mn^{2+} spin I onto the external magnetic field decreases, on average, by the value

$$\langle n(t) \rangle \approx I \frac{B_{ex}^2}{B^2 + B_{ex}^2} (1 - \cos \omega_L t) \tag{3}$$

where $\omega_L = \dfrac{\mu_B g_{Mn}}{\hbar}\left(\sqrt{B^2 + B_{ex}^2}\right)$ is the Larmor frequency of Mn spins in the effective magnetic field $\vec{B} + \vec{B}_{ex}$, μ_B is the Bohr magneton, g_{Mn} is the Mn^{2+} g-factor. The Zeeman energy of Mn^{2+} spins increases by $\Delta E_Z(t) = \mu_B g_{Mn} B\langle n(t)\rangle$. Meanwhile, the exchange energy $E_{ex} = -\mu_B g_{Mn} B_{ex} I_Z$ decreases by the same value. This process can be considered therefore as a *coherent energy transfer* from the exchange reservoir to the Zeeman reservoir of the spin system [3]. The first experimental evidence for this energy transfer has been obtained by observation of *multi-spin Raman paramagnetic resonance* (MRPR) [4]. In MRPR experiments, localised heavy-hole excitons serve as intermediate states for resonance Raman scattering of light. The energy dumped into the Zeeman reservoir due to spin precession during the exciton lifetime is ultimately taken from light, forming the Raman shift of the secondary-emission frequency. As in the absence of the exciton, i.e. both in the initial and in the final state of the scattering process, Mn^{2+} spins have fixed projections onto $\vec{B}$, the Raman shift takes discrete values proportional to the random integer n governed by the Poisson distribution. The mean value of n is equal to $<n(t)>$. To obtain the complete picture, one has to perform an additional averaging over the lifetimes of the intermediate state, which limits the actual time interval to the exciton relaxation time τ. A typical $<n>$ is of the order of 10 (the maximum number of observed Raman peaks was as large as 15 (Fig.2a)). This corresponds to a rather short relaxation time $\tau<1$ps. Typical recombination times of excitons in II-VI quantum wells are two orders of magnitude higher, so that the question arises what would happen beyond the subpicosecond times characteristic for Raman-scattering experiments. The answer is that various relaxation processes would eventually result in the formation of a correlated state of the hole spin and spins of all the ions covered by the hole orbit, which is usually referred to as a *magnetic polaron* (MP). The negative energy associated with the MP is just the exchange energy E_{ex}.

Magnetic polarons are known to be formed even in absence of an external magnetic field as a result of spin-spin and spin-lattice relaxation processes [5]. Moreover, magnetic fields usually suppress the magnetic polaron due to saturation of the susceptibility of the Mn^{2+} spin ensemble. But the above considerations suggests that in this case application of an in-plane magnetic field would greatly *enhance* the MP formation by providing a new channel of energy relaxation via energy transfer to the Zeeman reservoir. This scenario has been realised and experimentally observed [3] by measuring photoluminescence under selective excitation (the design of this experiment does not differ significantly from that of the Raman experiments discussed above) in QW structures where the relation between the MP formation time, τ_{MP}, and the exciton lifetime, τ_{ex}, is unfavourable for the MP formation ($\tau_{MP}>\tau_{ex}$). Involving the new mechanism of energy relaxation reverses this inequality, which results in a considerable increase of the MP energy measured from the Stokes shift of the photoluminescence line. (Fig.2b).

2) The Z-component of I, initially close to zero, increases. Besides, the signs of I_z and J_z coincide. In other words, there develops a correlation between the hole spin and the spins of magnetic ions. The magnetic-polaron energy expresses this correlation

quantitatively. Another important property of the correlated spin complex of the hole and magnetic ions is that it retains information about the initial polarisation of the hole. This feature shows up in experiments on optical orientation of localised excitons [6]. Normally, applying a transverse magnetic field suppresses the non-equilibrium spin polarisation of carriers [7]. This is due to the fact that the magnetic field provides the quantisation axis for the carrier spin, and splits states differing by their spin projection onto this axis. The carrier spins, initially oriented perpendicular to the field, rapidly depolarise as a consequence of quantum beats (the Hanle effect). The system under consideration, however, behaves in a different way. The point is that the hole does not precess in the magnetic field due to the zero in-plane g-factor. On the other hand, the quantisation axis for the magnetic-ion spins is inclined with respect to the external field direction because it is really governed by the effective field $\vec{B} + \vec{B}_{ex}$. Eventually, the spin system comes to the magnetic-polaron state with the magnetic-ion spin parallel to the effective field [3]. On the MP being formed, reorientation of the hole spin requires a simultaneous rotation of all the magnetic-ion spins covered by the hole orbit. Due to the structure anisotropy, this rotation implies surpassing a potential barrier of order the MP energy [8] and is for this reason strongly suppressed. The average polarisation of holes in the ensemble of magnetic polarons at these conditions is determined by the competition of the hole spin relaxation time and the MP formation time. As we could see above, this latter time decreases under in-plane magnetic fields. The rapid onset of spin correlation in the magnetic polarons being formed stabilises the spin orientation of holes. This was the reason why a paradoxical, at first glance, *increase* of the circular polarisation of luminescence with *transverse* magnetic field (Fig.2c) was observed under optical orientation in a DMS QW [6]. This effect was called "the anomalous Hanle effect".

There is one feature that is very important both for MP formation via energy transfer into the Zeeman reservoir and for the anomalous Hanle effect, but relatively unimportant for the multi-spin Raman paramagnetic resonance. Thus far we implicitly considered the exchange interaction to be of equal strength for all the magnetic ions involved. Usually, this is not the case due to the real shape of the hole wave function. In other words, the exchange field of the hole is not uniform within the volume where the hole is localised. Correspondingly, both the direction and the value of the effective field $\vec{B} + \vec{B}_{ex}$ vary from ion to ion. In photoluminescence experiments, where maximum effects are obtained at values of B close to the average B_{ex} (at these conditions the maximum projection of the Mn^{2+} total spin onto the structure axis can be reached), the variation of the effective field is of the order of its mean value. This results in a dephasing of the precession of different ion spins, and a relaxation of the transverse components of the total spin of the magnetic ions interacting with the hole, $\vec{I}$, which can be conveniently defined as $\vec{I} = \sum_i \vec{S}_i |\Psi_{hh}(\vec{r}_i)|^2$. As a consequence, $\vec{I}$ finds its way into the magnetic-polaron state within one revolution around the effective field [3]. To the opposite, Raman-scattering experiments are performed at high magnetic fields, in order to gain spectral resolution of Mn^{2+}-spin-flip peaks. In these conditions, the

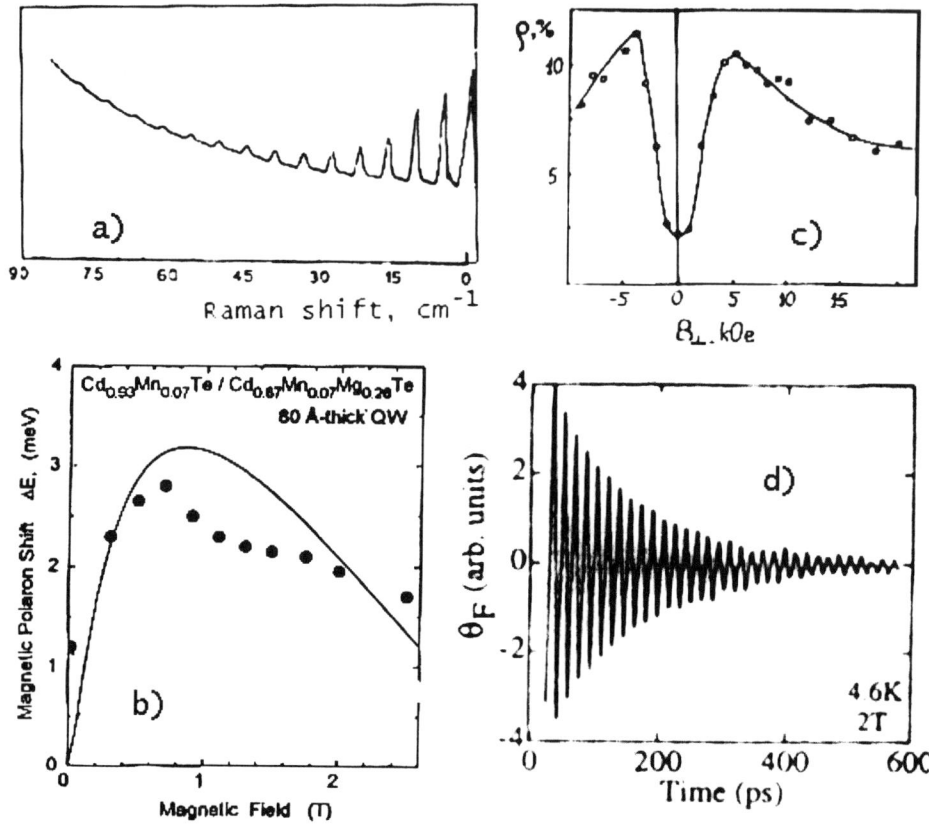

Figure 2. a) Multi-spin Raman paramagnetic resonance in a 18A (Cd,Mn)Te QW [4]; b) Magnetic-polaron energy vs. in-plane magnetic field [3]; c) Anomalous Hanle effect [6]; d) Free precession of Mn spins observed with time-resolved Faraday rotation [9].

relative variation of the precession frequency is of order $\dfrac{B_{ex}^2}{B^2} \ll 1$. This relation allows to neglect dephasing, at least within short time intervals typical of Raman scattering, and to describe the spin-system evolution in terms of the total ion spin $\vec{I}$, which has a fixed value and interacts with the hole and with the external magnetic field – a simplification especially important for a theoretical analysis of dynamic effects owed to the quantum properties of $\vec{I}$ (see later).

3) When the hole leaves the site (due to recombination with an electron, or by phonon-assisted tunnelling into another localised state), the magnetic-ion spins remain inclined at some angle to the external field, and immediately start to rotate around the field direction. If there are no other carriers, this precession is no longer influenced by the exchange interaction and can last as long as the spin-spin relaxation time T_{SS}. This

"free" precession of magnetic-ion spins has been observed [9] by means of time-resolved Faraday rotation, a pump-and-probe technique that allows subpicosecond time resolution. The design of this experiment is as follows. After excitation with a pulse of circularly polarised light, which creates a population of spin-polarised electrons and heavy holes, the QW structure is probed by another, linearly polarised pulse. The Faraday rotation of the probe polarisation plane is detected. The rotation angle is contributed to by the Z-components of all the magnetic moments in the structure, including non-equilibrium spin densities of electrons, holes and magnetic ions. By the use of this technique, an oscillating signal related to the Mn^{2+} precession has been observed, decaying with a characteristic time T_2 of order 200 ps (Fig.2d). As distinct from the Raman-scattering and photoluminescence experiments discussed above, in pump-and-probe experiments the excitation is introduced into the spin system by free holes that create a pulse of the exchange field with an amplitude proportional to the concentration and the mean spin of photoexcited holes. Free Mn^{2+} precession has been observed after relaxation of carrier spins that takes nearly 10 ps. This experiment has given experimental evidence of the collective precession of magnetic-ion spins induced by the exchange field of heavy holes.

3. The case of high transition rate between hole spin sublevels: Adiabatic approximation.

The model of spin precession in an effective field given above implies that there are no transitions between the heavy-hole spin states with J_Z equal to +3/2 and −3/2. In other words, both the in-plane g-factor and the spin relaxation rate of the hole should be zero. In reality, both statements hold but approximately. The range of their validity is confined to the most anisotropic structures, i.e. narrow or strained QWs, where the light hole – heavy hole splitting is large, and to extremely short times of spin evolution, met practically in Raman-scattering experiments only. Indeed, there are plenty of experimental observations, firmly supported by theory, that show evidence of the splitting of the heavy-hole spin doublet in sufficiently strong in-plane magnetic fields (see, for instance [10]). The reason is that magnetic-ion spins, polarised by an external magnetic field, produce an exchange field so strong that it is able to admix light-hole wave functions to the heavy-hole ones, and thus create a considerable in-plane g-factor of heavy holes. In-plane strain and/or an anisotropic localising potential result in a finite g-factor even in the zero-field limit. This has been demonstrated recently by the detection of a magnetic-field-induced linear polarisation of photoluminescence [11]. Spin-relaxation times of free holes in DMS QWs, while much longer than in bulk DMSs, can be nevertheless fairly short (near or even below 1ps) [9]. The most likely cause is the finite in-plane wave vector of the hole that violates the QW symmetry and produces non-zero matrix elements of the in-plane components of $\vec{J}$. Then, exchange scattering of the hole on magnetic ions provides relaxation of $<J_Z>$. In spite of all these facts, the in-plane components of the hole spin are much less than J_Z. For this reason, we can still describe the exchange interaction in terms of the hole exchange field directed along Z, but this field can no longer be considered as a constant vector. In many cases, it can be assumed that the hole spin either relaxes so fast (as compared with the period of Larmor precession of ion spins), or experiences quantum beats at such a high frequency, that magnetic-ion spins "see" its mean value, $<J_Z>$. This *adiabatic approximation* [3] reduces the problem of finding the time evolution of the spin system to the following self-consistent routine. First, $<J_Z>$ is found as a function of the total spin of the magnetic ions interacting with holes, $\bar{I}$. Then, $<J_Z>$ is used to calculate the exchange fields produced by the hole (or holes) at each ion, thus finding differential equations for all $\vec{S}_i$ as functions of time. Finally, an averaging over the ionic positions closes the loop, giving the rates of changing the components of $\bar{I}$ [3].

4. Spin dynamics in a DMS QW with a 2D hole gas.

Application of the formal recipe, given in the previous paragraph, to specific experimental situations is not, generally, an easy task, and usually requires further simplification of the problem. In the rest of the paper, the spin dynamics in DMS QWs is considered for two demonstrative cases: pump-and-probe experiments in the presence of a 2D hole gas, and multi-spin Raman paramagnetic resonance in wide QWs. In both cases, the external magnetic field is much stronger than the hole exchange field. In pump-and-probe experiments, the external field is of the order of 1 Tesla or more, while the exchange field that free holes can create does not exceed 0.1 Tesla [12]. As it has already been mentioned, the relation $B>>B_{ex}$ usually holds also for MRPR. Under these conditions, the direction of $<\bar{I}>$ can not move far away from the external field $\vec{B}$ during the spin-system evolution. This fact allows to exclude the component of $\bar{I}$ along $\vec{B}$ (let it be denoted as I_X), from the dynamical equations, because its relative variation is small. Thus, the three-dimensional rotation of the total-spin vector reduces to a two-dimensional motion in the (I_Y, I_Z) plane. Further simplification can be gained if we note that I_Y and I_Z are conjugate variables and enter the dynamical equations in the same way as the momentum and the coordinate of a point mass moving along a straight line. Indeed, the equations describing Larmor precession of I in the magnetic field B (without exchange interaction) are just the Hamiltonian equations of a harmonic oscillator:

$$\dot{I}_Z = \Omega I_Y$$
$$\dot{I}_Y = -\Omega I_Z \tag{4}$$

where $\Omega = \dfrac{\mu_B g_{Mn}}{\hbar} B$. Expansion of the Zeeman energy $E_Z = -\mu_B g_{Mn} B$ up to second powers of I_Y and I_Z near $\bar{I} \| \vec{B}$:

$$E_Z \approx -\hbar \Omega I + \left(\frac{1}{2} \hbar \Omega I_Z^2 + \frac{1}{2} \hbar \Omega I_Y^2 \right) \tag{5}$$

makes the analogy complete, if we assign the roles of potential and kinetic energy to $\frac{1}{2} \hbar \Omega I_Z^2$ and $\frac{1}{2} \hbar \Omega I_Y^2$, respectively.

Thus, any form of a mechanical oscillator (weight on a spring, pendulum, etc.) can serve as a model of the large spin in a magnetic field. The pendulum is especially demonstrative, because it is easily associated with the swinging spin vector. Besides, as we have chosen I_Z for the "coordinate", it is easy to modify the model so that it would describe the exchange interaction with the hole. This is done by making the pendulum into a device like a balance (see Fig.3). The cross beam attached to the pendulum shaft (Fig.3a) mediates mutual influence of the magnetic-ion spin, modelled by the pendulum itself, and the hole spin, whose energy sublevels corresponding to $J_Z=+3/2$ and $J_Z=-3/2$ are given by the heights at which the pans are hanging. When the hole occupies one of the spin sublevels, which in our picture is modelled by placing a weight onto one of the pans (Fig.3b), the equilibrium position of the balance changes. If the vertical position of the pendulum shaft corresponds to $\vec{I}\|\vec{B}$, the new equilibrium state naturally corresponds to $\vec{I}\|\vec{B}+\vec{B}_{ex}$. An oscillation of the device about the new equilibrium state is just a mapping of the precession of $\vec{I}$ around the effective field inclined to the QW plane.

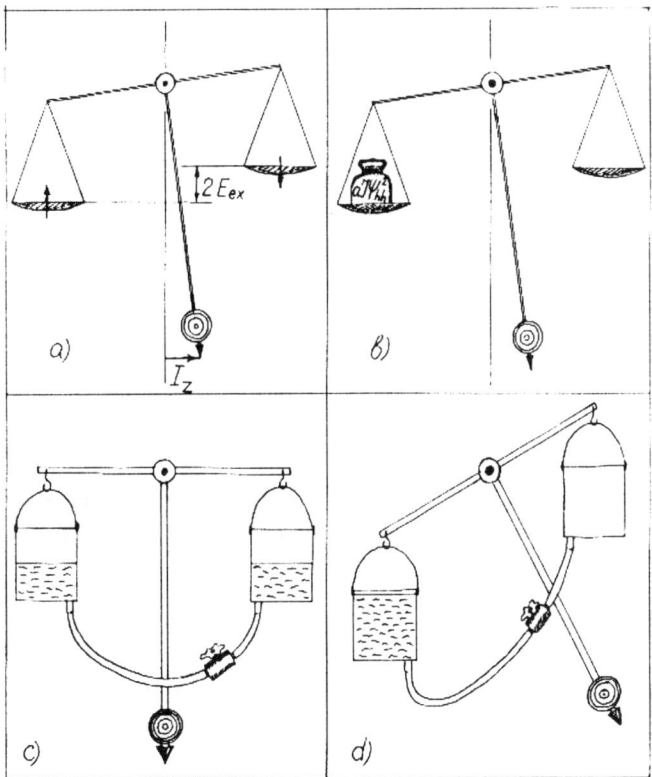

Figure 3. A mean-field model of the spin system of a DMS QW: a) Due to the anisotropy of the heavy-hole g-factor, only the Z-projection of $\vec{I}$ contributes to the giant Zeeman effect. b) Magnetic-polaron state in the in-plane magnetic field. c) Coupled spin system of a DMS quantum well with 2D hole gas. d) Carrier-induced ferromagnetic transition.

This toy mechanical model is, of course, unable to completely describe the quantum properties of the spin system. However, it is perfectly suited for an analysis of the spin dynamics using the adiabatic approximation. Indeed, within the adiabatic approximation the hole spin operator is replaced by its average, which in turn is determined by the spin state of magnetic ions. This is easily reflected by the model if we allow the weights on the pans to depend on the balance position. The device drawn in Fig.3c realises this idea for the case of a QW filled with free holes. The pans here are replaced with buckets connected by a flexible tube. Water levels in the buckets denote quasi Fermi levels in spin subbands. The cylindrical shape of the buckets reflects the step-like density of states in the 2D gas. The valve on the tube controls the spin relaxation rate. When it is closed, the water masses in the buckets do not change while the pendulum swings, which corresponds to the absence of spin relaxation. In this case the holes do not affect the dynamics of the Mn total spin modelled with the pendulum. When the valve is fully open, which corresponds to very fast spin relaxation, the water surface is kept at the same level in the two buckets. If the pendulum is not in the vertical position, the water masses in the buckets are different. As a result, an additional torque appears that tends to *increase* the inclination. It compensates partially the returning force created by the bob, so that the pendulum, once tilted, recovers its vertical position in a longer time. In other words, the frequency of small oscillations becomes *lower*. It is easy to express the renormalised frequency through the parameters of the model:

$$\omega = \Omega\sqrt{1-\zeta} \, , \qquad (6)$$

where $\xi = \dfrac{2l^2 A\rho}{Lm}$, L and l are respectively the shaft length and the arm of the crossbeam, m is the bob mass, A is the area of the water surface in the bucket, and ρ is the water density. The frequency goes to zero at $\xi = 1$, which corresponds to a stable equilibrium, when the balance can stay at any angle provided there is some water in both buckets. A further increase of ξ results in a spontaneous breaking of the symmetry. The overturning moment created by the buckets becomes greater than the returning moment due to the bob weight, the vertical position of the balance becomes unstable, and two new equilibrium states are formed (Fig.3d). In the new equilibrium state the upper bucket is empty. As a result, small oscillations near this state are not accompanied by water flow through the tube, and the eigenfrequency remains equal to Ω. This bifurcation is known as a carrier-induced ferromagnetic transition. The transition has been observed in modulation-doped QWs where it caused a zero-field splitting of the photoluminescence line [12]. The parameter ξ in reality is governed by the coupling constant, QW width, and the paramagnetic susceptibilities of Mn^{2+} spins and of the hole gas [13]. The temperature dependence of the Mn^{2+} susceptibility makes it possible to pass the transition by changing the temperature. A transition temperature of about 1.5K has been reported [12]. It is not clear, however, to what extent the transition can be described within this simple mean-field model. Though there is no theory of the transition going beyond the mean-field approximation, it has been supposed that the behaviour of the spin system near the transition is strongly affected by disorder resulted from structure imperfections [12]. An experimental observation of the precession-frequency shift near the transition would give new information that, on being compared with photoluminescence data, would help to elucidate the physics of the phase transition. This experiment would be also helpful for finding the concentration and temperature dependence of the paramagnetic susceptibility of the hole gas, χ_h. This value is known to be affected by the carrier-carrier exchange interaction [14], which is important for explanation of the observed transition temperature [12]. But a calculation of the exchange correction to the susceptibility at actual carrier concentrations (10^{11} cm^{-2}) is extremely difficult because the Coulomb energy is of the order of the Fermi energy, and there is no small parameter. The suggested method, being a frequency-domain technique, may provide a high-accuracy measurement of χ_h. A rigorous theory of the dynamical frequency shift, with an account for the realistic shape of the hole wave function, is given in Ref.13.

5. Single hole: quantum dynamics of a classically large spin.

Let us consider the balance model at $\xi > 1$. Its potential energy as a function of angle has two symmetric minima and a maximum in between. The maximum, corresponding to the vertical position of the pendulum, is a point of unstable equilibrium. A single localised hole interacting with magnetic-ion spins also forms a potential of this kind if the hole in-plane g-factor is large enough and the adiabatic approximation is valid [15]. The two minima in this case are magnetic-polaron states [8]. The process of the MP formation under an in-plane magnetic field with energy transfer to the Zeeman reservoir, discussed in paragraph 2, starts exactly from the central point, because the total spin of the magnetic ions initially lies in the QW plane. The derivative of the potential at this point is zero. The equilibrium is of course unstable, but a fluctuation is needed to push the system away from the dead point. If however we study the initial stage of the MP formation by means of Raman scattering, i.e. detect MRPR, all the extrinsic fluctuations (owed to interaction with phonons, spin relaxation etc) are excluded, because if there were such fluctuations we could have observed no discrete Raman lines. Nevertheless, MRPR has been observed in rather wide QWs also, where the in-plane g-factor is considerable [4]. This apparent contradiction has been resolved by taking into consideration quantum fluctuations of the magnetic-ion spin [15]. This can be done as follows. Let us consider the total magnetic-ion spin $\vec{I}$ in the initial state. According to Eqs.(4) and (5), it can be represented as a harmonic oscillator whose parabolic potential is associated with the magnetic field $\vec{B}$. As the Z and Y components of $\vec{I}$, which play the role of the coordinate and the momentum of the harmonic oscillator, do not commute, we are indeed dealing with a *quantum* oscillator. The interpretation of the discrete energy spectrum of the oscillator is straightforward: the equidistant levels are just the Zeeman sublevels of $\vec{I}$ in the magnetic field. The corresponding wave functions give fluctuations of the transverse components of $\vec{I}$ in states with definite I_X. Switching on the exchange interaction replaces the parabolic potential with the magnetic-polaron potential discussed above (Fig.4). As the wave function of the initial state is not an eigenfunction of this potential, it starts to change with time; furthermore, its width increases. Gradually, it spreads away from the central

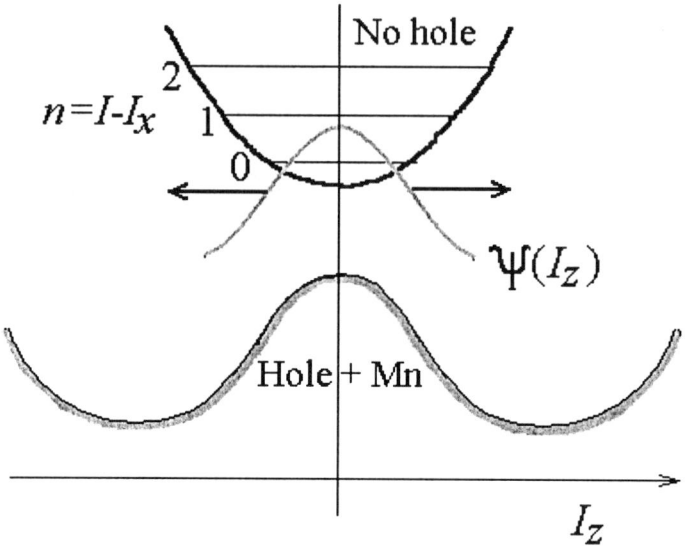

$n = I - I_x$

Figure 4. Quantisation of $\vec{I}$ in the external magnetic field and evolution of its wave function under the adiabatic (magnetic-polaron) potential.

maximum of the potential. This process results in a decrease of I_x, observed as multiple spin flips in the Raman spectrum. Exact calculations [15] yield a result that to some extent contradicts the common view on the quantum-to-classical divide in the mechanics of angular momenta, but is in qualitative agreement with the experimental data. Namely, the mean number of spin-flip peaks as a function of I: it first increases linearly in I according to Eq.(3), and then saturates at the level defined by the ratio of normal-to-plane and in-plane g-factor components of the hole, $\dfrac{g_{//}}{g_\perp}$ (one should just replace I in Eq.3 by $\dfrac{g_{//}^2}{2g_\perp^2}$). As long as the relation $\dfrac{g_{//}}{g_\perp} >> 1$ holds, $<n>$ can be still very large, which explains the observation of MRPR in wide QWs. The apparent paradox consists in the fact that a *macroscopic* change of I_x, entirely owed to *quantum* fluctuations of I, does not vanish in the limit of infinitely large I. In real experiments, I can be as large as several hundred. Such large angular momenta are believed to be purely classical – but this turns out to be not always right. Actually, fundamental laws are not violated because the *relative* change of I_x does vanish at very large I. The paradox is caused by the specifics of Raman-scattering experiments where numbers of quanta rather than classical quantities are measured. Nevertheless, it is a very instructive example of coherent dynamics of nearly classical objects. It should be noted that no detailed experimental study of this phenomenon has been reported so far.

6. Conclusion.

Before making the concluding remarks, some words should be added concerning the methods of experimental detection of the discussed effects. All the effects related to the changes in the spin state of holes are extremely sensitive to the component of the magnetic field along the structure axis, where the hole g-factor is large. If the external field is but slightly deflected from the plane, the Z-component of the magnetic-ion exchange field can completely polarise holes along Z and in this way restore the picture of the constant exchange field of holes. Estimated critical angles for realistic structures and magnetic fields are within a few degrees. This circumstance presents an additional difficulty for the experimentalist, but, on the other hand, gives a method of checking the validity of the results obtained. Both the oscillation frequency in QWs with a hole gas (paragraph 4) and the number of spin-flip Raman peaks in wide quantum wells (paragraph 5) should demonstrate sharp angular dependences.

The main message of this paper may be formulated in the following way: There are several spin-dynamical effects in diluted-magnetic quantum wells, explained within the model of spin precession in an effective field. This model, being very instructive, nevertheless oversimplifies the phenomena, since it neglects the influence of magnetic-ion spins on carriers. In fact, magnetic ions and holes form a strongly coupled spin system that can demonstrate peculiar dynamics. Studying these spin-dynamical effects can give new information on some problems of a general interest, including phase transitions and the quantum-to-classical divide. And, last but not least, this more complex physics still can be understood with simple models.

Acknowledgement.

I am grateful to everybody who contributed to understanding the phenomena touched on in this paper, especially to I.A. Merkulov, D.R. Yakovlev, J. Stuhler, G. Schaack, Yu.G. Kusrayev, and A.V. Koudinov, with whom I have had the pleasure to work in close contact. Partial support of this work by DFG (through SFB 410), RFBR (projects 96-16936 and 96-16941), and by the Volkswagen Foundation is acknowledged.

268

References

1. J.K.Furdyna, "Diluted magnetic semiconductors", J.Appl.Phys. **64**, R29 (1988).
2. M.I.Dyakonov, "Dynamic self-polarisation of magnetic impurity spins under optical pumping by unpolarized light in semiconductors", this conference.
3. D.R.Yakovlev, K.V.Kavokin, I.A.Merkulov, G.Mackh, W.Ossau, R.Hellmann, E.O.Gobel, A.Waag, and G.Landwehr, "Picosecond dynamics of magnetic polarons governed by energy transfer to the Zeeman reservoir", Phys.Rev.B **56**, 9782 (1997).
4. J.Stuhler, G.Schaack, M.Dahl, A.Waag, G.Landwehr, K.V.Kavokin, and I.A.Merkulov, "Multiple Mn^{2+}-spin-flip Raman scattering at high fields via magnetic polaron states in semimagnetic quantum wells", Phys.Rev.Lett. **74**, 2567 (1995).
5. D.R.Yakovlev and K.V.Kavokin, "Exciton magnetic polarons in semimagnetic quantum wells and superlatices", Comments on Condensed Matter Physics **18**, 51 (1996).
6. Yu.G.Kusrayev, A.V.Koudinov, K.V.Kavokin, and B.P.Zakharchenya, "Polarised Raman scattering and Hanle effect in CdTe/CdMnTe quantum wells", in Proceedings 23rd ICPS, Berlin, Germany, 1996, edited by M.Scheffler and R.Zimmermann (World Scientific, Singapore, 1996), p.2459.
7. *Optical Orientation*, edited by F.Meier and B.P.Zakharchenya (North-Holland, Amsterdam, 1984).
8. I.A.Merkulov and K.V.Kavokin, "Two-dimensional magnetic polarons: Anisotropic spin structure of the ground state and magneto-optical properties", Phys.Rev.B **52**, 1751 (1995).
9. S.A.Crooker, D.D.Awschalom, J.J.Baumberg, F.Flack, and N.Samarth, "Optical spin resonance and transverse spin relaxation in magnetic semiconductor quantum wells", Phys.Rev.B **56**, 7574 (1997).
10. B.Kuhn-Heinrich, W.Ossau, E.Banger, A.Waag, and G.Landwehr, "Zeeman pattern of semimagnetic (Cd,Mn)Te/(Cd,Mg)Te quanum wells in in-plane magnetic fields", Solid State Commun., **91**, 413 (1994)
11. Yu.G.Kusrayev, A.V.Koudinov, I.G.Aksyanov, B.P.Zakharchenya, T.Wojtowicz, G.Karczewski, and J.Kossut, "Extreme in-plane anisotropy of the heavy-hole g-factor in (001)-CdTe/CdMnTe quantum wells", Phys.Rev.Lett. **82**, 3176 (1999).
12. A.Haury, A.Wasiela, A.Arnoult, J.Cibert, S.Tatarenko, T.Dietl, and Y.Merle d'Aubigné, "Observation of a ferromagnetic transition induced by two-dimensional hole gas in modulation doped CdMnTe quantum wells", Phys.Rev.Lett., **79**, 511 (1997); P.Kossacki, "Ferromagnetic transition induced by 2D hole gas in CdMnTe quantum wells", this conference.
13. K.V.Kavokin, "Coherent dynamics of localised spins coupled with a two-dimensional hole gas in diluted-magnetic quantum wells", Phys.Rev.B **59**, 9822 (1999).
14. P.A.Wolff, "Spin susceptibility of an electron gas", Phys.Rev. **120**, 814 (1960)
15. K.V.Kavokin and I.A.Merkulov, "Multispin Raman paramagnetic resonance: Quantum dynamics of classically large angular momenta", Phys.Rev.B **55**, R7371 (1997).

NON-MAGNETIC MAGNETIC POLARON

I.A. MERKULOV
Ioffe Physico-Technical Institute RAS, 194021 St.Petersburg, Russia.

Abstract. The theoretical description of a novel magneto-polaron state with a total magnetic moment equal to zero is developed. This nonmagnetic magnetic polaron is formed by an electron in nanometer size heterostructures and is characterised by a sign-variable (antiferromagnetic) order of the magnetic ion spins. A sign-variable distribution of magnetic ion polarisation is related to the fact that the sign and the value of the exchange interaction between the electron and magnetic ions in such structures vary from point to point. The reason for such strong variations of the exchange parameter is an admixing of the amplitudes of the top valence band states to the Bloch amplitude of conduction band states. This admixing opens the possibility of a kinetic exchange interaction between the electron and magnetic ions.

1.

A magnetic polaron is a complex of spins of the localised charge carrier and of the magnetic ions surrounding it, with both parts correlated via the exchange interaction [1]. Usually, the total magnetic moment of such a system of correlated spins may be as great as hundreds of Bohr magnetons (see for example [2]). Such a macroscopically large value is the result of the interaction of the localised carrier's spin with a large number (>1000) of magnetic ions.

However, the spin correlation between a localised charge carrier and the surrounding magnetic ions is not necessarily accompanied by the presence of a large total magnetic moment. If the exchange field, $\mathbf{B}_p$, created at the surrounding ions by the carrier can change sign or direction, the correlation energy of this spin system may differ from zero, even with the total magnetic moment being equal to zero. For example, in the linear approximation $(\mathbf{M(B)} = (dM/dB)\mathbf{B})$ the polaron magnetic moment, $\mathbf{M}_p$, and energy, E_p, are given by:

$$\mathbf{M}_p = \int d^3 r \mathbf{M(r)} = \frac{dM}{dB} \int d^3 r \mathbf{B}_p(\mathbf{r}) \tag{1}$$

$$E_p = \int d^3 r \left(\mathbf{M(r)B(r)} \right) = \frac{dM}{dB} \int d^3 r B_p^2(\mathbf{r}) \tag{2}$$

As one can see, the average value of the exchange field $\langle \mathbf{B}_p(\mathbf{r}) \rangle$ (and $\mathbf{M}_p$) can easily become zero while the local value $\mathbf{B}_p(\mathbf{r})$ and the average field squared has a non-zero value and $E_p \neq 0$.

M.L. Sadowski et al. (eds.), Optical Properties of Semiconductor Nanostructures, 269–278.

In the following we will show that such a surprising magneto-polaron state with $M_p=0$ can emerge in quantum-size (~ nm) heterostructures. We will discuss in detail the magnetic polarons generated by two-dimensional electrons in a quantum well with a width of several nanometers. Specific calculations will be performed for semimagnetic heterostructures prepared from the solid solution $Cd_{1-x}Mn_xTe$, in which the magnetic polarons are detected by optical methods [3].

2.

In semimagnetic semiconductors, the exchange interaction of an electron with the magnetic ions is typically described by the Kondo Hamiltonian:

$$\hat{H}_{ex} = -\alpha \sum_n (s\mathbf{J}_n)\delta(\mathbf{r} - \mathbf{R}_n) \qquad (3)$$

where the parameter of exchange interaction α is positive. For holes, the exchange energy has the same form but the parameter of exchange interaction $\beta < 0$. In Eq (3) $\mathbf{s}$, $\mathbf{r}$ and $\mathbf{J}$, $\mathbf{R}$ are the spins and the radius vectors of, respectively, the electron and the magnetic ion with number n.

The value of the parameter α depends significantly on the electron kinetic energy [4]. The positive sign of α at the bottom of the conduction band (the Γ-point) results from the direct "potential" exchange of the conduction electron with d-electrons of Mn-ions. The negative sign of β is due to the dominant role of the "kinetic" exchange [5], which has been described in term of electron or hole capture in a virtual state in the d-shell [6]. For electrons at the Γ-point these transitions are forbidden by symmetry [5,7].

As the electron kinetic energy, E, increases, the Bloch amplitude of the electron wave function changes: $\Psi = (u_C C_C(\mathbf{k}) + u_V C_V(\mathbf{k}))\exp\{i(\mathbf{k}\mathbf{r})\}$. The Bloch amplitudes of the top valence band states (u_V) begin to admix to it ($C_V(0) = 0$). This admixing creates the possibility of virtual transitions from conduction band states to the magnetic ion d-shell states. Thus the kinetic exchange begins to contribute to the total value of the exchange parameter.

$$\alpha(\mathbf{k}) = \alpha(0)|C_C(\mathbf{k})|^2 + \beta(0)\gamma(\mathbf{k})|C_V(\mathbf{k})|^2 \qquad (4)$$

where the factor γ is given by the formula [4]:

$$\gamma(E) = \frac{(E_v + \varepsilon^+)(\varepsilon^- - E_v)}{[(E_v + E_g + E) - \varepsilon^+][\varepsilon^- - (E_v + E_g + E)]} \qquad (5)$$

and describes the resonant dependence of the kinetic exchange on the distance between the energy levels occupied by the electron and the acceptor ($\varepsilon^- - E_v$) and donor ($E_v - \varepsilon^+$) levels of the Mn^{2+} ion d-shell. For the solid solution $Cd_{1-x}Mn_xTe$ ($\varepsilon^- - E_v) \approx 2.5$ eV [4,10] and ($E_v - \varepsilon^+) \approx 305$ eV [7]. The position of these donor and acceptor levels in the quantum well and in the energy barriers is presented in Fig. 1.

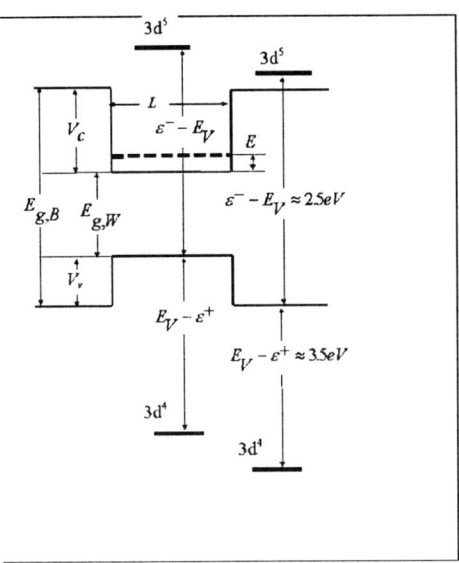

Figure 1. Schematic of the spatial distribution of the carriers' potential energy and of the Mn ion energy levels within the quantum well and the barriers.

The dashed line indicates the position of the electron level of dimensional quantisation; and ε^+ and ε^- are the donor and acceptor levels, respectively, of the Mn ion d-shell. These levels correspond to the presence of four and six electrons in the magnetic ion d-shell.

The kinetic exchange parameter for the states at the top of the valence band is negative, and the admixing of these states to those of the conduction band leads to a decrease of α. For a certain value of the electron kinetic energy (E_0) $\alpha(E_0) = 0$, and for $E > E_0$ α is negative. For electrons with kinetic energy E_0, the magnetisation of the magnetic ions by a uniform external magnetic field does not lead to the giant Zeeman splitting of spin levels: $\mathbf{B}_{ex} = (N_0 x \alpha(E_0)/\mu g_e)\langle \mathbf{J} \rangle = 0$. Here N_0 is the number of cations per unit volume, x is the concentration of Mn, μ is the Bohr magneton, g_e is the electron g-factor, and $\langle \mathbf{J} \rangle$ is the average value of the spin of the magnetic ions. At the same time, the average value of the exchange field created by a localised electron with kinetic energy E_0 at the surrounding magnetic ions also becomes zero: $\langle \mathbf{B}_p \rangle = \langle \mathbf{s} \rangle \alpha(E_0)/(V_l \mu g_{Mn}) = 0$. Here, V_l is the localisation volume of the electron, and $g_{Mn} \approx 2$ is the g-factor of the magnetic ion d-shell electrons. Thus, condition (1) is met for electrons with a kinetic energy of E_0. Now let us show that for these localised states the electron exchange field is not equal to zero at certain points of the localisation volume $(\mathbf{B}_p(\mathbf{r}) \neq 0)$. The localised electron wavefunction is a superposition of the standing waves formed by the states at the top of the valence band and the bottom of the conduction band: $\Psi(\mathbf{r}) = u_c \varphi_c(\mathbf{r}) + (\mathbf{u}_v \varphi_V(\mathbf{r}))$ and

$$\alpha(E, \mathbf{r}) = \alpha(0)|\varphi_c(\mathbf{r})|^2 + \beta(0)\gamma(E)|\varphi_V(\mathbf{r})|^2 \qquad (6)$$

In the Kane model framework, the envelope electron wave function corresponding to the upper states of the valence band ($\varphi_V(\mathbf{r})$) is directly proportional to the gradient of the envelope function corresponding to the states at the bottom of the conduction band ($\varphi_C(\mathbf{r})$): $\varphi_V(\mathbf{r}) \propto \nabla \varphi_C(\mathbf{r})$ [8]. At the point $\mathbf{r}_0$, where $\varphi_C(\mathbf{r})$ is maximum, $\varphi_V(\mathbf{r}) = 0$ and the exchange interaction is free of the kinetic part $\alpha_C(\mathbf{r}_0) > 0$. Near this point the electron exchange field at the surrounding magnetic ions is non-zero: $\mathbf{B}_p(\mathbf{r}_0) \neq 0$. Also, its sign is opposite to that of the electron exchange field $\mathbf{B}_p(\mathbf{r})$ at the points where the envelope function $\varphi_V(\mathbf{r})$, corresponding to the valence band states, is comparable with or exceeds $\varphi_C(\mathbf{r})$ ($|\beta(0)| > \alpha(0)$, $\gamma > 1$). Thus, the local values of the exchange field are actually not equal to 0. At the same time, $M_p = 0$ and $E_p \neq 0$ (see (1), (2)).

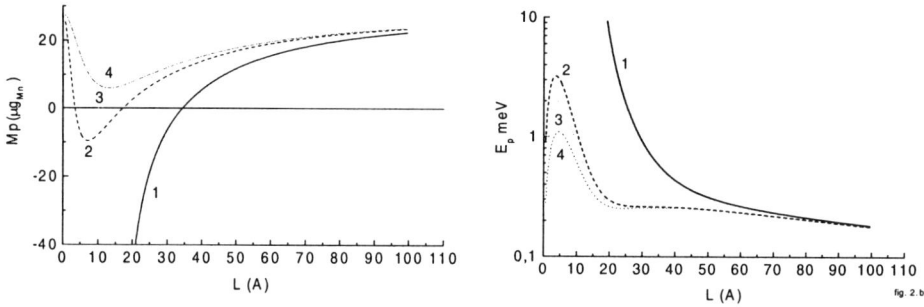

Figure 2. Dependence of magnetic moment (2a) and magnetic polaron energy (2b) on the quantum well width for heterostructures with a constant concentration ($x = 0.1$) of magnetic ions within the well and the barrier. The curves are calculated for the following quantum well depths:1 - $V_C = \infty$, 2 - $V_C = 600$ meV, 3 - $V_C = 555$ meV, 4 - $V_C = 500$ meV. For quantum wells with barriers above 555 meV, curve M_p (B, L, V_C) intersects the $M_p = 0$ line at certain values of the quantum well width. For these values the polaron magnetic moment is equal to zero but the magneto-polaron energy is not. However, no characteristic features occur in $E_p(L)$ at these points.

3.

Now let us consider the more specific case of a magnetic polaron created by a 2-dimensional electron localised at some non-homogeneity of a quantum well. Here we assume that the electron binding energy is small compared to the confinement energy. Also for simplicity we neglect the spin-orbital splitting of the valence band. For the conduction band states considered below, the error of this approximation is less than 10%.

Then, using the Kane model, one can write the wave function for the localised two-dimensional electron states in the following form [8,9]:

$$\Psi_{\pm 1/2}(\vec{\rho},z) = \begin{cases} \left[C_{C,B}|S,\pm 1/2\rangle + C_{V,B}|Z,\pm 1/2\rangle\right]\exp\{-Q(z-L/2)\}\dfrac{\psi(\vec{\rho})}{\sqrt{N}}, & L/2 < z \\[2mm] \left[C_{C,W}\cos(qz)|S,\pm 1/2\rangle + C_{V,W}\sin(qz)|Z,\pm 1/2\rangle\right]\dfrac{\psi(\vec{\rho})}{\sqrt{N}}, & -L/2 < z < L/2 \\[2mm] \left[C_{C,B}|S,\pm 1/2\rangle - C_{V,B}|Z,\pm 1/2\rangle\right]\exp\{Q(z+L/2)\}\dfrac{\psi(\vec{\rho})}{\sqrt{N}}, & z < -L/2 \end{cases} \quad (7)$$

where $\psi(\rho)$ is the envelope function of the localised electron state in the plane of the quantum well, L is the quantum well width, $|S,\pm 1/2\rangle$ and $|Z,\pm 1/2\rangle$ are the Bloch amplitudes of the electron states at the bottom of the conduction band and at the top of the valence band, which admix to each other when a z-component of the electron wave vector is nonzero ($\pm q$ and $\pm iQ$ within the quantum well and the barriers, respectively).

$$\left(\frac{\hbar^2\langle S|\partial/\partial z|Z\rangle q}{m_0}\right)^2 = E(E+E_g), \quad \left(\frac{\hbar^2\langle S|\partial/\partial z|Z\rangle Q}{m_0}\right)^2 = (V_C - E)(E+E_g+V_v) \quad (8)$$

$$\frac{C_{C,W}}{C_{V,W}} = -\sqrt{\frac{E_g+E}{E}}; \quad \frac{C_{C,B}}{C_{V,B}} = -\sqrt{\frac{E_g+V_V+E}{V_C-E}} \quad (9)$$

Here, E is the electron kinetic energy measured from the bottom of the conduction band in the quantum well, m_0 is the free electron mass, $\langle S|\partial/\partial z|Z\rangle = i\langle S|\mathbf{p}_z|Z\rangle/\hbar = P/\hbar$ is the matrix element of the gradient calculated using functions of the bottom of the conduction band and the top of the valence band, E_g is the energy gap in the quantum well, V_C and V_V are the energy barriers for the electron and the hole, respectively, and $V_C + V_V = E_{g,B} - E_g$, where $E_{g,B}$ is the energy gap within the barrier (see Fig. 1.). The boundary condition at the wall of the quantum well is determined by the requirement of the electron wave function being continuous upon the transition from the well into the barrier. Thus $C_{C,B} = C_{C,W}\cos(qL/2)$; $C_{V,B} = C_{V,W}\sin(qL/2)$ and at the same time

$$\tan\left(\frac{qL}{2}\right) = \sqrt{\frac{(E_g+E)\,(V_C-E)}{(E_g+E+V_V)\,E}} \quad (10)$$

$$C_{c,w} = \sqrt{\frac{E+E_g}{2E+E_g}}; \quad C_{v,w} = -\sqrt{\frac{E}{2E+E_g}}, \quad 2N = \left(L+\frac{1}{Q}\right) + \frac{E_g\,(Q\sin(qL)+q\cos(qL))}{(E_g+2E)qQ} \quad (11)$$

Using (3) we obtain the following expression for the dependence of the localised electron exchange field $\mathbf{B}_p(\mathbf{r})$ upon the position of magnetic ions:

$$
\mathbf{B}_p(z,\rho) = \begin{cases} 2\left(\tilde{\alpha}_{C,B} + \tilde{\beta}_{C,B}\right)\dfrac{|\psi(\bar{\rho})|^2}{N\mu g_{Mn}}\exp\left\{-2Q\left(z-\dfrac{L}{2}\right)\right\}<s> & \dfrac{L}{2}<z \\[12pt] \left[\left(\tilde{\alpha}_{C,W} + \tilde{\beta}_{C,W}\right)+\left(\tilde{\alpha}_{C,W} - \tilde{\beta}_{C,W}\right)\cos(2qz)\right]\dfrac{|\psi(\bar{\rho})|^2}{N\mu g_{Mn}}<s> & -\dfrac{L}{2}<z<\dfrac{L}{2} \\[12pt] 2\left(\tilde{\alpha}_{C,B} + \tilde{\beta}_{C,B}\right)\dfrac{|\psi(\bar{\rho})|^2}{N\mu g_{Mn}}\exp\left\{2Q\left(z+\dfrac{L}{2}\right)\right\}<s> & z<-\dfrac{L}{2} \end{cases} \quad (12)
$$

where $\langle s \rangle$ is the average spin of the localised electron, $\tilde{\alpha}_{C,n} = |C_{C,n}|^2 \alpha(0)/2$ and $\tilde{\beta}_{C,n} = \beta(0)\gamma(E)|C_{V,n}|^2/2$ are contributions to the electron exchange constant from the mixed states of the bottom of the conduction band and of the top of the valence band.

Substituting (12) into eq. (1) and (2), one obtains:

$$
\mathbf{M}_p = \frac{dM}{dB}\frac{\left[\left(\tilde{\alpha}_{C,W} + \tilde{\beta}_{C,W}\right)L+\left(\tilde{\alpha}_{C,W} - \tilde{\beta}_{C,W}\right)\sin(qL)/q\right]+2\left(\tilde{\alpha}_{C,B} + \tilde{\beta}_{C,B}\right)/Q}{\mu g_{Mn} N}\langle s \rangle \quad (13)
$$

$$
E_p = \frac{dM/dB}{8(\mu g_{Mn} N)^2 \Omega}\left[\begin{array}{l} L\left[3\left(\tilde{\alpha}_{C,W}^2 + \tilde{\beta}_{C,W}^2\right)+2\tilde{\alpha}_{C,W}\tilde{\beta}_{C,W}\right]+4\left(\tilde{\alpha}_{C,B} + \tilde{\beta}_{C,B}\right)^2/Q+ \\ +\left[4\left(\tilde{\alpha}_{C,W}^2 - \tilde{\beta}_{C,W}^2\right)\sin(qL)+\left(\tilde{\alpha}_{C,W} - \tilde{\beta}_{C,W}\right)^2 \sin(2qL)\right]/q \end{array}\right] \quad (14)
$$

where $\Omega = \left(\int \psi^4(\rho)d^2\rho\right)^{-1}$ is the localisation area in the quantum well plane.

In order to determine the dependence of the magnetic moment and the polaron energy on the quantum well width, we utilise a well-known formula for the magnetic susceptibility of a magnetic subsystem in a semimagnetic semiconductor [11]:

$$
dM/dB = 7J_0 x N_0 (\mu g_{Mn})^2 /[6k_B(T+T_0)]
$$

We also use the following values of the parameters for the solid solution $Cd_{1-x}Mn_xTe$: $\beta_{pot}N_0 = 0$, $\alpha_C(0)N_0 = 220$ meV, $\beta_{kin}N_0 = -880$ meV, $E_g = (1.606 + 1.592 x)$ eV, $P^2/m = 21$ eV, $T = 2$ K, $x = 0.1$, $N_0 \approx 7 \times 10^{22}$ cm^{-3} [11], $T_0 = 4.2$ K, $J_0 = 1.04$, $3V_C = 7V_V$. The localisation area Ω was taken to be 10^{-12} cm^2, which corresponds to a disk with a radius equal to the Bohr radius of the Coulomb donor.

Fig.2 shows the dependence of the magnetic moment M_p and energy E_p of a two-dimensional electron magnetic polaron upon the quantum well width. The calculations are performed for $Cd_{0.9}Mn_{0.1}Te$ quantum wells with equal Mn concentrations in the well and in the barrier and for the following values of the barrier height: - $V_C = 7V_V/3 = 500$ meV, 550 meV, 600 meV, as well as for an infinitely large barrier. One can see that for the infinitely large barrier E_p increases and M_p decreases monotonously with the decreasing width of the quantum well (L). For a large quantum well width ($L > 70\,\mathring{A}$) this dependence is determined by the usual law: the product of the polaron energy and polaron volume is constant. For $L < 70\,\mathring{A}$, the polaron energy increases more rapidly than L^{-1}. In this region the rate at which the antiferromagnetic exchange component $\left(\tilde{\alpha}_{C,W} - \tilde{\beta}_{C,W}\right)\cos(2qz)$ increases is higher than that at which the ferromagnetic exchange component $\left(\tilde{\alpha}_{C,W} + \tilde{\beta}_{C,W}\right)$ decreases. For $L \approx 34\,\mathring{A}$ $M_p(34\,\mathring{A}) \approx 0$ and $\left(\tilde{\alpha}_{C,W} + \tilde{\beta}_{C,W}\right) + \left(\tilde{\alpha}_{C,W} - \tilde{\beta}_{C,W}\right)\left\langle\cos(2qz)\right\rangle \approx 0$. For $L \approx 34\,\mathring{A}$ the correlator $\left\langle(\mathbf{M}_p\mathbf{s})\right\rangle$ changes sign.

For quantum wells with finite barriers, the dependence of polaron energy on the quantum well width is more complicated (curves 2, 3, 4 in Fig. 2a and 2b). In the limit of wide and narrow wells, the polaron magnetic moment is positive and its changes are governed by variations in the polaron volume. For $L > 70\,\mathring{A}$, a decreasing quantum well width leads to a decreasing polaron volume and increasing E_p. For $L < 8\,\mathring{A}$, a decreasing quantum well width leads to the repulsion of the electron wave function into the barrier and to an increase of the polaron volume. In the intermediate region $20\,\mathring{A} < L < 70\,\mathring{A}$, the polaron energy is practically independent of the quantum well width, although the polaron magnetic moment diminishes with decreasing L. (M_p=0 for $V = 550$ meV, $L = 10.7\,\mathring{A}$ and $V = 600$ meV, $L = 18\,\mathring{A}$). In this region, the increase of polaron energy with decreasing polaron volume is compensated by the decrease of this energy due to the diminished average value of the exchange constant. The diminution of this constant is manifest in the decrease of the polaron magnetic moment.

Fig. 3 shows the spatial distribution of magnetisation density of magnetic ions for polarons with a total magnetic moment equal to zero (nonmagnetic magnetic polaron). In the case of quantum wells with infinitely high barriers (Fig.3a), the electron wave function is localised within the well. In the centre of the well, spins of the magnetic ions are polarised along the direction of the localised electron spin while near the wall the electron exchange field changes its sign and the magnetic ions are polarised opposite to s.

The results of similar calculations for a quantum well with $V_c = 550$ meV and $L = 10.7\,\mathring{A}$ are presented in fig. 3b. In this case, the sign of the exchange constant within the quantum well does not change, but it is opposite to that of the exchange constant within the barrier. Thus, the magnetisations of Mn ions within the well and the barrier have opposite directions, and the total polaron magnetic moment becomes zero again.

Concluding this section we note that a similar situation will take place in the case of quantum dots with sizes of the order of nanometers. Here, in the case of the electron nonmagnetic magnetic polaron, spherical layers with opposite directions of magnetisation will occur.

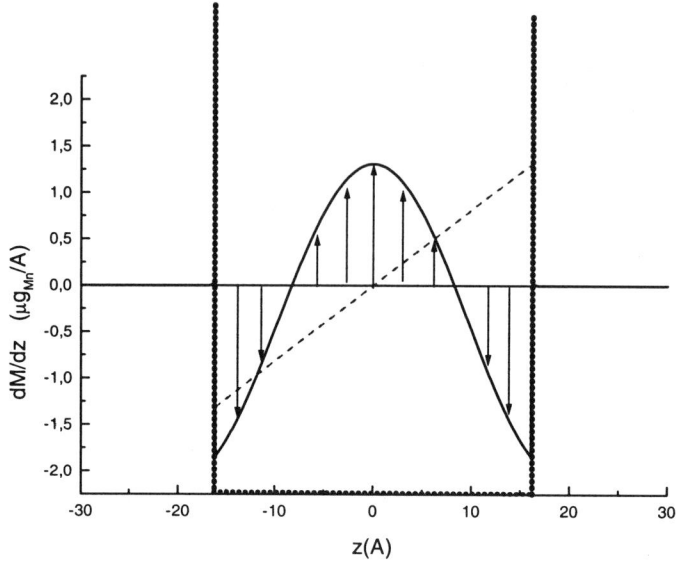

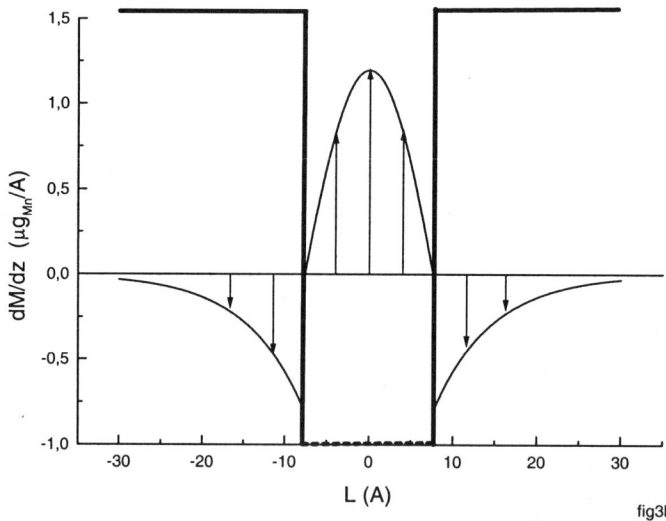

fig3b

Figure 3. Spatial distribution of magnetisation within a nonmagnetic magnetic polaron for relatively low ($Vc = 555$ eV) and extremely high ($V_C = \infty$) barriers. For $V_C = = 555$ meV the magntic moments of the ions in the well and the barrier are in opposite directions and consequently compensate each other. In the case of $V_C = \infty$, ψ is localised within the well only. Here the nonmagnetic magnetic polaron is characterised by the opposite polarisation of magnetic ions in the centre of the quantum well and at the barriers.

4.

Let us now discuss the possible experimental methods of detecting a nonmagnetic magnetic polaron. As can be seen from Fig. 2, the dependence of the polaron energy on the quantum well width does not exhibit any special features when the polaron magnetic moment is equal to zero. Therefore, this dependence does not provide any information about NMP formation.

From our point of view, the most direct experimental evidence for the existence of a nonmagnetic magnetic polaron can be obtained from Raman scattering experiments with electron spin flip [12,4]. In the absence of an external magnetic field the Stokes shift caused by the flip of the electron spin in the magneto-polaron state equals twice the polaron energy ($\hbar\Delta(0) = 2E_p$). In a weak magnetic field the slope of the curve $\hbar\Delta(B)$ yields double the polaron magnetic moment $d(\hbar\Delta(B))/dB \approx M_p$. Thus the dependence $\hbar\Delta(B)$ provides both the polaron energy and its magnetic moment. For an ordinary magnetic polaron, $E_p \propto M_p^2$: the energy of the polaron state decreases rapidly with decreasing M_p. In the case discussed above, of a nonmagnetic polaron independent of external magnetic field ($M_p = 0$), the Stokes shift will exceed 1 meV.

5.

In conclusion, we predict theoretically the existence of an unusual magneto-polaron state with a zero magnetic moment, which is formed by an electron in narrow and deep quantum wells and in other semimagnetic semiconductor quantum-size nanostructures. The appearance of this remarkable magneto-polaron state is due to the fact that the sign and the value of the exchange interaction constant of an electron and of the surrounding magnetic ions in such structures depend upon the magnetic ion position. The characteristic period of the sign-variable, antiferromagnetic component of the exchange parameter is of the same order of magnitude as the heterostructure size. The spatial distribution of the exchange parameter is a consequence of admixing components corresponding to the top valence band. A theoretical description of a novel magneto-polaron state with a total magnetic moment equal to zero is developed. This nonmagnetic magnetic polaron is formed by an electron in nanometer size heterostructures and is characterised by a sign-variable (antiferromagnetic) order of the magnetic ion spins. Sign-variable distribution of magnetic ion polarisation is related to the fact that the sign and the value of the exchange interaction between the electron and magnetic ions in such structures vary from point to point. The reason for such strong variations of the exchange parameter is admixing the amplitudes of the top valence band states to the Bloch amplitude of the conduction band states. This admixing opens the possibility of the kinetic exchange interaction between the electron and magnetic ions. Specific calculations are carried out for quantum wells based on a CdMnTe/CdMnMgTe solid solution. A magnetic polaron with a zero magnetic moment can be detected by Raman spin-flip spectroscopy.

278

Acknowledgements. The author is grateful to D. Yakovlev, K. Kavokin, W. Ossau, G.Landwehr, B.P.Zaharchenia and V.I.Perel' for fruitful discussions. This work was supported by the Volkswagen Stiftung and the Russian Foundation for Basic Research (grant 96-15-96392).

References

1. De Gennes P.G. (1960) Effects of double exchange in magnetic crystals *Phys. Rev.* **118**, 141-154;
   Nagaev E.L. (1992) Self-trapped states of charge carriers in megnetic semiconductors *J.Magn.Mat.* **110**, 39-60;
   Wolf P.A. (1988) Theory of bound magnetic polarons in semimagnetic semiconductors. In J.K.Furdyna and J.Kossut (eds.) *Semiconductors and Semimetals* **25** , Academic Press, London pp. 413-454.

2. Merkulov I.A., Yakovlev D.R., Kavokin K.V., Mackh G., Ossau W., Waag A., Landwehr G. (1995) Hierarchy of relaxation times in formation of an excitonic magnetic polaron in (CdMn)Te, *JETP Lett.*, **62**, 335-339;
   Merkulov I.A., Yakovlev D.R., Kavokin K.V., Mackh G., Kuhn-Heinrich B., Ossau W., Waag A., Landwehr G. (1997) Luminescence polarisation and spontaneous lowerny of symmetry caused by magnetic-polaron formation in semimagnetic-semiconductors quantum wells, *Phys.Solid State* **39**, 1859-1863.

3. Yakovlev D.R. and Kavokin K.V. (1996) Exciton magnetic polarons in semimagnetic quantum wells and superlattices *Comments Condens. Matter Phys* **18**, 51 81.

4. Merkulov I.A., Yakovlev D.R., Keller A., Ossau W., Geurts J., Landwehr G., Karczewski G., Wojtowicz T.and Kossut J. (1999) Kinetic exchange between the conduction band electrons and magnetic ions in quantum-confined structures. *Phys.Rev.Let.* to be published.

5. Bhattacharjee A.K., Fishman G. and Coqblin B. (1983) *Physica* **117&118**, 449.

6. Schrieffer J.R. and Wolff P.A. (1966) Relation between the Anderson and Kondo Hamiltonians *Phys.Rev.* **149**, 491-492.

7. Larson B.E., Hass K.C., Ehrenreich E. and Carlsson A.E., (1988) Theory of echange interactions and trends in diluted magnetic semiconductors. *Phys.Rev.B* **37**, 4137-4154.

8. Suris R.A.(1986) Surface states in heterostructures, *Sov. Phys. Semicond.* **20**, 1258-1261.

9. Ivchenko E.L. and Pikus G.E., (1995) Superlattices and Other Heterostructures, Springer-Verlag, Berlin.

10. Dietl T. (1994), Diluted magnetic semiconductors in S. Mahajan (ed*) Handbook of Semiconductors*, Vol. 3b, North-Holland,Amsterdam, pp.1252-1342.

11. Gaj J.A., Planel R. and Fishman G. (1979) Relation of magneto-optical properties of free electrons to spin alignment of Mn ions in (Cd,Mn)Te , Solid State Commun. **29**, 435-438.

12. .Ramdas A.K. and Rodrigues S. (1988) Raman scettering in diluted magnetic semiconductors, in J.K.Furdyna and J.Kossut (eds.) *Semiconductors and Semimetals* **25**, Academic Press, London pp. 345-412.

SEMIMAGNETIC SEMICONDUCTORS IN STUDIES OF CHARGED EXCITONS

P. KOSSACKI[a,b], J. CIBERT[c], Y. MERLE D'AUBIGNE[c],
A. ARNOULT[c], A.WASIELA[c], S. TATARENKO[c], AND J. A. GAJ[a]

*a) Institute of Experimental Physics, Warsaw University, 69 Hoża,
00-681 Warszawa, Poland*

*b) Department of Physics, Federal Inst. of Technology, CH-1015,
Lausanne, Switzerland*

*c) Laboratoire de Spectrométrie Physique, CNRS et Université Joseph
Fourier-Grenoble, B.P.87, 38402 Saint Martin d'Heres Cedex, France*

1. Introduction

Negatively or positively charged excitons in semiconductors, also known as trions, were studied theoretically by Lampert [1] already in the fifties. However, the first experimental observation of charged excitons by Kheng *et al.*[2] in modulation n-doped CdTe/CdZnTe multiple quantum wells (MQWs) occurred more than thirty years later. Formation of an experimentally detectable complex by three carriers was made possible due to the enhancement of their mutual interaction by confining them in a quantum well. The observation of Kheng *et al.* was rapidly followed by experimental evidence for negatively and positively charged excitons in III-V and II-VI semiconductor quantum structures [3-5].

Semimagnetic semiconductors in application to studies of charged excitons offer a unique opportunity of measuring precisely the effects of spin polarisation of carrier gas without perturbation of wavefunctions. Due to giant spin splittings of carrier states, characteristic for these materials [6], it is possible to control the occupation of spin sublevels by a small magnetic field [9], which does not significantly influence the orbital movement of the carriers. Thus one obtains pure information about population effects, which represent a fingerprint of optical transitions with a magnetically active initial state. Besides the case of charged excitons, effects of initial state spin polarisation occur for creation by light of excitons bound to neutral impurities [7]. When using higher magnetic fields, semimagnetic semiconductors can be useful to study the interplay between spin and orbital effects [5].

Another significant advantage of the use of semimagnetic semiconductors is the possibility of introducing a directly measurable significant variation of transition energies, again using a small magnetic field. In case of charged excitons, especially at higher carrier concentrations, this provides a tool for studying the influence of phase space filling on transition energies without changing the total carrier concentration (spin

279

M.L. Sadowski et al. (eds.), Optical Properties of Semiconductor Nanostructures, 279–289.

independent screening).

In the present paper we shall concentrate on the example of positively charged excitons in modulation doped CdMnTe quantum wells.

2. Samples and carrier concentration control

Quantum wells used for observation of charged excitons must be doped to provide initial state carriers. Modulation doping is often used to remove the impurities from the well and provide as narrow optical structures as possible. To control the carrier concentration Wojtowicz et al. [5] used a step-like thickness profile of the doping layer across the sample for X^- studies, and were able to study the influence of carrier concentration by varying the spot of light incidence on the sample. Another way of studying the influence of the carrier concentration is to control it by above-barrier illumination. We used this method on modulation p-doped CdMnTe quantum wells with CdZnMgTe barriers. The illumination suppresses the concentration of hole gas in the quantum well. The mechanism of this effect [8] is based on the diffusion of electrons created by light into the quantum well. These electrons neutralise the hole gas and under continuous illumination the system approaches a stationary state with reduced hole gas concentration.

The modulation doped structures studied in our work [9] consisted of a single 80Å quantum well (QW) of $Cd_{1-x}Mn_xTe$ embedded between $Cd_{0.66}Mg_{0.27}Zn_{0.07}Te$ barriers grown pseudomorphically on a [100] $Cd_{0.88}Zn_{0.12}Te$ substrate. Such a layout assures a large confinement energy for the holes in the quantum well, minimising at the same time the effects of lattice mismatch. A low Mn concentration in QW (x=0.0018) assured that the line broadening characteristic for mixed semimagnetic material was kept as low as possible, but was large enough to provide a significant Zeeman splitting (~5meV in a magnetic field of 5 T and at a temperature of 1.5 K). The nitrogen-doped region in the front barrier was located at a distance of 200Å. Furthermore, in order to reduce depleting effects, two additional nitrogen doped layers were placed at a distance of 1000Å from the QW on both sides. The nominal hole concentration in the doped structure, evaluated from a self-consistent solution of the Poisson and Schrödinger equations, was $2 \times 10^{11} cm^{-2}$.

The efficiency of the carrier concentration control was determined experimentally by standard magnetooptical measurements. We observed a step-like PLE structure characteristic for a degenerate hole gas, and representing transitions above the Fermi level. The distance between the well-resolved PL line and the energy of the absorption edge (Moss-Burstein shift) gives directly the hole gas concentration. Below a certain value of illumination (which corresponds to about $6 \times 10^{10} cm^{-2}$), the character of the absorption feature changed from step-like to a well-resolved line. This is due to excitonic effects, and this point also limits the applicability of the Moss-Burstein shift as a source of an accurate determination of the hole gas concentration. We achieved an extension of this range by polarising the hole gas in a small magnetic field (up to 1T), and measuring only the circular polarisation corresponding to the transition to the populated spin subband.

In order to obtain carrier concentrations in the range below the level for which sharp absorption lines develop, we applied a simple model describing the neutralisation of the hole gas by the photo-created carriers. We assume [10] that the current of photo-created electrons recombining in the quantum well (participating in the hole gas neutralisation) is proportional to the illumination, and at equilibrium it is equal to the current of holes tunnelling from acceptors into the quantum well. Therefore the hole concentration is determined by the probability of the hole tunnelling through the potential barrier between the quantum well and doped region. The model has two parameters defining the relation between the illumination and the electron current as well as the probability of the tunnelling from acceptor to the quantum well. Due to the complexity of the problem, the parameters were obtained empirically from fitting. The same parameters were used for the extrapolation of the hole gas concentration below the region of the applicability of the Moss-Burstein shift for the determination of carrier concentration.

3. Experimental methods

Photoluminescence (PL) is most frequently used in studies of charged excitons for its experimental simplicity. However, it is difficult to interpret, since it involves not only light-induced creation processes but also relaxation, and is sensitive to competitive recombination channels. Therefore we prefer to limit the use of photoluminescence to the basic characterisation of the structures studied, and rely mainly on absorption related measurements, which probe directly the density of optically accessible states. The information obtained from reflectivity measurements, although in principle equivalent to that provided by absorption, requires usually a more precise description of light interference in the sample, and is therefore less reliable.

Transmission measurements with simultaneous control of the carrier density by interband (blue) illumination present an additional difficulty: the transmission signal is confounded with photoluminescence excited by the blue light. In our experiments we separated the PL signal by switching off the source of transmitted light and subtracted the PL spectra to obtain unperturbed transmission (Fig. 1).

Among the absorption-related methods we would like to distinguish Faraday rotation. Usually it is assumed to originate from Zeeman splitting of the zero-field absorption spectra. However, particularly in the case of charged excitons, the magnetic field can strongly influence intensities of optical transitions, producing amplitude-type rotation spectra [11], similar in shape to those of the refractive index (asymmetric), and completely different from the Zeeman-type ones, resembling rather the derivative of the refractive index (symmetric). This property allows to separate the influence of magnetic field on the energy of a transition from that on its intensity. Due to their high sensitivity, Faraday effect measurements can be used to study very subtle effects.

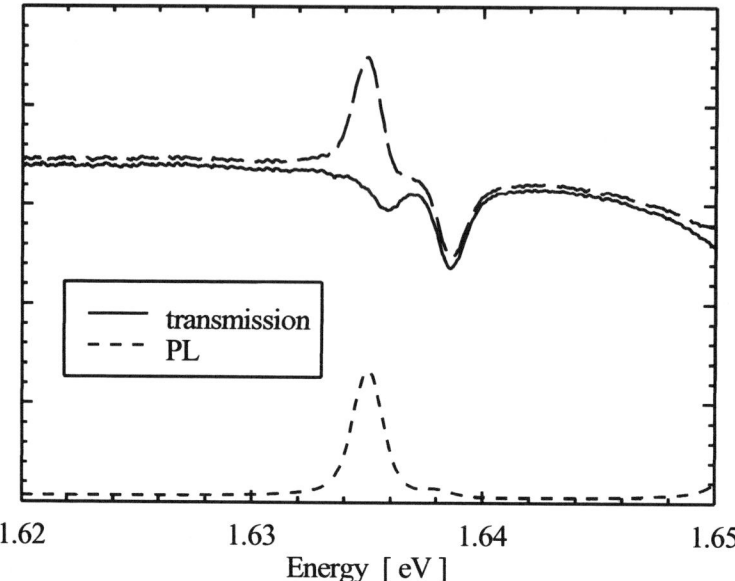

Figure 1. Transmission spectrum of a CdMnTe quantum well measured under additional blue light illumination (dashed line). The contribution of photoluminescence was measured directly without transmission light (dotted) and subtracted from the transmission signal producing a corrected spectrum (solid line).

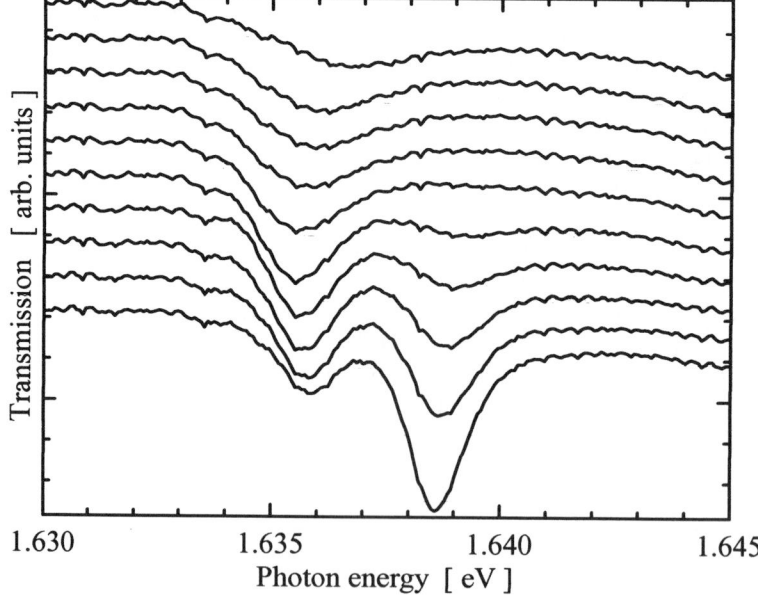

Figure 2. Zero-field transmission spectra for different hole gas concentrations (controlled by additional blue light illumination) varying from $2. x\ 10^{10}$ cm^{-2} (lowest curve) to $1\ x.6\ 10^{11}$ (upper curve). After Ref. [9].

4. Line intensities and their dependence on magnetic field

Figure 2 shows transmission spectra taken at various hole concentrations controlled by inerband illumination. The intensities of both lines vary as a function of hole concentration in the quantum well. The neutral exciton line decreases in intensity, while the X+ line first increases and then at higher concentrations develops into a step-like form characteristic of a Fermi edge singularity.

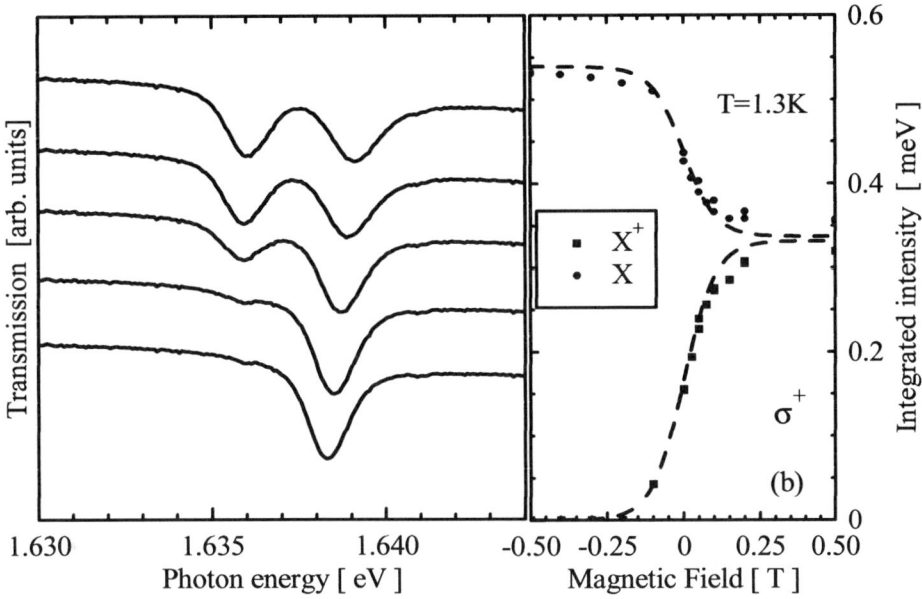

Figure 3. Transmission spectra of a CdMnTe quantum well with hole concentration p = 2·10^{10} cm^{-2}, taken at several magnetic fields (a) and dependence of integrated intensities of X and X$^+$ lines on magnetic field: experiments - points, solid lines - fit based on Eq. 1 (b). The sign of the magnetic field represents circular polarisation (σ$^+$ or σ$^-$). After Ref. [9].

Figure 3a shows transmission spectra at the lowest hole concentration, taken at several magnetic fields in both circular polarisations. We can see a characteristic population effect on the lower energy line: its intensity shifts from the sigma plus (red shifted X line) Zeeman component to the sigma minus (blue shifted X line) one. At a field of about 0.5 T the sigma plus component disappears completely, reflecting the absence of holes in the spin-up valence subband (in sigma plus polarisation the Pauli exclusion principle requires the presence of a spin-up hole for creation of a fundamental singlet X$^+$ state) [2]. This population effect is a fingerprint of spin polarisation in the initial state of the optical transition and represents a strong evidence for the identification of the lower energy line as being due to X$^+$ creation. For a quantitative test we analyse the intensities of both transitions by fitting two Gaussian functions to the optical density spectra (minus logarithm of transmission). The integrated optical density of the respective lines is represented versus magnetic field in Figure 3b. At constant total hole concentration p we expect the intensity of the X$^+$ line to be roughly proportional to the hole population in the appropriate hole spin subband (controlled by magnetic field). Using a simple thermodynamic (Boltzmann) distribution between Zeeman split subbands to describe the hole concentration we obtain for the integrated intensity:

$$A_{\pm}(\Delta) = A(\infty) \frac{1}{1 + \exp(\mp \Delta/k_B T)} \qquad (1)$$

where Δ denotes the valence band Zeeman splitting and is taken to be positive for one and negative for the other spin sub-band, T is the carrier gas temperature, and k_B is the Boltzmann constant. $A(\infty)$ is the maximum intensity value, achieved at total spin polarisation. The indices + and - denote the two circular polarisation's of light, used to select the spin polarisation. The valence band Zeeman splitting is obtained as 80% of the directly measured splitting of the exciton line in the magnetic field, assuming the ratio 1:4 of the conduction- to valence band splitting known for bulk heavy hole excitons [12]. The measured exciton splitting agrees well with that obtained with the use of the phenomenological expressions introduced for the bulk material [13].

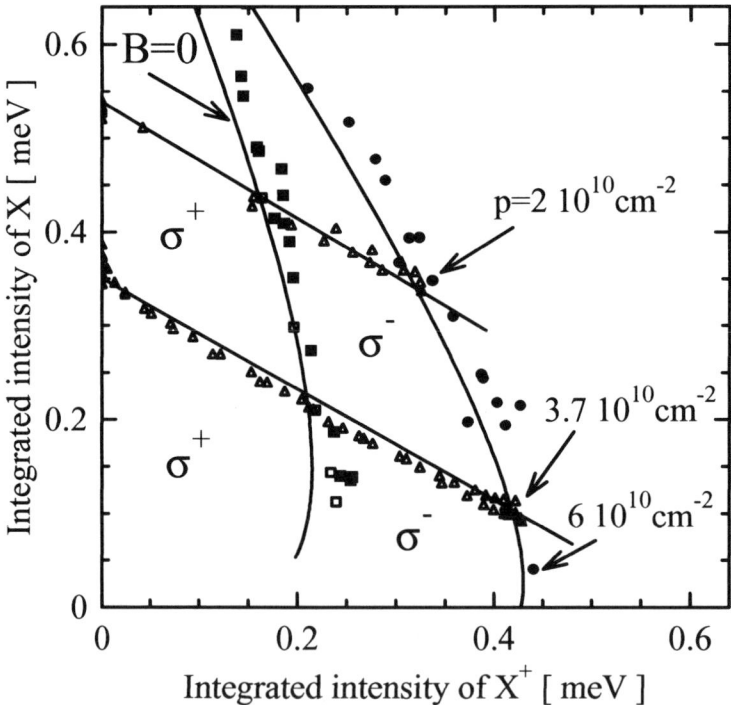

Figure 4. Integrated intensity of neutral exciton X versus that of charged exciton X⁺. Points - experiment, lines - empirical model described in text. After Ref. [9].

The temperature of the carrier gas and the maximum intensity $A(\infty)$ were treated as fitting parameters. Figure 3b shows the good applicability of Equation 1, confirming the proportionality of the X⁺ intensity to the hole population. The carrier gas temperature obtained from the fit is slightly higher than the sample temperature (by less than 0.5K). The obtained difference may be related to the fact that the sample remains under stationary conditions with a continuous diffusion of photoexcited electrons which neutralise the hole gas, but may also heat the carrier gas in the quantum well. One should also note that the Boltzmann distribution is inapplicable for the higher hole gas concentrations (degenerate carrier gas).

The variation of the X^+ intensity is accompanied by a significant opposite change of the X exciton line intensity, which can also be described by Eq. 1 with the same carrier gas temperature. It means that the two changes are proportional to each other. This can be observed directly in Figure 4, where X intensity is plotted vs. X^+ intensity. This effect cannot be explained as a result of the X line width variation as suggested for n-doped CdMnTe quantum wells [14], because the exciton line width remains almost constant in the experimental range of magnetic fields (LWHM=2 meV).

The presentation of X intensity vs. X^+ intensity eliminates the hole gas concentration, which is known with a poor precision, being extracted from a model of population control by light.

5. General empirical description of the influence of carriers on line intensities

Following [9], we may distinguish the following possible sources of the intensity variation:

- Polarisation independent screening (it is influenced by the total carrier concentration p, regardless of its distribution between the spin subbands). We will represent it by an intensity reduction factor $\eta(p)$.
- Phase space filling which is related to the occupation of the hole states participating in the exciton formation, and may play a significant role when the spin subband containing holes forming the exciton is occupied. This effect must vanish when the hole gas is polarised in the opposite spin subband. It may be represented by a reduction factor $\beta(p_\pm)$, where β is a function, in principle different for X and X^+. Of course, $\beta(0)=1$. This factor includes also any polarisation dependence of the screening.
- Presence of carriers in the initial state necessary for X^+ creation. It applies to the spin subband opposite to that participating in the exciton formation
- Intensity stealing - attenuation of the X line due to competition of X^+ creation [2,9]

Before we start an empirical discussion of these effects, we wish to introduce the following assumptions, which will allow us to achieve a simple quantitative approach:

1. The presence of the carriers in the initial state manifests itself in the form of linear proportionality of the (unscreened) X^+ intensity to the concentration of carriers in the opposite spin subband [2,9].
2. The intensity stealing represents a reduction of the X intensity proportional to the hole concentration in the opposite spin subband (number of holes contributing to the creation of X^+ in the same polarisation). We assume a coefficient σ for the relative reduction; it represents the area per hole inaccessible to free exciton creation.

Under the above assumptions the intensity can be written for the neutral exciton as

$$A_{\pm}^N(p, p_+, p_-) = A_0^N \eta_N(p) \beta_N(p_\pm)(1 - \sigma \qquad p_\mp)$$

and for the charged exciton as

$$A_{\pm}^C(p, p_+, p_-) = A_0^C \sigma \qquad p_\mp \eta_C(p) \beta_C(p_\pm)$$

The signs $\pm$ indexing the hole spin subbands in the above expressions refer to the two circular polarisations of light participating in the creation of both types of excitons. Of course, $p_\pm + p_\mp = p$. Introduction of the area per hole σ in the expression for X^+ intensity makes the coefficients A_0^N and A_0^C directly comparable.

We established that to a reasonable approximation the two intensities, measured at constant total hole concentration p, are linear as functions of $p_\pm$ (Figure 3). For the charged exciton it means that the phase space filling reduction factor $\beta(p_\pm)$ is constant, and therefore must be equal 1. For simplicity, we assume the same for the neutral exciton. As a consequence, the two intensities become

$$A_{\pm}^N(p, p_+, p_-) = A_0^N \eta_N(p)(1 - \sigma \qquad p_\mp)$$

and

$$A_{\pm}^C(p, p_+, p_-) = A_0^C \sigma \qquad p_\mp \eta_C(p)$$

Of course, the observed linear relation between the two intensities for a given total hole concentration follows directly from the above equations:

$$A_{\pm}^N(p, p_+, p_-) = A_0^N \eta_N(p) - \frac{A_0^N \eta_N(p)}{A_0^C \eta_C(p)} A_{\pm}^C(p, p_+, p_-)$$

6. Faraday rotation measurements

Figure 5 shows a Faraday rotation spectrum of the p-doped sample used for transmission measurements. The spectrum exhibits an excellent signal-to-noise ratio in spite of a low value of the magnetic field (0.04 T). For its interpretation the presence of two Lorentzian absorption lines was assumed with energy positions and width values determined from the zero-field transmission spectrum (the Lorentzian shape has been chosen for the computational simplicity). A fit of the rotation spectrum was performed, using two contributions for each of the lines: Zeeman- and amplitude-type ones. It is clear from the figure that the amplitude-type contribution to the Faraday rotation is dominant for both lines. The sign of this contribution is opposite for the two lines in agreement with the transmission results: the variation of the X^+ intensity with magnetic field is accompanied by an opposite variation of the X intensity. The much smaller Zeeman-type contributions are also opposite in sign. This result indicates an interesting behaviour of the dissociation energy of the charged exciton as a function of hole concentration distribution between the spin sublevels. We shall discuss it below in more detail.

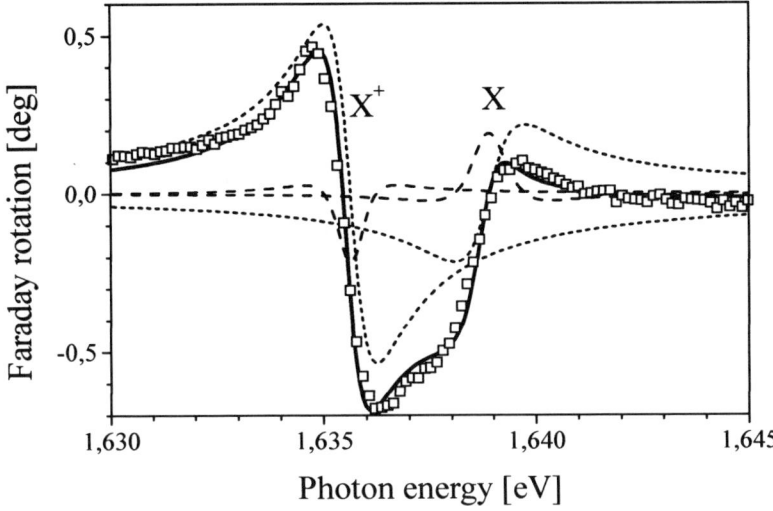

Figure 5. Faraday rotation spectrum of the modulation doped $Cd_{0.998}Mn_{0.002}Te$ quantum well used in [9] with hole concentration $p = 3.x \ 10^{10} \ cm^{-2}$, taken at $T = 1.8$ K and $B = 0.04$ T. Experimental points are fitted by a combination (solid line) of two Lorentzian lines, each containing a Zeeman-type (dashed) and an amplitude type (dotted) contribution.

7. Transition energies

Faraday rotation results reveal a variation of charged exciton energy with magnetic field opposite to that of the neutral exciton. These variations can be studied directly using transmission results (Figure 6).

At relatively high magnetic fields in sigma minus polarisation the energy of X^+ varies with field almost identically as that of the neutral exciton, while the X^+ dissociation energy increases with hole concentration. Observation of the low field region shows that this increase is reduced when the hole concentration in the spin subband containing carriers necessary for X^+ creation decreases. The dependence of the dissociation energy on subband concentration, shown in Figure 7, comes out to be approximately equal for all the values of total hole concentration (some deviation appears at the highest hole concentration, when the X^+ line transforms into a Fermi edge singularity and where the determination of the hole gas polarisation is less accurate).

An extrapolation to a vanishing subband concentration gives a value of 2.5 meV, reasonably comparable with 2.7 meV reported by Haury et al. [4]. The observed increase of the X^+ dissociation energy with subband hole concentration requires a deeper analysis. Theoretical results of Hawrylak et al. [15] predict in fact such an effect. In a simple intuitive picture it may be understood in terms of kinetic energy of the carrier in the initial state of the optical transition.

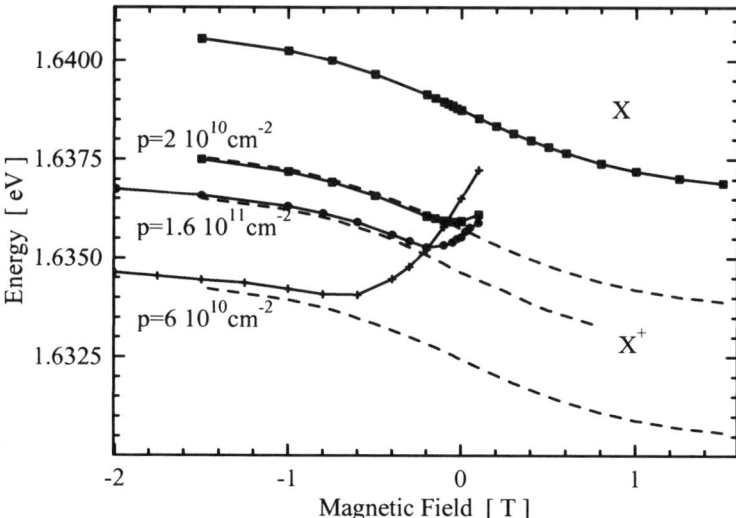

Figure 6. Energies of the absorption lines versus magnetic field measured under different blue light illumination. Sign of the field represents circular polarisation. Dashed lines represent "bulk CdMnTe" Zeeman effect. After Ref. [9].

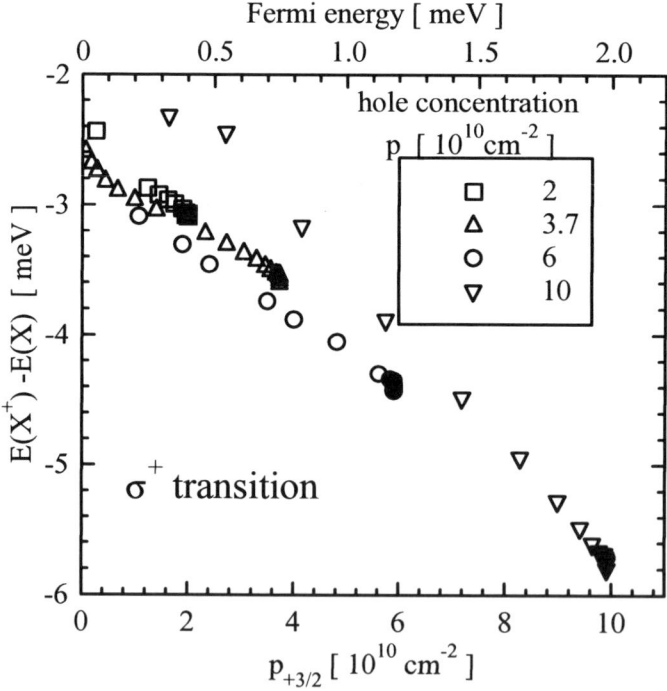

Figure 7. Charged exciton dissociation energy (difference between the X and X^+ energy) as a function of the concentration in the spin subband containing holes present in the initial state of X^+ formation. After Ref. [9].

8. Conclusions

- Application of semimagnetic semiconductors in modulation doped quantum wells allows one to control the distribution of the carrier concentration between the spin subbands by a small magnetic field, negligibly perturbing the electronic wave functions, and therefore to analyse population effects, as opposed to the direct influence of the magnetic field, present in most studies of similar systems.
- By controlling also the total carrier concentration (by means of light or wedge doping) one can distinguish the influence of spin-independent effects (screening) from spin-dependent ones (phase space filling and intensity stealing).
- Faraday rotation represents a sensitive tool to study effects of magnetic field on line intensities and energies.
- The positively charged exciton dissociation energy increases linearly with respective spin subband concentration.

Acknowledgements. This work has been supported in part by Polish Committee for Scientific Research (grant PBZ 28.11) and the French – Polish collaboration program Polonium.

References

1.   M.A. Lampert, Phys. Rev. Lett. **1**, 450 (1958).
2.   K. Kheng, R.T. Cox, Y. Merle d'Aubigné, F. Bassani, K. Saminadayar, S. Tatarenko, Phys. Rev. Lett. **71**, 1752 (1993).
3.   G.Finkelstein, H.Shtrikman, I.Bar-Joseph, Phys. Rev. Lett. **74**, 976 (1995);
     A.J. Shields, J.L. Osborne, M.Y. Simmons, M. Pepper, D.A. Ritchie, Phys. Rev. **B 52**, R5523 (1995).
4.   Haury, A. Arnoult, V.A. Chitta, J. Cibert, Y. Merle d'Aubigné, S. Tatarenko, A. Wasiela, Superlattices and Microstruct. **23**, 1097 (1998).
5.   T. Wojtowicz, M. Kutrowski, G. Karczewski, J. Kossut, F.J. Teran, and M. Potemski, Phys. Rev. B **59**, R10437 (1999).
6.   J.K. Furdyna, J. Appl. Phys. **64**, R29 (1988).
7.   Kudelski, J.A. Gaj, T. Wojtowicz, G. Karczewski, and G. Cywiński, to be published in Solid State Commun.
8.   A.J. Shields, J.L. Osborne, M.Y. Simmons, D.A. Ritchie, M. Pepper, Semicond. Sci. Technol. **11**, 890 (1996).
9.   P. Kossacki, J. Cibert, Y. Merle d'Aubigné, A. Arnoult, A.Wasiela, S. Tatarenko and J.A. Gaj, to be published.
10. E.F. Schubert and K. Ploog, Phys. Rev. B 29, 4562 (1984).
11. W. Maślana et al., to be published.
12. J. A. Gaj, R. Planel, and G. Fishman, Solid State Commun. **29**, 435 (1979).
13. J.A. Gaj, W. Grieshaber, C. Bodin-Deshayes, J. Cibert, G. Feuillet, Y. Merle d'Aubigné, A. Wasiela, Phys. Rev. B **50**, 5512 (1994).
14. D.R. Yakovlev, V.P. Kochereshko, R.A. Suris, H. Schenk, W. Ossau, A. Waag, G. Landwehr, C.M. Christianen, J.C. Maan, Phys. Rev. Lett. **79**, 3974 (1997).
15. P. Hawrylak, Phys. Rev. B **44**, 3821 (1991); S. A. Brown, J.F. Young, J. A. Brum, P. Hawrylak, Z. Wasilewski, Phys. Rev. B **54**, R11082 (1996).

FAR INFRARED RESPONSE OF CdTe STRUCTURES ON GaAs SUBSTRATES

M. L. SADOWSKI[1], M. GRYNBERG[2], AND G. MARTINEZ[1]

[1] *Grenoble High Magnetic Field Laboratory, Max-Plank Institut für Festkörperforschung and Centre National de la Recherche Scientifique Boîte Postale 166, 38042 Grenoble Cedex 9, France*

[2] *Institute of Experimental Physics, Warsaw University, Hoża 69, PL-00-681 Warsaw, Poland*

Abstract. Placing CdTe structures on a GaAs substrate offers the advantage of being able to work close to, or even in, the Reststrahlen band of CdTe. This permits interesting studies of resonant polaron effects. However, we show that far-infrared photoconductivity measurements performed on conducting CdTe layers grown on a semiinsulating GaAs substrate yield unexpected results due to the substrate itself.

1. Introduction

Investigations of phenomena occurring close to the Reststrahlen band in semiconductors are difficult from the point of view of optical spectroscopy. Since light does not penetrate into a semiconductor at frequencies close to the Reststrahlen, it is usually not possible to observe e.g. resonant polaron effects in the most interesting region, close to the resonant frequency. Placing a thin layer of the active semiconductor on a substrate whose Reststrahlen band is well removed in energy from that of the layer creates a possibility of getting around this difficulty. Using photoconductivity instead of e.g. absorption has made it possible to follow the resonant polaron effect in CdTe layers on GaAs all the way up to the resonance [1]. The photoconductive method has the additional advantage that its response at low temperature is much more sensitive than absorption measurements, as the photo-generation of even a relatively small number of conducting

M.L. Sadowski et al. (eds.), Optical Properties of Semiconductor Nanostructures, 291–298.

electrons can lead to large changes in the resistance of a sample. Thus, the combination of such two materials and photoconductivity is extremely appealing. However, care should be taken in interpreting the photoconductivity spectra of thin conducting samples on semiinsulating substrates. Near the Reststrahlen region, which falls in the far-infrared for most semiconductors, the transmission of light through a multilayered semiconductor structure is a complicated process. The optical properties of the component layers change rapidly with the wavelength, and the effects due to consecutive reflections, absorptions and transmissions can lead to quite startling phenomena in optical spectra. Such effects are known to significantly affect line positions or lineshapes [2, 3, 4] in transmission measurements. In this paper we show results of photoconductivity measurements performed on CdTe layers on GaAs substrates and suggest a physical mechanism to explain the strange features found in the spectra. We also show that the mechanism is quite universal and is encountered in different structures, of different materials.

2. Measurements

The samles were equipped with electrical contacts. A constant voltage was applied to these contacts, and changes of the samples resistivity as a function of incident light wavelength were recorded in the spectral range 50 cm^{-1} - 350 cm^{-1} using a Fourier-transform spectrometer (Bruker IFS 113v). Typical PC spectra at zero magnetic field for three different samples are shown in figure 1. The samples, which are described in detail elsewhere [5], consist of a layer of n-doped CdTe on a semiinsulating GaAs substrate. The thickness of the CdTe layer is different for the different samples used. It may be seen that the shape of the spectrum depends strongly on the sample - that is, on the thickness of the CdTe layer. A common feature of all the spectra is the dip around 150 cm^{-1}, which corresponds to the Reststrahlen band in CdTe. Very little light penetrates into the sample at this wavelength, because of the strong reflection coefficient. A small but distinct peak can be seen in some samples at 144 cm^{-1} - the energy of the TO phonon in CdTe. The second feature occurs at energies between 270 - 300 cm^{-1}, which in turn is the Reststrahlen band of GaAs. There is either a peak in the general region of the TO phonon of GaAs (sample 6275C) or very near the LO phonon energy (samples 6225A, 6235B). It should be noted that the positions of these features do not change with magnetic field up to 13 T.

3. The model

A semiconductor lattice absorbs energy resonantly at the TO phonon frequency. This resonance of course has a finite width, moreover we are dealing with a system of three different lattices and four interfaces at which reflections occur. This absorption of light by the lattice results in an increase of the temperature of the sample On the other hand, we know that the conductivity of an n-type semiconductor layer at low (helium) temperatures is usually strongly temperature dependent. In fact, transport measurements performed on the same samples as those used in these experiments have shown [6] that at low temperatures their resistivity is temperature-activated with an activation energy $\triangle E \approx 1\text{meV}$:

$$R = R_0 \exp\left(\frac{\triangle E}{kT}\right) \tag{1}$$

This means that with a resistance of the order of 100MΩ at 4.2K, which is the case here, an increase of the temperature by as little as 10mK will lead to a relative change of the resistance of about 0.8%, which is readily detectable. A doped layer, placed on a semi-insulating substrate, will therefore detect changes in the temperature of the substrate - and of the layer itself. The whole structure - layer and substrate - acts as a bolometer, quite independently of other photoconductive processes which might take place in the layer. The bolometric response of the structure, while not being strictly speaking a photoconductive process in the generally accepted sense of the term, will always be present and should be duly accounted for.

Taking into account only the lattice absorption, neglecting all electronic transitions, we calculated the absorptance of a sample composed of a CdTe layer, a ZnTe layer and a GaAs substrate. To do this, we used a simple model based on the complex refractive index $n^*(\omega)$ of a material, which is connected to the (complex) dynamic dielectric function ϵ^* by the relation

$$(\epsilon^*)^{1/2} = n^* = n + ik. \tag{2}$$

The lattice contribution to the dielectric function for CdTe and GaAs may be described by a single oscillator model. Taking into account the Lyddane-Sachs-Teller relation [7] we have the following expression for ϵ^*

$$\epsilon_l^*(\omega) = \kappa_\infty + \frac{\omega_{TO}^2(\kappa_0 - \kappa_\infty)}{\omega_{TO}^2 - \omega^2 + i\gamma\omega} \tag{3}$$

where ω_{TO} is the frequency of the transverse optical phonon, κ_0 and κ_∞ are the low frequency and optical (high frequency) dielectric constants, respectively, and γ is a damping parameter.

We can now write an electromagnetic wave in the form of $\vec{E}(x) = Ee^{-i(\frac{\omega}{c})nx}$ and, setting appropriate boundary conditions at the interfaces, desribe the response of a semiconductor system composed of N layers by means of a complex 2x2 matrix [8] linking the incident ($\vec{E}_I$), transmitted ($\vec{E}_T$) and reflected ($\vec{E}_R$) waves by means of the relation

$$\begin{pmatrix} \vec{E}_I \\ \vec{E}_R \end{pmatrix} = \mathcal{M} \begin{pmatrix} \vec{E}_T \\ \vec{0} \end{pmatrix} \qquad (4)$$

The transfer matrix $\mathcal{M}$ of the whole system is a product of the transfer matrices of the component layers, $\mathcal{M} = \mathcal{M}_1\mathcal{M}_2...\mathcal{M}_N\mathcal{M}_{N+1}$. The expressions for these may be derived from a consideration of the boundary conditions for the electromagnetic waves at the interfaces. Assuming the incident wave to be monochromatic, the layers to be homogeneous and the response function to be local, the expressions for the individual matrices can be written as:

$$\mathcal{M}_i = \begin{pmatrix} \frac{(n_i^*+n_{i-1}^*)}{2n_{i-1}^*}e^{-i(\frac{\omega}{c})n_{i-1}^*d_i} & \frac{(n_i^*+n_{i-1}^*)}{2n_{i-1}^*}e^{-i(\frac{\omega}{c})n_{i-1}^*d_i} \\ \frac{(n_i^*+n_{i-1}^*)}{2n_{i-1}^*}e^{-i(\frac{\omega}{c})n_{i-1}^*d_i} & \frac{(n_i^*+n_{i-1}^*)}{2n_{i-1}^*}e^{-i(\frac{\omega}{c})n_{i-1}^*d_i} \end{pmatrix} \qquad (5)$$

The index i denotes the $i-th$ layer, and d is the thickness of the layer. The complex reflection and transmission coefficients of the system (from equation 4) are written as $r = \mathcal{M}_{21}/\mathcal{M}_{11}$ and $t = 1/\mathcal{M}_{11}$, respectively. The absorptance, i.e. the energy absorbed in the sample, is then given by

$$A = 1 - rr^* - tt^* \qquad (6)$$

This value should be distinguished from the absorption coefficient α, usually defined in terms of the imaginary part of the refractive index as $\alpha = 4\pi\omega k/c$. The wavelength-dependent absorption coefficient describes the probability of absorbing a photon at a given wavelength, while the absorptance gives the total energy absorbed by the system, and depends on the sample dimensions and its reflectivity. Thus, the spectral dependence of the two may be completely different. This approach takes into account the multiple reflections and refractions of light in the multi-layered structure.

The intensity spectrum of the incident light was approximated by a Gaussian function, fitted to the response of the spectrometer measured with a DTGS detector, a mercury lamp and the same beamspiltter as used in the measurements. The absorption spectrum $A(\omega)$ was then calculated from eq.(6), for layer thicknesses corresponding to those of the measured samples. No adjustable parameters at all were used.

TABLE 1. Material constants used in the calculations

Material	κ_0	κ_∞	ω_{TO} (cm^{-1})	γ (cm^{-1})
CdTe	10.23	7.21	144.7	2
ZnTe	9.86	7.28	180.0	2
GaAs	12.41	10.60	173.1	2

The material parameters used in the calculations are given in table 1. The data for CdTe and ZnTe are taken from [9], while those for GaAs are given after [10].

4. Results

The PC spectra for all samples, together with the calculated absorption spectra, are shown in figure 1. It can be observed that in most cases the calculated absorption spectra recreate the characteristic features of the photoconductivity spectra in such a way as to validate the mechanism of signal generation described in the previous section. It should be noted here that the relationship between the absorptance and the photoconductive response is not expected to be direct and straightforward, for several reasons:

- other photoconductive mechanisms, e.g. photothermal ionisation of shallow donors, are also present - the calculation takes into account only the absorption by the lattice and the resultant heating;

- the bolometric response does not necessarily have to be linear, therefore the relative signal strength does not have to follow the calculated curve exactly;

- the calculation takes no account of the imperfections which must exist at the interfaces.

- we calculate the total energy absorbed in the whole sample; the CdTe layer may reasonably be expected to be more sensitive to temperature changes in its immediate vicinity.

In spite of all this, the shapes of the calculated absorptance spectra reproduce the behaviour of the photoconductive response in a very reasonable way. A dip is observed in the Reststrahlen band of CdTe, usually preceded by a small peak, also visible in most PC spectra. Its energy is close to that of the TO phonon in CdTe. For samples with a $1\mu m$ thick layer of CdTe (6235B) the spectrum is then flat up to the Reststrahlen band in GaAs, where another dip is visible. A peak whose energy usually coincides closely with that of the LO phonon in GaAs follows. Sample 6275C, with a CdTe

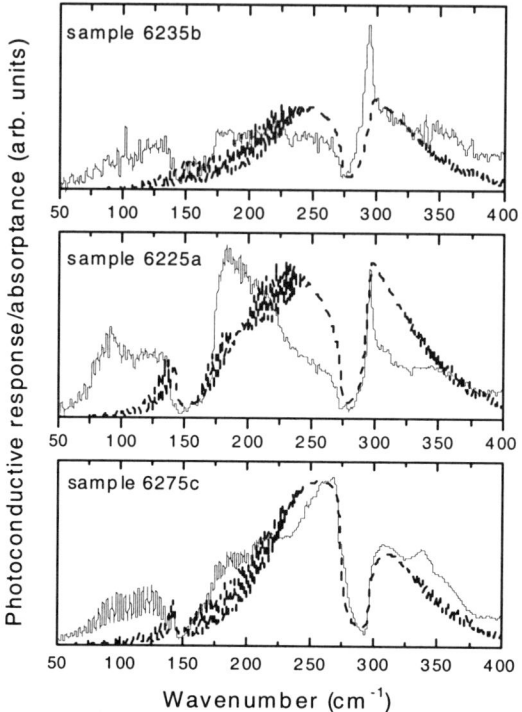

Figure 1. Far-infrared photoconductivity spectra of CdTe samples (solid lines) and absorptance calculated for each sample (dotted lines)

thickness of $3\mu m$, shows a maximum at about 270 cm^{-1} followed by a dip in the Reststrahlen band of GaAs. The thickest sample (6225A) exhibits an increase of absorption, then a dip at the GaAs Reststrahlen band and a peak at the GaAs LO phonon energy.

Deviations from the remarkable similarity between the calculated absorption curves and the measured PC signal may be observed for sample 6275B, where the shape of the spectrum is similar but the position of the maximum is different. The position of the main maximum in the spectrum is rather sensitive to the thickness of the CdTe layer, which might be different from the nominal one, although it must be admitted that there is no thickness which allows the experimental position to be reproduced in the calculation. A possible explanation is that in sample 6275B, which has no buffer ZnTe layer, the lattice mismatch and resulting strain strongly influence the local lattice parameters. Misfit strain is known to change the phonon energies in semiconductor structures [11]. The photoconductive signal at around 100 cm^{-1}, which is larger than the calculated absorptivity for all samples, is due to the electronic shallow donor 1s-2p$_+$ transitions, as demonstrated by magnetic field measurements [1].

5. Conclusions

We propose a physical mechanism to explain the complex photoconductivity spectra of In-doped CdTe layers grown on GaAs substrates. According to this model, changes observed in the photoconductivity of the CdTe:In layer arise from a bolometric-type detection of the heating of the structure due to lattice absorption by the structure components. The local increase of temperature affects the conductivity of the doped layer, which acts as a thermometer integrated with the sample. The intricate and surprising shape of the observed spectra are due to multiple reflections in the multilayered structure. Simple calculations of the absorptance of such a system, performed in the formalism of the dynamic dielectric function, show good qualitative agreement with observed photoconductivity spectra for most of the samples, indicating that the model suggested to explain the origin of the signal is correct. It should also be added that a similar peak in the photoconductivity at the LO phonon energy, has also been observed in PC measurements of thin weakly n-type GaAs layers [14].This feature can perfectly well be recreated using the model presented above [15]. The model is also very efficient in recreating the extremely complicated PC spectra of n-type GaN layers on sapphire substrates [16]. It appears therefore that the bolometric effect described above is a general phenomenon which should always be taken into account when analyzing far-infrared photoconductivity experiments performed on thin conducting layers.

References

1. M. Grynberg, S. Huant, M.L. Sadowski, J. Kossut, T. Wojtowicz,G. Karczewski, J.M. Shi, F.M. Peeters, J. Devreese, Shallow Level Centres in Semiconductors, edited by C.A.J. Ammerlaan and B. Pajot (World Scientific Publishing Co., 1997) p. 1 (1997)
2. M. von Ortenberg, Sol. State Comm. **17**, 1335 (1975)
3. T.A. Kennedy, R.J. Wagner, B.D. McCombe and J.J. Quinn, Sol. State. Comm. **18**, 275 (1976)
4. K. Karraï, S. Huant, G. Martinez and L.C. Brunel, Sol. State Comm. **66**, 355 (1988)
5. M. Grynberg, S. Huant, G. Martinez, J. Kossut, T. Wojtowicz, G. Karczewski, J.M. Shi, J.M. Peeters and J. Devreese, Phys. Rev. B **54**, 1467 (1996)
6. J. Łusakowski, K. Karpierz, M. Grynberg, G. Karczewski, T. Wojtowicz, S. Contreras, O. Callen, Acta Physica Polonica A92, 911 (1997)
7. R.H. Lyddane, R.G. Sachs and E. Teller, Phys. Rev. **59**, 673 (1941)
8. O.S. Heavens, Optical Properties of Solid Thin Films (London: Butterworths Scientific Publications) (1955)
9. Landolt-Börnstein Numerical Data. The Physics of II-VI and I-VII Compounds vol **17b** (Berlin: Springer) (1982)
10. A. Mooradian and G.B. Wright, Sol. State Comm. **4**, 431 (1966)

11. B. Jusserand, P. Voisin, M. Voos, L.L. Chang, E.E. Mendez and L. Esaki, Appl. Phys. Lett. **46**, 678 (1985)
12. F. Gervais and B. Piriou , J. Phys. C: Sol. St. Phys. **7**, 2374 (1974)
13. A.M. Witowski, M.L. Sadowski, K. Pakuła and P. Wyder, Phys. Stat. Solidi (b) **210**, 385 (1998)
14. Zh. Chen, P. Liu, Zh. Chen, X. Shi, G. Shi, S.C. Shen, B. Yang, Zh. Wang, L. Lin, Phys. Rev. Lett. **79** 1078, (1997)
15. M.L. Sadowski, G. Martinez and M. Grynberg, Phys. Rev. Lett. **81**, 2834 (1998)
16. M.L. Sadowski, M. Grynberg, A.M. Witowski, S. Huant, G. Martinez, Phys. Rev. B **60**, 10908 (1999)

COMBINED EXCITON-ELECTRON PROCESSES IN MODULATION DOPED QUANTUM WELL STRUCTURES.

V.P. KOCHERESHKO, D.R. YAKOVLEV, G.V. ASTAKHOV, R.A. SURIS
A.F.Ioffe Physico-Technical Institute, RAS, 194021 St.Petersburg, Russia
J. NÜRNBERGER, W. FASCHINGER, W. OSSAU, G. LANDWEHR
Physikalisches Institut der Universität Würzburg, 97074 Würzburg, Germany
T. WOJTOWICZ, G. KARCZEWSKI, AND J. KOSSUT
Institute of Physics, Polish Academy of Sciences, PL-02608 Warsaw, Poland

Abstract: Combined exciton-electron processes were studied in modulation doped quantum well structures in magnetic fields. In the case of such combined processes an incident photon creates an exciton and induces a transition of an additional electron between two Landau levels. In the present work we found the combined process in which the incident photon creates a trion (negatively charged exciton-electron complex) and also promotes the inter-Landau level transition of an additional electron. Such combined processes are found and studied in CdTe/CdMgTe and ZnSe/ZnMgSSe modulation-doped quantum wells containing two dimensional electron gas of low and moderate density.

Until recently the prevailing opinion has been that the exciton-electron interactions in the presence of a two dimensional electron gas (2DEG) reduce solely to the screening of excitons by free electrons. Thus, the system of excitons and 2DEG was viewed as a homogeneous system without any internal structure. It is clear that within such a picture there is no room for any interesting physical phenomena. Here we show that, while such a view might be valid in the case of a dense 2DEG, in a 2DEG of low and moderate density, i.e., when $E_F \ll Ry$ (E_F being the Fermi energy of the 2DEG, and Ry being the exciton Rydberg), the internal structure of the system of excitons coexisting with electrons is quite rich.

A new type of electron-hole bound state, namely a *negatively charged exciton* or trion was experimentally found in 1993 in quantum well (QW) structures with 2DEG [1]. Since that time many papers devoted to trions have been published. Trions were observed in a number of QW structures based on different semiconductor compounds, such as e.g. CdTe/CdZnTe, CdTe/CdMgTe, CdTe/CdMnTe, GaAs/AlGaAs, and ZnSe/ZnMgSSe [2-4]. Negatively charged trions as well as positively charged trions, related to heavy holes as well as to light holes, were found and studied experimentally

M.L. Sadowski et al. (eds.), Optical Properties of Semiconductor Nanostructures, 299–308.
© *2000 Kluwer Academic Publishers. Printed in the Netherlands.*

[5,6]. A trion triplet state, in addition to the normally observed singlet state, was observed in high magnetic fields [5]. Pico- and femtosecond trion dynamics in the presence of magnetic fields were studied [7]. In support of the experimental studies, theoretical papers devoted to trions started to appear [8,9].

In addition to bound exciton-electron states, resonances involving three-particle states have been observed in the presence of magnetic fields [10]. Such states were termed *Combined Exciton Cyclotron Resonances* (ExCR), meaning that in an external magnetic field an incident photon creates an exciton in its ground state and simultaneously excites one of the resident electrons from the lowest to one of the higher Landau levels. The energy of these transitions is equal to the sum of the exciton energy and a multiple of cyclotron energy.

In the present paper we report the observation of yet another new type of combined exciton-electron processes. These processes involve four particles: one hole and three electrons and are similar to the ExCRs. During these new processes an incident photon creates a trion and simultaneously excites one of the resident electrons from the lowest to one of the higher Landau levels.

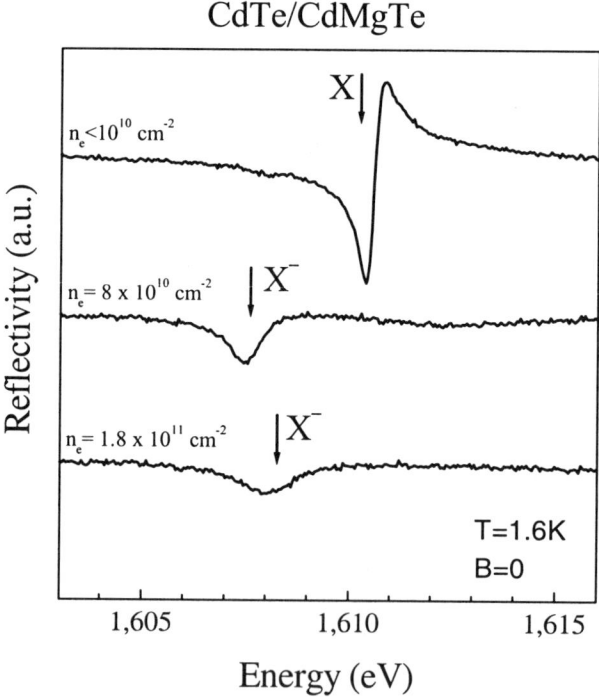

Figure 1. Reflectivity spectra taken at 1.6K in modulation-doped $CdTe/Cd_{0.7}Mg_{0.3}Te$ 80 Å wide single QW structures with different electron concentrations n_e in the QW.

In this study we performed measurements on two different types of structures which were grown by molecular-beam epitaxy on (100)-oriented GaAs substrates. We used modulation-doped CdTe/(Cd,Mg)Te and ZnSe/(Zn,Mg)(S,Se) QW structures with a 2DEG of low and moderate density (the electron density varied from $n_e = 5 \times 10^{10}$ to 2×10^{12} cm^{-2}). A typical structure contains an 80Å single QW and is modulation doped in the barriers at a 100 Å distance from the QW. A special design of the structures made it possible to control the electron concentration while keeping all other QW parameters (QW width, barrier height, background impurity concentration, etc.) constant with a high accuracy [11]. We studied reflectivity, photoluminescence and photoluminescence excitation spectra in magnetic fields applied in the Faraday geometry in the σ^+ and σ^- circular polarisations.

Figure 1 shows the reflectivity spectra of 80 Å wide CdTe/Cd$_{0.7}$Mg$_{0.3}$Te single QWs with different electron concentrations taken at 1.6 K. The upper spectrum corresponds to a nominally undoped structure with the electron concentration in the QW less than 10^{10}cm^{-2}. The lowest spectrum in Fig. 1 corresponds to an electron concentration of 1.8×10^{11} cm^{-2}. Only the exciton line X was revealed in the spectrum taken in the undoped structure. With increasing electron concentration, a negatively charged exciton complex line X$^-$ appears in the spectra. Such states, which are bound exciton-electron complexes, can be observed in modulation-doped QWs even with relatively low electron concentrations.

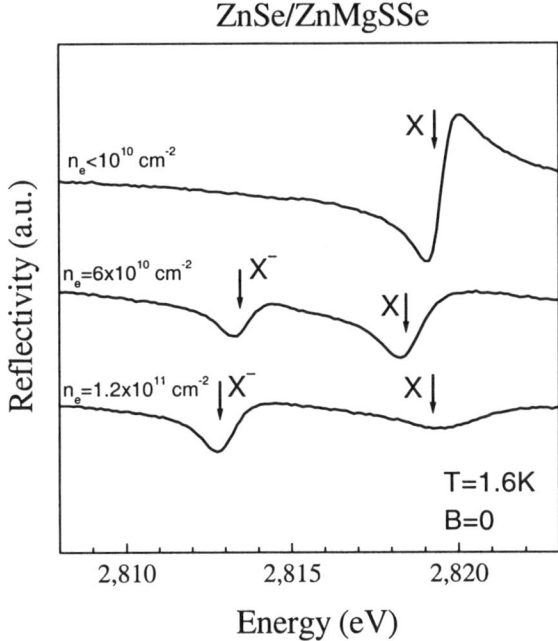

Figure 2. Reflectivity spectra taken at 1.6K in modulation doped ZnSe/Zn$_{0.89}$Mg$_{0.11}$S$_{0.18}$Se$_{0.82}$ 80 Å single QW structures with different electron concentrations n_e in the QW.

We have also observed similar spectra in ZnSe- based QW structures. Figure 2 shows a set of reflectivity spectra taken in modulation doped QW structures of ZnSe/Zn$_{0.89}$Mg$_{0.11}$S$_{0.18}$Se$_{0.82}$ with different electron concentrations. Due to the larger Coulomb interaction energy, the effects of the exciton-electron interactions are even more pronounced as compared to CdTe-based QW structures.

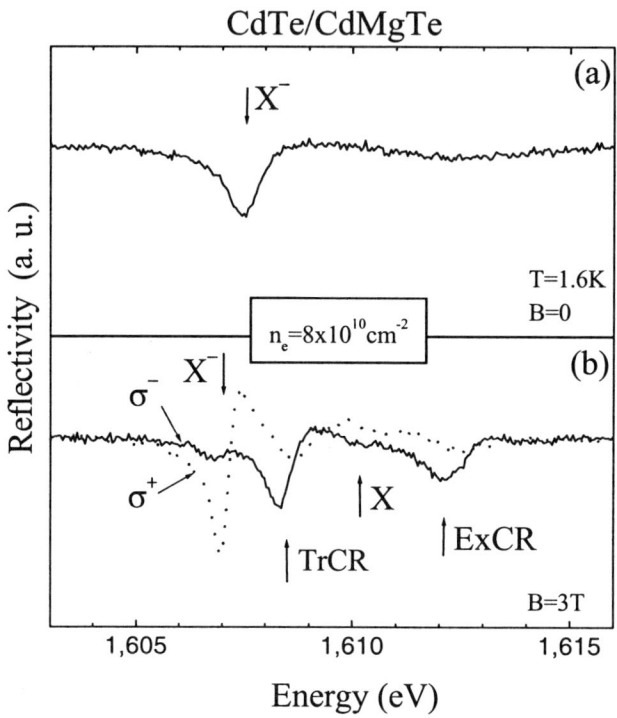

Figure 3. Reflectivity spectra taken in a CdTe/Cd$_{0.7}$Mg$_{0.3}$Te SQW with an electron concentration of 8×10^{10}cm^{-2} (a) at zero magnetic field and (b) at a magnetic field of 3T in σ⁺ (dotted) and σ- (solid) circular polarisations.

Strong modifications of these spectra are observed in magnetic fields. Figure 3 shows these modifications for the QW structure with an electron concentration of 8×10^{10}cm^{-2} for the field increasing from 0 T (Fig.3.a) to 3 T (Fig.3.b). The trion reflectivity line is completely circularly polarised in the presence of the magnetic field. It can be observed in the σ⁺ polarisation only. This strong polarisation of the line reflects the fact that the trion ground state is a singlet. Consequently, in the magnetic field, in which all the background electrons are spin-polarised, the trion can be created for one circular polarisation of the incident photon only, namely, for that in which the background electron and the electron in the exciton have opposite spins.

In addition to the exciton (X) and the trion (X⁻) reflectivity lines, two new lines appear in magnetic fields, i.e., the Combined Exciton Cyclotron Resonance line (ExCR) and the Combined Trion Cyclotron Resonance line (TrCR). The ExCR line appears in the presence of a magnetic field at energies higher than that of the exciton ground state and the TrCR line is observed in between the exciton and the trion reflectivity features.

A similar behaviour is observed in the case of ZnSe based QW structures. Figure 4 shows the reflectivity spectra taken in an 80 Å ZnSe/ZnMgSSe QW structure with an electron concentration of $1.2 \times 10^{11} cm^{-2}$ in the magnetic field B = 4.5T (Fig.4.a) and B = 6T (Fig.4.b). The exciton (X), the trion (X⁻) and the TrCR line were revealed in these spectra.

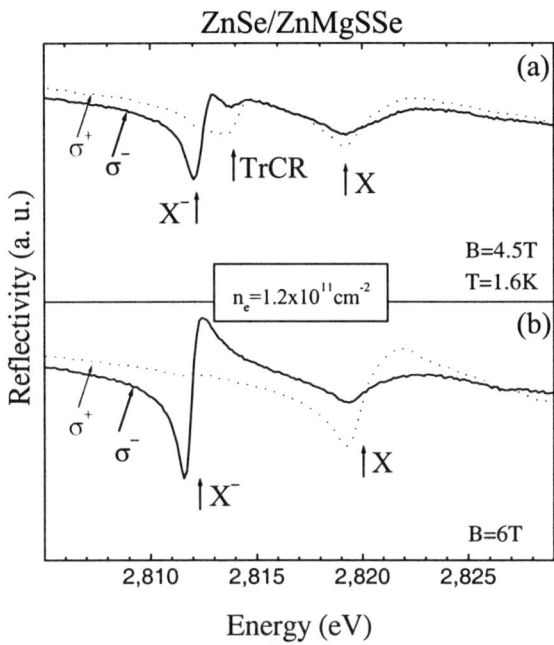

Figure 4. Reflectivity spectra taken from a ZnSe/Zn₀.₈₉Mg₀.₁₁S₀.₁₈Se₀.₈₂ single QW with an electron concentration of $1.2 \times 10^{11} cm^{-2}$ at (a) a magnetic field of 4.5T and (b) a magnetic field of 6T in σ⁺ (dotted) and σ⁻ (solid) circular polarisations.

We have used the effect of strong circular polarisation of the trion reflection line in the magnetic field to determine the electron concentration in the QW. The idea is based on the fact that the circular polarisation of the trion is controlled by the spin polarisation of the 2DEG. No polarisation of the trion reflection line occurs if there is no polarisation of the 2DEG. On the other hand, the polarisation of the 2DEG appears when the lowest Landau level crosses the Fermi energy, i.e., at filling factors $\nu < 1$ (the filling factor defined as $\nu = n_e ch/eB$, with e being the electron charge, B - the magnetic field, and n_e – the electron concentration). Thus, the polarisation of the trion feature in the reflectivity appears under the same conditions. Figure 5 shows the experimentally measured dependence of the trion line circular polarisation $P_{cir} = (I^+ - I^-)/(I^+ + I^-)$, with I^+, I^- representing the trion reflectance line amplitudes in σ⁺ and σ⁻ circular polarisations. The result of a fit of the calculated polarisation (assuming a Fermi-Dirac distribution of the electrons between spin-split levels) to these experimentally observed dependences is shown in Fig.5 by solid lines. Such fits allowed us to determine the electron concentration in the QW.

Photoluminescence excitation (PLE) spectra show even more remarkable modifications in the presence of magnetic fields. Figure 6a shows the photoluminescence (PL) and PLE spectra for a structure with an electron concentration $n_e=8\times10^{10}cm^{-2}$ in the absence of a magnetic field. Two lines were observed in the PLE spectrum - the exciton absorption line X and the trion absorption line X⁻. These two lines have similar intensities in the PLE. The exciton oscillator strength normalised to the unit cell volume is nearly equal to the trion oscillator strength per one electron.

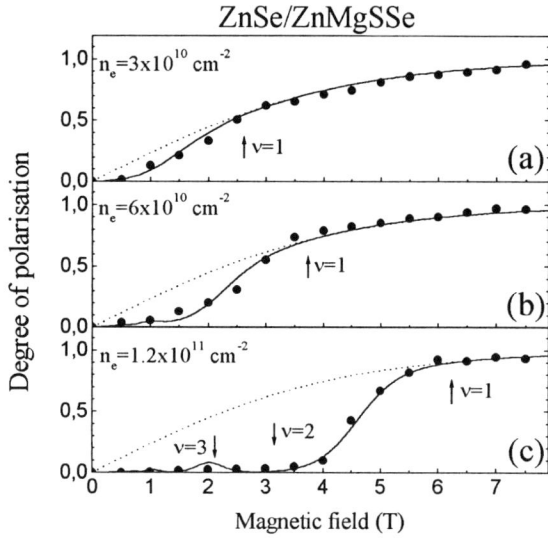

Figure 5. Magnetic field dependences of the circular polarisation of the trion reflectivity line for ZnSe/Zn$_{0.89}$Mg$_{0.11}$S$_{0.18}$Se$_{0.82}$ SQW at different electron concentrations. Circles are the experimental data, the solid line is the result of the fitting.

Thus, at a sufficiently high electron concentration the trion and the exciton contributions to the absorption can be comparable.

In the PL only the trion line can be observed. This fact can be connected to the fast exciton thermalisation and trapping to the trion state.

In the presence of a magnetic field two new absorption lines ExCR and TrCR appear in the PLE spectra (see Fig.6).

Figures 7a and 7b show the magnetic field dependences of the energy position of the exciton X, the trion X⁻, and the combined resonances ExCR and TrCR in the reflectivity for two electron concentrations in the QW for CdTe based structures. The line corresponding to the combined exciton cyclotron resonance ExCR (Fig.4.a) shows a linear shift to higher energies as the magnetic field increases. Its position approaches the exciton ground state when the field is reduced to zero. The slope of this dependence is described by the relation $E_{ExCR}=(1+m_e/M)\hbar\,\omega_c$ [10], where m_e is the electron effective mass, M is the exciton mass, ω_c-the cyclotron frequency. Such combined exciton cyclotron transitions have been reported for the first time in Ref. [10].

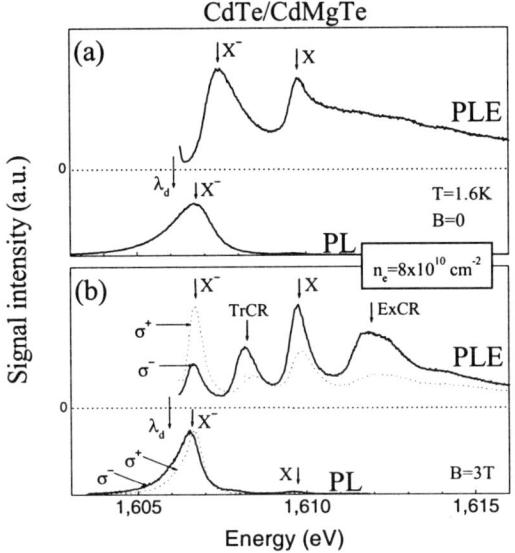

Figure 6. Photoluminescence (PL) and Photoluminescence excitation (PLE) spectra taken in a CdTe/Cd$_{0.7}$Mg$_{0.3}$Te SQW with an electron concentration of 8×10^{10}cm^{-2} at zero magnetic field - a) and at a magnetic field of 3T in σ^+ (dotted) and σ^- (solid) circular polarisations.

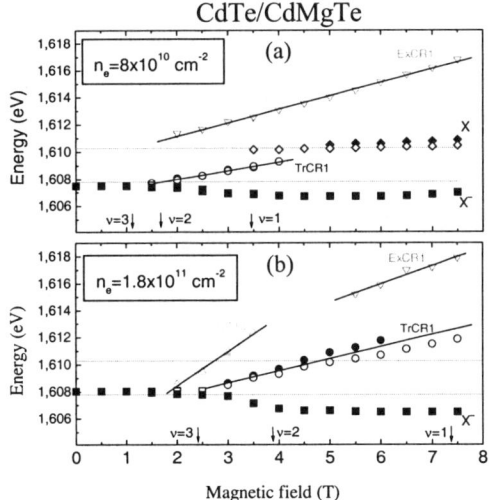

Figure 7. Magnetic field dependences of the energy position of the exciton X, the trion X$^-$, ExCR and TrCR lines for a CdTe/Cd$_{0.7}$Mg$_{0.3}$Te single QW with an electron concentration (a) n$_e$= 8×10^{10}cm^{-2} and (b) n$_e$ = 1.8×10^{11}cm^{-2}. Arrows indicate the values of filing factor. Open symbols correspond to σ^- and closed to σ^+ polarisation.

306

These processes could be explained in the following fashion: an incident photon creates an exciton in its ground state and promotes an excitation of an additional electron from the 0-th to a higher Landau levels.

The combined trion-cyclotron resonance line TrCR shows a different behaviour. This line can only be observed at magnetic fields corresponding to filling factors between 3 and 1. The position of the line tends to the trion energy as the magnetic field vanishes. The line energy depends linearly on the magnetic field. The slope of this dependence is describes by the relation $E_{TrCR} = 1/2\,\hbar\,\omega_c$.

We connect this line with four-particle processes, which involve three electrons and one hole. An incident photon creates a trion and causes another electron to be excited between the Landau levels. In Fig. 7b one can see two transitions of this type: TrCR1 and TrCR2. Thus, in the initial stage we have two electrons in the 2DEG and one photon, then the photon creates an exciton, which binds one electron, and in the final stage we have one trion and one electron on one of the higher Landau levels. The energy of this combined transition is

$$1/2\,\hbar\,\omega_c + 1/2\,\hbar\,\omega_c + ph \rightarrow Tr + 3/2\,\hbar\,\omega_c.$$

The expected energy of the line is thus $E_{TrCR} = E_{Tr} + 1/2\,\hbar\,\omega_c$

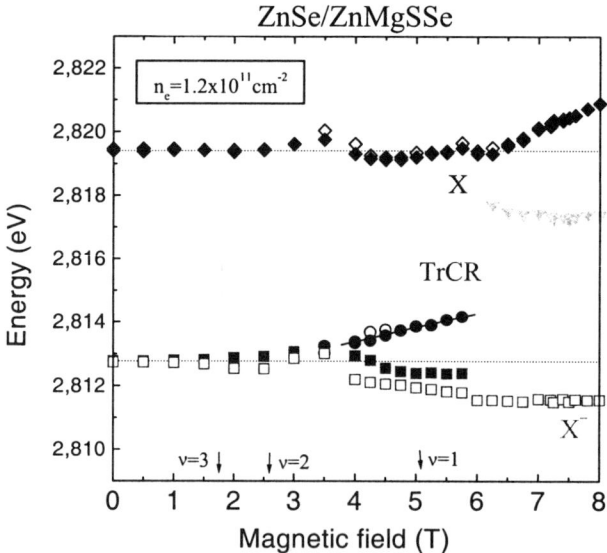

Figure 8. Magnetic field dependences of the energy position of the exciton X, trion X⁻, ExCR and TrCR lines for a ZnSe/Zn$_{0.89}$Mg$_{0.11}$S$_{0.18}$Se$_{0.82}$ SQW with an electron concentration of 1.2×10^{11}cm^{-2}. Arrows indicate the values of the filling factor. Open symbols correspond to σ^- and closed to σ^+ polarisation.

The same transitions were also observed in ZnSe QW structures, see Fig.8. The main feature of these trion-combined transitions is the linear dependence of their energy

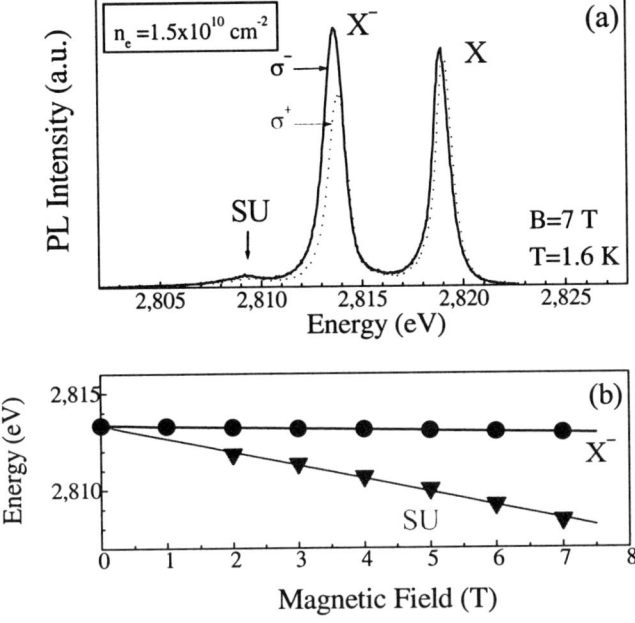

Figure 9. Shake-up processes (SU): (a) – the photoluminescence spectrum taken in a ZnSe/Zn$_{0.89}$Mg$_{0.11}$S$_{0.18}$Se$_{0.82}$ SQW at B= 7T. (b) – The magnetic field dependence of the SU line position.

position, varying as $1/2\,\hbar\,\omega_c$, and the existence of the lines only in the magnetic field range corresponding to filling factors 3 and 1. The fact that the line disappears at filling factors ν<1 also supports the idea that the process requires two additional electrons. Only at filling factors higher than 1 can one find two electrons at the same point in the sample.

The combined processes could be observed not only in the reflectivity or in the absorption but also in PL. Figure 9 shows PL spectra of a ZnSe-based QW structure with 1.5×10^{10}cm^{-2} electrons. On the low energy side of the trion PL line X$^-$ one can see an additional weak line labelled SU. This line is connected with shake-up processes [12] for which the exciton annihilates and simultaneously one 2DEG electron is excited from zero to one of higher Landau levels. The line shifts toward lower energies with the magnetic field as $\hbar\,\omega_c$. The shake-up process is apparently an opposite of the ExCR process.

In conclusion, in CdTe and ZnSe based QW structures, we found a new type of combined exciton processes, which involve four particles: one hole and three electrons. These combined processes were studied in magnetic fields up to 7.5T in reflectivity and PLE spectra.

Acknowledgements: This work was supported in part by RFBR grant 98-02-18219, mutual grant RFBR-DFG No.98-02-04089, NATO Linkage grant LG-974702 and SFB410. The work in Poland was supported by PBZ 28.11

References:

1. K.Kheng, R.T.Cox, Y.Merle d'Aubigné, F.Bassani, K.Saminadayar and S.Tatarenko, Phys.Rev.Lett. **71**, 1752, (1993)
2. K.Kheng, R.T.Cox, V.P.Kochereshko, K.Saminadayar, S.Tatarenko, F.Bassani and A.Franciosi, Superlatt Microstruct. **15**, 253, (1994)
3. G.Finkelstein, H.Shtrikman and I.Bar-Joseph, Phys.Rev.Lett. **74**, 976, (1995)
4. A.J.Shields, M.Pepper, D.A.Ritchie, M.Y.Simmons and G.A.C.Jones, Phys.Rev. B **51**, 18049 (1995)
5. G.Finkelstein, H.Shtrikman and I.Bar-Joseph, Phys.Rev.B **53**, R1709, (1996).
6. A. Haury, A.Arnoult, V.A.Chitta, J.Cibert, Y.Merle d'Aubigné, S.Tatarenko and A.Wasiela, Superlatt. Microstrict. **23**, 1097, (1998)
7. G.Finkelshtein, V.Umansky, I.Bar-Joseph, V.Ciulin. S.Haacke, J.-D.Ganière and B.Deveaud Phys. Rev. B **58**, 12637, (1998)
8. B.Stébé, E.Feddi, A.Anane, F.Dujardin Phys. Rev. B **58**, 9926, (1998).
9. A.Dzubenko to be published in Phys. Rev. Lett.
10. D.R.Yakovlev, V.P.Kochereshko, R.A.Suris, H.Schenk, W.Ossau, A.Waag, G.Landwehr, P.C.M.Christianen, and J.C.Maan, Phys. Rev. Lett. **79**, 3974 (1997).
11. T.Wojtowicz, M.Kutrowski, G.Karczewski, J.Kossut, Acta Physica Polonica A **94**, 199, (1998)
12. K.J.Nash, M.S.Skolnick, M.K.Saker, and S.J.Bass, Phys. Rev. Lett. **70**, 3115 (1993).

A NEW OPPORTUNITY OF PROBING LATERAL DISTORTIONS IN QUANTUM WELLS

A.V. KOUDINOV, YU.G. KUSRAYEV, I.G. AKSYANOV,
B.P. ZAKHARCHENYA
A.F. Ioffe Physical Technical Institute, 194021 St. Petersburg, Russia

T. WOJTOWICZ, G. KARCZEWSKI, AND J. KOSSUT
Institute of Physics, Polish Academy of Sciences, Warsaw

Abstract. The in-plane distortions of quantum well structures as revealed by polarised luminescence are studied. It is shown that C_{2v} distortions induce an extreme anisotropy of the in-plane g factor of quantum confined heavy holes. The anisotropy of the linear polarisation of the photoluminescence turns out to be much less pronounced in structures with a special buffer layer introduced to stop spreading misfit dislocations.

1.

A strong s,p-d exchange interaction in semiconductor quantum well (QW) structures based on diluted magnetic semiconductors (DMS) leads to an unusual sensitivity of the band states to external magnetic fields [1]. On the other hand, the structural properties of such structures are usually quite similar to their nonmagnetic counterparts. The DMS-based QW structures can, therefore, be considered as model systems for the study of magnetic field- and deformation-induced band mixing as well as of the influence of symmetry reduction on the spin structure and spin dynamics of confined states.

In the present paper we report our recent results on the anisotropy of magneto-induced linear polarisation of the luminescence from QWs. First, we discuss the nature of the fourth angular harmonic contents in the angular dependence of the polarisation, which cannot be accounted for by assuming only the anisotropy of the hole g factor [2]. Second, we show that the quality of the buffer layer strongly influences the amount of the second harmonic present in the angular dependence of the PL polarisation from QWs. Third, we show that there exists also a sizeable anisotropy of the polarisation of the emission from the barrier material.

M.L. Sadowski et al. (eds.), Optical Properties of Semiconductor Nanostructures, 309–313.

2.

The structures under study were deposited by standard molecular beam epitaxy on (001)-GaAs substrates, with the exception of sample W, which was deposited on a (001)-CdTe substrate. Sample W contained QWs of $Cd_{0.93}Mn_{0.07}Te$ placed between barrier layers of (Cd,Mg,Mn)Te. Each of the remaining samples contained a set of isolated QWs of CdTe of width ranging from 2 to 10 nm embedded in (Cd,Mn)Te barriers. The manganese content in the barrier material was 11% (sample A), 30% (samples B and C), and 50% in sample D. Samples B and C differed by the buffer layer, separating the QW structure from the substrate. In sample B, the buffer consisted of ZnTe, CdTe and (Cd,Mn)Te layers of 3 µm thickness, while in sample C the substrate was covered with an aperiodic superlattice (10 alternating layers of ZnTe, CdTe and (Cd,Zn)Te of thickness 3-20 nm) followed by a 3 µm layer of CdTe and an 0.7 µm layer of (Cd,Mn)Te.

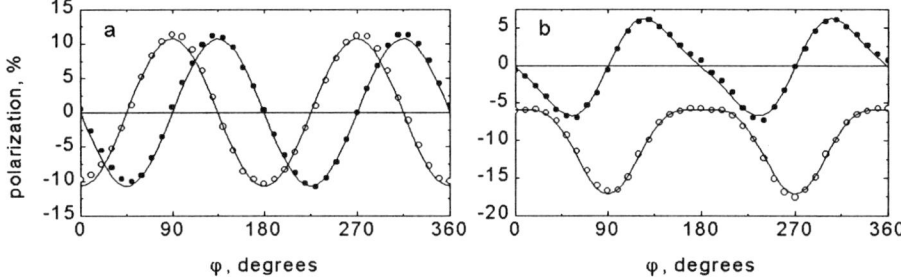

Figure 1. Luminescence polarisation degrees ρ_0 (open circles) and ρ_{45} (closed circles) vs angle φ between magnetic field and [110] crystal axis in $CdTe/Cd_{0.7}Mn_{0.3}Te$ (structure B): (a) QW width L=60 Å, (b) L=20 Å. The solid lines are the results of symmetry-based calculations.

The experiments were performed with standard spectroscopic equipment, including a photoelastic modulator of polarisation and two-channel photon counting electronics. The samples were placed in a helium bath. The magnetic field was perpendicular to the sample growth axis (z-axis). PL was excited by an Ar or a He-Ne laser and detected along the z-axis. We measured the intensities of the luminescence polarised along (I_x) and perpendicular to (I_y) the magnetic field, and also along the directions rotated by 45° ($I_{x'}$ and $I_{y'}$). The two pairs of intensities were used to calculate the polarisation degrees $\rho_0 = (I_x-I_y)/(I_x+I_y)$ and $\rho_{45} = (I_{x'} - I_{y'})/(I_{x'} + I_{y'})$. The experiment consisted of measurements of ρ_0 and ρ_{45} while rotating the sample about z.

Figure 1 shows the dependences of ρ_0 and ρ_{45} on the angle between the magnetic field and [110] crystal axis. The case presented in Fig.1a corresponds to an extreme anisotropy of the g factor of the heavy holes: $g_{xx} = - g_{yy}$ [2]. Such behaviour of the polarisation (when both $\rho_0(\varphi)$ and $\rho_{45}(\varphi)$ contain only second angular harmonics) indicates that the symmetry of the hole states in our QW is C_{2v}. Moreover, the in-plane g factor of the holes is completely induced by an axial in-plane perturbation. In the case of Fig.1b the contribution of the second harmonics remains the dominant one, but,

a   b

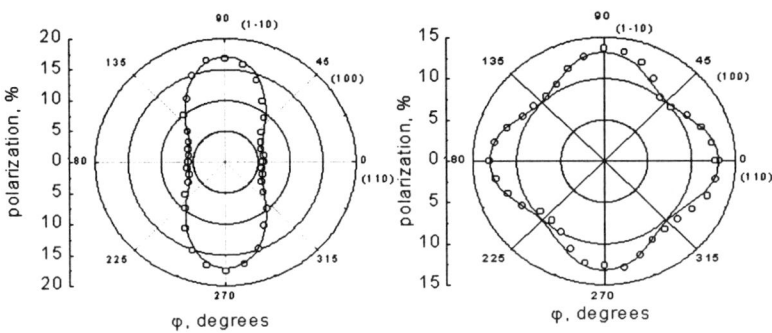

Figure 2. Polar plots of linear polarisation (parameter ρ_0) as a function of angle φ for two QWs with the same width and barrier height: (a) structure B, L=20 A (see fig. 1b), (b) structure C, L=20 A. The solid lines are the results of symmetry-based calculations.

surprisingly, the contributions of the fourth harmonics are also noticeable.

Figure 2 shows a comparison of the symmetry of the ρ_0 signal from the QWs having the same width in samples B and C. The data plotted in Fig.2a are just another representation of the case shown in Fig.1b. The data in Fig.2b correspond to the structure where special measures were taken to prevent spreading of the misfit dislocations originating at the interface with the substrate. One can see that in contrast to previous case, the angular dependence shown in Fig.2b is invariant with respect to rotations by 90°, i.e., the symmetry of radiative states is higher (D_{2d}).

Figure 3 shows the anisotropy of the magnetic-field-induced PL polarisation from the barrier layers in sample B (the anisotropy of bulk (Cd,Mn)Te was reported in Ref. [3].). It is clear that the crystal axes [110] and $[\bar{1}\,10]$ are not equivalent. This fact seems to indicate that the in-plane distortion leading to a reduction of the symmetry exists not only in the vicinity of the heterointerfaces but extends through the whole structure.

3.

An axial (tetragonal) symmetry of the physical properties of QWs grown on (001) substrates would be a natural property. Our experiments show, however, that the application of a magnetic field along the [110] and $[\bar{1}\,10]$ directions leads to different results. This means that the QW structures have orthorhombic and not tetragonal symmetry. Possible reasons of this symmetry reduction are: i) uniaxial in-plane strain

[4], ii) QW interface anisotropy [5], iii) anisotropy of the exciton localisation potential. The expressions for $\rho_0(\varphi)$ and $\rho_{45}(\varphi)$ obtained on the basis of symmetry considerations [2] contain the zero-th, second and fourth angular harmonics. The experimental results for different QWs as well as for the barrier luminescence can be represented by these

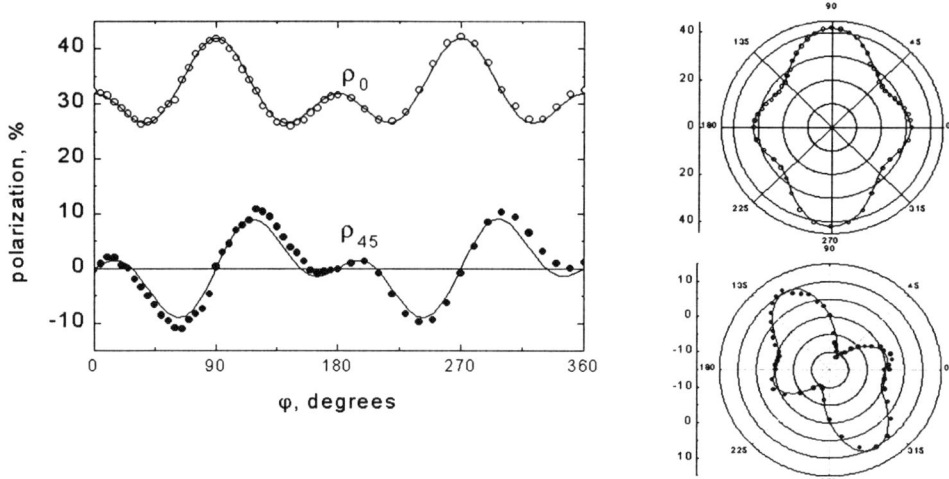

Figure 3. Angular dependences of the degree of linear polarisation ρ_0 (open circles) and ρ_{45} (closed circles) for the barrier luminescence in structure B. On the right hand side the polar plots of the same dependencies are given.

expressions very well (see Figs. 1, 2 and 3). Figure 2 demonstrates that the special buffer in structure C strongly influences the quality of the structure. Such buffers ("aperiodic superlattices") are known to efficiently stop the spreading of misfit dislocations. As a result, our experiment on the 20 Å wide QW in structure C (in contrast to the 20 Å QW in structure B) can be fitted without the second harmonics. Nevertheless, according to our experience, the absence of the second harmonic is rather exceptional. Even in sample C, the signal from wider QWs is not free from the second harmonic content, although it is relatively less intense than in samples A, B, and D. Observation of the anisotropy of the PL polarisation from the barriers indicates that the symmetry reducing perturbation is present in the entire structure. This observation favours the hypothesis of "dislocation" (strain) origin of C_{2v} distortions of the QW states, and contradicts the "interfacial" hypothesis. Because of the differences in dislocation mobilities along [110] and $[\bar{1}10]$, the strain leads to a C_{2v} symmetry [6]. Recent high resolution X-ray diffraction measurements for homoepitaxial CdTe films grown on (001)-CdTe [6] or GaAs [7] substrates, in fact, revealed a difference between the lattice constants of the substrate and the layer.

The presence of the second harmonics in the angular dependences can be explained in terms of the model developed in [4] as being due to the anisotropy of the hole g factor, while the explanation of the presence of the fourth harmonics requires additional assumptions. One possible explanation involves magnetic polarons whose effects in narrow QWs are significant [8]. Namely, the electron spin can align with the magnetic moment of the polaron rather than with the external field. The two directions may differ due because of hole g factor anisotropy. This alignment leads to the appearance of the fourth harmonic. An alternative explanation assumes an anisotropy of the electron g factor, apart from that of the hole g-factor. Further experiments are necessary to clarify this point.

Acknowledgements. Work in Poland was supported by KBN, contract PBZ 28.11, in Russia – by the program "Physics of the Solid State Nanostructures".

References

1. *Diluted Magnetic Semiconductors*, ed. by J. Furdyna and J. Kossut, Ser.: *Semiconductors and Semimetals*, eds. R.K. Willardson and A.C. Beer, Vol. 25 (New York, Academic Press, 1988).
2. Yu.G. Kusrayev, A.V. Koudinov, I.G. Aksyanov, B.P. Zakharchenya, T.Wojtowicz, G. Karczewski and J. Kossut. *Extreme in-plane anisotropy of heavy-hole g factor in (001)-CdTe/CdMnTe quantum wells.* Phys. Rev. Letters, **82**, 3176 (1999).
3. A.V. Koudinov, Yu.G. Kusrayev, B.P. Zakharchenya, V.N. Yakimovich. *Anisotropy of cubic semimagnetic Cd1-xMnxTe solid solutions and magnetic polaron energy from polarised luminescence measurements,* Fiz. Tverd. Tela, **40**, 894 (1998) [English transl. Phys. Solid State, 40, 823 (1998)].
4. C. Gourdon, P. Lavallard. *Fine structure of heavy excitons in GaAs/AlAs superlattices,* Phys. Rev.B **46**, 4644 (1992).
5. E.L. Ivchenko, A.Yu. Kaminskii, I.L. Aleiner. *Exchange splitting of exciton levels in type I and type II superlattices,* Zh. Eksper. Teor. Fiz., **104**, 3401 (1993) [English transl. JETP, **77**, 609 (1993)].
6. H. Heinke, A. Waag, M.O. Moller, M.M. Regnet and G. Landwehr. *Unusual strain in homoepitaxial CdTe(001) layers grown by molecular beam epitaxy.* J. Crystal Growth, **135**, 53-60 (1994).
7. J. Domagała, J. Bąk-Misiuk, J. Adamczewska, Z.R. Żytkiewicz, E. Dynowska, J. Trela D. Dobosz, E. Janik, M. Leszczyński. *Anisotropic misfit strain relaxation in thin epitaxial layers,* Phys. Status Solidi A, **171**, 289 (1999).
8. D.R. Yakovlev, W. Ossau, G. Landwehr, R.N. Bicknell-Tassius, A. Waag, S. Schmeusser, I.N. Uraltsev. *Two dimentional exciton magnetic polaron in CdTe/CdMnTe quantum well structures,* Solid State Communs., **82**, 29 (1992).

MAGNETO-REFLECTION IN CdMnTe/CdTe MULTIPLE QUANTUM WELLS

R. BRAZIS[1], R. NARKOWICZ[1], L. SAFONOVA[1], AND
T. WOJTOWICZ[2]

[1]*Semiconductor Physics Institute, Goštauto 11, Vilnius 2600-LT, Lithuania*
[2]*Institute of Physics, Polish Academy of Sciences, Al. Lotnikow 32/46, Warszawa, Poland*

Magnetic Mn^{++} ions substituting host atoms in A_2B_6 compound crystals, as in CdMnTe, are known to enhance an external magnetic field acting on electron and hole spins. This results in the giant Faraday rotation of the transmitted wave polarisation plane [1]. Unfortunately, light transmission at the exciton lines is strongly reduced. Reflection measurements including the Kerr effect have been recently performed in quantum-well structures in the Faraday geometry with a constant magnetic field normal to the layer plane [2]. Quite different behaviour of excitons is expected in quantum wells in the case of in-plane orientation of the constant magnetic field because the rotational symmetry is broken due to the quantum size effects. Photoluminescence excitation spectroscopy data have been reported in this case in the Voigt geometry [3,4].

This work presents the first observations of non-reciprocal reflection in the CdTe/CdMnTe MQW structures in such geometry.

MQW structures were grown by molecular beam epitaxy on a [100] ZnTe single crystal substrate with a 1μm thick $Cd_{0.95}Mn_{0.05}Te$ buffer layer reducing the lattice mismatch-induced strain. The MQW spatial period implied the 10 nm thick CdTe well layer and the 10 nm $Cd_{0.95}Mn_{0.05}Te$ barrier layer. The total thickness of the structure was 0.3 μm. The light emitted by a halogen lamp was reflected from the crystal at an angle of $\vartheta = 45°$ and analysed with the use of a polariser selecting the p-polarised component of the light beam. The reflected beam spectrum was recorded with the use of a monochromator and a CCD camera. The sample temperature was stabilised at 2 K in a cryostat with a superconducting coil.

P-polarised light reflection from the MQW side of the layered structure is found to exhibit the heavy- and light-hole exciton lines below 1.64 eV related to excitons in the MQW (Fig.1a). A noticeable difference between the reflection coefficients is observed upon magnetic field reversal: $R(\mathbf{B}) \neq R(-\mathbf{B})$. Exciton lines in the difference spectrum (Fig.1b) are better resolved than in the conventional reflection.

Further improvement of resolution is obtained by means of attenuated total reflection (ATR). The ZnTe substrate is shaped in this case so as to obtain a prism. The light-hole exciton line in the MQW structure is found to be stronger than the heavy-hole one in the ATR of p-polarised light. The ATR spectrum in the vicinity of the light-hole exciton line is shown in Fig. 2. Attenuation in this figure is defined as the logarithm of

M.L. Sadowski et al. (eds.), Optical Properties of Semiconductor Nanostructures, 315–318.
© *2000 Kluwer Academic Publishers. Printed in the Netherlands.*

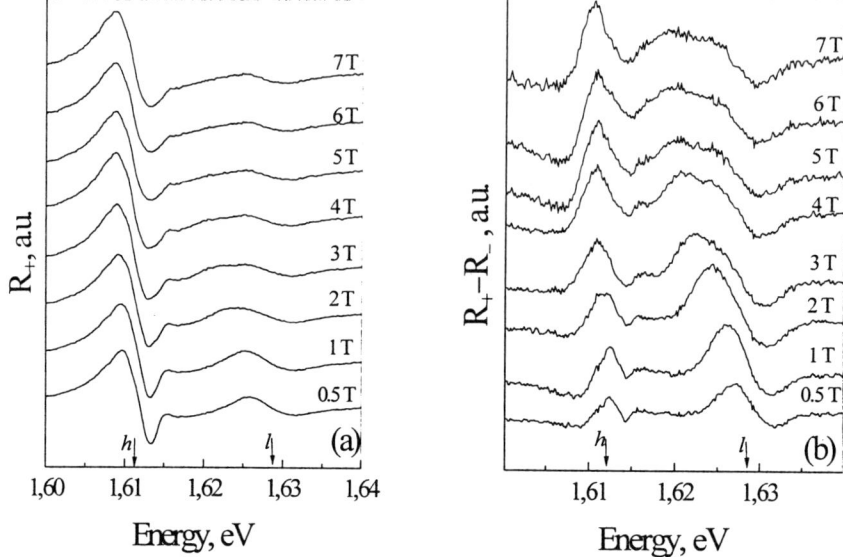

Figure 1. P-polarised light reflection coefficient (a) and the difference of R$_+$–R$_-$ for the CdTe/Cd$_{0.95}$Mn$_{0.05}$Te MQW structures at $\vartheta = 45°$. Arrows at the energies of 1.6123 eV and 1.6284 eV denote the heavy-hole (h) and the light-hole (l) exciton transitions in the wells.

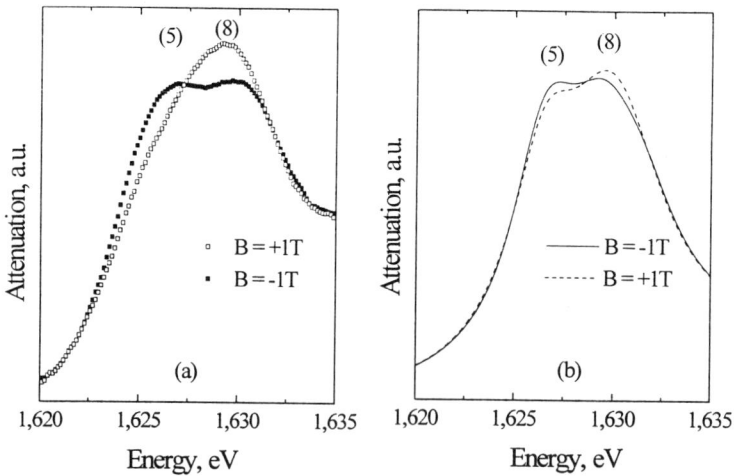

Figure 2. P-polarised light ATR for the CdTe/Cd$_{0.95}$Mn$_{0.05}$Te MQW structures. (a) – experiment, (b) – theory. The prism is the ZnTe substrate, $\vartheta = 45°$

the reflection coefficients R$_\pm$. The light-hole exciton line is seen to start splitting at a specific direction of the constant magnetic field. Therefore the non-reciprocity effect is useful in seeking a better resolution of the Zeeman splitting at low fields. Note that the s-polarised light did not exhibit any non-reciprocity. Experimental results have been modelled using the complex reflection coefficient in the recurrent form [5]:

$$r_j = \frac{\left(\eta_{j-1}^- - \eta_j^-\right)\exp(-2ik_{x,j-1}d_{j-1}) + \left(\eta_{j-1}^+ - \eta_j^-\right)r_{j-1}}{\left(\eta_j^+ - \eta_{j-1}^-\right)\exp(-2ik_{x,j-1}d_{j-1}) + \left(\eta_j^+ - \eta_{j-1}^+\right)r_{j-1}},$$

where the subscript j runs from 1 for the "dark" half-space to p for the illuminated half-space, d_j is the layer thickness, and the j-th material characterising coefficients are

$$\eta_j^\pm = \left(\mathbf{M}_j \times \mathbf{k}_j\right)_x f_j(B^2) \pm g_j(B^2),$$

$\mathbf{M}_j = (0, 0, M_j(B))$ is the magnetisation, $\mathbf{k}_j = (k_{xj}, k_y, 0)$ is the light wave vector, $f_j(B^2)$ and $g_j(B^2)$ are the even functions of the magnetic field determined by the background and the exciton oscillator contributions to the material polarisation. The magnetisation, which arises due to the Mn^{++} ion spin orientation, is an odd function of the steady magnetic field. Therefore, the power reflection coefficient of $R = r_p r_p{}^*$ is subject to change upon the field reversal. Modelling results (Fig. 2b) are in reasonable agreement with experiments. The fitting procedure enables one to obtain the values of exciton parameters in magnetic fields. They are evaluated from the average values of $(R_+ + R_-)/2$ and summarised in Fig. 3. The radiative damping is actually the measure of the oscillator strength [6].

The excitonic transitions (1) and (3) originate from the valence band state $|h-3/2\rangle$, (2) and (4) come from the $|h-1/2\rangle$ state, and those labelled with (5) and (6) come from the states $|l+1/2\rangle$, whereas (7) and (8) belong to $|l+3/2\rangle$. Selection rules impose the observability of the transitions (2, 3, 6 and 7) in the s-polarisation, whereas the other transitions occur in p-polarised light. The notation of states identifies their behaviour at zero magnetic field (h and l refers to heavy and light holes) [3].

There are at least three types of modes responsible for the observed effects. They are classified according to the direction of their dipole moment with respect to the plane of incidence x-z [7]. Transverse (T) modes with the dipole moment perpendicular to the plane of incidence are excited by the s-polarised light. Longitudinal (L) modes refer to the dipole moment orientation along the layer in the plane of incidence, and Z-modes refer to the dipole orientation along the layer normal. They are excited at an oblique incidence of p-polarised light. The modes L and Z are coupled in the presence of steady magnetic field.

In the case of p-polarised light the (5-8) transition energies are seen to be essentially the same for the L and T-modes (Fig. 3a). However, the Z-mode, which can be observed only in the case of the oblique reflection, is remarkably shifted with respect to L-mode, and the observed L-Z energy difference depends on the magnetic field. The radiative damping of the Z-mode, i.e., the corresponding oscillator strength at zero magnetic field is approximately 3 times larger for the Z-mode compared to L and T modes (Fig. 3b), because the Z-mode is directly affected by size quantisation. With the rise of magnetic field the Z-mode related to the transition (5) becomes weaker and comparable with the L-mode of transition (8), whereas the intensity of the transition (8) Z-mode increases.

The non-zero lateral momentum of photons in ATR results in a ~1meV shift of the L-mode with respect to its energy obtained in normal transmission experiments. This cannot be related to the influence of Z-modes, because this transition is forbidden for the heavy hole exciton [7]. No heavy hole line splitting was observed up to 7 T.

318

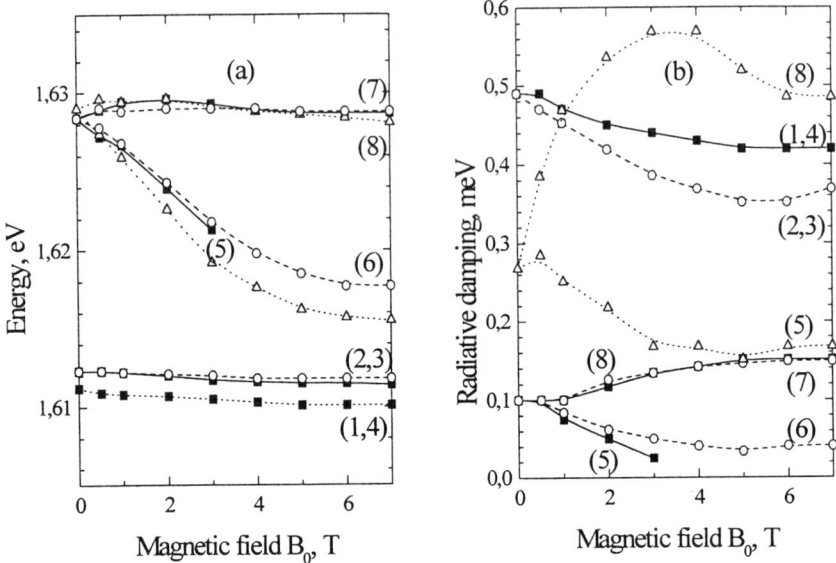

Figure 3. Zeeman splitting (a) and radiative damping (b) in CdTe\Cd$_{0.95}$Mn$_{0.05}$Te MQW in the Voigt configuration obtained from ATR experiments. Solid curves and squares denote L – modes, dotted curves and up-triangles – Z-modes in p–polarisation, dashed curves and open circles - T-modes in s-polarisation. Curves are drawn as guides for the eye.

It seems that the magnetic field is not sufficiently strong to reorient the heavy hole spins towards the field direction.

In conclusion, magnetoreflection measurements at an oblique incidence provide information on the Z-modes which is not accessible in the normal transmission and reflection. The reflection non-reciprocity enables one to refine the exciton line splitting resolution at low magnetic fields.

Acknowledgements. We thank Prof. A. Mycielski for the ZnTe substrates, Dr G. Karczewski for valuable suggestions concerning the multiple QW structure technology, and Prof. J.K. Furdyna for encouraging discussions. The work is partially supported by the Committee for Scientific Research (Poland) through grant PBZ-28-11.

References

1. Eunusoon, Oh, Ramdas, A.K., and Furdyna, J.K. (1992) Magneto-optic phenomena in II-VI diluted magnetic semiconductors: the Faraday and the Voigt effect, *Journ. Lumin.* **52**, 183-191.
2. Testelin, C. and Rigaux, C. (1997) Resonant magneto-optic Kerr effect in CdTe/Cd$_{1-x}$Mn$_x$Te quantum-well structures, *Phys. Rev.* B**55**, 2360-2367.
3. Peyla, P., Wasiela, A., Merle d'Aubigné, Y., Ashenford, D. E. and Lunn, B. (1993) Anisotropy of the Zeeman effect in CdTe/Cd$_{1-x}$Mn$_x$Te multiple quantum wells, *Phys. Rev.* B**47**, 3783-3789.
4. Kuhn-Heinrich, B., Ossau, W., Bagnert, E., Waag, A. and Landwehr, G. (1994) Zeeman pattern of semimagnetic (CdMn)Te/(CdMg)Te quantum wells in inplane magnetic fields, *Sol. State Commun.* **91**, 413-418.
5. Brazis, R. and Kunigelis , A. (1992) Hole cyclotron resonance in germanium in DC electric and magnetic fields: E$_0$‖B$_0$‖⟨100⟩, *Lithuanian Journ. Phys.*, **32**, 147-157.
6. Narkowicz, R., Brazis, R. and Safonova, L. (1998) Exciton-polaritons in CdTe/CdMnTe multiple quantum well structures, *Solid State Commun.* **108**, 229-233.
7. Tassone, F., Bassani, F. and Andreani, L. C. (1992) Quantum-well reflectivity and exciton-polariton dispersion, *Phys. Rev.* B**45**, 6023-6030.

OPTICAL PROPERTIES OF QUANTUM DOTS

PAWEL HAWRYLAK
Institute for Microstructural Sciences
National Research Council of Canada
Ottawa, Canada K1A 0R6

Abstract. Theoretical investigations of optical properties of quantum dots are reviewed. Emission, absorption and inelastic light scattering as probes of the ground and excited states of quantum dots charged with electrons are discussed. Special attention is paid to Hund's rules and their manifestation in the emission spectrum from self-assembled charged quantum dots. Band-gap renormalisation, shake-up, and ground state emission are discussed. It is shown that partially filled shells lead to break-down of the optically created hole in the electronic state at the bottom of the filled band and a splitting of the emission line. The electronic structure and emission spectrum of quantum dots filled with excitons is reviewed. Hidden symmetries replace Hund's rules and excitons condense into correlated multiplicative states. The emission spectrum is related to the number of excitons in these excitonic artificial atoms.

1. Introduction

Recent advances in the growth and processing of quantum dots [1] in which both electrons and valence holes can be confined opened up the possibility of optical investigation of their electronic properties. This progress in growth is coupled with progress in "single dot spectroscopy", where individual quantum dots can be interrogated optically [2, 3, 4, 5, 6]. It is therefore neccessary to relate different optical spectroscopies to the electronic properties of quantum dots.

Modulation doped quantum dots contain electrons while optically or electrically pumped dots contain electrons and holes. Their electronic properties are determined by the number and structure of bound single particle levels, the form of the Coulomb interaction among carriers, and many par-

M.L. Sadowski et al. (eds.), Optical Properties of Semiconductor Nanostructures, 319–336.
© *2000 Kluwer Academic Publishers. Printed in the Netherlands.*

ticle configurations which depend strongly on the number of carriers. The single particle levels of electrons and holes in smooth confining potentials are described by the coupled harmonic oscillator spectrum with characteristic degeneracies and shell spacing ω_0. The electronic structure depends on the ratio of the quantisation of the kinetic energy ω_0 to the interaction energy $V_0 \approx \sqrt{\omega_0}$. For large confining energies, electronic quantum dots satisfy Hund's rules[7] while excitonic quantum dots are controlled by "hidden symmetries" [8, 9]. For small confining energies many different correlated phases are expected [10, 11, 12, 13, 14]. Quantum dots containing free carriers are interesting examples of strongly correlated electron and electron-hole systems. For confined electrons one might expect that their electronic structure, and in particular electronic correlations, are a function of the confining potential. However, this is not the case. In smooth confining potentials, the centre of mass motion does not depend on the confining potential [12] and the electron-electron interaction contribution to the energy, and hence correlations, is the same as in a translationally invariant system. Hence quantum dots offer a unique insight into correlated electronic states, both already existing in extended quasi-two-dimensional systems, and those particular to finite systems. This is particularly true in strong magnetic fields, where these correlated states are analogues of incompressible liquids encountered in the FQHE [10, 11, 14], or spin excitations encountered in Quantum Hall ferromagnets [14, 15, 16]. The second obvious property of quantum dots is that they are finite electronic systems and therefore have edges. In a strong magnetic field the edge excitations of quantum Hall droplets are examples of chiral Fermi and Luttinger liquids [17, 18]. We attempt here to relate some of these electronic properties of quantum dots to what one might observe, or has observed, in emission, absorption, and Raman scattering experiments.

In Chapter 2 we investigate optical probes of the electronic structure of modulation doped quantum dots. For example, we show that the dependence of the ground state energy on the number of electrons and/or the magnetic field can be probed by measuring emission on acceptors [19, 20]. The spectral function of the "hole"(vacancy) in modulation doped quantum dots can be measured by the recombination of valence holes with electrons. The deviations from the single particle result in this spectral function are due only to electron-electron interactions [7, 21]. The two-particle spectral function, excitation spectrum and its evolution with magnetic field can be obtained from resonant Raman spectra [22, 23, 24, 25, 26]. Absorption/emission spectra can probe correlated electronic states [27] in quantum dots in a strong magnetic field, and in particular edge excitations, and their chiral Fermi and Luttinger liquid character.

In Chapter 3 we investigate factors which control the optical properties of quantum dots filled with N excitons. Calculations of the electronic structure of these excitonic artificial atoms are reviewed. The role of "hidden symmetry" in the emission spectrum as a fingerprint of the number of excitons in a quantum dot is analysed in connection with experiments.

2. Spectroscopy of charged quantum dots

At present, detailed information about the structural parameters of quantum dots is not yet fully available. Hence it is natural to focus on a general model of a quantum dot. Any soft confining potential can be approximated by a parabola and we shall focus on quantum dots with parabolic confinement. This model is supported by a number of studies of self-assembled and gated quantum dots [1, 28]. The single particle Hamiltonian corresponds to a particle moving in a parabolic potential, and in the presence of the magnetic field. It can be exactly diagonalised [12, 19], with single-particle energies $E_{mn}^e = \Omega_+^e(n + \frac{1}{2}) + \Omega_-^e(m + \frac{1}{2})$, and eigenstates $|m, n >$ of two harmonic oscillators. The frequencies are $\Omega_\pm^e = [\Omega \pm \omega_c]/2$ ($\hbar = 1$ for the rest of this work). ω_c is the cyclotron energy, $l_0 = 1/(m^*\omega_c)^{1/2}$ is the magnetic length, m^* is the effective mass, and $\Omega = \sqrt{\omega_c^2 + 4\omega_0^2}$. The kinetic energy $\sim \Omega_-$ decreases with the magnetic field while the Coulomb energy increases with the magnetic field. The Coulomb energy is measured in units of exchange energy $E_0 = Ry\sqrt{2\pi}a_0/l_{eff}$, where Ry is the effective Rydberg, a_0 is the effective Bohr radius, and $l_{eff} = l_0/(1+4\omega_0^2/\omega_c^2)^{1/4}$ is the effective magnetic length. The same applies to valence band holes photo-excited in a quantum dot. The valence band holes are characterised by characteristic frequencies $\Omega_\pm^h = \beta\Omega_\pm$ and energies $E_{mn}^h = -[\Omega_+^h(n+\frac{1}{2})+\Omega_-^h(m+\frac{1}{2})]$ (with the semiconductor gap E_G set to zero). Holes have the opposite charge to electrons and the angular momentum of holes is opposite to the angular momentum of electrons.

The N electron quantum dot system couples to radiation via interband polarisation operators P^+, P^-. Absorbing/annihilating a photon creates/annihilates interband polarisation. Denoting the creation (annihilation) operators for electrons (holes) in states $|i >$ by c_i^+ (c_i), h_i^+ (h_i), the polarisation operator can be written as an operator creating electron-valence hole pairs $P^+ = \sum_i c_i^+ h_i^+$. The final state Hamiltonian describes $N + 1$ electrons and one valence hole:

$$H = \sum_i E_i^e c_i^+ c_i + \sum_l E_l^h h_l^+ h_l$$

$$+ \frac{1}{2} \sum_{i_1 i_2 i_3 i_4} < i_1, i_2|V_{ee}|i_3, i_4 > c_{i_1}^+ c_{i_2}^+ c_{i_3} c_{i_4}$$

$$+ \sum_{i_1 l_2 l_3 i_4} < i_1, l_2 |V_{eh}| l_3, i_4 > c_{i_1}^+ c_{i_4} h_{l_2}^+ h_{l_3}, \qquad (1)$$

where $< i_1, i_2 |V_{ee}| i_3, i_4 >$ are the electron-electron Coulomb matrix elements[29] and $< i_1, l_2 |V_{eh}| l_3, i_4 >$ are electron-hole Coulomb matrix elements[30]. The conservation of angular momentum in the Coulomb scattering of electrons guarantees that $i_1 + i_2 = i_3 + i_4$ (if $i = m$) and for electrons scattered by valence holes $i_1 - l_2 = i_4 - l_3$.

The final state Hamiltonian describes $N + 1$ electrons interacting with a single valence hole. The system before optical excitation contained N electrons. We wish to know the properties of this "initial state" N electron system. Let us assume that we know the exact eigenstates of the initial $|u >_N$ and final $|v_{N+1} >$ electronic systems. The final state droplet is negatively charged and attracts valence holes. To account for this interaction we must form many-particle Wannier states $|f >= \sum_{i,v} A_{i,v}^f h_i^+ |v_{N+1} >$ as a sum of the products of many-electron and valence hole states with amplitudes $A_{i,v}^f$ determined by interactions. This linear combination means that the hole entangles many-electron states and increases the number of degrees of freedom. The absorption spectrum, given by:

$$A(\omega) = \sum_f | < f|P^+|u_N > |^2 \delta(\omega - (E_f(N + 1) - E_u(N))), \qquad (2)$$

describes the addition of an electron-valence hole pair. The transition probability matrix element $< f|P^+|u_N >= \sum_{i,v} A_{i,v}^f < v_{N+1}|c_i^+|u_N >$ differs from the matrix element associated with just adding an electron $< v_{N+1}|c_i^+|u_N >$ into a single particle state i. The interference effects, contained in the squares of the transition matrix element $| < f|P^+|u_N > |^2$, lead to nontrivial effects, such as Fermi edge singularity (FES) [31], even in the absence of e-e interactions. Little is known about the effect of electron-electron interactions on the FES.

In the recombination process we assume that the system has relaxed to the lowest many-electron valence hole state $|G >= \sum_{i,v} A_{i,v}^G h_i^+ |v_{N+1} >$. The transition probability involves the matrix element $< u_N|P^-|G >= \sum_{i,v} A_{i,v}^G < u_N|c_i|v_{N+1} >$. This matrix element involves the matrix element describing the probability $< u_N|c_i|v_{N+1} >$ of removing an electron from the $N + 1$ electron system. If holes are strongly localised in some state $i = 0$, the emission spectrum describes very well a hole spectral function of the removed electron:

$$E(\omega) = \sum_u | < u_N|c_0|v_{N+1} > |^2 \delta(\omega - (E_G(N + 1) - E_u(N)). \qquad (3)$$

Hence in emission we may hope to measure directly the spectral function of a "hole", i.e. an electron removed from the lowest single particle level of a quantum dot. Since we are forced to remove an electron from the lowest level, the removed electron does not know how many electrons there are in the dot unless there are interactions. The emission process in the noninteracting system does not depend on the number of electrons, and all the deviations from this result must come from interactions.

The electronic Raman scattering process is much more complicated as it couples to the electronic excitations through virtual excitonic transitions. While some preliminary work on the cross section has appeared [32] it suffices to say that the Raman spectrum is proportional to the spectrum of collective excitations [14].

2.1. QUANTUM HALL DROPLETS

Quantum Hall droplets describe electrons confined in the lateral direction and subject to a strong perpendicular magnetic field B. When the cyclotron energy is the largest energy, electrons occupy lowest angular momentum orbitals $|m>$ similar to orbitals in the lowest Landau level. In a certain magnetic field range all lowest orbitals $m>$ can be occupied and electrons can form a compact spin polarised droplet corresponding to filling factor e.g. $\nu = 1$.

When the magnetic field is increased further, the droplet with a small number of electrons undergoes a series of magnetic field induced phase transitions between different total angular momentum states [12]. In Ref [27] the effect of these transitions on the absorption spectrum was investigated. It was concluded that new transition lines should appear as a result of electronic transitions and accompanying charge redistribution in the dot. In a similar way, the edges of larger quantum Hall droplets reconstruct [11, 17, 33], and this reconstruction can be seen optically [27]. A quantum Hall droplet at $\nu = 1$ provides an interesting example of a chiral spin polarised Fermi liquid. Excitonic absorption in the presence of a Fermi liquid transforms into a Fermi edge singularity [31]. The role of electron-electron interactions and of the finite valence hole mass on the FES is not well understood. Hence investigation of simpler systems, such as compact droplets, might help in our understanding of more complex problems.

2.1.1. *Fermi edge singularity in a chiral Fermi liquid*

We will summarise below some essential physical processes which control absorpton into a spin polarised compact droplet while detailed results can be found in Ref. [27]. These processes are summarised in Fig.1.

Fig.1a shows electrons occupying several lowest orbitals m up to a Fermi

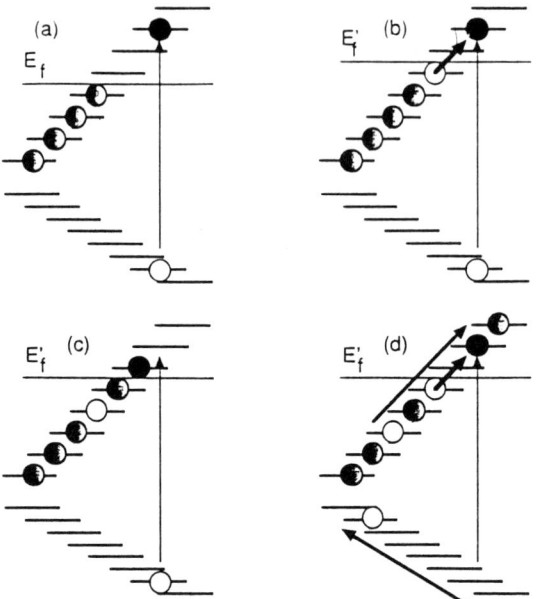

Figure 1. Occupied states and absorption processes involving a photo-excited electron and valence hole pair and free carriers in a compact spin polarised droplet.

level E_f. The valence band orbitals are also shown. In an absorption process, an additional electron and a valence hole are created. This optically created electron should be viewed as a part of the $N + 1$ electron system. This is shown in Fig.1b, where we have presented our extra electron as an electron-hole pair excitation from the ground state of the $N+1$ electron system. Hence our excited state consists of a valence hole and an electron-hole pair excitation. This optically created electron-hole pair excitation mixes with other electron-hole pair excitations with the same angular momentum. An example of a different pair, mixing with the photo-excited pair is shown in Fig.1c. This illustrates how a photo-excited electron immediately mixes with other electrons already present. Processes (a)(b) and (c) all take place in the space of the same total electronic angular momentum. One can also couple different electronic subspaces due to valence hole recoil and account for multiple pair excitations. We can, for example, add an additional electron-hole pair to the photo-excited one by compensating the increase of angular momentum of the electronic system by the momentum of the valence hole. Such a configuration involving two excitations of the electronic system and a scattered valence hole is shown in Fig.1d. All these processes were accounted for in the calculations of Ref.[27]. The final result was that the enhancement of absorption at the Fermi edge survived e-e interactions and valence hole recoil. More studies of the effect of spin degrees of freedom is needed.

2.1.2. *Recombination on acceptors*

Optical spectroscopy of modulation doped quantum dots using purposely introduced acceptors has been investigated theoretically in Ref.[19, 27] and experimentally by Plaut and co-workers in Refs.[20]. When valence holes localise on negatively charged acceptors, they form a neutral charge complex. This complex is a weak perturbation of the initial electronic state. When the spatial extent of the electronic charge is much larger than the average separation between acceptors, emission spectra average over different acceptor configurations. It has been shown in Ref.[19] that the mean emitted photon energy $< \omega >$ can be directly related to the mean kinetic and mean interaction energy per electron in a given ground state of the dot: $< \omega >=< T > + \gamma < V >$, where $0 < \gamma < 2$ is a measure of final state corrections. In particular, when acceptors are in the plane of the dot $\gamma = 0$ and the recombination process measures the kinetic energy directly. The magnetic field causes transitions between ground states of a quantum dot with different total angular momenta i.e. jumps in kinetic energy. These transitions should lead to discontinuous oscillations of the emission line. Unfortunately, these predictions have not yet been verified experimentally.

2.1.3. *Inelastic light scattering*

Inelastic light scattering measures the excitation spectrum of a QD [22, 23, 24, 25] and, in principle, can provide direct evidence of the discrete nature of excitations in zero-dimensional (0D) systems. The electron-electron interactions play a significant role in determining these excitations. For example, in a strong magnetic field the gap in the excitation spectrum of the Hall droplet is determined by electron-electron interactions. Changing the magnetic field leads to a collapse of the gap at special values of the magnetic field i.e. to a compressible Hall droplet. One could in principle observe this behaviour as soft modes in Raman spectra[22]. However, the wavelength of the perturbation neccessary to excite electrons of the droplet across the gap has to be much smaller then the physical size of the droplet. In typical Raman experiments on 2DEG this wavelength was of the order of a few thousand angstroms, i.e. much larger than the size of the droplet. Nevertheless, a number of groups[23, 24, 25, 26] have undertaken inelastic light scattering experiments in quantum dots. Lockwood et al. [24] carried out investigation of electronic excitations in modulation doped quantum dots in a magnetic field. Their result supported the notion of electronic shells even in dots with ≈ 100 electrons.

Recently, Potemski et al. [26] carried out resonant inelastic light scattering studies of InGaAs self-assembled quantum dots embedded in GaAlAs. Because of a very high Al content, these dots had two well energetically separated subbands for motion perpendicular to the plain of the disk, superimposed with weaker quantisation of the lateral motion. The two subbands allowed for the separation of the Raman signal from the resonantly excited higher subband from the PL signal originating from the lower subband. The observed broad Raman band was interpreted as originating from the scattering of quantum dot excitons by acoustic phonons. The very presence of this signal implies a significant degree of coherence of interband polarisation in quantum dots.

2.2. EMISSION IN CHARGED SELF-ASSEMBLED QUANTUM DOTS

Emission from charged self-assembled quantum dots has been investigated experimentally by Schmidt et al.[21] and theoretically by Wojs et al. [7]. The purpose of these investigations was to establish whether one can determine the number of electrons in the dot N from the emission spectrum. This is a valid question since the recombination process involves the recombination of valence holes from the $|0,0>$ topmost valence level with the lowest electron level $|0,0>$ of the dot. This process leaves an empty electron state at the bottom of the occupied band. In the absence of electron-electron interactions the emission energy does not depend on the number N of electrons in the dot, and hence PL is an ideal candidate to measure the effect of electron-electron interactions. In self-assembled dots with large kinetic energy quantisation ω_0 the electronic configurations are filled according to their increasing kinetic energy. The only difficulty arises for partially filled degenerate shells. Detailed numerical diagonalisation studies[7] showed that partially filled shells are filled according to Hund's rules, i.e. electrons tend to maximise their total spin. Hund's rules for quantum dots are summarised in Fig.2, where the total spin and total angular momentum of the calculated ground state of a quantum dot as a function of N is shown. We see that closed shells have zero spin and partially filled shells have maximum spin. The rules of occupying single particle orbitals influence the magnetic field dependence of the energy of adding an electron. Below the total angular momentum behaviour of the ground state we show also a qualitative dependence of the addition energy on the magnetic field. This characteristic behaviour of the addition energy has been observed in vertical tunneling quantum dots [34]. Far infrared spectroscopy of charged quantum dots [35, 36] has also shown features specific to a number of particles in the dot [39]. Let us qualitatively describe the results of our investigation of emission from charged quantum dots by considering in detail the emission from the $N = 4$ electron dot. The $N = 4$ electron dot has a filled s-shell and a partially filled p-shell. The two p-electrons are in a triplet state. In the initial state we have the 4-electron complex and a valence hole with two possible spin orientations, as shown in Fig.3a and

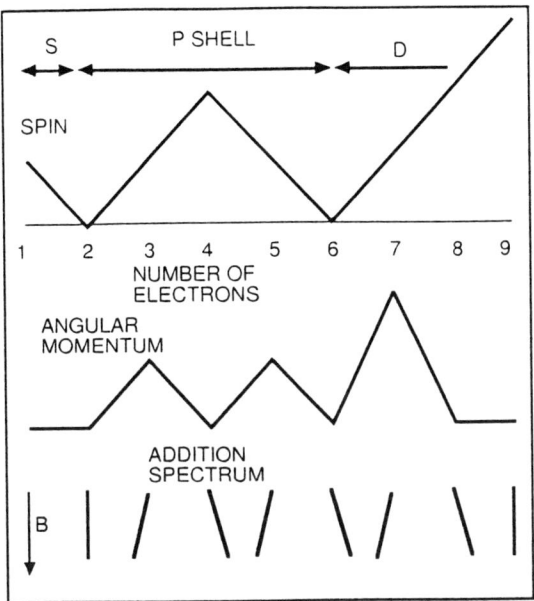

Figure 2. Ground state angular momentum and total spin as a function of the number of electrons in a quantum dot. The occupation of successive orbitals leads to the magnetic field behaviour of the addition spectrum as shown at the bottom.

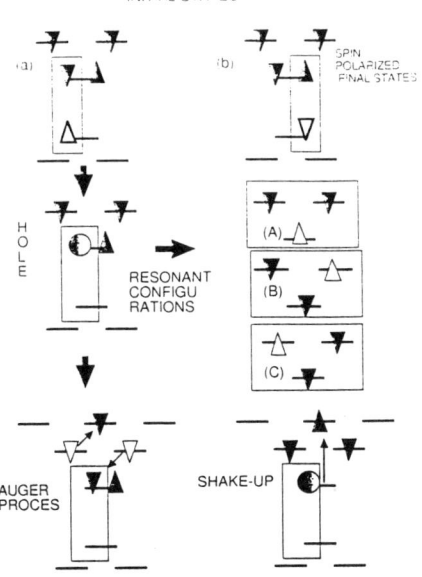

Figure 3. Initial and final configurations in the emission from a 4 electron quantum dot

b. The specific spin orientation of the valence hole selects an electron with a specific spin for the recombination process, and hence leaves a different spin configurations of the electron left in the partially filled s-shell. For the initial spin configuration (b) there is only one final state with all electrons with parallel spins. Since we removed an electron with a spin opposite to those of all the other electrons, the energy of recombination for this spin configuration is exactly the same as the energy of a single exciton $E(1)$. The same is true for $N = 1 - 4$. For the spin configuration shown in Fig.3a, the final configuration involves a spin down electron in the s-shell and two electrons with parallel spins in the p-shell. This configuration belongs to a group of 4 degenerate configurations. The most important group A,B,C is obtained by simultaneously flipping the spin of a p-shell electron and the s-shell electron. Alternatively, we can think of a spin down electron circulating on a plaquette of three sites[12]. The three degenerate states are strongly mixed but in the end only two of them have a finite overlap with the optically created configuration A. Due to resonant configuration mixing the optically created configuration ceases to exist and breaks into two configurations with energies $E^1 = E(A) + 0.46E_0$ and $E^2 = E(A) - 0.27E_0$ and transition probabilities of 0.63 and 0.37, respectively. The splitting of the two peaks is very large i.e. $\approx 70\%E_0$. E_0 equals the binding energy of the s-shell electron-hole pair. In addition to spin related resonant configuration mixing there are resonant Auger processes, illustrated in Fig.3. This process involves one electron jumping down and annihilating an optically created hole in the s-shell, and a second electron jumping up to the unoccupied d-level. This process involves final state electrons in s and d shells and contributes much less to the breakdown of the optically created hole. In addition to resonant configurations, we also have excited final state configurations coupled to the optically created configuration. These shake-up configurations involve an electron excited by twice the electron kinetic energy $2T = 2\omega_0$ and an emission line shifted down in energy by the same amount. In general, electron-electron interactions tend to lower the energy of emitted photons, an effect called band-gap renormalisation(BGR). In a quantum dot a new effect, an emission at higher energy, appears when the d-shell becomes populated. The initial and final configurations for $N = 9$ are shown in Fig.4. The three electrons in a d-shell are needed to populate the $|1, 1 >$ state with angular momentum zero. The configuration with an optically created hole is mixed by Coulomb interactions with a configuration with a hole relaxed to the zero angular momentum state in the d-shell. This is clearly a lower energy configuration and emission from this configuration will appear at higher energy, shifted approximately by $2T$ from the main emission line. In Fig.5 we summarise the effect of electrons on the emission spectrum of a quantum dot. The effect of free carriers is to broaden and

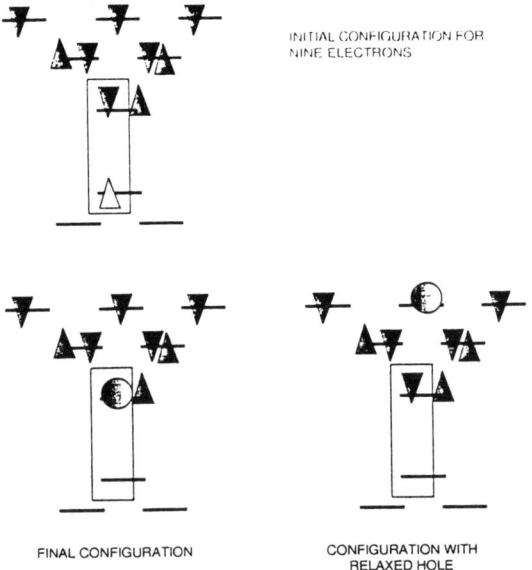

Figure 4. Initial and final configurations in the emission from a 9 electron quantum dot

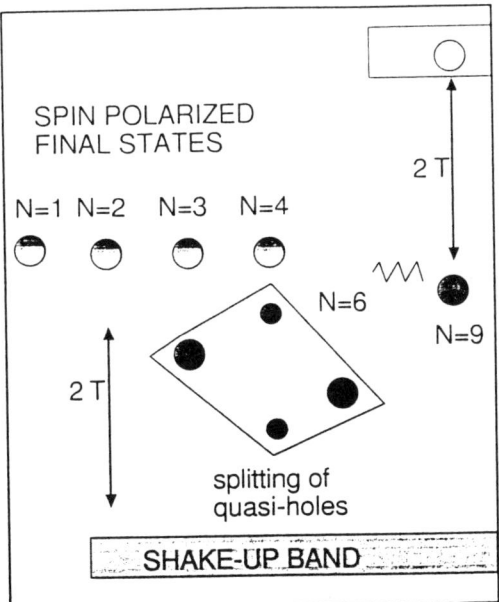

Figure 5. Summary of characteristic features in emission spectrum from a modulation doped quantum dot as a function of the number of electrons.

shift the emission spectrum to lower energy. The main emission spectrum is accompanied by a shake-up band below the main emission line, at energy $\approx$ twice the kinetic energy quantisation for electrons. When the zero angular momentum channel of higher shells is populated, the photo-created hole in the electronic system can relax to the Fermi level and an emission at higher energies (approximately even multiples of the electron kinetic energy) should be observable. A preliminary observation of this high energy band has been reported in Ref.[21]. Perhaps the most startling and interesting phenomenon is associated with partially filled shells. Due to a large number of spin degrees of freedom of a partially filled shell, the simple minded picture of a photo-created hole at the bottom of the occupied band breaks down. This quasi-particle breakdown manifests itself in the splitting of the emission line for a partially filled shell and a return to a single emission line for closed shells (broadened by Auger processes). Hence the width of the recombination spectrum should oscillate as a function of the filling factor of successive shells. The physical picture of invoking spin degrees of freedom of partially filled shells of degenerate levels is identical to the picture invoked to explain anomalies and oscillations in the emission line from 2DEG at odd and even filling factors [40].

3. Spectroscopy of excitonic complexes in quantum dots

A number of groups investigated radiative recombination from a single quantum dot filled with electrons and holes, the population of which was controlled by the power of optical excitation in the barrier material [2, 3, 4, 6]. Filling dots with electron-hole pairs creates multi-exciton complexes. In a dot with up to five shells one can store up to $N = 30$ excitons. One therefore hopes to realise excitonic artificial atoms. These atoms would have emission spectra characteristic of an artificial excitonic atom with a single exciton, bi-exciton, tri-exciton, and so on. In our opinion existing experiments are not conclusive as yet and both experiments and a detailed theoretical fingerprint of the exciton removal (emission) spectrum as a function of the number N of excitons is needed.

We summarise here the results of our theoretical investigation of the electronic structure of excitonic artificial atoms. Calculations of many-exciton states have been carried out in Refs [37, 3, 8, 9, 38] using exact diagonalisation techniques. In Ref.[38] a model asymmetric quantum box was investigated. The box allowed two non-degenerate orbital levels per electron/hole, and could accommodate up to $N = 4$ excitons. A similar calculation has been carried out in Ref.[3]. In Refs.[8, 37, 9] lens shaped quantum dots with degenerate electronic shells were investigated for up to $N = 20$ excitons. The key physics was found to be associated with excitons occupying degenerate shells according to "hidden symmetry". This symmetry is particularly useful in designing quantum dynamics, and possibly quantum information processing properties of quantum dots [41].

3.1. HIDDEN SYMMETRY AND EXCITONIC CONDENSATES

Symmetries are associated with commutation properties of relevant operators. For excitons in quantum dots the relevant operators are the set of interband polarisation operators P^+ (P^-) which create (annihilate) electron-hole pairs $P^+ = \sum_i c_i^+ h_i^+$ ($P^- = \sum_i h_i c_i$) by annihilating (creating) photons [8, 9]. The third component $P_z = \frac{1}{2}(N^e + N^h - N_g)$ measures population inversion, i.e. the number of excitons N in N_g of the single particle levels. P satisfies the commutation relations of a 3D angular momentum: $[P^+, P^-] = 2P_z$, $[P_z, P^\pm] = \pm P^\pm$. The total polarisation $P^2 = \frac{1}{2}(P^+ P^- + P^- P^+) + P_z^2$ commutes with P^+.

The general Hamiltonian of the interacting electron-hole system (we omit spin degrees of freedom σ for simplicity) treats electrons and holes on the same footing:

$$
\begin{aligned}
H &= \sum_i E_i^e c_i^+ c_i + \sum_i E_i^h h_i^+ h_i - \sum_{ijkl} < ij|V_{eh}|kl > c_i^+ h_j^+ h_k c_l \\
&+ \frac{1}{2} \sum_{ijkl} < ij|V_{ee}|kl > c_i^+ c_j^+ c_k c_l + \frac{1}{2} \sum_{ijkl} < ij|V_{hh}|kl > h_i^+ h_j^+ h_k h_l. \tag{4}
\end{aligned}
$$

The commutator of the polarisation operator with the Hamiltonian can be written in a suggestive way as [8, 9]:

$$
\begin{aligned}
[H, P^+] &= \sum_i (E_i^e + E_i^h) c_i^+ h_i^+ - \sum_{ijk} < ij|V_{eh}|kk > c_i^+ h_j^+ \\
&+ \frac{1}{2} \sum_{ijkl} (< ij|V_{ee}|kl > - < ik|V_{eh}|jl >)(c_i^+ h_l^+ c_j^+ c_k - c_i^+ h_k^+ c_j^+ c_l) \\
&+ \frac{1}{2} \sum_{ijkl} (< ij|V_{hh}|kl > - < ik|V_{eh}|jl >)(c_i^+ h_i^+ h_j^+ h_k - c_k^+ h_i^+ h_j^+ h_l). \tag{5}
\end{aligned}
$$

The commutator $[H, P^+]$, or the dynamics of the interband polarisation $i\dot{P}^+$, requires knowledge of both two-particle and four-particle operators, the dynamics of which has to be sought and truncated at some level of approximation. This coupling to many degrees of freedom is responsible for fast dephasing in semiconductors. To avoid this we must design our energy levels and matrix elements.

First let us observe that if the single particle energy levels are degenerate i.e. $(E_i^e + E_i^h) = (E^e + E^h)$ and the Coulomb matrix elements $< ij|V_{eh}|kk >$ satisfy $< ij|V_{eh}|kk >= \delta_{ij}V_0$ the first two terms $\sum_i (E_i^e + E_i^h)c_i^+ h_i^+ - \sum_{ijk} < ij|V_{eh}|kk > c_i^+ h_j^+ = E_X P^+$ turn out to be proportional to the polarisation P^+. We are still left with four body terms. Fortunately, the four body operators are proportional to the difference of matrix elements e.g. $< ij|V_{ee}|kl > - < ik|V_{eh}|jl >$. Hence, if this difference can be made very small, the four body terms vanish. The smallness of the four body terms was demonstrated for lens shaped quantum dots via numerical calculations of Coulomb matrix elements [8] with recent extensive discussion in Ref. [9]. As was demonstrated in Ref. [8], the degeneracy of single particle levels and the symmetry of (ee), (hh), and (eh) interactions cause a remarkable cancellation of the four particle contribution and leads to very simple commutation properties of the interband polarisation operator operating on a degenerate shell,

$$[H, P^+] = E_X P^+. \tag{6}$$

This commutation relation allows us to construct "multiplicative" states $|N >= (P^+)^N |v >$ with energies $E(N) = N E_X$. These states are exact eigenstates of the interacting, spinless on-shell Hamiltonian. They turn out to be a good approximation to true eigenstates of the full Hamiltonian. As a result, the energy necessary to add/subtract an exciton to/from a degenerate shell does not depend on the population N of this shell.

3.2. RECOMBINATION SPECTRA OF FEW-EXCITON COMPLEXES

Recombination spectra of few-exciton complexes have been discussed in detail in Refs.[8, 9, 37]. There are three different energy scales which enter the problem in the following order of importance: (a) the kinetic energy quantisation for electrons and for holes T, (b) the direct, exchange, and correlation contribution from the electron-electron, electron-hole and hole-hole interactions E_0, (c) electron-hole exchange interaction δ which modifies the attractive electron hole interaction. The typical scales of these effects are of the order $T \approx 50 - 100 meV$, $E_0 \approx 10 - 30 meV$, and $\delta \approx 0.1 - 1.0 meV$ [5]. For obvious reasons, the calculations discussed here neglect the electron-hole exchange interaction. These calculations were carried out in optically active zero angular momentum and zero spin Hilbert spaces of the electron hole systems. As an illustration let us discuss qualitatively the emission from a $N = 4$ exciton complex in a simple model of a quantum dot with only s and p shells. The $N = 4$ exciton complex has two total spin arrangements of the partially filled p-shell electrons and holes, the triplet-triplet, the singlet-singlet configuration, and the singlet-triplet configurations. There are nine possible degenerate triplet-triplet states, only three of them optically active. Let us first discuss the triplet-triplet configuration, $|t> = (c^+_{10\downarrow}h^+_{10\uparrow})(c^+_{01\downarrow}h^+_{01\uparrow})|XX>$. Here $|XX>$ describes a filled s-shell. The triplet-triplet state has been written explicitly as a product of two electron-hole pairs. The energy of each pair is the energy E_p. Two electrons and two holes have parallel spins and lower their respective energy by exchange. The ground state energy, measured from the energy of the filled s-shell, is:

$$E_{4Xt} = [2T - (V^{sp,x}_{ee} + V^{sp,x}_{hh}) - V^{pp,d}_{eh}] - (V^{pp,x}_{ee} + V^{pp,x}_{hh}), \qquad (7)$$

where the short hand notation for the differen Coulomb matrix elements has been adopted e.g. $V^{pp,x}_{ee}$ is the exchange matrix elements for electrons in the two p orbitals. Because exchange interaction and electron-hole scattering interactions are equal and attractive, $V^{pp,x}_{ee} = V^{pp}_{eh}$, the energy of the two additional excitons in a p-shell of the four exciton complex turns out to be exactly twice the energy of a single exciton in a three exciton complex, $E_{4Xt} = 2E_{3X}$ [9], in accordance with "hidden symmetry".

When both electrons and holes are in singlet configurations, the total number of possible configurations increases. There are three possible singlet-singlet configurations:

$$
\begin{aligned}
|a> &= (c^+_{10\uparrow}c^+_{10\downarrow})(h^+_{10\downarrow}h^+_{10\uparrow})|XX> \\
|b> &= \frac{1}{\sqrt{2}}(c^+_{10\uparrow}c^+_{01\downarrow} + c^+_{01\uparrow}c^+_{10\downarrow})\frac{1}{\sqrt{2}}(h^+_{10\downarrow}h^+_{01\uparrow} + h^+_{01\downarrow}h^+_{10\uparrow})|XX> \\
|c> &= (c^+_{01\uparrow}c^+_{01\downarrow})(h^+_{01\downarrow}h^+_{01\uparrow})|XX>.
\end{aligned} \qquad (8)
$$

Electron-hole scattering can move an electron-hole pair from configuration (b) to either (a) or (c). This matrix element $< a|H|b >$ equals the electron-hole scattering matrix element V^{pp}_{eh} times the product of the normalisation factors of states $|a>$ and $|b>$, equal to 2. Hence the effective scattering from a to b is twice the single pair scattering. The energy of configurations (a) and (c) is just the sum of the pair energy, i.e. $2E_p$. The energy of configuration (b) is increased by the repulsive exchange energy $2E_p + (V^{pp,x}_{ee} + V^{pp,x}_{hh})$ of singlet p-shell electrons and holes. The resulting Hamiltonian for the singlet-singlet states can be written as:

$$
\begin{pmatrix}
2E_p & -2V^{pp}_{eh} & 0 \\
-2V^{pp}_{eh} & 2E_p + 2V^{pp,x}_{ee} & -2V^{pp}_{eh} \\
0 & -2V^{pp}_{eh} & 2E_p
\end{pmatrix}
$$

334

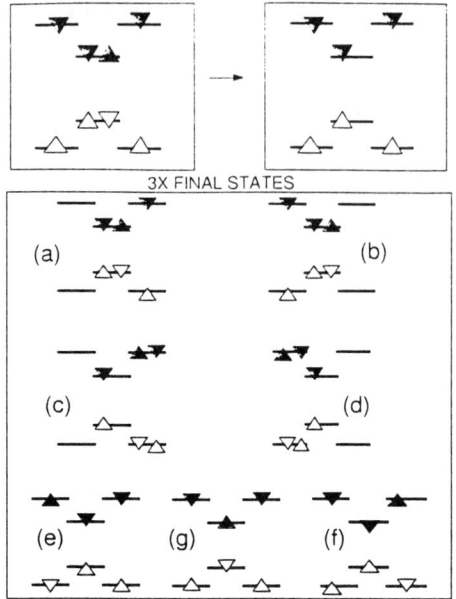

Figure 6. Initial and final configurations in the emission from a 4 exciton quantum dot

We again utilise the fact that exchange interaction, repulsive here, equals in magnitude the attractive electron-hole scattering to obtain the three eigenvalues expressed solely in terms of electron-hole matrix elements:

$$E_{4Xs}(1) = 2E_p - 2V_{eh}^{pp}, \quad E_{4Xs}(2) = 2E_p, \text{ and } E_{4Xs}(3) = 2E_p + 4V_{eh}^{pp}.$$

We see that $E_{4Xs}(1) = E_{4Xt}$ i.e singlet-singlet and triplet-triplet ground state energies are degenerate. This is to be contrasted with Hund's rules for the four-electron artificial atom, which predict the triplet to be the ground state. This degeneracy may be removed in favour of the singlet-singlet states by considering an admixture of excited states.

We can conclude here that there are few states for spin polarised configurations (one triplet-triplet in this example). Their energy is reduced by attractive exchange interaction among like-particles. For spin unpolarised configurations there are many more available configurations (three singlet-singlet configurations in this example). The energy of these individual configurations is increased by exchange, but correlations due to mixing of these states by electron-hole scattering more than compensate for the repulsive exchange energy, and reduce the ground state energy.

In the emission process from either a triplet or singlet $N = 4$ exciton ground state, excited states of the $N = 3$ exciton are created. These states are shown in Fig.6. One final state configuration corresponds to spin polarised electrons and holes. The emission line corresponding to this configuration has the same energy as the emission line from a single exciton.

The rest of the configurations involve spin degrees of freedom. The first two configurations (a) and (b) correspond to the removal of an electron-hole pair from the p-shell. The energy of this emission line is the same for a spin triplet initial state as well as for the $N = 3$ exciton state. The remaining degenerate states correspond to the removal of an electron-hole pair from an s-shell. The last three states (e),(g), and (f) are equivalent to the spin resonant configuration states responsible for the splitting of the emission line in the $N = 4$ electron complex. The spin states (e),(g), and (f) of electrons and of holes each have zero angular momentum and we must make a linear combination of all of them. This generates a band of nine excited states, plus two configurations (c) and (d) for a total of 11 degenerate states. These degenerate states form a band of states with a width proportional to the strength of Coulomb interactions. Of course there is a direct analogy between the spectral function of a "hole" in the electronic droplet and a spectral function of a hole (missing exciton) in the excitonic droplet. The band of resonant states leads to the breakdown of the "hole" spectral function in the excitonic droplet. For results of detailed calculations of the emission spectrum as a function of the number of excitons N see Ref.[9].

4. Summary

We reviewed our work on the optical properties of quantum dots with electrons and of quantum dots with excitons. We have shown that the radiative emission spectrum carries detailed information about the number of carriers in the dot, and correlations among them.

References

1. For reviews and references see S. Fafard, Z. R. Wasilewski, C. N. Allen, D. Picard, P. G. Piva, J. P. McCaffrey, Superlattices and Microstructures **25** . . . (1999); L. Jacak, P. Hawrylak, A. Wojs, "Quantum Dots", Springer Verlag (1998); R. C. Ashoori, Nature **379**,413 (1996); P. M. Petroff and S. P. Denbaars, Superlattices and Microstructures **15**, 15 (1994); M. Kastner, Physics Today,**24**, January 1993; T. Chakraborty, Comments in Cond.Matter Physics **16**,35(1992);
2. A. Zrenner, M. Markmann, A. Paassen, A. L. Efros, M. Bichler, W. Wegscheider, G. Bohm and G. Abstreiter Physica B **256**, 300 (1998).
3. E. Dekel, D. Gershoni,E. Ehrenfreund, D. Spektor, J. M Garcia and P. M. Petroff, Phys. Rev. Lett.**80** 4991 (1998).
4. L. Landin et al., Science **280** 262 (1998).
5. M. Bayer, A. Kuther, A. Forchel, A. Gorbunov, V. B. Timofeev, F. Schafer, J. P. Reithmayer, T. L. Reinecke, and S. N. Walck, Phys. Rev. Lett.**82**,1748 (1999).
6. M. Bayer, T. Gutbrod, A. Forchel, V. D. Kulakovskii, A. Gorbunov, M. Michel, R. Steffen, and K. H. Wang, Phys. Rev. B, **58**, 4740 (1998).
7. A. Wojs and P. Hawrylak, Phys. Rev. B **55**, 13066 (1997).
8. A. Wojs and P. Hawrylak, Solid State Comm.**100**, 487 (1996); P. Hawrylak and A. Wojs, Semic. Sci. Tech. **11**, 1516 (1996).
9. P. Hawrylak, Phys. Rev.B, July 15 1999.

336

10. P. A. Maksym, Physica **B184**, 385 (1993); P. A. Maksym and T. Chakraborty, Phys. Rev. Lett.**65**, 108 (1990); P. A. Maksym and T. Chakraborty, Phys. Rev.**B45**, 1947 (1992).
11. A. H. MacDonald, S. R. Eric Yang, and M. D. Johnson, Aust. J. Phys. **46**, 345 (1993).
12. P. Hawrylak, Phys. Rev. Lett. **71**, 3347 (1993).
13. R. C. Ashoori, H. L. Stormer, J. S. Weiner, L. N. Pfeiffer, K. W. Baldwin, and K. W. West, Phys. Rev. Lett. **71**, 613 (1993). R. C. Ashoori, Nature**379**, 413 (1996).
14. Arkadiusz Wojs and Pawel Hawrylak , Phys. Rev. B **56**, 13227 (1997).
15. J. H. Oaknin, L. Martin-Moreno, and C. Tejedor, Phys. Rev. **B54**, 16 850 (1996).
16. P. Hawrylak, C. Gould, A. Sachrajda, Y. Feng, Z. Wasilewski, Phys. Rev.**B59**,2801 (1999).
17. X. G. Wen, Phys. Rev. B, **41**, 12 838 (1990); C. de Chamon and X.-G. Wen, Phys. Rev.**B49**, 8227 (1994).
18. J. J. Palacios and A. H. MacDonald, Phys. Rev. Lett.**76**, 118 (1996).
19. P. Hawrylak and D. Pfannkuche, Phys. Rev. Lett. **70**, 485 (1993)
20. S. Patel, A. S .Plaut, P. Hawrylak, H. Lage, P. Grambow, D. Heitmann, K. von Klitzing, J. P. Harbison and L. T. Florez, Solid State Comm.**101**,865 (1997).
21. K. H. Schmidt, G. Medeiros-Ribeiro, P. M. Petroff, Phys. Rev. B, **58**, 3597 (1998).
22. P.Hawrylak, Solid State Commun.**93**, 915(1995).
23. R.Strentz *et al.*, Phys. Rev. Lett. **73**, 3022 (1994).
24. D. J. Lockwood, P. Hawrylak, P. D. Wang, C. M. Sotomayor-Torres, A. Pinczuk, and B. S. Dennis, Phys. Rev. Lett.**77**, 354 (1996).
25. C. Schuller, G. Biese, K. Keller, C. Steinebach, D. Heitmann, P. Grambow, K. Eberl, Phys. Rev. **B54**, 17304 (1996).
26. P. Hawrylak, M. Potemski, H. Labbe, D. J. Lockwood, J. Temmyo , and J. Tamamura, "Inelastic light scattering from self-assembled quantum disks", Physica E, 1998.
27. P. Hawrylak, A. Wojs, and J. A. Brum, Phys. Rev. B, **54**, 11 397 (1996).
28. A. Wojs, P. Hawrylak, S. Fafard, L. Jacak; Phys. Rev. **B54**, 5604 (1996).
29. P. Hawrylak, Solid State Comm. **88**, 475 (1993).
30. A. Wojs and P. Hawrylak, Phys. Rev. B **51**, 10 880 (1995).
31. J. A. Brum and P. Hawrylak,Comm. on Cond. Matter Physics **18**, 135 (1997).
32. Maura Sassetti and Bernhard Kramer, Phys. Rev. Lett. **80**, 1485 (1998).
33. O. Klein, S. de Chamon, D. Tang, D. M. Abusch-Magder, U. Meirav, X.-G. Wen, M. A. Kastner, and S. J. Wind, Phys. Rev. Lett. **74**, 785 (1995).
34. S. Tarucha, D. G. Austing, T. Honda, R. J. van der Hage, and L. P. Kouvehoven, Phys. Rev. Lett **77**, 3613 (1996).
35. H. Drexler, D. Leonard, W. Hansen, J. P. Kotthaus, and P. M. Petroff, Phys. Rev. Lett., **73**, 2252 (1994).
36. M. Fricke, A. Lorke, J. P. Kotthaus, G. Medeiros-Ribeiro, and P. M. Petroff, Europhys. Lett. **36**, 197 (1996).
37. S. Raymond, P. Hawrylak, C. Gould, S. Fafard, A. Sachrajda, M. Potemski, A. Wojs, S. Charbonneau, D. Leonard, P. M. Petroff, and J. L. Merz, Solid State Comm.**101**,883 (1997).
38. A. Barenco and M. A. Dupertuis, Phys. Rev.**B 52**, 2766 (1995)
39. A. Wojs and P. Hawrylak, Phys. Rev. B **53**, 10 841 (1996)
40. L. Gravier, M. Potemski, P. Hawrylak, and B. Etienne, Phys. Rev. Lett.**80**,3344 (1998).
41. P. Hawrylak, S. Fafard, Z. Wasilewski, Condensed Matter News (in press)

THE STARK EFFECT AND ELECTRON-HOLE WAVEFUNCTIONS IN InAs-GaAs SELF-ASSEMBLED QUANTUM DOTS

M.S. SKOLNICK[1], P.W. FRY[1], I.E. ITSKEVICH[2,1], D.J. MOWBRAY[1], J.A. BARKER[3], E.P. O'REILLY[3], M. HOPKINSON[4], M. AL-KHAFAJI[4], A.G. CULLIS[4], G. HILL[4], AND J.C. CLARK[4]
[1]Department of Physics, University of Sheffield, Sheffield S3 7RH, UK
[2]Institute for Solid State Physics, RAS, Chernogolovka, Moscow district 142432, Russia
[3]Department of Physics, University of Surrey, Guildford GU2 5XH, UK
[4]Department of Electronic and Electrical Engineering, University of Sheffield, Sheffield S1 3JD, UK

Abstract. New information on the electron-hole wavefunctions in InAs-GaAs self-assembled quantum dots is obtained from study of the quantum confined Stark effect. From the sign of asymmetry observed in the Stark effect, it is deduced that the dots have a permanent dipole moment directed from base to apex, implying that holes are localised above the electrons in the dots. This highly unexpected electron-hole alignment is opposite to that predicted by all previous theories. We explain our results by comparison with strain/electronic structure modelling, and are able to deduce that the nominally InAs dots contain significant amounts of gallium, and have a strongly truncated shape. In the light of these results most if not all previous modelling of the electronic structure of InAs self-assembled quantum dots needs to be re-examined. The mechanisms involved in the photocurrent process employed to observe the Stark effect, and the significance of photocurrent techniques to measure absorption spectra in quantum dots are discussed.

1. Introduction

In spite of the large amount of work on self-assembled InAs-GaAs quantum dots (QDs) very little experimental information is available on the forms of the electron and hole wavefunctions in such structures. Such QDs provide nearly ideal examples of zero-dimensional semiconductor systems [1] and have very favourable opto-electronic properties. As a result they have generated very considerable interest for both fundamental studies and for device applications. However because of the uncertainty regarding the nature of the wavefunctions, and because precise structural information is very difficult to obtain, as we show reliable modelling of the electronic structure is very likely still lacking.

In this paper we show that study of the quantum confined Stark effect in such

M.L. Sadowski et al. (eds.), Optical Properties of Semiconductor Nanostructures, 337–346.
© *2000 Kluwer Academic Publishers. Printed in the Netherlands.*

338

structures reveals very surprising and hitherto unsuspected information regarding the alignment of the electron-hole wavefunctions. From the sign of the asymmetry we observe in the Stark effect, we show that the electrons are localised below the holes in the dots, in contrast to the predictions of all previous theoretical studies [2-5]. By comparison with our own modelling we are able to show that the results can only be explained if the nominally InAs dots contain significant amounts of gallium, graded from base to apex, and have strongly truncated shapes, not pyramidal as frequently assumed in the literature. We are also able to deduce the effective height of the dots. We are thus able to deduce important new information on the shape and composition of QDs from purely optical studies. In view of our new results, we believe that most previous electronic structure calculations of InAs QDs need to be re-examined.

We employ photocurrent spectroscopy to measure absorption-like spectra in the dots and to study the response of the wavefunctions to applied electric fields (F). We show that photocurrent measurements provide a highly sensitive means to study the very weak absorption from QDs (absorption fractions $<10^{-5}$ are easily measured). As well as the information on the vertical extent and ordering of the wavefunctions discussed above, we are also able to show that the excited state splittings in the spectra arise from lateral quantisation.

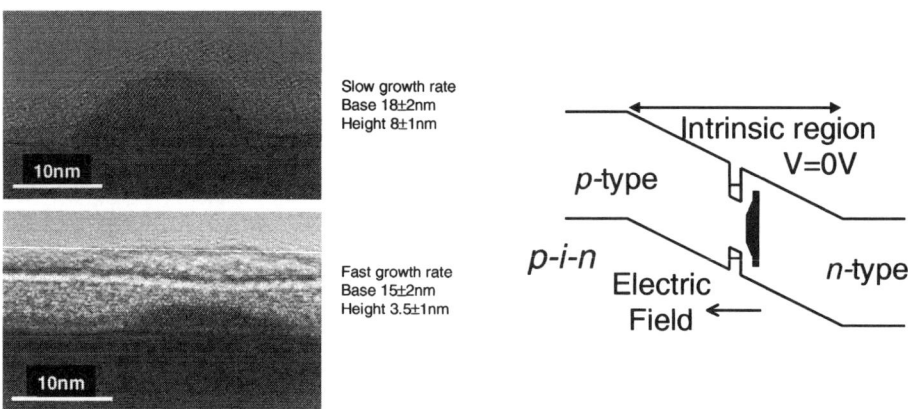

Figure 1. Transmission electron micro-graphs of uncapped slow (0.01 monolayers per second) and fast growth rate samples (0.4 monolayers per second) in Figs 1a and b respectively.

Figure 2. Schematic band diagram of a p-i-n device under applied reverse bias V.

2. Experimental details

Two types of dots were studied, both grown by MBE on (001) GaAs substrates at 500°C. The first type (samples A-C) was deposited at 0.01 monolayers per second (ML/s) to give a density $\sim 1.5 \times 10^9 \text{cm}^{-2}$, base size 18nm and height 8.5nm (Fig. 1a), as determined from transmission electron microscopy. The second type (sample D) had a higher deposition rate of 0.4 ML/s, resulting in a density $\sim 5 \times 10^{10} cm^{-2}$ and size 15nmx3.5nm (Fig. 1b). The asymmetric shaped QDs, sitting on a $\sim$1 ML thick wetting layer, have their apex oriented along the growth direction. Attention is focussed in this paper on samples containing single layers of dots contained within the intrinsic region (width d) of either p-i-n or n-i-p structures, allowing fields (F) up to 300kV/cm to be applied either parallel or anti-parallel to the growth direction. Results for samples containing five dot layers will also be mentioned briefly. Applying a reverse bias to a p-i-n structure (p region at surface) results in F pointing from substrate to surface as shown in Fig. 2. For an n-i-p structure the field direction is reversed. The total electric field is given by $F=(V+V_{bi})/d$, where V is the externally applied voltage and V_{bi} is the built in voltage ($\approx$1.5V) (d=0.3μm for samples A-C and 0.5μm for sample D). Monochromated white light ($\sim$3mW/cm^2, corresponding to extremely low dot carrier occupancies ($<<$1)) from a projector lamp was used for excitation. The photocurrent was measured using lock-in techniques from 10 to 300K using 400μm diameter, annular contact mesas.

3. Experimental results and analysis

Photocurrent spectra, recorded as a function of reverse bias from 0 to 8V (total field F from 50 to 320kV/cm), are shown in Figs. 3a and b for sample A (p-i-n) at temperatures of 5 and 200K respectively. At 200K, four well-defined features are seen in the range 1.1 to 1.3eV, arising from inter-band transitions between the confined electron and hole levels in the dots. Similar spectra are seen at 5K, though at low fields the interband transitions are quenched, and a photocurrent signal is only observed from the InAs wetting layer (1.41eV) and the GaAs barriers (1.5eV) (see inset to Fig 3b). The temperature dependence of the spectra is discussed later in this section. The most important point in Fig 3 is that with increasing field, at all temperatures, all the QD transitions shift strongly to lower energy (by 30meV at 8V ($\equiv$300kV/cm)), the signature of the quantum confined Stark effect [6], but without any qualitative change in the form of the spectra.

The ground state transition energies at 200K for all samples are plotted as a function of F, calculated using V_{bi}=1.5V, in Fig. 4. We focus attention on samples B (p-i-n) and C (n-i-p) [7]. Positive field corresponds to reverse bias for the p-i-n structure. The transition energy exhibits a marked asymmetry about F=0, with the maximum obtained for a non-zero field of -90kV/cm. This asymmetry implies that the QDs have a permanent dipole moment (p). A non-zero Stark shift at F=0 of the ground state transition has also been observed in the PL of InAlAs-AlGaAs QDs [8] over the limited field range of $\pm$60kV/cm since the PL is unavoidably quenched at high fields.

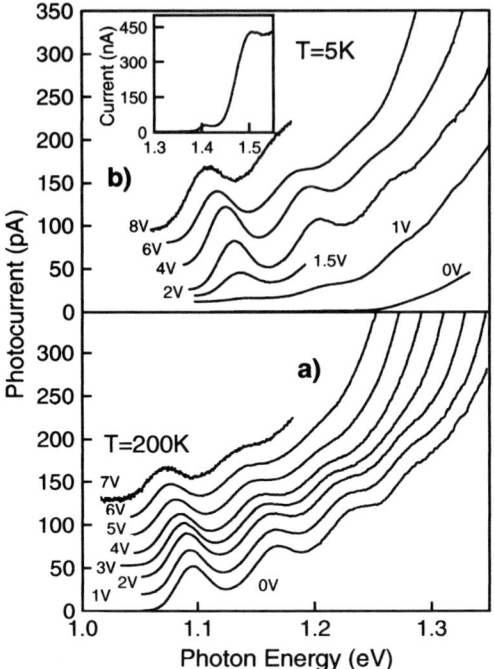

Figure 3. Photocurrent spectra of sample A, as a function of applied reverse bias, for temperatures of 5K and 200K. The inset shows a photocurrent spectrum (V=0V) to higher energy exhibiting wetting layer and GaAs band edge features.

The field dependence of the transition energies (E) in Fig. 4 can be described to a very reasonable approximation by the expression $E = E_0 + pF + \beta F^2$ (the solid lines in Fig. 2), where E_0 is the energy at $F=0$, the second term arises from the non-zero dipole moment (p), and the third term from polarisation of the dots in the applied field (the quantum confined Stark effect). By fitting to the experimental data for samples B and C a value of $p=(7\pm2)\times10^{-29}$ Cm^{-1} is determined, corresponding to an electron-hole separation of r=4.0±1Å (from $p=er$).

Self-assembled QDs are expected to possess a permanent dipole moment as a result of their asymmetric shape [2]. In general the lack of inversion symmetry means that the eigenfunctions of the system do not have definite parity, the dots are expected to have a permanent dipole moment and the first order perturbation due to an applied electric field will be non-zero. For the same reason asymmetric molecules such as CO are polar and possess a permanent dipole moment whereas symmetric molecules such as CO_2 have no dipole moment [9]. Thus although the existence of a permanent dipole moment for self-assembled quantum dots is not surprising, the experimentally determined direction of the observed dipole is highly unexpected. The maximum transition energy occurs for F in the direction from apex to base, corresponding to electron (hole) attraction to the apex (base) of the dots. This implies that the electron

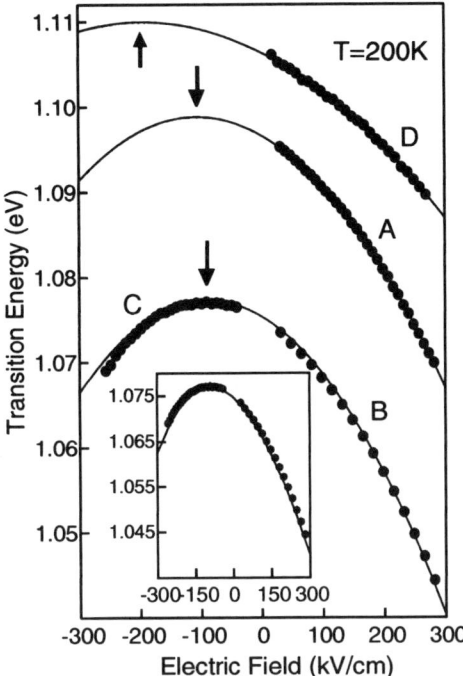

Figure 4. Transitions energies as a function of electric field for samples A-D at 200K. The solid curves are a fit to $E = E_0 + pF + \beta F^2$. The data for sample C have been rigidly shifted by 7meV. The inset shows a theoretical fit to the data for samples B and C.

charge density distribution lies below that of the hole at $F=0$, with the dipole pointing from base to apex. This result, to the best of our knowledge, is in contrast to that predicted by all accurate modelling [2-5] of InAs QDs, which predicts hole localisation towards the base of the dots, below the electrons. This alignment, which is predicted for pure InAs dots of *any shape* for which the lateral size decreases from base to apex (e.g. the pyramidal shape in Refs. [2-5]) results from the strain-induced form of the valence band edge profile [3] which leads to hole localisation in the widest region of the dots, and from the ratio of the effective masses along the growth direction ($m^*_{hh} \gg m^*_e$) which means that the electrons tend to spread out to fill the dots whereas the (heavy) holes are localised in the regions of deepest potential.

To determine dot parameters necessary to reverse the relative alignment of the electron and hole, and to fit to the experimental data, envelope function calculations, treating the electrons and holes with separate one-band Hamiltonians, have been carried out. The strain distribution for a given dot shape was calculated using a Green's function technique which provides an analytical expression in the form of a Fourier series for the strain tensor [10]. The band gaps and offsets were calculated using model solid theory [11], including hydrostatic strain effects; the heavy-hole Hamiltonian included the spatial variation of the biaxial strain deformation potential and the directional dependence of the heavy-hole mass. Carrier effective masses, determined using 3-band **k.p** theory, and band offsets were assumed to vary linearly with composition.

For pure InAs pyramidal dots [12] our calculations give good agreement with previous theories [2-5], with the hole wavefunction always below that of the electron, confirming the accuracy of our approach. Reversal of this alignment is only possible for dots having a graded $In_{1-x}Ga_xAs$ composition, with x decreasing from base to apex (the holes tend to be localised in the region with the largest In composition). This tends to shift the localisation of the holes away from the base. However, in addition it is also necessary to severely truncate the pyramid to achieve a dipole of the correct sign, since strain effects continue to localise the hole strongly below the electron until the truncation factor is greater than ≈ 0.6.

The best fit to the experimental data for samples B and C is shown in the inset to Fig. 4. This was obtained using a pyramid of base length (w) 15.5 nm, height 22 nm, of which the top 75% is truncated (see Fig 5 for diagrams of the shape employed) to give an actual height (d) of 5.5 nm, and an x varying linearly from 50 % at the base to zero at the top surface. The quadratic shift is determined principally by d, the dipole by the grading and the degree of truncation, and the absolute energy mainly by d and w. Although other parameter sets can also give a good fit, we emphasise that the shape employed represents a good approximation to that obtained from structural measurements (Fig. 1a). Furthermore pyramids without truncation, and any shape wider at the base than the apex but without grading, never have a dipole of the correct sign. We thus obtain new information by optical methods on the structure and composition of buried quantum dots which is very difficult to obtain from structural studies alone.

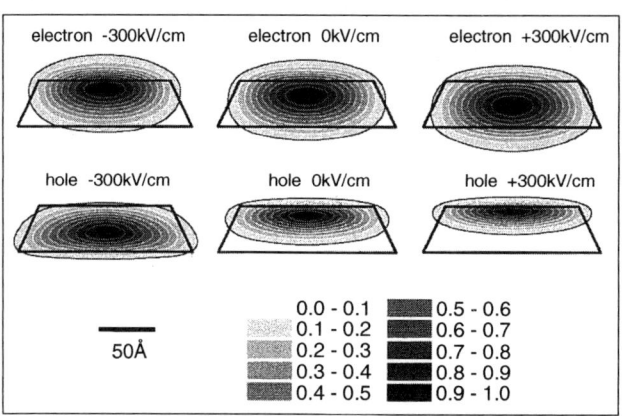

Figure 5. Calculated electron and hole probability densities using parameter set which gives the best fit to the data of the fig 4 inset, at zero, -300kV/cm and 300kV/cm. The localisation of the electron below the hole at zero electric field is clearly visible.

Contour plots of the calculated electron and hole probability densities are presented in Fig. 5, for the parameter set that gives the best fit to the experimental Stark shift data. At zero electric field, the localisation of the electron below the hole is seen very clearly. For negative fields the hole distribution moves towards the base and the electron to the apex, with the dipole moment being zero for fields of -90kV/cm in very good agreement with the results of Fig 4. For positive fields by contrast the hole is attracted further to the apex and the electron to the base with a consequent increase of the dipole moment.

Our conclusions regarding the composition of *buried* InAs dots are consistent with results of previous STM investigations on *uncapped* dots [13,14]. For the growth temperature used for the present QDs (500°C) the total dot volume was found in ref [13] to be greater than that of the deposited InAs, a finding that could only be explained by in-diffusion of Ga. The grading required to explain the sign of the observed dipole has also been observed in STM studies of $In_{0.3}Ga_{0.7}As$ QDs and attributed to In segregation during growth [14].

Data for devices A and D (slow and fast growth rates respectively) in Fig. 4 also exhibit a maximum transition energy for a negative field, showing that a permanent dipole with hole above electron is a general property of our QDs. In addition the size of the quadratic component of the Stark shift for the slow growth rate samples (A-C) is larger than that of the faster growth rate sample D, consistent with the smaller height of the dots in these samples as seen by inspection of the TEM images in Figs 1a, b. The same sign of asymmetry of the Stark effect has been reported by Raymond et al for AlInAs/AlGaAs self-assembled QDs [8] supporting the generality of our findings.

In Fig 1, the energy separations and relative intensities of the transitions are independent of field, in strong contrast to the behaviour for quantum wells [6]. This is consistent with the excited state splittings arising from lateral quantisation since, to first order, the response to F along the growth direction (z) is determined by quantisation along z. Between 30 and 300kV/cm the magnitude of the ground state photocurrent decreases by ~35%. This arises from a reduction in oscillator strength due to the field-induced separation of the electron (e) and (h). Our calculations of the e-h overlap predict a 30% reduction over this field range, in excellent agreement with experiment. Furthermore an increase of ~10% in the ground state photocurrent was observed from – 30 to ~–100kV/cm for the n-i-p structure, before decreasing for more negative electric fields. The peaking of the photocurrent signal at –100kV/cm is fully consistent with expectations since at this electric field the dipole is cancelled to zero and the electron-hole overlap is maximum.

The above analysis has been performed on 200K data where the QD interband transitions are observed at all fields. At low temperature (T) and F the transitions are quenched (see Fig1a). However as the temperature is raised, the peaks are observed at increasingly lower field. In Fig. 6, the variation of transition energies with F from 240 to 5K is plotted. At low temperature, and F less than ~120kV/cm, there is increasing departure from the monotonic behaviour of Fig. 4. These effects arise from the competition between two carrier escape mechanisms from the dots: tunnelling and thermal emission. At low temperature the photo-excited carriers can only escape by tunnelling. In this regime, at low F we observe selective tunnelling from the inhomogeneous distribution of dot

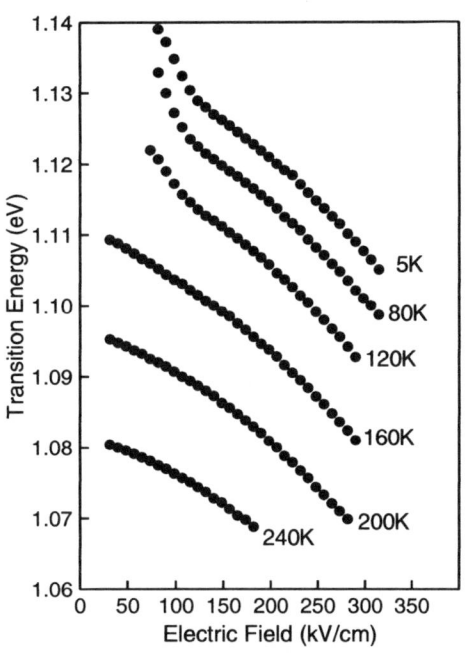

Figure 6. Transition energies as a function of field for sample A at temperatures from 5 to 240K

energies; only carriers excited in the highest energy part of the distribution can escape, and the transition energy occurs above that expected by extrapolation from higher field. As F is increased tunnelling becomes possible from the lower energy parts of the distribution, until at 120kV/cm all carriers escape and the observed energy corresponds to the mean of the distribution. Increasing temperature results in thermal emission from the QDs, which at high temperature leads to field independent escape of all carriers from the QDs. We are thus able to achieve selective carrier escape from within the inhomogeneous distribution of dot energies by variation of the applied electric field.

4. Significance of photocurrent spectroscopy for investigation of electronic states of self-assembled quantum dots

Absorption spectra provide the most fundamental optical probe of the electronic properties of bulk and low dimensional semiconductors. However absorption spectra of QDs are very difficult to obtain due to their very small absorption strengths. The two reports of absorption spectra in the literature previously required the use of very difficult experimental techniques. Calorimetric absorption spectroscopy [15] requires very low temperatures and resulted in broadened spectra, and direct absorption spectroscopy required the development of very sensitive detection techniques and long integration times [16]. Photoluminescence excitation spectroscopy, commonly employed to measure absorption spectra in quantum wells, is dominated by phonon related features in self assembled quantum dots as a result of the discrete density of states in a 0d system, and the influence of competing non-radiative recombination paths [17]. Photocurrent spectroscopy by contrast suffers from none of these drawbacks permitting absorption spectra to be obtained with very good signal to noise in short experimental times (~5minutes).

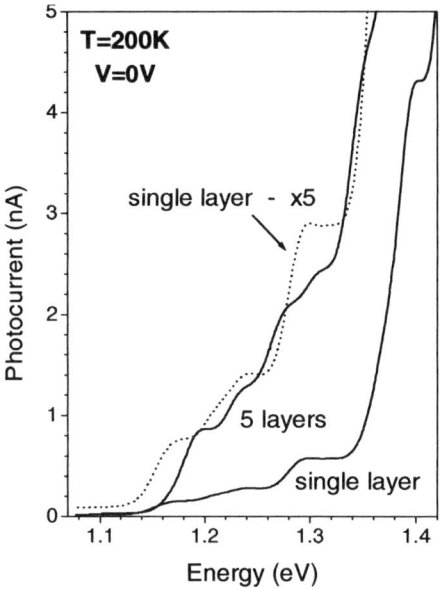

Figure 7. Comparison of photocurrent spectra from one and five dot layer samples. The photocurrent signals are found to scale with the number of dot layers to a very good approximation.

At sufficiently high fields all carriers escape from the dots [18] and their absorption strength (A) can be determined directly from the magnitude of the photocurrent (I) and the relationship $I=APe/h\nu$ where P is the optical power at frequency ν. A value of $A=1.3\times10^{-5}$ is obtained, in good agreement with a previously published value after correcting for the different dot density. In Fig 7 we report photocurrent spectra from samples containing five dot planes, and compare to the spectrum for a single layer sample. It is seen that the absorption strength measured is, to a very good approximation, five times greater than that for a single layer sample, providing further confirmation of the reliability of the technique.

One feature of the spectra of Figs 3 and 7 remains unexplained, the presence of the broad background signal underlying particularly the excited state spectra [19]. Its origin remains uncertain at the present time, although it most likely arises from band tail absorption from either the GaAs band edge or from the InAs wetting layer, some enhancement occurring in applied electric field due to Franz-Keldysh effects. It is interesting to note that a similar underlying background was observed in the absorption measurements of ref [16].

5. Conclusion

In summary, we have demonstrated from Stark effect spectroscopy that InAs-GaAs self-assembled quantum dots possess a permanent dipole moment. The dipole moment is expected from the asymmetric shape of the dots but its sign is found to be opposite to that predicted by all recent calculations. In combination with theoretical modelling we have deduced the effective height of the dots and the electron-hole vertical separation, and have demonstrated that the nominally InAs dots contain significant concentrations of gallium. New information on the structural properties of buried quantum dots, of key importance for modelling of electronic structure has thus been deduced from optical spectroscopy. The power of photocurrent spectroscopy to determine absorption spectra of quantum dots has been demonstrated.

Acknowledgements. We thank A D Andreev for useful discussions and acknowledge EPSRC and the Russian Foundation for Fundamental Research (grant 97-02-17802) and GNTP 'Nanostructures' (grant 97-1068) for financial support.

References

1. D Bimberg, M Grundmann and N N Ledentsov, Quantum Dot Heterostructures, Wiley, (1998).
2. M Grundmann, O Stier and D Bimberg, Phys. Rev. *B*. **52**, 11969, (1995)
3. M Cusack, P R Briddon and M Jaros, Phys. Rev. B. **54**, R2300, (1996)
4. C Pryor, Phys. Rev. B. **57**, 7190, (1998). This reference (8 band k.p theory) provides a summary of most earlier theoretical work.
5. J Kim, L-W Wang and A Zunger, Phys. Rev. B. **57**, R9408, (1998)
6. D A B Miller, D S Chemla, T C Damen, A C Gossard, W Wiegmann, T H Wood and C A Burrus, Phys. Rev. B. **32**, 1043, (1985)
7. A small shift of 7meV to higher energy has been applied to the results of sample C, in order to obtain a continuous variation of the peak positions between positive and negative electric fields. Samples B and C were grown consecutively in order to obtain the minimum possible run-to-run variation in dot parameters between samples. The observed energy difference of 7meV corresponds to only a ~2.5% variation in dot base size.
8. S Raymond, J P Reynolds, J L Merz, S Fafard, Y Feng and S Charbonneau, Phys. Rev. B **58**, R13415, (1998)
9. see e.g. B Bleaney and B I Bleaney, Electromagnetism (Oxford University Press) p477, (1965)
10. A.D. Andreev and E.P. O'Reilly, MRS Internet J. Nitride Semicond. Res. 4S1, G6.45, (1999); also in S.J. Pearton, C. Kuo, T. Uenoyama and A.F. Wright (eds.), GaN and related alloys, MRS vol. **537** (1999)
11. M P M C Krijn, Semicond. Sci. Technol. 6 27, (1991)

12. S Ruvimov et al Phys. Rev. B. **51**, 14766, (1995). This paper reported pyramidal shapes for InAs quantum dots from transmission electron microscopy and provided the basis for much subsequent modelling. However other reports (e.g. D S L Mui, D Leonard, L A Coldren and P M Petroff, Appl. Phys. Lett. **66**, 1620, (1995)) including our own in figure 1 of this paper, find the dots more closely approximate to lens shapes.

13. P B Joyce, T J Krzyzewski, G R Bell, B A Joyce and T S Jones, Phys. Rev. B **58**, R15981, (1998)

14. N Grandjean, J Massies and O Tottereau, Phys. Rev. B. **55**, R10189, (1997)

15. M Grundmann, J Christen, N N Ledenstov, J Bohrer, D Bimberg et al, Phys Rev Lett, **74**, 4043, (1995)

16. R Warburton, C S Durr, K Karrai, J P Kotthaus, G Mideiros-Ribeiro and P M Petroff, Phys. Rev. Lett. **79**, 5282, (1997).

17. M J Steer, D J Mowbray, W R Tribe, M S Skolnick, M D Sturge, H Hopkinson, A G Cullis, C R Whitehouse and R Murray, Phys Rev B **54**, 17738 (1996), R Heitz, M Grundmann, N N Ledentsov et al, Appl Phys Lett **68**, 361, (1996)

18. This is known since the magnitude of the dot photocurrent is independent of field and temperature, at sufficiently high fields and temperatures.

19. The background is very weak at the energy of the QD ground state transition and thus has a very small effect on the absorption strength deductions.

INTERBAND OPTICS OF CHARGE-TUNABLE QUANTUM DOTS

R. J. WARBURTON, C. SCHÄFLEIN, H. PETTERSSON,
D. HAFT, F. BICKEL, C. S. DÜRR, K. KARRAI,
J. P. KOTTHAUS,
Center for NanoScience and Sektion Physik, LMU,
Geschwister-Scholl-Platz 1, 80539 München, Germany

G. REBEIROS-RIBEIRO, J. GARCIA, W. SCHOENFELD,
P. M. PETROFF,
Materials Department and QUEST, University of California,
Santa Barbara, California 93106, USA

AND

N. CARLSSON, W. SEIFERT, AND L. SAMUELSON
Division of Solid State Physics,
Lund University, S-221 00 Lund, Sweden

Abstract. Self-assembled quantum dots have been incorporated into a field-effect structure allowing the dots to be filled sequentially with electrons. Results of interband spectroscopy on charge-tunable structures are presented. For InAs dots in GaAs and for InAs dots in InP characterisation of the devices with capacitance and transmission spectroscopies yields detailed energy level diagrams. Furthermore, we have measured the oscillator strengths of the various interband transitions. Both these systems have a band gap in an awkward spectral range for single dot experiments. Instead, we present photoluminescence experiments on single quantum rings. Charging of the rings with single electrons is detected by pronounced shifts in the photoluminescence energy.

1. Introduction

Self-assembled quantum dots in semiconductors confine electrons in all three directions and can therefore be thought of as artificial atoms. The

M.L. Sadowski et al. (eds.), Optical Properties of Semiconductor Nanostructures, 347–363.

properties of ordinary atoms oscillate with electron number as described by the periodic table. The fundamental reason for the periodicity is the existence of atomic orbitals which are filled according to Hund's rules. An equivalent periodicity has been observed in a lithographically defined quantum dot by transport experiments [1]. Similar results exist also for self-assembled quantum dots [2]. These dots have a strong confinement in one direction, the vertical direction, but a relatively weak lateral confinement and so are essentially two-dimensional in character. This leads to large gaps for electron filling 2, 6, 12 ..., and not fillings 2, 8, 18 ... as in ordinary atoms. It should be noted of course that the energy scales in the two systems are completely different, as are the forms of the confinement potentials.

A major difference between real atoms and artificial atoms in semiconductors is the existence of the valence band in semiconductors which has no analogue in atomic physics. Strong optical transitions are allowed between valence and conduction band states but they have not been extensively explored as a function of electron occupation. There are two challenges experimentally. First, a quantum dot system must be designed with which the dots can be successively filled with electrons. Secondly, for a full understanding experiments should be performed on a single quantum dot. This typically demands microscopy with a high spatial resolution, quite possibly smaller than the wavelength of the interband transition. The point is that although self-assembled quantum dots are remarkably homogeneous, with dot to dot fluctuations of only $\sim 10\%$, interband transitions from a dot ensemble are inhomogeneously broadened such that all effects on a meV energy scale are obscured.

In this paper, interband optics are presented on quantum dots which are embedded in a metal-insulator-semiconductor transistor. It is shown how the occupation of the dots with electrons can be controlled simply by the gate voltage. Interband absorption is detected through changes in the intensity of light transmitted through the sample. Results on two different systems are presented, InAs dots in GaAs and InAs dots in InP. As in other groups, we have not yet managed to perform single dot spectroscopy on these low band gap dots. However, we do present results on single quantum rings which show very clearly shifts on charging with single electrons.

2. Charge-tunable dots

We have designed and optimised MISFET structures with which self - assembled dots can be loaded with electrons [3, 4]. The layer structure and band diagram are shown in Fig. 1. The dots are situated between a highly doped GaAs layer, the back contact, and a metallic gate on the sample

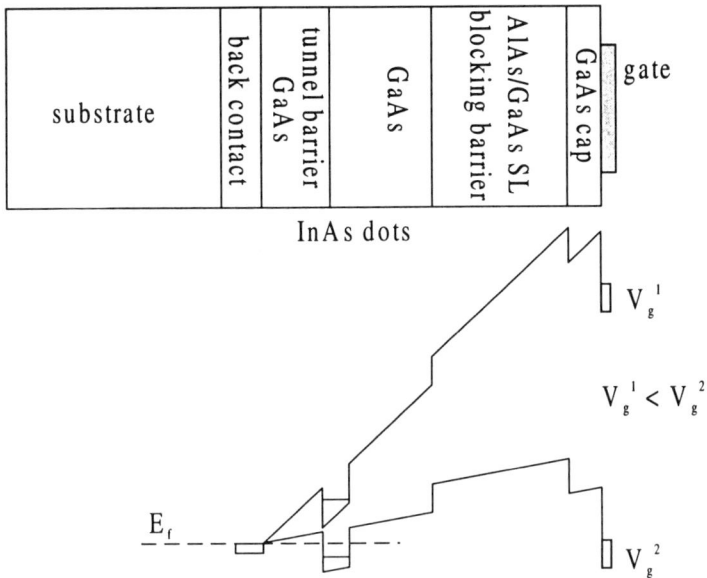

Figure 1. The upper part of the figure shows the layer structure of a metal-insulator-semiconductor device. The dots are separated by 25 nm from the back contact; the back contact is 175 nm away from the surface. The lower part of the plot shows the conduction band edge for two different applied potentials showing how it is possible to change the electron occupation of the dots with a voltage applied between gate and back contact. In the upper case, the potential is so large and negative that the dot levels lie above the Fermi energy. In the lower case, a more positive voltage has been applied so that the dot levels lie beneath the Fermi energy.

surface. When a large negative voltage is applied to the gate, the ground state of the dots lies well above the Fermi energy defined by the high doping in the back contact and the dots are not occupied. When the voltage is made more positive, the ground state in the dots becomes resonant with the Fermi energy in the back contact, and electrons can tunnel into and out of the dots. A further increase in gate voltage essentially traps electrons in the dots. These loading processes can be monitored with capacitance spectroscopy: a dc voltage and small ac voltage are applied to the gate. When tunnelling takes place, the oscillating current between back contact and dots can be measured as a capacitive signal.

We present here results on three systems, each of which is embedded into a field-effect device with the same geometry. First, InAs dots in GaAs from Santa Barbara [5] have been investigated. The dots are lens-shaped, approximately 20 nm in base diameter and 6 nm in height. Secondly, InAs dots in InP from Lund have been studied. These dots are also roughly speaking lens-shaped but they have a pronounced asymmetry in the plane, with a diameter of 45 nm in the [-110] direction, but only 35 nm in the [110] direction. Finally, InGaAs rings have been investigated. The rings are grown exactly as the dots but with a growth pause of 3 minutes after 1 nm of the GaAs cap layer has been deposited [6]. This induces a remarkable transformation in the form of the dots: In diffuses laterally, generating a ring of InGaAs. For the present work, the detailed nature of the confinement potential is actually not crucial. The main point is that the rings have a smaller extent in the growth direction than the dots and this pushes up the confinement energy.

3. Experimental details

The transmission experiments were performed at 4.2 K with a Fourier transform spectrometer. The output of the spectrometer was focused into one end of a 200 μm multi-mode fibre. The other end of the fibre was positioned a few tens of microns above the sample surface. The transmitted light was detected with a p-i-n diode placed immediately behind the sample. We employed a Ge diode for the InAs dots in GaAs (band gap 1.05 eV), an InAs diode for the InAs dots in InP (band gap 0.7 eV), and a Si diode for the InGaAs rings in GaAs (band gap 1.3 eV). The current generated by the p-i-n diode was converted into a voltage by a commercial pre-amplifier. There are two requirements for the transmission experiments. First, the signal:noise ratio must be at least 10^5. The system was optimised and the noise is actually limited by the dynamic range of the pre-amplifier. The noise is random and scales as $1/\sqrt{n}$ where n is the number of mirror scans in the Fourier transform spectrometer. The spectra presented here represent the average over 10,000 scans and typically took 2-4 hours to record. Secondly, it has to be possible to distinguish the dot absorption from features in the energy response of the whole system. We do this by recording at each gate voltage of interest a reference spectrum at a large and positive gate voltage where the dots are fully occupied with electrons and the dot absorption is essentially turned off. Presented are ratios of the spectrum at a particular gate voltage to the reference spectrum.

Photoluminescence on single rings was recorded using a 822 nm laser diode as pump source and a multi-channel, cooled Si CCD as detector. A grating spectrometer with resolution 0.2 nm was employed. We achieved a high spatial resolution either with a low temperature near-field microscope or with a confocal microscope. In the former, a near-field tip with diameter $\sim$ 150 nm was positioned $\sim$ 5 nm from the sample surface using quartz tuning fork technology [7]. The sample was excited through the tip and the photoluminescence was collected and focused into a multi-mode fibre with an ellipsoidal mirror. The spatial resolution is determined by the exciton diffusion length in this experiment, typically 0.5 to 1 μm for these samples. The confocal microscope consists of optics designed to image two single-mode fibres onto the same point on the sample surface. One fibre was used for excitation, the other for collection. Excitons which diffuse out of the excitation region cannot contribute to the measured photoluminescence. In this way, exciton diffusion does not worsen the resolution. Instead, the resolution is determined only by the diffraction limit for Gaussian optics. We used a microscope objective with numerical aperture 0.55, giving a spot size of ~ 1 μm. This is very similar to the resolution achieved with the near-field microscope. The data presented here were all recorded with the confocal microscope as its collection efficiency is higher. Similar spectra were obtained however also with the near-field microscope.

The pump power was ~ 1 μW. Assuming that the quantum efficiency of the samples is 100%, that the recombination time is 1 nS, and that the wetting layer absorbs 1% of the incident excitation, the rings are occupied on average with a single exciton for a power density ~ 1 μW/μm^2 for 10^{10} rings per cm^2. In other words, the pump intensity is sufficiently small that we can be confident that the rings are not occupied by more than a single electron-hole pair. This is important as it has been demonstrated that the decay of a dot occupied by multiple excitons consists of a several lines, with also a broad background at high occupancy [8]. These are obviously interesting effects, but we are concerned here with the recombination of rings with a given number of electrons and one hole and so we need to need to exclude multiple exciton occupation.

4. Ensemble measurements of InAs dots in GaAs

Fig. 2 shows a capacitance spectrum of an ensemble of InAs dots in GaAs. The first peak corresponds to tunneling into the first state of the quantum dots which we label s in analogy with atomic physics. The peak is split into two by Coulomb blockade. The remarkable feature is that Coulomb blockade is visible even on this ensemble of about 100 million quantum dots. It is therefore possible to set the gate voltage such that the vast majority of dots has occupation $N = 1$ or occupation $N = 2$. At higher gate voltages, tunneling into the first excited state, the p state, occurs. For large ensembles the broadening is too large for each charging peak to be observed. Occupancy $N = 4$ for example cannot be achieved exactly: there is inevitably a distribution in occupancy from dot to dot. At still larger voltages, the capacitance rises rapidly. This is a general feature of all the samples and corresponds to tunneling into the wetting layer.

Transmission spectra of InAs dots in GaAs are shown in Fig. 3. At large negative voltages there are three peaks corresponding to the valence

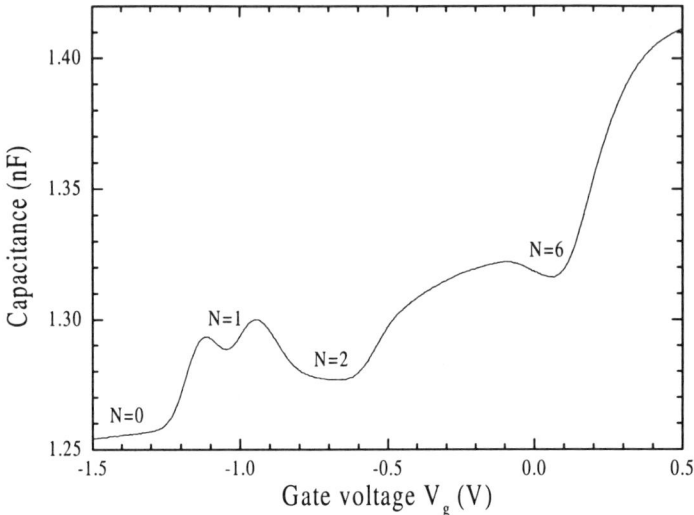

Figure 2. Capacitance versus gate voltage at 4.2 K for InAs dots in GaAs. Charging of the electron levels in the dots corresponds to peaks in the capacitance enabling the electron occupancy per dot, N, to be determined.

band—conduction band transitions $s - s$, $p - p$ and $d - d$. On increasing the gate voltage, the peaks disappear exactly as expected from Coulomb blockade: at $N = 2$, the $s - s$ transition is blocked; at $N = 6$, the $p - p$ transition is blocked. The $d-d$ transition seems also to disappear at positive gate voltages, and this may be because the electron d state is no longer bound for higly charged dots. The electron quantisation energy and the effective length of the electron s wave function have been measured with far infrared spectroscopy to be 50 meV and 4.7 nm, respectively [9].

Assuming that the confinement potentials for both electrons and holes are parabolic, we can use the positions of the $s-s$, $p-p$ and $d-d$ interband transitions to determine the properties of the valence band [10]. The exciton binding energies were estimated by perturbation theory, and have to be included as the exciton binding energy decreases from $s - s$ to $p - p$ to $d - d$. We find that the hole quantisation energy is 25 meV, and that the effective length of the s hole wave function is 2.5 nm.

A close examination of Fig. 3 shows that the peaks not only disappear according to Pauli blocking but also shift in energy. The shifts on occupation are all to lower energy. For instance, the $p-p$ transition shifts by -10 meV at $N = 1$ and by -18 meV at $N = 2$. The shifts are summarised in Fig. 4, a plot of the shifts against electron occupation for the $s - s$, $p - p$ and

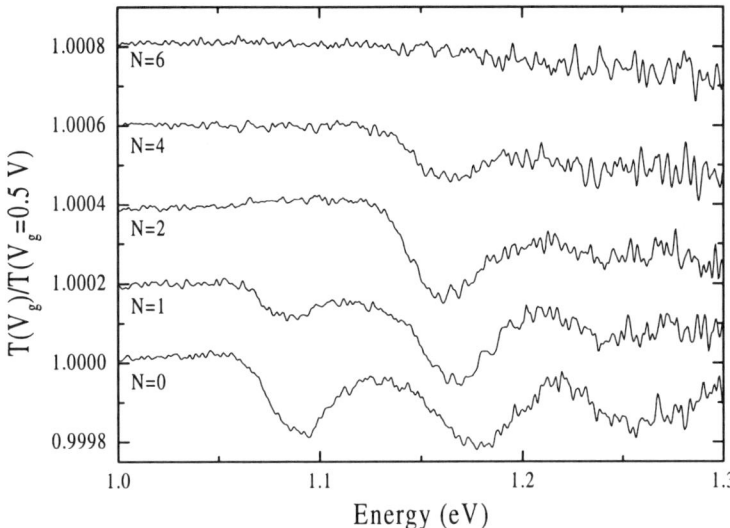

Figure 3. Interband transmission of InAs dots in GaAs for various gate voltages at 4.2 K. In each case, a ratio to a background spectrum taken at 0.5 V is plotted. The curves are offset from 1 for clarity.

$d - d$ transitions. The precision with which we can determine these shifts is limited by the inhomogeneous broadening to a few meV.

The $s - s$ transition is obviously blocked for $N > 2$. The point at $N = 5$ corresponds to photoluminescence of a fully occupied dot [11]. All the shifts correspond to an interaction between the stored electrons and the excited electron-hole pair. Generally speaking, the interaction of the excited hole with the stored electrons is larger than the interaction between the excited electron and the stored electrons because the hole is more localised than the electrons. This explains the shifts to the red [12]. In order to understand the shifts quantitatively, we have calculated the Coulomb energies between the particles for a given configuration with perturbation theory [10, 13]. Using the dot parameters as described above, we calculate the shifts as shown by the solid lines in Fig. 4. It can be seen that the agreement with the experimental data is good. These model calculations also give an excellent description of capacitance spectroscopy on small dot ensembles where changes in the single-particle configurations in high magnetic field are observed [13].

A particular feature of transmission spectroscopy is that it provides a direct measurement of the integrated absorption which itself can be related to the oscillator strength of the dots. We relate the integrated absorption to

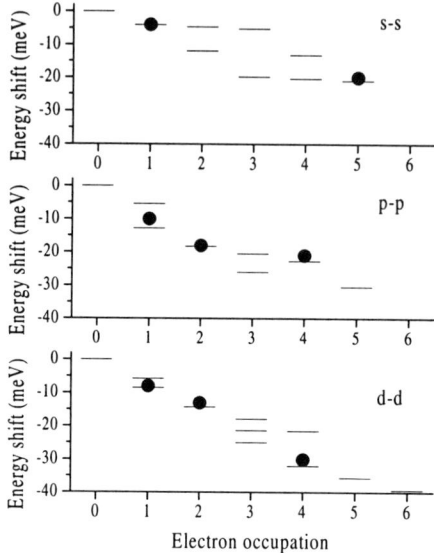

Figure 4. The shifts of the $s - s$, $p - p$, and $d - d$ interband transitions against electron occupation for the InAs dots in GaAs. The points are the results of the experiments; the solid lines the results of perturbation theory calculations. The electron occupation refers to the number of electrons in the dots before optical excitation of an electron-hole pair takes place.

an integral of the real part of the conductance with simple electrostatics, and we use a semi-classical model for the conductivity [10, 13]. The dot density was determined from the capacitance by integrating over the first two charging peaks. We find from the data of Fig. 3 that the oscillator strength of the $s - s$ and $p - p$ transitions are $f_{s-s} = 10.9$ and $f_{p-p} = 1.7 f_{s-s}$. The perturbation theory calculations assume that the dots are in the strongly confined limit, namely that the quantisation energies are larger than the Coulomb energies. In the same spirit, we can calculate the oscillator strength essentially as the overlap between the electron and hole wave functions. The picture is that the electron and hole are confined primarily by the parabolic potential and are not coherently bound to each other. Using the dot parameters deduced from the optical spectroscopy, we find $f_{s-s} = 10.7$ and $f_{p-p} = 1.7 f_{s-s}$, in excellent agreement with the experimental results. f_{p-p} is less than twice as large as f_{s-s} because the electron wave function becomes delocalised more quickly than the hole wave function on going from s to p.

An alternative way of interpreting the experiment is to consider the minimum in transmission as a result of a scattering process. A scattering cross-section can be deduced from the integrated absorption. We find that

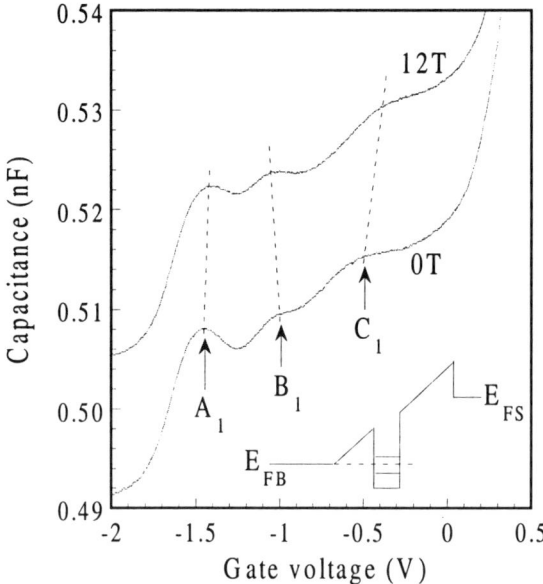

Figure 5. Capacitance versus gate voltage for InAs dots in InP at 4.2 K. The inset is a schematic of the band diagram of the device.

the scattering cross-section for a single quantum dot is 1.0×10^{-15} cm^2. A useful comparison is the scattering cross-section for a single rhodamine 6G dye molecule 2×10^{-16} cm^2 [14]. The conclusion is that quantum dots are effective scatters of radiation, even in the strongly confined limit. (The oscillator strength of a quantum dot increases as one goes into the weakly confined limit when the exciton becomes a coherent entity [15].)

5. Ensemble measurements of InAs dots in InP

Fig. 5 shows a capacitance trace of a field-effect structure containing InAs dots in InP. The first peak, A_1, corresponds to tunnelling into the ground state of the dots. A splitting of the peak into two cannot be observed because the broadening is too large. However, the capacitance peaks corresponding to tunnelling into the p-like states (B_1 and C_1) are clearly split into two.

This must mean that the p-like single-particle states of the dots are split. Confirmation that the states are p-like comes from the magnetic field dependence of the tunneling peaks. Fig. 5 shows how the peaks move apart in magnetic field as expected. It is possible to estimate the splitting between the p states from the capacitance trace by converting the voltage into an energy with the lever arm and by estimating the contribution of the charging

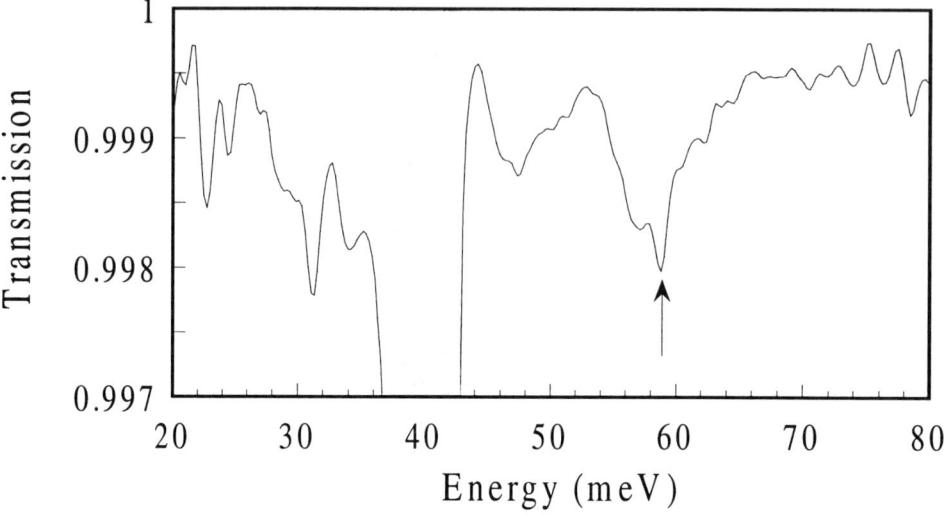

Figure 6. Far infrared absorption of InAs dots in InP at 4.2 K. The resonance marked with an arrow is thought to arise from an $s - p$ transition in the dots. The other, weaker resonance at 42 meV is probably related to a phonon interaction. The marked area corresponds to the Reststrahlenband where the sample is essentially opaque.

energies to the splitting. In the absence of any detailed structural information, we assume that the confinement potential is parabolic, but with different curvatures in the x- and y-directions. The charging energies can be expressed in terms of elliptical integrals in the perturbation theory approach [16]. We find that $\hbar\omega_x = 60$ meV and $\hbar\omega_y = 35$ meV. An obvious way to measure these energies directly is with far infrared spectroscopy. A spectrum of the InAs dots in InP is shown in Fig. 6. A resonance can be observed at 57 meV supporting the interpretation of the capacitance trace. The weak resonance at 42 meV does not move with magnetic field (not shown here) and is probably related to an interaction of the electrons with an interface phonon as it lies very close to the Reststrahlenband [9]. A second quantum dot-related resonance cannot be discerned above the noise in Fig. 6. It could well be the case that the resonance is simply too close to the Reststrahlenband to be detected.

The simplest explanation for the split p states is that the lateral asymmetry in the shape gives two different quantisation energies. One would expect that $L_y : L_x = \sqrt{\hbar\omega_x} : \sqrt{\hbar\omega_y}$ where L_x and L_y are the dot dimensions along the minor and major axes, respectively. This relation is quite accurately fulfilled in practice. An alternative explanation is that the strain

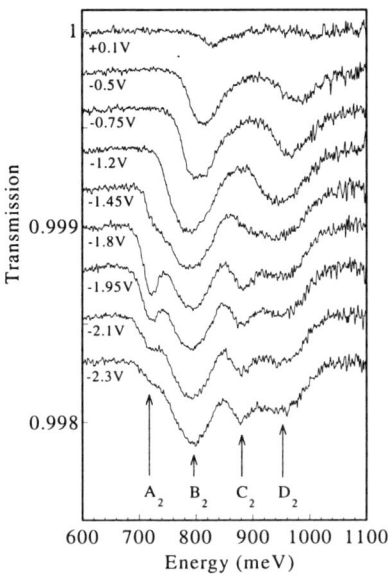

Figure 7. Interband transmission of InAs dots in InP for various gate voltages at 4.2 K. In each case, a ratio to a background spectrum taken at 0.5 V is plotted. The curves are offset from 1 for clarity.

field is asymmetric. In fact, calculations have predicted that the splitting works against the splitting from the shape alone [17]. However, these calculations assume a highly idealised shape and composition, and predict also a polarisation dependence of the interband absorption which we have not observed. We feel that the most likely explanation for the split p states lies simply in the asymmetry in the shape. Linearly polarised far infrared spectroscopy is required to resolve this point definitively.

Fig. 7 shows the interband absorption for the InAs dots in InP for a variety of gate voltages. The first peak, A_2, is the $s - s$ transition and has a complicated dependence on the gate voltage, being weak for $V_g < -2$ V, stronger around -1.8 V and then disappearing altogether at -1.2 V. At -1.2 V the transition is blocked because the s state in the conduction band is fully occupied. The explanation for the behaviour at -2 V is that the valence band state fills with holes, blocking the transition. The idea is that the electrons tunnel rapidly out of the dots, but the holes, owing to their large effective mass, do not and so positive charge tends to accumulate in the dots. This process has been investigated by DLTS, showing that holes can be trapped for several hours at low temperatures [17]. Similar effects have also been observed in other quantum dot systems [18, 19, 20]. At intermediate voltages, the electrons have a longer tunnelling time so that hole storage is less effective, allowing the $s - s$ transition to be partially observed.

The transition B_2 is the $p - p$ transition as is clear both from its energy and oscillator strength. It can be seen how the lower half disappears on increasing the gate voltage from -2.3 V to -0.75 V, and that the upper half disappears on further increasing the gate voltage. This is consistent with the idea that the p states are split. An interesting and unexpected feature of Fig. 7 is that there are two further transitions, labelled C_2 and D_2, at higher energy. The intensities of peaks A_2 and C_2 behave the same with gate voltage. The implication is that the final state is the same. The same statement can be made about peaks B_2 and D_2. There are therefore states in the valence band, 160 meV from the ground state, which have the same lateral character as the states close to the ground state. The data suggest an adiabatic picture: the higher states have the same lateral wave functions but a different vertical wave function. Such states have never been observed in InAs dots. The important point is that the InAs dots in InP have a particularly high valence band offset which can bind more than one state. All this information is summarised in the energy level diagram in Fig. 8.

6. Spectroscopy of single rings of InGaAs in GaAs

The two systems discussed so far, InAs dots in GaAs and InAs dots in InP, both have band gaps smaller than the low-energy cut-off of Si detectors which are able to detect single photons. This makes experiments on single dots extremely demanding. Instead, we have investigated single InGaAs rings in GaAs which emit at 1.3 eV. A capacitance curve of the device is shown in Fig. 9. Consistent with the higher band gap is the fact that the first tunnelling peak occurs at -0.3 V and not at -0.9 V as for the InAs dots. Three charging peaks are visible for the rings before occupation of the wetting layer begins, implying that the rings can be filled with at least three electrons.

In the confocal arrangement, we measure typically ~ 50 rings simultaneously. However, individual rings can be distinguished by their emission wavelength. This works particularly well on the long wavelength side of the ensemble peak where we can always find single ring emission which is reasonably isolated in wavelength from the other rings. Spectra were recorded as a function of wavelength and are displayed as a grey-scale plot in Fig. 10. The x-axis is the emission wavelength, the y-axis the gate voltage, and the level of darkness represents the intensity of the emission. A number of effects are apparent.

First, at large and negative voltages the sample is dark: no photoluminescence could be detected. Emission from single rings emerges around

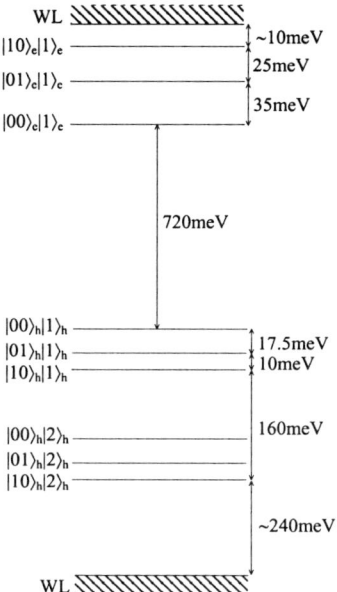

Figure 8. An energy level diagram for InAs dots in InP as deduced from capacitance measurements, intraband and interband transmission experiments.

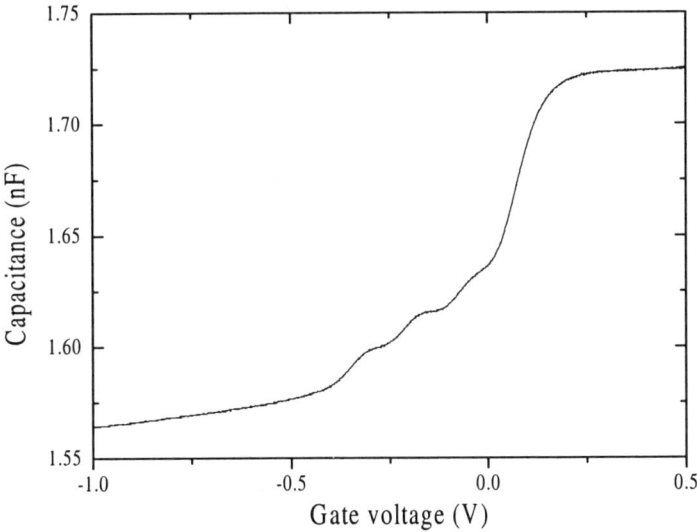

Figure 9. Capacitance versus gate voltage for InGaAs rings in GaAs at 4.2 K.

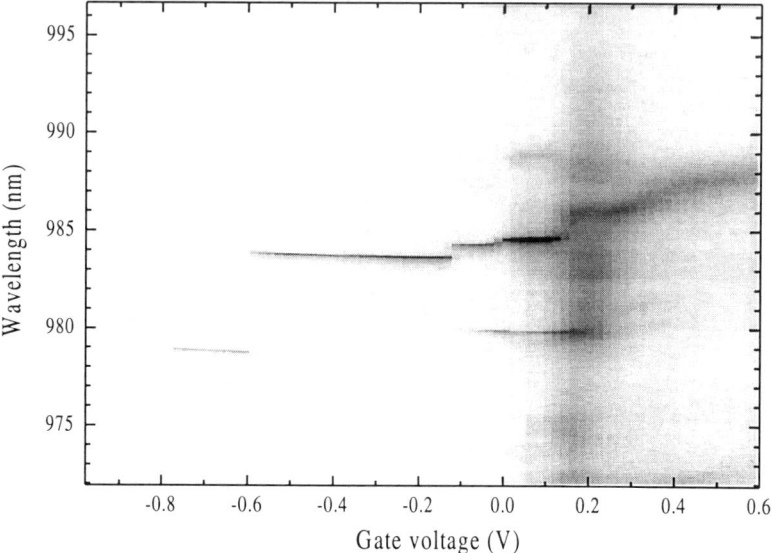

Figure 10. A grey-scale plot of the photoluminscence of a $\sim 1~\mu\mathrm{m}^2$ area of the sample with InGaAs rings. The intensity is logarithmically related to the grey-scale.

−0.8 V. Our explanation is that the excitons are ionised by the field at particularly large voltages. The voltage at which the rings turn on fluctuates from ring to ring, but there is a tendency for the rings with a small emission wavelength (high emission energy) to turn on at a more positive voltage than those with a large emission wavelength. This suggests that the excitons relax into the rings, but that the electron tunnels out at high electric field before recombination takes place. The dependence of the critical voltage on emission wavelength is then simply a consequence of the changing confinement energy and hence tunnelling time.

Secondly, the photoluminescence for negative gate voltages is sharp: we cannot resolve the line with our spectral resolution of $\sim 250~\mu\mathrm{eV}$. The homogeneous linewidth of an exciton in a quantum dot is not understood in any detail even for samples containing only buried dots. We simply note here that we have sharp lines even in our more complicated sample geometry.

Thirdly, each ring emission shifts to longer wavelength at $V_g \sim -0.6$ V. The shifts occur because the gate voltage is reached at which it is energetically favourable for an electron to be stored in the ring. The emission is then from a singly charged exciton, X^-, and not from a neutral exciton, X. The shift, 6 meV (fluctuating slightly from ring to ring) is exactly the binding energy of an X^- in this system. A further shift occurs around −0.1 V, and our interpretation is that this shift corresponds to the transition from an X^- to an X^{--}. The shift is much smaller, only 0.6 meV, and this may suggest that we have a weakly confined system: the circumference of the rings is larger than the excitonic Bohr radius so that in some sense an exciton can form and move around in this rather complex potential landscape. By analogy with quantum wells, the X^- forms a complex which is bound. The X^{--} however is unbound in a quantum well; it is bound here by the confinement potential of the ring, but only weakly. In the strong quantisation limit, the picture of an exciton moving as a composite particle in a potential landscape is not accurate and one would expect that the X^- and X^{--} have comparable binding energies. Both transitions, from X to X^-, and from X^- to X^{--}, occur in a range of gate voltage of ~ 20 mV, corresponding to an electrostatic energy of ~ 3 meV at the rings. Exactly this width was measured in capacitance spectroscopy on very small dot ensembles for tunneling into single dots [21], suggesting that there are some fluctuations in the sample. It is very unlikely that they are thermal in origin. It could be the case that the broadening arises from fluctuations in the back contact as tunneling takes place or even from fluctuations arising from dot-dot interactions.

Fourthly, the photoluminescence becomes much stronger and much broader around 0.1 V. This gives rise to the large black band visible in Fig. 10. The capacitance trace of Fig. 9 shows that the wetting layer starts to be occupied at this voltage and we suggest that this is the key to explaining the behaviour of the photoluminescence. Around 0 V, the Fermi energy becomes degenerate with unconfined resonances of the rings. The rings are then effectively coupled with each other and with the wetting layer leading to the massive increase in broadening. A resonance behaviour is supported by the data at higher gate voltage where the photoluminescence from each ring becomes shaper and weaker. A residual interaction with the Fermi sea in the wetting layer can however be identified: the photoluminescence is still broad compared to the behaviour at $V_g < 0$ when the wetting layer is unoccupied, and there is a continuous red-shift in the emission wavelength as the gate voltage increases beyond 0 V.

7. Conclusions and outlook

We have embedded various forms of quantum dots in a field-effect structure so that the occupation of the dots with electrons can be controlled simply by tuning the voltage applied to a surface gate. Generally speaking, we believe that this is an important technique for probing the electronic structure and many-electron effects in quantum dots.

For InAs dots in GaAs, the dots are sufficiently homogeneous that we can observe Coulomb blockade in the vertical transport properties of an ensemble of several million dots. The absorption spectrum of an ensemble of quantum dots has been measured with transmission spectroscopy and we have argued that this experiment leads to direct measurements of the energies and oscillator strengths of the transitions. The results of both transmission and capacitance spectroscopies can be understood with a two-dimensional, parabolic confinement potential treating the Coulomb interactions in perturbation theory. The dots are in the strong confinement limit: the quantisation energies are larger than the Coulomb energies in this system. An equivalent statement is that the extent of the electron and hole wave functions is smaller than the excitonic Bohr radius. InAs dots in InP are shown to have two p-like states, each with degeneracy 2. It is argued that the splitting arises from the pronounced anisotropy in the dots' shape. These dots are also in the strong confinement limit. Experiments on single quantum dots are presented on InGaAs rings in GaAs where the energy gap, 1.3 eV, lies in an experimentally amenable region. Relative to the small linewidth, we see an enormous shift in photoluminescence energy as the emission changes from recombination of a neutral exciton to recombination of a charged exciton. We also observe a field-ionisation of the excitons at high electric field. Furthermore, the occupation of the wetting layer with electrons leads to an enormous increase in the intensity and broadening of the emission.

A number of questions remains however. It is still unclear what the optical response is of the strongly confined InAs dots in GaAs on a meV scale. On charging, a splitting of the main features has been predicted [22]. These effects arise through Coulomb couplings of various final state configurations. It is also unclear what the linewidth is for single-dot absorption and emission. This would be particularly interesting in transmission experiments where the measurement represents a minimal perturbation to the system. The rings were employed here because they have a larger band gap than the ordinary dots. But there is more to it than this: initial indications are that the electronic structure of the rings is completely different from that of the dots. All these issues need to be addressed by further experiments.

Acknowledgements. We would like to thank A. Lorke, R. J. Luyken, A. Govorov and D. Pfannkuche for stimulating discussions. The work was funded by the Deutsche Forschungsgemeinschaft (SFB 348), and by QUEST, a National Science and Technology Center (grant no. DMR 20007).

References

1. S. Tarucha, D. G. Austing, T. Honda, R. J. van der Hage, and L. P. Kouwenhoven, Phys. Rev. Lett. **77**, 3613 (1996).
2. B. T. Miller, W. Hansen, S. Manus, R. J. Luyken, A. Lorke, J. P. Kotthaus, S. Huant, G. Medeiros-Ribeiro, and P. M. Petroff, Phys. Rev. B **56**, 6764 (1997).
3. H. Drexler, D. Leonard, W. Hansen, J. P. Kotthaus, and P. M. Petroff, Phys. Rev. Lett. **73**, 2252 (1994).
4. G. Medeiros-Ribeiro, F. G. Pikus, P. M. Petroff, and A. L. Efros, Phys. Rev. B **55**, 1568 (1997).
5. D. Leonard, K. Pond, and P. M. Petroff, Phys. Rev. B **50**, 11687 (1994).
6. J. M. Garcia, G. Medeiros-Ribeiro, K. Schmidt, T. Ngo, J. L. Feng, A. Lorke, J. P. Kotthaus, and P. M. Petroff, Appl. Phys. Lett. **71**, 2014 (1997).
7. K. Karrai and R. G. Grober, Appl. Phys. Lett. **66**, 1842 (1995).
8. E. Deckel, D. Gershoni, E. Ehrenfreund, D. Spektor, J. M. Garcia, and P. M. Petroff, Phys. Rev. Lett. **80**, 4991 (1998).
9. M. Fricke, A. Lorke, J. P. Kotthaus, G. Medeiros-Ribeiro, and P. M. Petroff, Europhys. Lett. **36**, 197 (1996).
10. R. J. Warburton, C. S. Dürr, K. Karrai, J. P. Kotthaus, G. Medeiros-Ribeiro, and P. M. Petroff, Phys. Rev. Lett. **79**, 5282 (1997).
11. R. J. Warburton, C. Bödefeld, C. S. Dürr, K. Karrai, J. P. Kotthaus, G. Medeiros-Ribeiro, and P. M. Petroff, Festkörperprobleme/Advances in Solid-State Physics **38**, 183 (1998).
12. Ph. Lelong and G. Bastard, Solid State Commun. **98**, 819 (1996).
13. R. J. Warburton, B. T. Miller, C. S. Dürr, C. Bödefeld, K. Karrai, J. P. Kotthaus, G. Medeiros-Ribeiro, P. M. Petroff, and S. Huant, Phys. Rev. B **58**, 16221 (1998).
14. A. L. Huston and C. T. Reimann, Chem. Phys. **149**, 401 (1991).
15. W. Que, Phys. Rev. B **45**, 11036 (1992).
16. H. Pettersson, R. J. Warburton, J. P. Kotthaus, N. Carlsson, W. Seifert, M.-E. Pistol, and L. Samuelson, unpublished.
17. H. Pettersson, C. Pryor, L. Landin, M.-E. Pistol, N. Carlsson, W. Seifert, and L. Samuelson, unpublished.
18. G. Yusa and H. Sakaki, Appl. Phys. Lett. **70**, 345 (1997).
19. J. J. Finley, M. Skalitz, M. Arzberger, A. Zrenner, G. Böhm, and G. Abstreiter, Appl. Phys. Lett. **73**, 2618 (1998).
20. M. C. Bödefeld, R. J. Warburton, K. Karrai, J. P. Kotthaus, G. Medeiros-Ribeiro, and P. M. Petroff, Appl. Phys. Lett. **74**, 1839 (1999).
21. B. T. Miller, W. Hansen, S. Manus, R. J. Luyken, A. Lorke, J. P. Kotthaus, G. Medeiros-Ribeiro, and P. M. Petroff, Physica B **249-251**, 257 (1998).
22. A. Wojs and P. Hawrylak, Phys. Rev. B **55**, 13066 (1997).

OPTICAL CHARGING OF SELF-ASSEMBLED InAs QUANTUM DOTS

D. HEINRICH[†], J. FINLEY[‡], M. SKALITZ, J. HOFFMANN,
A. ZRENNER, G. BÖHM, AND G. ABSTREITER.
Walter Schottky Institut, Technische Universität München,
Am Coulombwall, D-85748, Garching, Germany.

Abstract. In this paper we describe the results of spectrally resolved photo-resistance investigations of optically induced charge storage in self-assembled InAs quantum dots. The results obtained demonstrate that, following resonant photo-excitation of the dots, excitons can be selectively ionised leaving *either* electrons or holes stored. This charge is sensed remotely using a density tuneable 2D electron – hole system and is shown to remain stored over very long timescales (>8 hours) at elevated temperature (~150K). By analysing the temporal dependence of the charge storage effect, the optical absorption strength of the quantum dots is estimated to be ~3.5×10^{-5}. The potential operation of the devices investigated as highly sensitive photo-transistors or a basic optical memory element is suggested.

1. Introduction

Self-assembled quantum dots (QDs) have been shown to be almost ideal examples of quasi-zero dimensional electronic systems [1] and as such have attracted considerable attention during recent years. This interest has been primarily stimulated by the possibilities this system provides for both fundamental physics studies and potential novel device applications. Recently the use of self-assembled QDs for data storage has been proposed [2] in which the information is recorded as a small number of stored electrons or holes within the deep trapping potential of the QDs. These charges may be electrically injected and indeed numerous devices have recently been demonstrated [3,4]. Alternatively, the quantum dots may be charged optically [5,6,7], an approach which ,combined with wavelength selective writing or local probe techniques, could approach the situation where each quantum dot stores a *single* bit of information. Such an ideal quantum dot optical memory element promises to combine both ultra-dense storage capacities (~1Tbitcm^{-2} [2]) with very low switching energies and thus represents a highly desirable technological goal.

Recently, a number of workers have investigated the effects of a self-assembled

---

[†] *EMAIL – DORIS.HEINRICH@WSI.TU-MUENCHEN.DE*
[‡] *Present Address : Department of Physics & Astronomy, The University of Sheffield, Hounsfield Road, Sheffield, S3 7RH, U.K. EMAIL : J.J.FINLEY@SHEFFIELD.AC.UK*

M.L. Sadowski et al. (eds.), Optical Properties of Semiconductor Nanostructures, 365–377.

InAs QD layer on the lateral transport properties of a two-dimensional electron gas (2DEG) in MODFET type structures [4,7,8,9]. This work has shown that the QDs can efficiently act as controllable scattering centres, which can be used to effectively tailor the transport properties of the 2DEG. Furthermore, several authors [4,7] have demonstrated that such devices are strongly photo-sensitive forming the foundation for a basic QD optical memory element or highly sensitive QD photo-transistor.

In this paper, we spectrally investigate such optically induced charging of self-assembled InAs QDs using such optically gated FETs in which a layer of InAs quantum dots are embedded ~50nm from an electron (section 3) or hole (section 11) 2D system. By separating the 2D channel and QD layer using a wider bandgap blocking barrier, photo-created excitons can be selectively ionised leaving electrons (or holes) stored preferentially within the QDs.

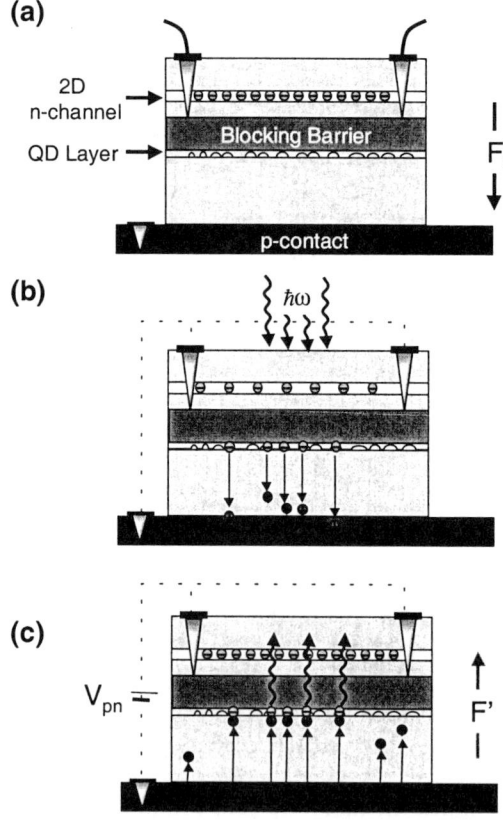

Figure 1. Schematic device structure and operating principles of the FET structure investigated (a) natural state of the device ; p-contact maintained at the same potential as the n-channel by the external circuit. (b) After resonant optical excitation of the dots, electrons are stored, the holes removed by the electric field in the p-i-n junction. (c) Injecting holes into the dot layer from the p-contact erases stored charge.

2. General device operation.

Figure 1 illustrates schematically the fundamental operating principles of the devices investigated. Whilst the following discussion pertains to an electron storage device, the principles for holes are entirely analogous.

The devices consist of a two-dimensional (2D) electron channel that was spatially separated by a blocking barrier from a layer of self-organised quantum dots (figure 1a). The QD layer is embedded within the intrinsic region of a vertical p-i-n junction. By fabricating separate ohmic contacts to both the 2D n-channel and buried p-contact, the electric field (F_0) in the i-region and conductivity of the n-channel (σ) can be tuned by varying the DC bias applied to the p-i-n junction (V_{pn}). Following resonant optical excitation of the quantum dots, the photo-created holes escape by thermal activation (T>100K) or by tunnelling (T<100K) [10]. By contrast, the electrons remain preferentially stored due to the presence of the wider bandgap-blocking barrier (Figure 1b). The stored electrons within the QD layer then selectively deplete the 2DEG, resulting in a reduction of its lateral conductivity (σ) which can be measured electrically. The magnitude of this conductivity change ($\Delta\sigma$) is expected to reflect the charge density stored in the QDs (Figure 1b). The stored electrons are then removed from the QDs by forward biasing the vertical p-i-n junction. Holes are then injected into the QD layer where they become trapped by the dots, recombining radiatively with the stored electrons. The device then switches back to its initial state and another *write-detect-reset* cycle can be performed.

3. Electron storage devices

3.1. SAMPLE & EXPERIMENTAL DETAILS

The device investigated was grown on a [100] orientated semi-insulated GaAs substrate and nominally consisted of the following epitaxial layers: 500nm p(Be)=2.10^{18} cm^{-3} back contact followed by a 240nm un-doped (u.d.) GaAs spacer. Then a 2.25ML layer of InAs was deposited at 530°C / 0.04MLs^{-1} which self-assembles into quantum dots. The QD layer was capped with 0.57nm GaAs before growth of the blocking barrier. This consisted of 30nm un-doped $Al_{0.3}Ga_{0.7}As$ and a five period 2nm(AlAs) / 2nm(GaAs) short period superlattice to inhibit impurity segregation in the 2D electron channel. After this, the 10nm thick $In_{0.1}Ga_{0.9}As$ electron channel was grown followed by a 40nm wide modulation doped $Al_{0.3}Ga_{0.7}As$ barrier region. The modulation doping was incorporated into two narrow (1.5nm) δ-doped (~2 x 10^{12} cm^{-12}) GaAs quantum wells in an effort to inhibit persistent photo-conductivity associated with carrier excitation from DX centres in the $Al_{0.3}Ga_{0.7}As$ modulation doped region [11]. Finally, the sample was capped with 10nm of n-doped (4 x 10^{18} cm^{-3}) GaAs. A reference sample was also grown which was nominally identical to the above but without the QD layer. In the following discussion we refer to the sample containing dots as sample-A whilst the reference sample is termed sample- B. A schematic band-structure for the device investigated is shown in figure 2.

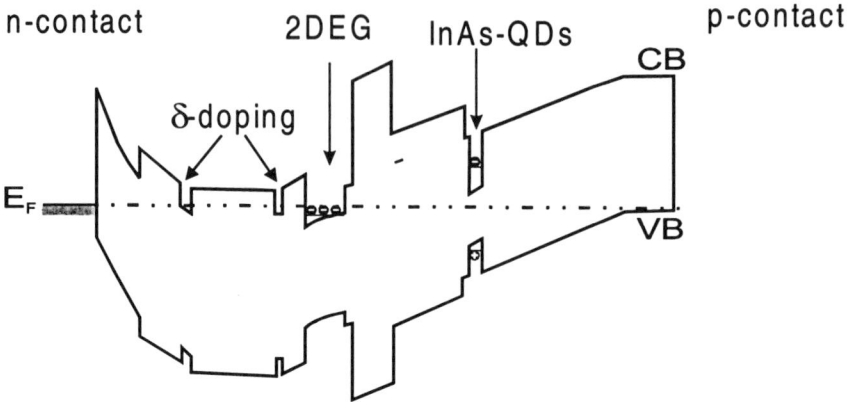

Figure 2. Schematic bandstructure of the electron storage sample. The band-structure is depicted for the situation when the p & n contact are maintained at the same potential by the external circuit (V_{pn}=0.0V)

Cross sectional TEM measurements taken on similar quantum dots [12] reveal the average height and lateral dimension of the dots to be ~10nm and ~30nm, respectively, while their density is typically $1.5\pm0.5\times10^{10}$ cm^{-2}.

The wafers were processed into a Hall bar geometry using standard optical lithographic techniques and separate ohmic contacts were established to both the 2D-electron channel and the back p-contact (figure 3a(inset)). The lateral conductivity (σ) of the n-channel is plotted versus V_{pn} in figure 3a. Over the range $-3<V_{pn}<0$ the channel conductivity can be tuned from ~2-0.2 mS □ whilst the leakage current (I_{pn}) remains less than 10nA confirming the high electrical quality of the devices fabricated. This corresponds approximately to an order of magnitude change in the carrier density in the n-channel.

TABLE 1 – Magneto transport data for QD containing (A) and reference (B) samples

Sample	n_e ($x10^{11}$ cm^{-2})	μ_e (cm^2 / Vs)
A	3.4	8400
B	5.1	22000

The 2D-electron channel of both samples was characterised using magneto-transport techniques at 4.2K, the results of which are summarised in table I. Both samples exhibited no parallel conduction, even at elevated temperatures, with typical carrier densities in the range ~3-8x10^{11} cm^{-2} as the temperature increases from 4-200K. We note that this value is approximately an order of magnitude greater than the dot density and consequently for low electron charging occupancies we expect the stored charge density (N.) to be much smaller than the channel density (n_e). This enables us to simply model the electrostatic effects of the charged dot layer on the conductivity of the electron channel and reconcile the photo-conductance data with the absorption strength of the quantum dots.

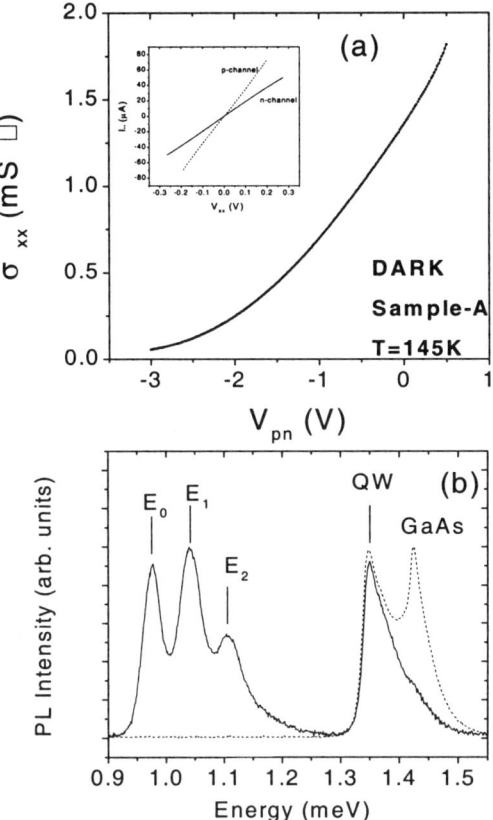

Figure 3. (a) Electrical characterisation of the charge storage device illustrating back gating of the conductivity 2D electron channel (inset) Lateral transport characteristics of the ohmic n-channel & buried p-contact. (b) Photoluminescence spectra obtained from samples A (full line) & B (dashed line) at 293K.

3.2. EXPERIMENTAL RESULTS

Most of the experimental results presented in this section were obtained at lattice temperatures between 140-200K. Over this temperature range the observed photo-response of our devices arises solely from optically induced charge storage with the quantum dots [13].

Figure 3b shows far field PL spectra obtained from samples A (full line) & B (dashed line). Excitation was performed using the 488nm line of an Ar^+ laser at an incident power density of $5\pm1Wcm^{-2}$. Over the energy range from 950 - 1200meV a series of well resolved peaks labelled E_n, n=0,1,2 and 3 are observed which are separated by 64 ± 3meV and exhibit a linewidth (FWHM) of ~35meV. The energy at which these features arise together with their characteristic dependence on the excitation power density identifies them as arising from radiative recombination of ground and

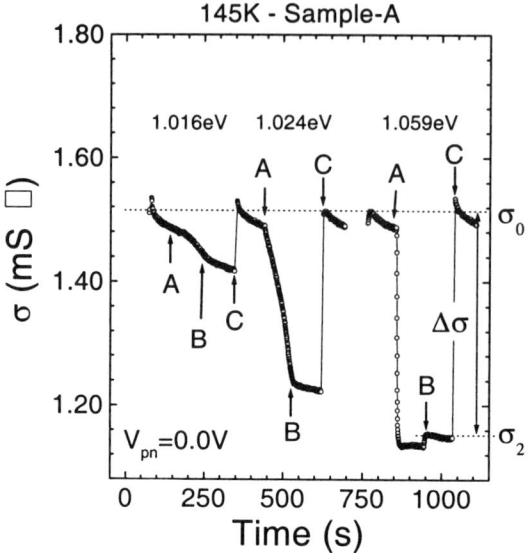

Figure 4. Temporal dependence of the n-channel conductivity for sample-A as a series of 100s illumination, recovery and reset cycles are performed as described in the text. The measurements were taken with excitation energies of $E_{ex}=1016$ (leftmost), 1024 and 1059meV (rightmost) respectively. The arrows labelled A & B denote the points at which the illumination is applied and removed, respectively, the sample is reset at C. Sample B showed a much weaker photo-response over the excitation range 900<E_{ex}<1200meV.

excited state excitons localised within the quantum dots. To higher energy, additional PL peaks are observed which arise from recombination in the $In_{0.1}Ga_{0.9}As$ quantum well (QW) and bulk GaAs (GaAs) regions of the device. As expected, sample B (dashed line) generates no luminescence over the energy range 800-1200meV confirming the absence of optically active centres.

3.3 PHOTO-CONDUCTIVITY RESULTS

In order to investigate optically stimulated charge storage effects in our samples (section 3) we performed a series of spectrally resolved photo-conductivity measurements. These consisted of a 100s illumination period after which the temporal dependence of the sample conductivity - $\sigma(t)$ was monitored before a reset electrical pulse was applied to the p-i-n junction [14]. The excitation energy was then changed and cycle repeated.

For these measurements the samples were held at T~145K in a stabilised environment designed to eliminate parasitic effects arising from ambient illumination. Optical excitation was achieved via a shutter using power stabilised mono-chromated white light from a Tungsten Halogen lamp (P_{max}~1mWcm^{-2}, FWHM~5meV) and the effect of exciting different quantum dots within the ensemble was studied by tuning the excitation energy (E_{ex}) through the dot absorption spectra. The conductivity was measured using a high sensitivity AC technique with low modulation currents (typically ~0,5-1nA) to eliminate effects arising from the presence of DC conduction along the 2D n-channel.

Figure 4 shows σ(t) data obtained from sample-A as a series of three illumination, recovery and reset phases were performed with E_{ex}=1016meV, 1024meV and 1059meV respectively. The onset of illumination is denoted by the label **A**, with **B** showing the point at which the shutter is closed. The reset pulse is applied to the sample at **C**. The channel conductivity prior to illumination is denoted by σ_o, which for this sample was measured to be 1.5±0.2 mS □. As can be seen clearly in figure 4, for E_{ex}=1016meV the sample responds only weakly to illumination. By contrast, after increasing E_{ex} the photo-effect becomes much stronger until for E_{ex}>1020meV a rapid switching of σ is observed following illumination. We focus now on the illumination cycle at E_{ex}=1059meV [15]. After closing the shutter (B) σ recovers weakly, saturating at a new level σ_2. Without resetting the sample, σ remained at σ_2 over timescales longer than *eight hours* at 145K whilst recovering towards σ_0 by less than 10%. This directly reflects the potential for extremely long charge storage times in the deep trapping potential of the quantum dots, even at elevated temperatures. Sample B, which did not contain quantum dots, exhibited little or no photo-response independent of E_{ex} over the energy interval 900meV<E_{ex}<1200meV [6].

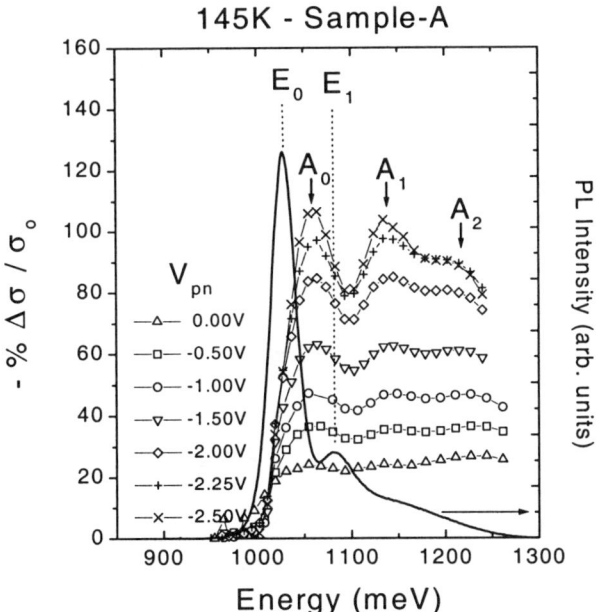

Figure 5. Spectral dependence of the photo-conductive effect for sample-A as a function of the reverse bias applied to the vertical p-i-n junction from V_{pn}=0 to -2.5V. The thick line shows a PL spectrum for comparison.

Figure 5 shows the spectral dependence of the quantum dot charging over the energy range 950<E_{ex}<1260meV and for different reverse biases applied to the vertical p-i-n junction (up to V_{pn}=-2.5V). The bold line shows the PL spectrum obtained at 145K for comparison. The onset of the photo-response discussed above is very clearly observed for $E_{ex} \sim 1015meV$, firmly establishing the correspondence between the observed photo-effect and charging of ground states of selected quantum dots.

With increasing V_{pn} the magnitude of the photo-effect increases strongly (up to $-\Delta\sigma/\sigma_0 \sim 100\%$ for $V_{pn}=-2.5V$) reflecting the electrostatic depletion of the n-channel (figure 3a). Over the same energy range, only a very weak photo-effect ($\Delta\sigma/\sigma_0 < 5\%$) is observed from sample – B. To higher energies, a series of peaks labelled A_{0-2} are observed. Before discussing these features, we note that the data plotted is the *saturation* value of the photoresponse. Thus, the magnitude of the charge storage observed includes *all* possible QD excitation mechanisms regardless of the relative timescales over which they occur. The energy separation between A_0 and A_1 is $75\pm5meV$, much larger than the splitting between ground and excited quantum dot peaks ($E_1-E_0=60\pm5meV$). Furthermore, these resonances are shifted from E_0 by $33\pm5meV$, $115\pm10meV$ and $172\pm20meV$ for A_0, A_1 and A_2 respectively. This indicates that they do not simply reflect *direct* optical charging of the ground and excited states of the quantum dots. Previously we have tentatively suggested that they may reflect the participation of inelastic processes [6] over long timescales involved. Following direct excitation of the quantum dots excited states, stored electrons are expected to relax rapidly [16] to the ground state and multiple charging of the quantum dots may occurr. Therefore, one must also consider the possibility that such features may reflect Coulomb renormalisation of the quantum dot energy levels [17]. These comments aside, the observations of such resonances clearly indicate that the quantum dots have an enhanced *capacity* for charge storage close to A_0, A_1 and A_2. Below, we show that similar resonances are also evident in the quantum dot charging rate at these energies, reflecting a resonance of the optical absorption strength.

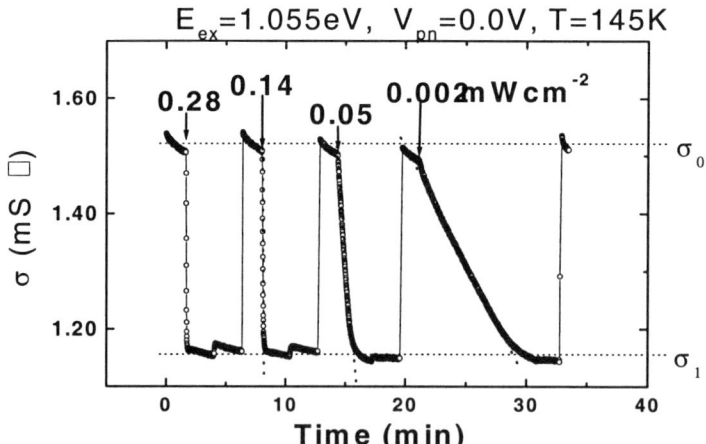

Figure 6. Excitation power dependence of the photo-conductive effect for sample-A at $V_{pn}=0.0V$ and for excitation power densities in the range $0.28 – 0.002mW$ cm^{-2}

Next we investigated the excitation power dependence of the quantum dot charging. Figure 6 shows a number of illumination-recovery-reset cycles obtained with excitation power densities (P_{ex}) in the range ~0.25mWcm^{-2}-50μWcm^{-2}. The excitation energy used was 1055meV, close to the peak of the A_0 peak in the charging spectrum (figure 5). As P_{ex} decreases, the dots become charged more slowly and clear linear transients, marked by the dotted lines on figure 6, are observed. We note that the saturation level of the photo-effect (marked by σ_1) remains independent of P_{ex}, consistent with qualitative expectations for charging a fixed density of quantum dots. The gradient of the observed linear transient ($d\sigma / dt$) is plotted versus P_{ex} in figure 7a for E_{ex} around the A_0 resonance (1022meV<E_{ex}<1077meV). Over this range of energy $d\sigma / dt$ increases linearly with P_{ex} at a rate which exhibits a pronounced resonance close to ~1050meV. This resonance is more clearly observed in figure 7b, which depicts the $d\sigma / dt$ vs E_{ex} cross section for fixed P_{ex}. We now proceed by developing a simple model for the QD charging, with which these observations can be qualitatively understood.

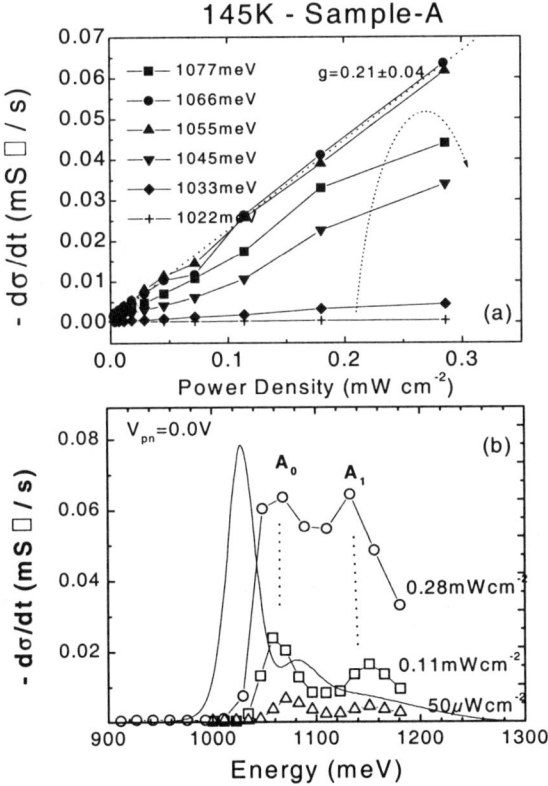

Figure 7. (a) Excitation power dependence of $d\sigma / dt$ for 1022<E_{ex}<1066meV showing the strong linear relationship predicted by eqn. 3. (b) Spectral dependence of $d\sigma / dt$ at fixed values of P_{ex}.

3.4. SIMPLE MODEL OF QD CHARGING

Assuming that the quantum dots are non-interacting we may model the temporal and excitation power dependence of the photo-response using simple state filling arguments. In this case, the density of uncharged QDs (N_0) which are addressed by the optical excitation decreases with time after first illuminating, reflecting the bleaching of the absorption in the charged dots. In this case, N_0 decreases exponentially with time according to

$$N_0 = N_{QD} \exp\left(- \frac{P_{ex}\alpha}{\hbar\omega N_{QD}} t \right) \tag{1}$$

where N_{QD} is the total density of QDs addressed by the optical excitation, P_{ex} the optical power density at $E_{ex}=\hbar\omega$, and α is the mean quantum dot absorption strength. Eqn. 1 assumes that the photo-generated holes escape from the dots over timescales which are short compared with the radiative recombination lifetime (~1ns). The validity of this assumption is supported by recent temperature and electric field tuned photo-current measurements performed on similar InAs QDs [10]. These measurements clearly show that for elevated temperature (T>100K) and electric field (F>60 kV cm^{-1}) both electrons and holes escape from the quantum dots before recombining. The density of charged dots is then $N_- = N_{QD} - N_0$, and $N_-(t)$ follows simply from eqn. 1. We now proceed by relating N_- to the change in conductivity of the n-channel ($\Delta\sigma$). Following charging of the dots, the carrier density in the electron channel (n_{2D}) is reduced by $n_{2D} = n_{2D}^0 - fN_-$, where f is an electrostatic leverage factor and n_{2D}^0 is the channel density before illumination. For the present sample, this factor is f~0.8 and the 2DEG is depleted by approximately one electron for each stored electron in the QD layer [18]. The rate of change of σ can be written

$$\frac{d\sigma}{dt} = e\frac{dn_{2D}}{dt}\left(\mu + n\frac{\partial\mu}{\partial n_{2D}} \right) \tag{2}$$

Which, providing the mobility of the 2DEG is not strongly perturbed by the charged QD layer [19], can be using eqn. 1 rewritten as

$$\frac{d\sigma}{dt} = -\alpha\left(\frac{ef\mu}{\hbar\omega} \right)P_{ex}\exp\left(-\frac{\alpha P_{ex}}{\hbar\omega.N_{QD}}t \right) \approx -\alpha\left(\frac{ef\mu}{\hbar\omega} \right).P_{ex} \tag{3}$$

Where the approximation corresponds to small times ($t\langle\langle\frac{N_{QD}}{\alpha\phi_{ph}}$) after the onset of illumination. Eqn. 3 predicts a linear dependence of $d\sigma/dt$ on P_{ex} as observed in figure 7a. According to eqn 3 the gradient (g) of the $d\sigma/dt$ vs P_{ex} curves is proportional to the optical absorption strength (α) of the QDs. From figure 7a, we measure g=0.21±0.04 at E_{ex}~1055meV from which we extract [20] α~3.5x10^{-5}. This value is comparable with previously published results obtained using more direct methods [21], further confirming that the observed photo-response arises from optically induced charge storage in the QD layer. Furthermore, the extremely slow experimental charging rate observed (figure 6) is now clearly reconciled with the absorption strength of the quantum dots, being due to the very low photon fluxes employed in the present experiment.

4. Hole storage devices.

We have recently performed similar experiments on inverted devices in which *holes* are optically stored in the QDs and sensed using a 2D $In_{0.1}Ga_{0.9}As$ *p*-channel. In all other respects the device operating principles and fabrication are similar to the electron storage sample discussed above. Besides being interesting from the fundamental physics viewpoint, hole storage may enable charge storage up to higher temperatures as a consequence of the larger effective mass [22]. Here, we present only selected preliminary results of such experiments, in order to demonstrate that the principles describe above are fully applicable to hole storage devices. More complete results and fuller discussion will appear in a future publication.

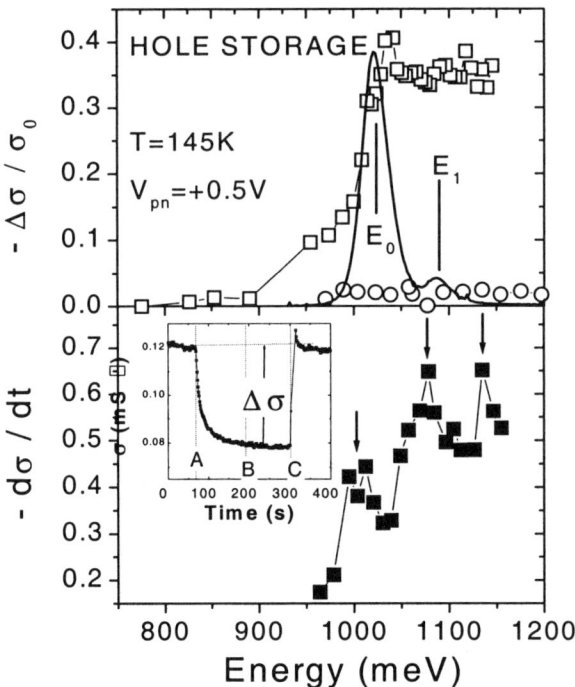

Figure 8. (inset) Persistent photo-conductivity arising from optically stimulated hole storage in the QD layer (top panel) QD charging spectrum for sample with (open squares) and without (open circles) the QD layer. (lower panel) Switching rate spectrum (dσ /dt) for the hole storage sample at P_{ex}~0.02mWcm^{-2}.

Similar to the electron storage sample discussed above (section 3.1) the modulation doped p-channel was separated by 50nm from the InAs QD layer which was embedded within the intrinsic region of a vertical n-i-p junction. In addition, the growth conditions for the QD layer were nominally identical to the electron sample discussed above.

Figure 8 (inset) shows the photo-response observed following illumination of the sample at $E_{ex}=1050$meV, just above the peak of the ground state PL emission (figure 8 - bold line, top panel). As for the electron sample, a strong decrease ($\Delta\sigma$) in the channel conductivity was observed following illumination (**A**), which remains persistent for many hours after closing the shutter (**B**). After resetting the device by forward biasing the n-i-p junction (**C**) the conductivity, once again, returns to the pre-illumination level (σ_o). A similar photo-response is not observed in a separate reference sample, which did not contain quantum dots. Figure 8 (top panel) shows the spectral dependence of the induced photo-effect $\Delta\sigma/\sigma_0$ for both QD (open squares) and reference (open circles) samples respectively. Since this data shows the saturation level of the photo-response, care should be exercised whilst interpreting the detailed form of the charging spectrum. However, a clear onset is observed close to ~1030meV confirming that the photo-effect arises from *hole* storage within the quantum dots. Furthermore, the peaked structure observed in the $\Delta\sigma/\sigma_0$ data provides evidence for *selective* storage of holes within the QDs. The filled squares in figure 8 (lower panel) show the spectral dependence of the device switching rate ($P_{ex}=0.05$mW/cm^{-2}) *immediately* after illumination for comparison. A series of peaks, marked by the bold arrows on the figure, are clearly observed in dσ / dt arise close to the ground and excited state PL emission from the QDs. Whilst it is believed that these features reflect the absorption spectrum of the quantum dots (section 0) a more detailed interpretation remains, at present, incomplete. The applications of the device principles to hole storage are, nevertheless, successfully demonstrated.

5. Summary

In summary we have investigated optical charging of self-assembled quantum dots using a novel quantum dot "optically gated" FET. Resonant optical excitation of the quantum dots results in a strong modulation of the conductivity of the 2D conducting channel which has been shown to arise from charge storage within the QD layer. The excitation power dependence of the charging effect enables us to estimate the absorption strength of the quantum dots, being found to be ~3.5×10^{-5}. By engineering the device geometry, we have shown that either electrons or holes can be preferentially stored in the quantum dots. The operation of the devices investigated as a highly sensitive optical switch or basic QD memory element is demonstrated.

Acknowledgements. The authors would like to thank D.J. Mowbray & I. Larkin for useful discussions during the preparation of this manuscript. The work was supported financially by the DFG via SFB-348.

One of us (*J. J. Finley*) gratefully acknowledges financial support from The Royal Society within the framework of the European Scientific Exchange Scheme.

References

1. *For example* : D. Leonard, M. Krishnamurty, C. M. Reaves, S. P. DenBaars, and P. M. Petroff, Appl. Phys Lett. **63**, 3203 (1993).

2. S. Muto. Jpn. J. Appl. Phys. **34**, L210 (1995).

3. L. Zhuang, L. Guo and S.Y. Chou, Appl. Phys. Lett. **72**, 1205 (1998).

4 H. Yusa and H. Sakaki, Appl. Phys. Lett. **70**, 345 (1997).

5 K Imamura, Y. Sugiyama, Y. Nakata, S. Muto, and N Yokoyama, Jpn. J. Appl. Phys **34**, L1445 (1995).

6 J.J.Finley, M Skalitz, M Arzberger, A Zrenner, G Böhm and G Abstreiter, Appl. Phys. Lett. **73**, 2618, (1998).

7. A. J. Shields, M. P. O'Sullivan, I. Farrer, D. A. Ritchie, K. Cooper, C L Foden and M Pepper, *"Optically Induced Bistability in the Mobility of a 2DEG Coupled to a Layer of Quantum Dots"* Submitted to Phys. Rev. B (1999).

8. H. Sakaki, G. Yusa, T. Someya, Y. Ohno, T. Noda, H. Akiyama, Y. Kadoya, and H Noge, Appl. Phys. Lett. **67**, 3444 (1995).

9. E. Ribeiro, E. Miller, T. Heinzel, H. Auderset, K. Ensslin, G. Medeiros-Ribeiro, and P M Petroff, Phys. Rev. **B58**, 1506 (1998).

10. P. W. Fry, I. E. Itskevitch, D. J. Mowbray, M. S. Skolnick, J. A. Barker, E. P. O. O'Reilly, L. R. Wilson, I. A. Larkin, P. A. Maksym, M. Hopkinson, M. Al-Kafaji, J. P. R. David, A. G. Cullis, G. Hill, and J. C. Clarke, Submitted to Phys Rev. Lett.

11. D. J. Chadi and K. J. Chang, Phys. Rev. Lett. **61**, 7 873 (1988).

12. L. Chu, M. Arzberger, G. Böhm and G. Abstreiter, Journal of Applied Physics **85**, 2355 (1999).

13. For T<100K, both samples-A and B exhibit an additional persistent photo-response, the contribution of which becomes stronger as the temperature decreases. These effects are completely absent at T=145K, and the observed photo-response arises solely from charge storage in the QD layer.

14. The samples were typically reset by passing between I_{pn}=1~10μA for times between 100-300ms.

15. We note that the form of the excitation cycle for E_{ex}=1059meV is representative of the measured response to higher E_{ex}, up to 1250meV whilst the E_{ex}=1016meV cycle typifies the lower excitation energy response

16. S. Raymond, P. J. Poole, S. Fafard, A. Wojs, P. Hawrylak, S. Charbonneau, D. Leonard, P. M. Petroff, and J. L. Merz, Phys. Rev. **B54**, 11548 (1996).

17. R. J. Warburton, B. T. Miller, C. S. Dürr, C. Bödefeld, K. Karrai, J. P. Kotthaus, G. Medeiros-Ribeiro, P. M. Petroff, and S. Haunt, Phys. Rev. **B58**, 16221 (1998).

18. Using simple electrostatic arguments, this factor may be related simply to the separation from the 2D electron channel of the dot layer (ℓ_1) and p-contact (ℓ_2) respectively i.e. $f = \left(1 - \dfrac{\ell_2}{\ell_1}\right)^{-1}$.

19. For our sample we expect that the n-channel mobility is controlled by optical phonon and ionised impurity scattering at 145K. Furthermore, the QD layer is separated by 50nm from the n-channel and has in ref 8, for a similar structure, been shown not to strongly perturb the channel mobility. Thus, the strongest effect on the channel conductivity is expected to arise from the reduction of the n_{2D}.

20. Using f=0.8 and μ=8000 cm^2 / Vs.

21. R. J. Warburton, C. S. Dürr, K. Karrai, J. P. Kotthaus, G. Medeiros-Ribeiro, and P. M. Petroff. Phys Rev. Lett. **79**, 5282 (1997).

22. For the electron storage sample, the strength of the observed photoeffect reduces for T>170K, and is not observed for T>230K.

OPTICAL PROPERTIES OF COLOIDALLY SYNTHESISED II–VI SEMICONDUCTOR NANOCRYSTALS

ANDREY L. ROGACH
Physico–Chemical Research Institute, Belarussian State University,
220050 Minsk, Belarus, & Institute of Physical Chemistry, University of
Hamburg, 20146 Hamburg, Germany

1. Introduction

In recent years there has been much activity covering the growth of semiconductor nanocrystals (often called quantum dots in physics related literature). In the quantum dots, 'artificial' atom-like electronic energy level systems are formed due to charge carrier confinement in three dimensions. The optical properties of semiconductor nanocrystals can be controlled not only by the elemental composition but also by the particle size via the so-called quantum confinement effect (see e.g. [1]). Nanometer sized particles of III-V and II-VI semiconductors are of great technological interest in different optoelectronical applications [2–6]. There are two possible routes to making semiconductor nanocrystals: in the "top down approach" they are grown by molecular beam techniques, making use of non-uniform layer growth (island-like) of certain materials on semiconductor substrates. In the "bottom up approach", they are formed by colloidal particle growth in a solvent medium. In the framework of the last approach, nanocrystals of different III-V and II-VI semiconductor materials with narrow size distribution, high crystallinity and size-dependent absorption and photoluminescence features have been obtained on a gram scale which can be handled like ordinary chemical substances. Among others, wet chemical synthetic routes have been developed for cadmium chalcogenide nanocrystals in non-aqueous solutions using TOP/TOPO (tri–octylphosphine/tri–octylphosphine oxide) capping [7] and in aqueous solutions using thiols as stabilising agents [8,9]. The latter is the growth technique that we have used to synthesise CdSe, CdTe, CdHgTe and HgTe nanocrystals. In this paper, the structural and optical properties of the nanocrystals are reported, and perspectives of the applications of the luminescent nanocrystals incorporated in light–emitting structures, optical amplifiers and photonic crystals are discussed.

2. Experimental

2.1 SYNTHESIS OF NANOCRYSTALS.

All the chemicals used were of analytical grade or of the highest available purity. They

M.L. Sadowski et al. (eds.), Optical Properties of Semiconductor Nanostructures, 379–393.

were obtained from Sigma, Merck, Aldrich, Alfa, and Fluka and used as received. The solutions of 0.05 M NaHSe or NaHTe were prepared according to the following procedure [9]: 100 mL of 0.05 M NaOH solution was titrated with H_2Se or H_2Te (generated by the reaction of Al_2Se_3 or Al_2Te_3 with 10% H_2SO_4) in an N_2 atmosphere. Al_2Se_3 (Al_2Te_3) was used in a 1.7-fold excess to the calculated quantity. After titration, the solution was bubbled with N_2 for 30 min to remove any traces of H_2Se (H_2Te) present in the solution.

Colloidal solutions of CdSe, CdTe and HgTe nanocrystals were prepared either in aqueous or in dimetylformamide (DMF) solutions in the presence of thiols as effective stabilising agents following, in general, the methods introduced in Refs. [9–11]. Freshly prepared oxygen-free NaHSe or NaHTe solutions or H_2Te gas were added to N_2-saturated $Cd(ClO_4)_2 \cdot 6H_2O$ or $Hg(ClO_4)_2$ solutions at pH 11.2 in the presence of thiols (RSH, 2-mercaptoethanol, 1-thioglycerol or thioglycolic acid) or dithiols (2,3–dimercapto–1–propanol). The molar ratio of $Me^{2+}:E^{2-}:RSH$ (Me = Cd, Hg; E = Se, Te) was, generally, 1:0.5:2.4. The particle size and chemical composition was controlled by the duration of the subsequent heat treatment (refluxing from 30 minutes up to 14 days depending on the solvent and stabiliser), through post-preparative size-selective fractionation, and by the nature of the solvent and stabilising agent.

Aqueous colloidal solutions of CdHgTe nanocrystals were prepared starting with the formation of CdTe nanocrystal precursors as described above. At the first preparation stage, mercury ions were subsequently substituted for cadmium at the surface of the CdTe nanocrystals: an alkaline solution of CdTe nanocrystals (5×10^{-4} M CdTe, pH 11) was treated with a small amount of 1 M mercury perchlorate solution, the net amount of mercury ions added being equivalent to 20–120% of the cadmium ion solution used to form the precursor CdTe nanocrystals. On the second preparation stage, the diameter of the particles was subsequently increased by growing a further shell of cadmium onto the surface of the nanocrystals by a treatment of hydrogen telluride gas which combines with Cd ions still present in the solution, these having been previously displaced by mercury ions during the first synthesis stage. The first and the second stages can be repeated in order to increase the diameter of the CdHgTe nanocrystals and/or introduce more Hg–ions into the nanocrystals.

2.2 CHARACTERISATION OF NANOCRYSTALS

Structural characterisation of the nanocrystals involved powder X–ray diffraction (XRD), high resolution transmission electron microscopy (HRTEM) and energy dispersive x-ray analysis (EDX). XRD spectra were taken on a Philips X'Pert diffractometer (Cu K_α-radiation, variable entrance slit, Bragg-Brentano geometry, secondary monochromator). Samples for these measurements were prepared by placing finely dispersed powders of CdTe nanoparticles on standard PVC supports. HRTEM and EDX were performed on a Phillips CM–300 microscope operating at 300 kV. TEM samples were prepared by dropping diluted solutions of nanoparticles onto 400-mesh carbon-coated copper grids with the excess solvent immediately evaporated.

3. Results and Discussion

The wet chemical synthesis of extremely small and highly monodisperse semiconductor nanocrystals is based on the principle of "arrested precipitation" [12], i.e. the surface of the growing nuclei should be sufficiently terminated to prevent spontaneous formation of the macroscopic substance. A choice of a suitable stabiliser is, therefore, the key point in the "bottom up approach". Stabilisers prevent nanocrystals from coagulation and regulate the size of growing particles; moreover, organic capping on the surface of semiconductor nanocrystals make them soluble and re-soluble in a variety of solvents. The use of polyphosphate [13] is an example of the stabilisation of nanocrystals based on the kinetic effects through electrostatic and steric repulsive forces. Another class of stabilisers working thermodynamically belongs to ligands almost covalently bonded at the nanocrystal surface. We used different thiols (mercaptoalcohols, mercaptoacids, dithiols) to prepare thermodynamically stable metal chalcogenide nanocrystals by the wet chemical route.

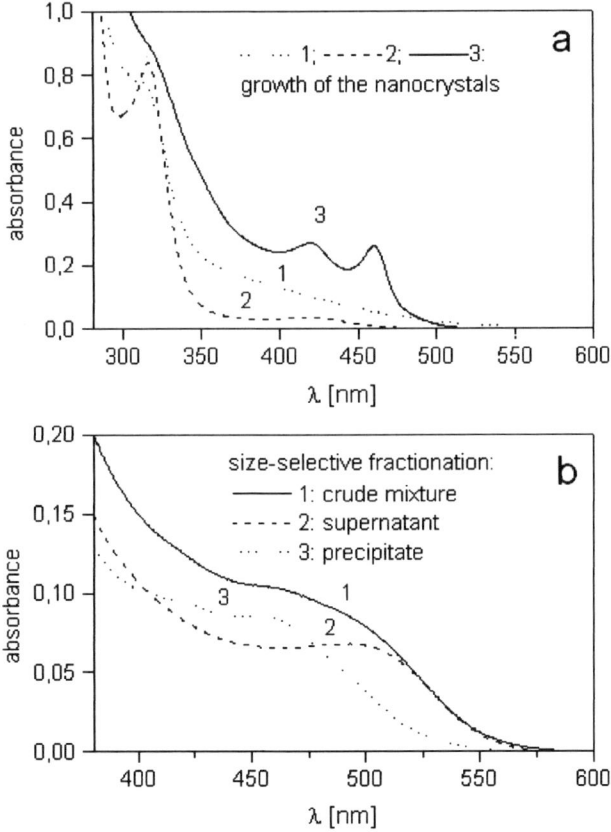

Figure 1. UV–vis absorption spectra demonstrating the growth (a) and size–selective fractionation (b) of CdTe nanocrystals synthesised in aqueous solution (taken from Ref. [9]).

Figure 1a shows a set of absorption spectra recorded during the synthesis of CdTe nanocrystals stabilised with 2–mercaptoethanol in aqueous solutions. During the earliest stage of the synthesis, directly after addition of the NaHTe solution into the solution containing Cd^{2+}-ions and thiol molecules, a spectrum evolves exhibiting an absorption shoulder in the UV and an unstructured absorption in the whole visible region. With time, and especially fast during the heating of the colloidal solution, the absorption at wavelengths longer than 330 nm decreases, and a sharp absorption peak at 315 nm belonging to the smallest nanocrystals is formed. During further heating, the band at 315 nm becomes less pronounced and the absorption intensity at longer wavelengths increases. New bands appear first at 420 nm and then at 425 and 460 nm, suggesting the formation of new nanocrystals. This growth can occur either due to additive reactions of reagents present in the colloidal solution which were up to now not involved in the reaction, or via the mechanism of Ostwald ripening when the larger particles grow on the account of smaller ones.

The CdTe nanocrystals capped with 2–mercaptoethanol with two well pronounced absorption maxima shown in Figure 1a belong to a thermodynamically favourable structure for which a structural formula $Cd_{54}Te_{32}(SR)_{48}^{4-}$ has been proposed based on EXAFS data [14,15]. Using nother thiols as stabilisers, or applying a prolonged heat treatment to the colloidal solution of CdTe nanocrystals allows a further shift of the absorption spectrum to longer wavelengths, which is indicative of further growth of the nanocrystals since the size of the particles is directly related to the absorption spectrum of quantum sized particles. The genesis of the absorption shown in Figure 1a is typical for metal chalcogenides growing in aqueous solutions in the presence of thiols.

In order to obtain highly monodisperse nanocrystal samples, post–preparative size–selective fractionation was applied to crude colloidal solutions. The technique is based on the different solubility of nanocrystals with different sizes/masses in a solvent/nonsolvent mixture [7,16]. For cadmium chalcogenide nanocrystals synthesised in aqueous solutions and capped with different thiols, a gradual addition of alcohols (ethanol, 2–propanol) leads to the precipitation of monodisperse species. Typically, larger nanocrystals precipitate first, but there are exceptions to this rule. Figure 1b shows the absorption spectrum of the crude solution (spectrum 1) containing two types of CdTe nanocrystals. During the size-selective precipitation the largest clusters (fraction 2 with an electronic transition in the UV-vis absorption spectrum at longer wavelengths than those of fraction 3) were obtained as a supernatant. This fact contradicts usual observations, where the larger particles precipitated first. This allows us to assume that the solubility of the nanoclusters in the aqueous-organic mixtures might be dependent not only on the cluster size but also on the cluster structure, e.g. the number and type of neighbouring atoms in the cluster and, most probably, the capping groups binding at the cluster surface.

Small- and wide-angle powder X-ray diffractometry was performed on powders of isolated nanocrystal fractions. The degree of crystallinity, crystalline structure, and the mean nanocrystal size can be derived from the XRD patterns. Figure 2 shows X-ray diffraction patterns of some selected fractions of CdSe, CdTe and HgTe nanocrystals. The diffractograms confirm the crystallinity of the nanoparticles; for all samples the positions of the wide-angle diffraction peaks correspond to the cubic modification of CdSe (zinc blende phase), CdTe (zinc blende phase) and HgTe (coloradoite phase).

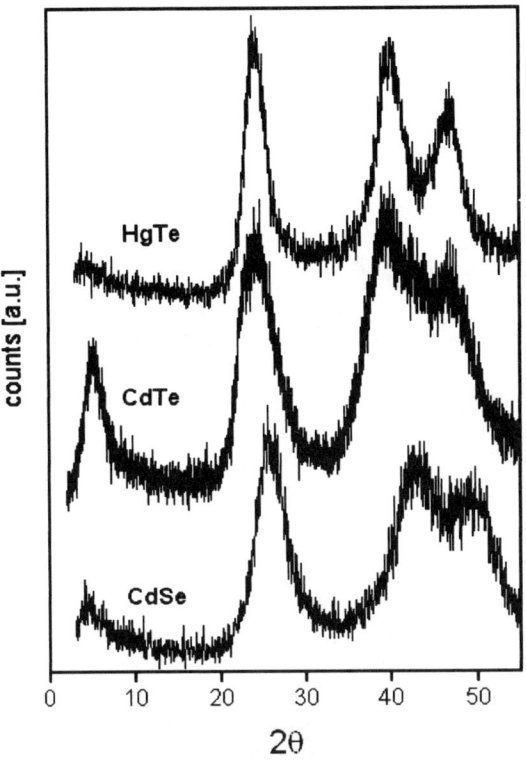

Figure 2. Powder X-ray diffractograms of CdSe (a), CdTe (b) and HgTe (b) nanocrystals.

The wide-angle diffraction peaks are broadened due to the small particle size. The mean nanocrystal sizes obtained from the full width at half-maximum intensity of the (111) zinc blende reflection according to the Scherrer equation are about 2.1 nm for CdSe, 1.9 nm for CdTe and 3.5 nm for HgTe nanocrystals. Reflection maxima appear also in the small-angle region of the X-ray diffraction patterns of CdSe and CdTe nanocrystals due to the periodicity of their arrangement. The appearance of such a peak provides further confirmation for the narrow size distribution of the particles. Using the Bragg equation, the peak angle maxima can be converted to the nearest neighbour distances of the nanoparticles in the powdered samples, which can be used as a measure of the mean particle size. The values are 2.1 nm for CdSe and 2.4 nm for CdTe nanocrystals.

Thiol-stabilised CdTe nanocrystals belonging to the cubic (zinc blende) crystalline phase are formed in aqueous solution in the presence of thiols as stabilisers by reaction of Cd^{2+} and Te^{2-} ions. A prolonged refluxing of the aqueous CdTe colloidal solutions in the presence of an excess of thiols is needed to promote further growth of the nanocrystals. Under these conditions, the partial hydrolysis of thiols causes an incorporation of the sulphur from the thiol molecules covalently bound at the

384

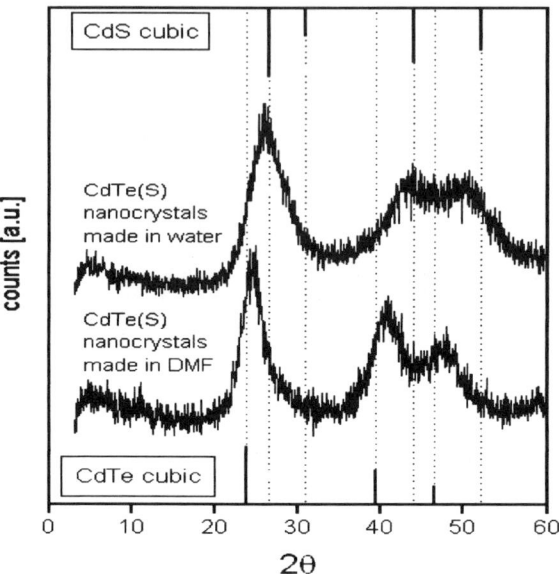

Figure 3. Powder X-ray diffractogram of CdTe(S) nanocrystals prepared in DMF and in water. The line spectra show the bulk CdTe and CdS zinc blende reflections with their relative intensities.

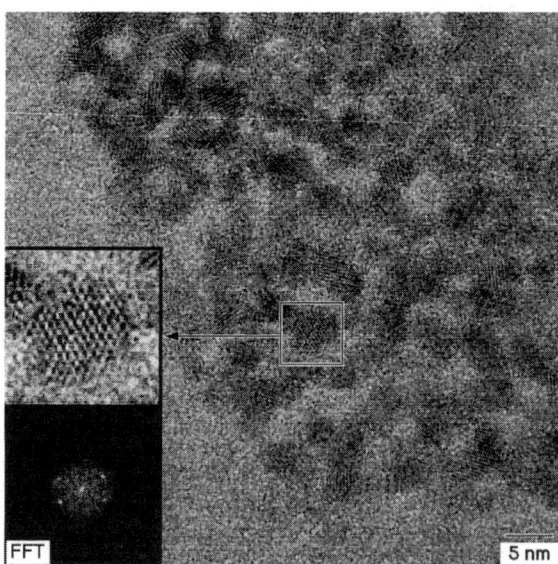

Figure 4. High-resolution TEM image of HgTe nanocrystals. Inserts show a single HgTe particle with the corresponding FFT. (Taken from Ref. [11]).

nanoparticle surface into the growing nanocrystals which results in the formation of mixed CdTe(S) nanocrystals under prolonged refluxing. Figure 3 shows XRD patterns of nanocrystals with a size of about 4 nm and absorption maxima positioned at 590 nm, obtained under prolonged refluxing of aqueous or DMF colloidal solutions of CdTe nanocrystals. The nanocrystals belong to the cubic (zinc blende) structure; however, the positions of the XRD peaks are intermediate between the values of the cubic CdTe and the cubic CdS phases. Using DMF instead of water as the solvent and reaction medium allowed us to partially prevent the hydrolysis of thiols and, thus, the incorporation of sulphur into the growing CdTe nanocrystals. It is, however, impossible to completely prevent the incorporation of sulphur into largely grown CdTe nanocrystals, either by the choice of solvent or by varying the stabilisers.

The determination of the exact composition of the CdTe(S) particles prepared under different synthetic conditions will be the subject of future investigations. In this paper, the term "CdTe" is used for the nanoparticles synthesised under moderate conditions of heat treatment, when the XRD and HRTEM data indicate that predominantly CdTe is formed. The term "CdTe(S)" is used for nanoparticles synthesised under prolonged refluxing, when the XRD and HRTEM indicate a formation of the mixed CdTe(S) phase.

HRTEM allows to determine the size, shape and crystal structure of the nanocrystals. Figure 4 shows, as an example, a typical overview image of thioglycerol–stabilised HgTe nanocrystals synthesised in aqueous solution [11] together with a micrograph of an individual HgTe particle with the corresponding Fast–Fourier–Transform (FFT). HRTEM indicates, along with the X-ray diffraction data, a rather broad size distribution of HgTe nanoparticles with sizes ranging from 3 to 6 nm. The existence of lattice planes on the HRTEM images of the particles indicates a high crystallinity of the sample. All images of single particles exhibited an interplanar distance of 3.73 Å belonging to the (111) lattice plane of the coloradoite HgTe phase which, again, is consistent with X-ray diffraction data.

Figure 5 shows absorption spectra of CdSe, CdTe and CdTe(S) nanoparticles of different sizes. The size of the nanocrystals determined by HRTEM varies between 2.5 and 5 nm. The pictures are typical for a variety of stabilisers. All nanocrystals are in the size quantisation regime and show a well-developed size–dependent maximum near the absorption onset, which is ascribed to the first excitonic transition. This maximum shifts to higher energies with decreasing particle size, indicating the increase of the nanocrystal's bandgap. Higher electronic transitions were also clearly resolved for the smallest nanocrystals.

Both CdTe and CdTe(S) nanocrystals capped with 1–thioglycerol, thioglycolic acid or 2,3–dimercapto–1–propanol possess a sharp emission band near the absorption onset ("excitonic" photoluminescence) which is tunable with particle size through the visible spectral range (Figure 6). The quantum yield of this excitonic emission at room temperature is in the range of 3–6 % and can be enhanced, for CdTe and CdTe(S) nanoparticles stabilised with thioglycolic acid, up to 18 % by making the colloidal solution acidic (inset of Figure 6). In the acidic range a shell of cadmium-thiol

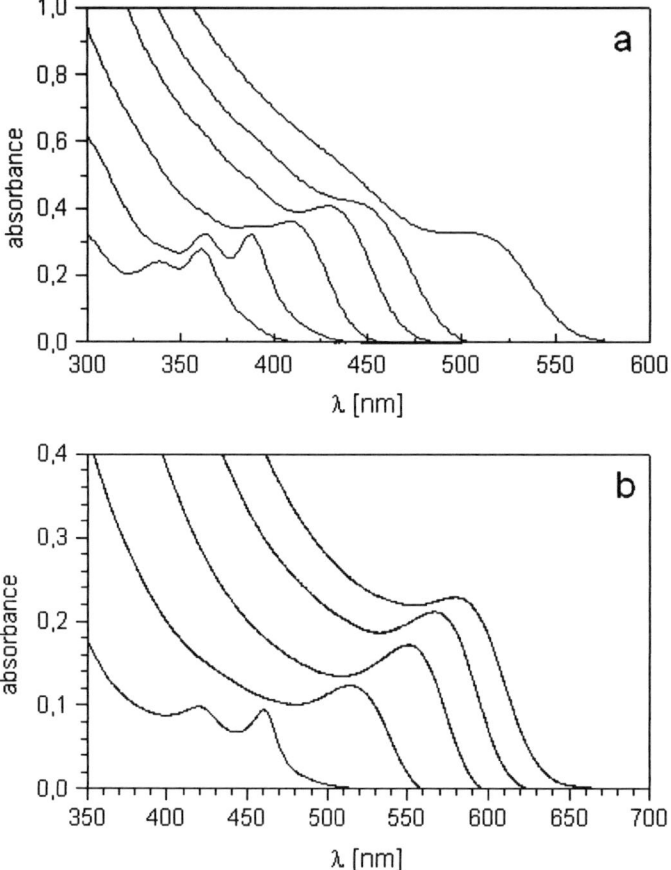

Figure 5. UV–vis absorption spectra of CdSe (a) and CdTe and CdTe(S) (b) nanocrystals of different sizes.

complexes is formed around the CdTe core, which removes the local trap sites from the surface and thus significantly increases the quantum yield of the exitonic emission [18]. In order to obtain semiconductor materials luminescent in the near IR, CdHgTe mixed nanocrystals have been synthesised. The alloy $Cd_{1-x}Hg_xTe$ is a well-known material for long wavelength IR detector technologies [18,19], and is obtained mainly by molecular beam epitaxy growth. Here, we present a wet chemical route to CdHgTe nanocrystals starting from CdTe nanoparticles. Two types of CdTe nanocrystals with narrow size distributions and extremely small particle sizes (1.9 nm and 2.4 nm respectively as determined by XRD) stabilised with either 2-mercaptoethanol (ME) or 1-thioglycerol (TG) have been used. XRD and HRTEM indicate the cubic (zinc blende phase) CdTe crystalline structure. Pronounced electronic transitions in the absorption

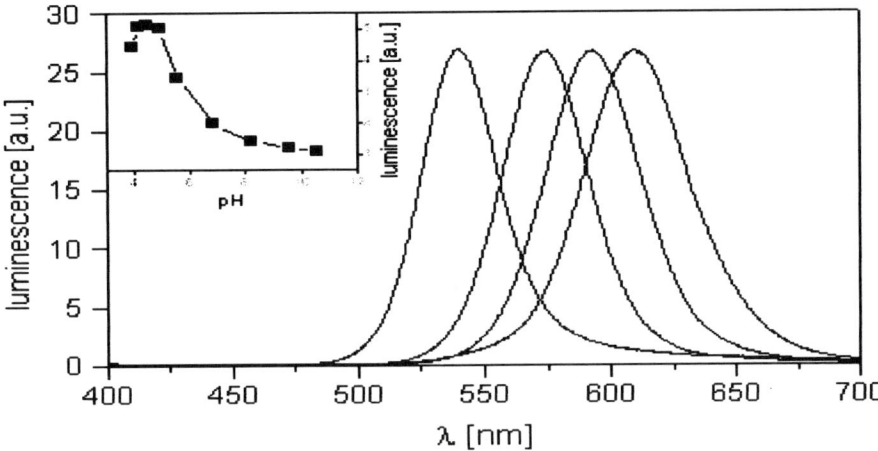

Figure 6. Photoluminescence spectra of CdTe and CdTe(S) nanocrystals of different sizes. Inset: luminescence intensity of CdTe nanocrystals capped with thioglycolic acid vs the pH of the colloidal solution.

spectra of CdTe nanocrystals are shifted dramatically to higher energies in comparison with bulk CdTe as a consequence of the strong quantum confinement (Figure 7). Two kinds of photoluminescence are observed: a broad "trapped" emission with a maximum at about 620 nm and a quantum efficiency of about 0.2% at room temperature for smaller CdTe nanocrystals stabilised with ME, and a narrow "excitonic" emission with a maximum at 540 nm and a quantum efficiency of about 3% at room temperature for larger CdTe nanocrystals stabilised with TG (Figure 7).

A modification of the CdTe nanocrystal surface with mercury ions was performed to obtain CdHgTe nanocomposites. Hg^{2+} ions added into aqueous solutions of CdTe nanocrystals substitute Cd^{2+} ions in the CdTe core due to a large difference in the solubility products of CdTe and HgTe. The preparation scheme can be described by the following equations:

$$CdTe \xrightarrow[-Cd^{2+}]{(1)\ Hg^{2+}} CdHg_xTe \xrightarrow{(2)\ Cd^{2+},\ Te^{2-}} CdHg_xCdTe \xrightarrow{(1)\ Hg^{2+}} CdHg_xHg_yTe \xrightarrow{(2)\ Cd^{2+},\ Te^{2-}} \ldots$$

and includes two main stages: (1) substitution of cadmium at the surface of the CdTe nanocrystals by mercury ions and (2) a subsequent increase of the diameter of the particles by growing a further shell of cadmium onto the surface of the nanocrystals by a treatment with hydrogen telluride gas, which combines with Cd ions still present in the solution, these having been previously displaced by mercury ions during the first synthesis stage. The subsequent repetition of the Hg^{2+} and Te^{2-} addition leads to an increase of both the diameter of the CdHgTe nanocrystals and the Hg content. Note that the Hg indices in the preparation scheme and in Figure 7 do not indicate a real Hg content in the CdHgTe mixed crystals, but show the percentage of mercury ions added during subsequently repeated stages into the CdTe colloidal solutions in comparison with the initial Cd–content in this solution

The absorption and photoluminescence of CdHgTe nanocomposites was found to be nearly independent of the size of the initial CdTe nanocrystals but to depend strongly on the amount of added Hg^{2+} ions. The maximum of the photoluminescence band shifts to longer wavelengths, up to about 820 nm, with an increase of the amount of added Hg^{2+} ions if only the first stage of Cd/Hg substitution is made (Figure 7b), and the luminescence intensity increases gradually with increasing Hg content. Consequent repetition of the first and second preparation stages leads to a further shift of the photoluminescence band into the IR and to an increase of the luminescence intensity (Figure 7d). The photoluminescence quantum yield of the CdHgTe nanocomposites reaches a value of about 15% at room temperature for $CdHg_{1.2}Te$ and about 40% for $CdHg_{0.4}Hg_{0.4}Hg_{0.4}Te$ nanocrystals.

The telecommunications industry is interested in the near infrared wavelengths of around 1.3 and 1.55 microns, which are the two windows used in optical fibre systems. A material with strong photoluminescence at these wavelengths could be the basis of a broadband optical amplifier, and by having control of the nanocrystal size distribution, it should also be possible to completely tailor the optical properties of the system. We have carried out a synthesis of HgTe nanocrystals since in the bulk, this material is a semi-metal with an inverted band structure [20], and hence an effectively zero band-gap material. The blue shift caused by the quantum confinement effect should therefore yield luminescence in the near infrared.

HgTe nanocrystals have been synthesised in aqueous solution using 1-thioglycerol as a stabiliser [11]. Figure 8 shows the absorption and the room

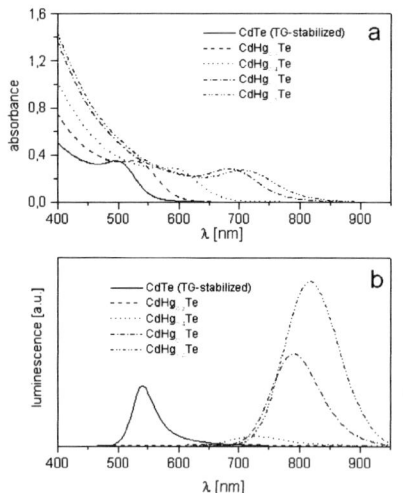
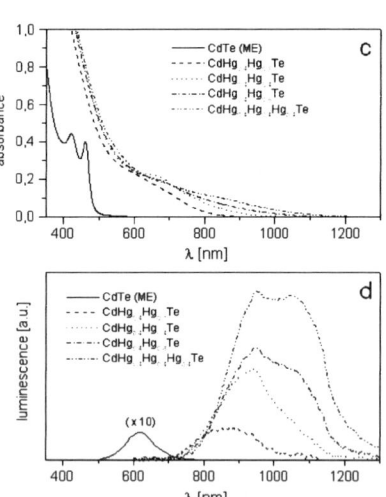

Figure 7. (a, c) UV–vis absorption and (b, d) photoluminescence spectra of CdHgTe nanocrystals vs amount of Hg^{2+}–ions added to the CdTe colloidal solution. See text for preparation routes and nomenclature.

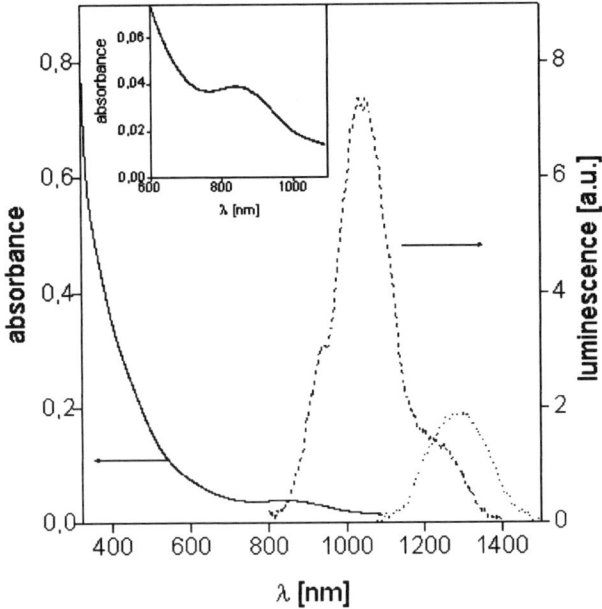

Figure 8. Absorption of HgTe nanocrystals in water (solid lines) and photoluminescence of HgTe nanocrystals in water (dashed line) and pyridine (dotted line).

temperature photoluminescence spectra of the sample directly after preparation. The initial absorption spectrum shows a long tail extending into the IR with a well–pronounced maximum at about 850 nm. The tail shifts further into the infrared over a period of several days, and the maximum becomes less pronounced, probably due to particle growth. The room temperature PL emission spectrum of HgTe particles in aqueous solution as prepared covers the spectral region from 800 to 1400 nm with a maximum located at about 1050 nm. Numerical integration of the peaks in the 800-1400 nm region of the PL spectrum for the HgTe nanoparticles indicates an overall quantum efficiency of approximately 50% for the sample. The minima seen in the PL spectra of HgTe particles in aqueous solution are due to water re-absorption. They were not observed when measuring the photoluminescence of thin films of HgTe nanocrystals on glass supports. The position of the emission maximum shifts into the infra–red with time, which coincides with the red–shift of the absorption tail, and the PL intensity decreases.

For both amplifier and optical source applications it is highly desirable to be able to include luminescent semiconductor nanocrystals in an organic solvent having low optical absorption and scattering losses rather than using aqueous solutions. Fig. 8 shows the room temperature PL emission spectrum of HgTe particles precipitated from the aqueous solution and then redispersed in pyridine. Although the quantum efficiency of HgTe nanoparticles decreases, a further shift of the PL maximum is achieved so that it is located exactly over the 1.3 μm telecommunications window. The absorption spectrum of HgTe particles in pyridine does not differ markedly from that in aqueous solution.

As indicated above, XRD and HRTEM data show that HgTe nanocrystals synthesised in aqueous solution have a sufficiently broad size distribution (from 3 to 6 nm). By using the size-selective precipitation technique it is possible to narrow this broad distribution of sizes, although this may not be significant for broadband applications. By using other thiol stabilisers it is possible to move the photoluminescence band of HgTe nanocrystals further into the IR, so that another important telecommunication window at 1.55 μm can be covered. The most exciting feature of this novel material is the extremely high quantum yield of their room temperature infrared luminescence of around 50%. We are currently adopting HgTe nanocrystals into telecommunication fibre devices.

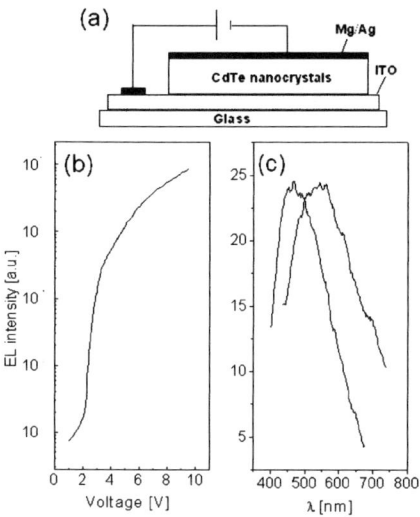

Figure 9. (a) Schematic cross-sectional diagram and (b) typical EL–voltage dependence of a light–emitting diode based on a film of closely packed CdTe nanocrystals; (c) EL spectra of light–emitting diodes based on films of closely packed CdTe nanocrystals with different sizes.

Another important technological direction which is currently developing is the integration of luminescent semiconductor nanocrystals into light-emitting diodes (LEDs) [2–4,21].The attractivity of nanocrystals is based on the tunability of the emitted light by simply varying the size of the nanoparticles. CdTe nanocrystal/polyaniline composite films as well as films of closely packed CdTe nanocrystals were used to fabricate LEDs [21] with low turn-on voltages (approx. 2.5 V for a device with a Mg cathode and an ITO anode) and the emitted colour depending on the size of the nanocrystals. Figure 9 shows the schematic cross-sectional diagram, a typical electroluminescence (EL) intensity – voltage dependence of the diode with a CdTe active layer, and EL spectra of the diode structures containing CdTe nanocrystals with two different sizes.

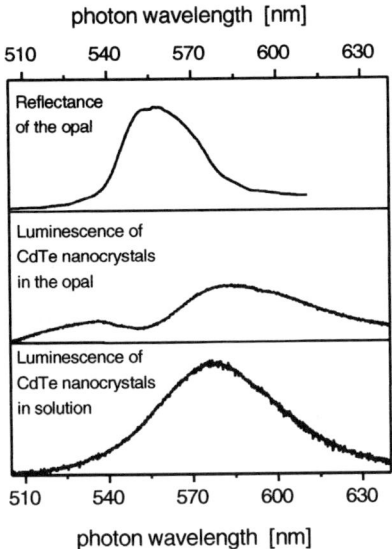

photon wavelength [nm]

510   540   570   600   630

Reflectance
of the opal

Luminescence of
CdTe nanocrystals
in the opal

Luminescence of
CdTe nanocrystals
in solution

510   540   570   600   630

photon wavelength [nm]

Figure 10. Modification of the luminescence spectrum of CdTe nanocrystals embedded in a photonic crystal (artificial opal). (Taken from Ref. [22]).

The impregnation of artificial opals with photoluminescent CdTe nanocrystals whose emission spectrum overlap the photonic stop gap provides an intriguing opportunity of controlling the spontaneous emission using both electronic (semiconductor nanoparticles) and photonic (artificial opals) states. Novel mesoscopic structures with separately controllable densities of photon and electron states can be developed. CdTe nanocrystals embedded in a photonic crystal from colloidal aqueous solution exhibited a noticeable change of the luminescence spectrum when the latter overlapped with the photonic stop gap of the opal [22]. The modification of the spontaneous emission is due to a modified density of photon states in the gap region. It is known that the spontaneous decay of an excited state of every quantum system is not an intrinsic property of the system but a result of its interaction with the electromagnetic vacuum. The density of photon states redistributes in a photonic crystal as compared to free space. It vanishes within the gap and increases near the gap edges. Therefore the spontaneous decay rate, which is directly proportional to the photon density of states, is expected to be inhibited within the gap and enhanced in the close vicinity outside the gap. The experiments clearly showed a dip in the emission spectrum correlating with the spectral position of the gap (Figure 10). Although a modification of the emission spectrum occurred at the edges of the gap as well, additional studies are necessary to draw unambiguous conclusions on the enhanced emission at the gap edges. They are currently carried out using colloidal solutions of luminescent CdTe nanocrystals in DMF.

4. Conclusions

CdSe, CdTe, CdHgTe and HgTe nanocrystals capped on the surface with different thiols (mercaptoalcohols, mercaptoacids, dithiols) have been synthesised by a wet chemical route and were obtained in the form of redispersible powders on a gram scale. They were characterised by UV-Vis absorption and photoluminescence spectroscopy, X-ray diffractometry and high-resolution transmission electron microscopy. The nanocrystals belong to the cubic phase with sizes ranging from 2.5 to 6 nm, tunable on the preparative and post-preparative stages. The size-quantisation effect together with the surface chemistry determine the optical properties of the nanocrystals. Room–temperature photoluminescence with its maximum depending on the nanocrystal size and composition was observed in the visible (CdTe; quantum efficiency 3–6%, at appropriate conditions up to 18%) and near–IR (CdHgTe, HgTe; quantum efficiency up to 50%) spectral ranges. Luminescent semiconductor nanocrystals are perspective candidates for use in broad-band optical amplifiers, light–emitting diodes, and novel structures with separately controllable densities of photon and electron states based on photonic crystals.

Acknowledgement. This work was supported by the Volkswagen Foundation, Hannover, and partially by a Collaborative Research Grant of the NATO Scientific and Environmental Affairs Division and an INTAS–Belarus research grant (project 97–0250). Thanks are due to all colleagues and co–authors of the articles cited contributing to the topic of this paper. Special thanks are due to Dr. A. Eychmüller for carefully reading the manuscript.

References

1. Gaponenko, S.V. (1998) *Optical Properties of Semiconductor Nanocrystals,* Cambridge University Press, Cambridge.
2. Colvin, V.L., Schlamp, M.C., and Alivisatos, A.P. (1994) Light-emitting diodes made from cadmium selenide nanocrystals and a semiconducting polymer, *Nature* **370**, 354–357.
3. Dabbousi, B.O., Bawendi, M.G., Onitsuka, O., and Rubner, M.F. (1995) Electroluminescence from CdSe quantum-dot/polymer composites, *Appl. Phys. Lett.* **66**, 1316–1318.
4. Gao, M., Richter, B., Kirstein, S., and Möhwald, H. (1998) Electroluminescence studies on self-assembled films of PPV and CdSe nanoparticles, *J. Phys. Chem. B* **102**, 4096–4103.
5. Klein, D.L., Roth, R., Lim, A.K.L., Alivisatos, A.P., and McEuen, P.L. A single-electron transistor made from a cadmium selenide nanocrystal, *Nature* **389**, 699–701.
6. Feldheim, D.L. and Keating, C.D. (1998) Self-assembly of single electron transistors and related devices, *Chem. Soc. Rev.* **28**, 1–12.
7. Murray, C.B., Norris, D.J., and Bawendi, M.G. (1993) Synthesis and characterization of nearly monodisperse CdE (E = S, Se, Te) semiconductor nanocrystallites, *J. Am. Chem. Soc.* **115**, 8706–8715.
8. Vossmeyer, T., Katsikas, L., Giersig, M., Popovic, I.G., Diesner, K., Chemseddine, A., Eychmüller, A., and Weller, H. (1994) CdS nanoclusters: synthesis, characterization, size dependent oscillator strength, temperature shift of the excitonic transition energy, and reversible absorbance shift, *J. Phys. Chem.* **98**, 7665–7673.
9. Rogach, A.L., Katsikas, L., Kornowski, A., Su, D., Eychmüller, A., and Weller, H. (1996) Synthesis and characterization of thiol-stabilised CdTe nanocrystals, *Ber. Bunsenges. Phys. Chem.* **100**, 1772–1778.
10. Rogach, A.L., Kornowski, A., Gao, M., Eychmüller, A., and Weller, H. (1999) Synthesis and characterization of a size series of extremely small thiol-stabilised CdSe nanocrystals, *J. Phys. Chem. B* **103**, 3065–3069.

11. Rogach, A.L., Kershaw, S.V., Burt, M., Harrison, M., Kornowski, A., Eychmüller, A. and Weller, H. (1999) Colloidally prepared HgTe nanocrystals with strong room-temperature infrared luminescence, Adv. Mater. 11, 552–555.

12. Steigerwald, M.L. and Brus, L.E. (1990) Semiconductor crystallites: a class of large molecules, Acc. Chem. Res. 23, 183–188.

13. Spanhel, L., Haase, M., Weller, H. and Henglein, A. (1987) Photochemistry of colloidal semiconductors. 20. Surface modification and stability of strong luminescing CdS particles, J. Am. Chem. Soc. 109, 5649–5655.

14. Rockenberger, J., Tröger, L., Rogach, A.L., Tischer, M., Grundmann, M., Eychmüller, A. and Weller, H. (1998) The contribution of particle core and surface to strain, disorder and vibrations in thiolcapped CdTe nanocrystals, J. Chem. Phys. 108, 7807–7815.

15. Rockenberger, J., Tröger, L., Rogach, A.L., Tischer, M., Grundmann, M., Weller, H. and Eychmüller, A. (1998) An EXAFS study on thiolcapped CdTe nanocrystals. Ber. Bunsenges. Phys. Chem, 102, 1561–1564.

16. Chemseddine, A. and Weller, H. (1993) Highly monodisperse quantum sized CdS particles by size selective precipitation, Ber. Bunsenges. Phys. Chem. 97, 636–637.

17. Gao, M., Kirstein, S., Möhwald, H., Rogach, A.L., Kornowski, A., Eychmüller, A. and Weller, H. (1998) Strongly photoluminescent CdTe nanocrystals by proper surface modification, J. Phys. Chem. B 102, 8360–8363.

18. Rogalski, A. (1994) New trends in semiconductor infrared detectors, Opt. Eng. 33, 1395–1412.

19. Balcerak, R., Gibson, J.F., Guiterrez, W., and Pollard, J.H. (1987) Evolution of a new semiconductor product: mercury cadmium telluride focal plane arrays, Opt. Eng. 26, 191–200.

20. Landolt-Börnstein, (1982) Numerical Data and Functional Relationships in Science and Technology: New Series. Vol 17b : Semiconductors, Springer-Verlag Berlin.

21. Gaponik, N.P., Talapin, D.V. and Rogach, A.L. (1999) A light-emitting device based on a CdTe nanocrystal/polyaniline composite, PCCP 1, 1787–1790.

22. Gaponenko, S.V., Bogomolov, V.N., Kapitonov, A.M., Prokofiev, A.V., Eychmüller, A. and Rogach, A.L. (1998) Electrons and photons in mesoscopic structures: quantum dots in a photonic crystal, JETP Letters, 68, 142–147.

PHOTOLUMINESCENCE FROM InGaAs/GaAs QUANTUM DOTS IN A HIGH ELECTRIC FIELD

A. BABIŃSKI
Institute of Experimental Physics, Warsaw University, ul. Hoża 69,
00-681 Warszawa, Poland

1.Introduction

Growing interest in semiconductor quantum dots (QDs) results both from their peculiar properties and their prospective applications in optoelectronics. Semiconductor laser devices with QDs have been shown to have lower threshold current density, higher differential gain, and higher thermal stability as compared to their quantum well-based counterparts [1-2]. The 3 dimensional confinement of electronic motion in QDs results in a δ-like density of states, which is crucial to QD properties [3]. The energy distribution of these levels depends on the size of a particular dot. Therefore the QD photoluminescence (PL) spectra obtained under a non-resonant excitation, which influences a large ensemble of QDs, have the form of Gaussian peak with a broadening factor depending on the size distribution of the QDs [4-5]. With a decrease of the number of excited QDs, well reproducible individual peaks of ground states can be observed in the PL spectrum. The most successful way of fabricating QDs makes use of the Stranski-Krastanow mode of growth. Deposition of a mismatched material layer (such as InGaAs on GaAs) proceeds in two steps: first a very thin layer of highly strained material is grown, forming the so called wetting layer (WL), then the Stranski-Krastanow growth mode transition takes place, resulting in the formation of 3-dimensional highly strained islands [6-8]. The areal density and shape strongly depends on the substrate orientation and growth conditions.

Very intense investigations of QDs in recent years enable researchers to understand their basic properties. One of the important questions raised is the effect of an electric field on the properties of QDs. Electric field influences quantum confinement and exciton dynamics in semiconductor structures [9-10]. Extensive work on the effect of an electric field on the optical properties of QDs has been performed using semiconductor nanocrystals [11-12]. Microphotoluminescence measurements investigating the effect of electric field on excitons localised on the quantum well fluctuations [13], interdiffussed QDs [14] or self-assembled QDs [15-16] have also been performed. The effect of electrically induced charging of QDs on the PL from the InAs/GaAs self-assembled QDs was also investigated [17]. In this work the influence of electric field on the PL from the large array of self-assembled InGaAs/GaAs QDs in a field-effect structure is investigated both at room temperature and at low temperature.

M.L. Sadowski et al. (eds.), Optical Properties of Semiconductor Nanostructures, 395–404.
© 2000 *Kluwer Academic Publishers. Printed in the Netherlands.*

2. Samples

The samples investigated in this work were grown using low pressure Metal Organic Vapour Phase Epitaxy. Two samples were grown in the same growth procedure on a [100] - oriented SI GaAs substrate (sample A) and on a 2 deg off [100] towards [110] - oriented SI GaAs substrate (sample B). The investigated structures consisted of 200 nm of Si doped ($n = 2 \times 10^{18}$ cm^{-3}) n$^+$GaAs layer acting as a back contact, followed by 20 nm of undoped GaAs tunnelling barrier, the In$_{0.6}$Ga$_{0.4}$As layer with QDs grown in the Stranski-Krastanow mode, a 20 nm undoped GaAs spacer, a 31 nm Al$_{0.3}$Ga$_{0.7}$As barrier, and a 25 nm GaAs cap. A schematic sample structure is shown in Fig. 1. The In$_{0.6}$Ga$_{0.4}$As layer was deposited at a growth temperature of 550° C, GaAs was grown at 650° C and Al$_{0.3}$Ga$_{0.7}$As at 750° C. The surface topography of QDs was examined using an Atomic Force Microscopy (AFM) on a sample with uncapped QDs (for an AFM image of QDs on [100] oriented substrate see Fig. 2.). It was found that lens shaped islands of average basal size 890 nm^2 ± 20%, (820 ± 22%), with an average surface density 7.1×10^9 cm^{-2} (1.7×10^{10} cm^{-2}), average island height 7 ± 2 nm (9 ± 2 nm) were formed in the sample A (sample B) [19].

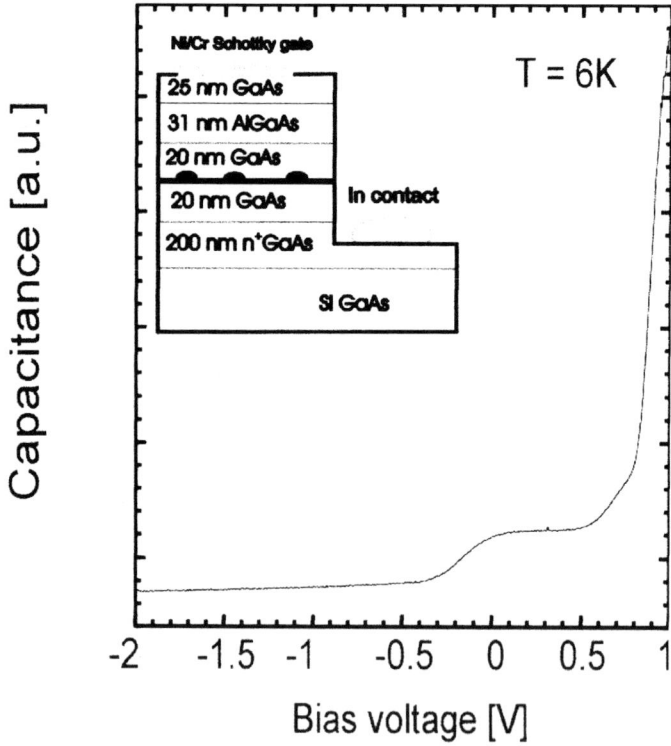

Figure 1. A sample structure and capacitance-voltage characteristic of sample A at *T*=6K.

For the bias dependent PL measurements, a circular (1.5 mm diameter) semitransparent (30 nm thick) Ni/Cr Schottky contact was evaporated through a shadow mask on the top of the structure. The back gate was formed by alloying an In contact on top of the n^+ GaAs layer after part of the sample had been electro-chemically etched off.

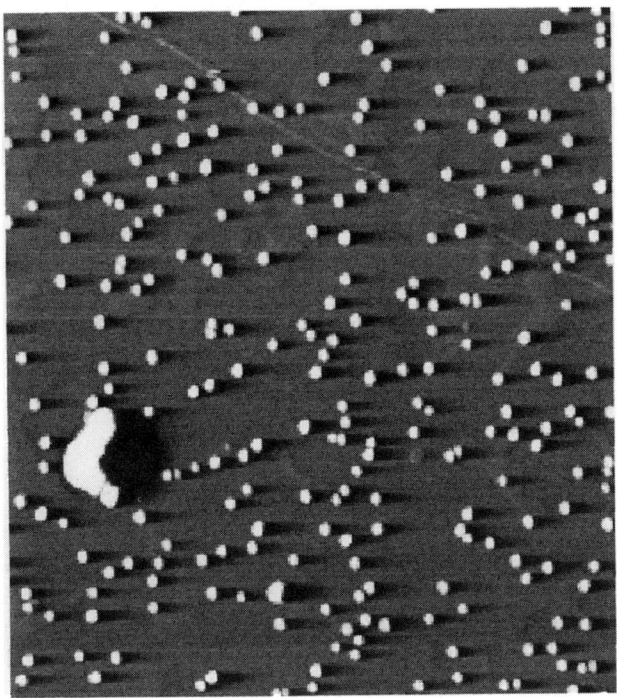

Figure 2. Atomic Force Microscopy image of QDs grown on a [100] oriented GaAs substrate (sample A). The area shown is 2μm x 2μm. [18]

Contacting the back n^+ GaAs layer prevents an undesired short-circuit of the structure. A negative bias denotes in this paper a reverse polarisation of the Schottky structure. Bias dependent PL measurements were taken with a laser illumination through a Schottky contact.

The PL measurements were performed with the sample cooled in a continuous flow Oxford CF-1204 cryostat. Laser illumination (λ=780 nm) was used for a non-resonant PL excitation. PL spectra were dispersed by a SPEX spectrometer and detected with a liquid nitrogen-cooled Ge p-i-n diode. The PL spectra of the investigated samples measured at RT are presented in Fig.3. A peak at 1.283 eV (1.264 eV) dominates the PL spectrum for sample A (sample B). We relate this PL peak to the electron-hole recombination within the QDs. The difference in PL lineshape between two samples can be understood as follows. Because of the strong confinement in the growth direction, the absolute energy position of the QD PL is determined by the thickness of the QDs [20]. Therefore in sample A (with QDs of lower height) the PL peaks at higher energy than in sample B. The PL FWHM reflects the morphological size distribution of QDs, therefore it is larger in sample B, with a broader QDs size distribution, than in sample A. The relatively high energy of the QDs PL peak has to be noted. We relate this to the effect of In interdiffusion through the growth cladding material at elevated temperature. It is known that the QDs morphology and therefore their electrical and optical properties can be tuned by post-growth annealing [21-23]. It is believed that a change in the dot composition makes the confining potential shallower and results in a blue-shift of the QD emission [24-25], reduction of the QDs interlevel spacing [26] and reduction of the PL FWHM [27]. Quenching of the WL emission after interdiffussion was also observed [24], which explains also that no WL emission can be observed in our samples. A weak spectral feature due to the WL absorption can be observed in electroreflectance and photovoltaic measurements at 1.35 eV.

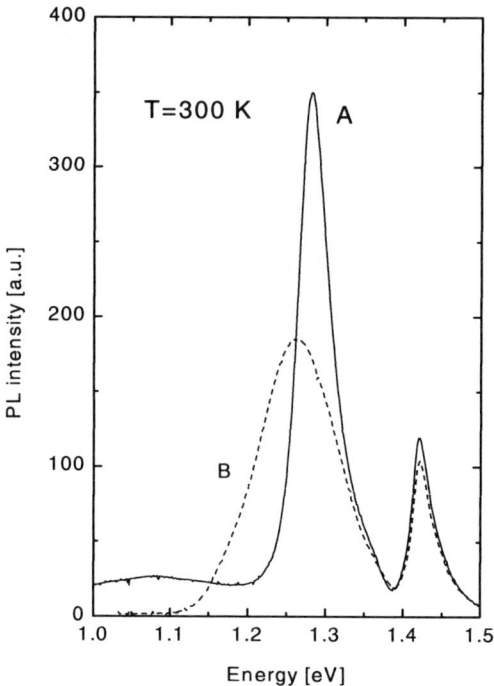

Figure 3. Photoluminescence at room temperature from sample A ([100] oriented substrate) and sample B (2 deg off [100] toward [110] oriented substrate).

3. Room temperature photoluminescence

The PL spectra at RT in biased structures are presented in Fig. 4. It can be seen that the QDs PL can be completely quenched in a high electric field in both investigated samples. In order to investigate more specifically the effect of bias on the QDs PL lineshape an electrically modulated (e-m) PL measurement was performed [28]. A field-effect structure in such an experiment is continuously illuminated with a laser beam, and it is biased with a sum of DC bias U_0 and AC modulation signal ΔU. The AC voltage is also used as a reference for a lock-in amplifier. The e-m PL signal depends on the PL in phase with an applied bias and it can be regarded as a derivative of the PL intensity on the bias. The e-m PL experiment spectra measured at room temperature in samples A and B are presented in Figs. 5a and 5b, respectively. The asymmetric e-m PL lineshape in sample A may result from transitions involving heavy and light holes in

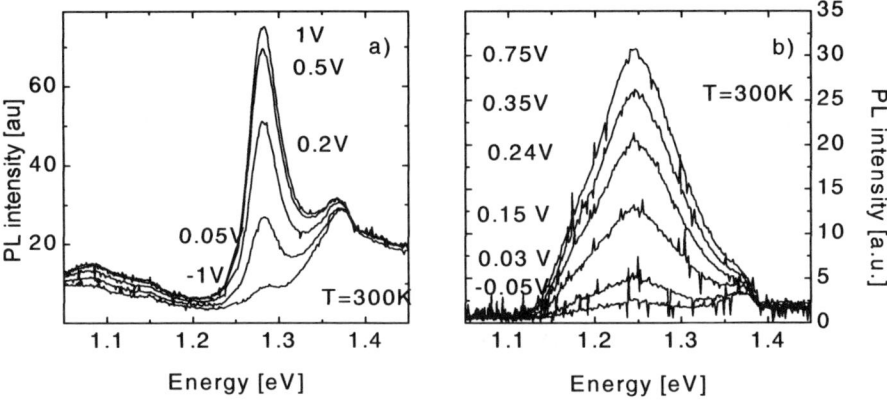

Figure 4. Photoluminescence at room temperature from sample A (a) and sample B (b), as a function of bias.

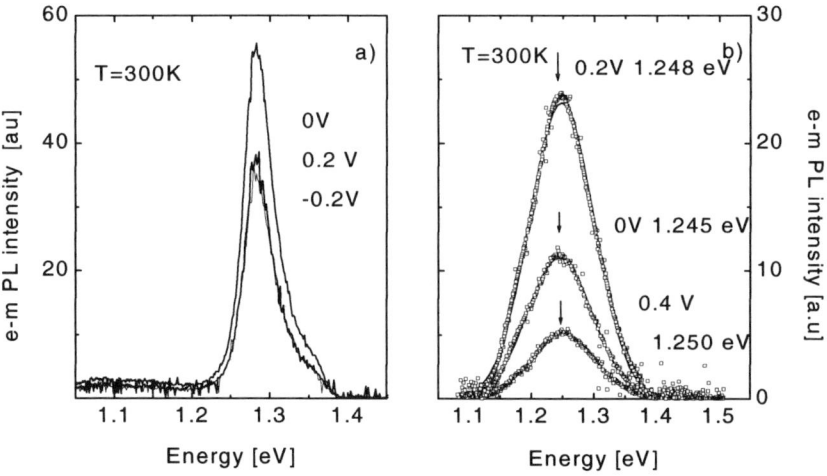

Figure 5. Electrically modulated photoluminescence at room temperature from field effect structures on sample A (a) and sample B (b) as a function of bias. The e-m PL lineshape fits with Gaussians are also presented (see continuous line in Fig. 5b). The e-m PL peak positions at bias $U_0 = 0V$, 0.2V, 0.4V were found to be equal to 1245± 0.5 meV, 1248± 0.5 meV, and 1250 ± 0.5 meV respectively.

the QDs [30]. No observable bias dependence of the PL lineshape was observed in the e-m PL from sample A (see Fig. 5a). The e-m PL spectra in sample B were fitted with single Gaussians (see lines in Fig 5b). As a result, a very weak redshift of the PL peak energy (approx. 5 meV) accompanying the quenching of the PL was found.

The quenching of the non-resonantly excited PL in the electric field is attributed to the separation of photoexcited carriers [13,16] before their relaxation into the QDs. This process is enhanced at room temperature by very effective thermal evaporation of electrons from QDs to the continuum states due to a relatively small energy separation between the QDs ground state and the WL states (approx. 100 meV in our samples) [30]. The PL intensity can be correlated with bias-controlled steady electron occupancy of the QDs. The electric charge on the QDs screens the field, which increases the probability of radiative recombination within the dot.

The PL peak redshift accompanying the PL quenching in sample B results most likely from the quantum confined Stark effect (QCSE) [9]. Spatial separation of electron and hole wavefunctions in tilted conduction and valence bands leads to a decrease of the transition energy within the QDs. This redshift strongly depends on the confinement in the electric field direction. The average QD height in sample B (9 ± 2 nm) is larger than in sample A (7 ± 2 nm), which might explain why no shift is observed in sample A.

4. Low temperature photoluminescence

A more complicated dependence of the PL lineshape on bias can be observed in the field-effect structure at low temperature [31]. The bias dependence of the QDs PL at $T = 4.2$K in sample A is shown in Fig. 6. Spectra shown in Fig. 6 were obtained after subtracting the „background" PL measured at $U_0 = -1.5$V. The PL intensity continuously increases with bias up to $U_0 = 0.3$V. At bias $U_0 > 0.3$V an additional feature (marked with an arrow) becomes apparent on the high energy tail of the QDs PL and the PL peak intensity decreases. The bias dependence of the PL peak position is shown in Fig. 7. The PL peak energy can be related to the electron occupancy of the QDs. Charging of QDs with electrons is reflected in capacitance-voltage (CV) characteristics (see Fig. 1). At large negative bias the QDs are empty and the structure capacitance reflects the geometrical capacitance of the fully depleted region between the back n$^+$ GaAs contact and the Schottky gate. The QDs PL intensity decreases with increasing negative bias in this field region. It was found that the PL decay time substantially decreases with increasing electric field (from 740 ps at $U_0 = -0.5$V to 550 ps at $U_0 = -1.5$V) [32], which indicates that carrier tunnelling from the QDs is effective in this electric field region [10]. The reason for the relatively high PL peak energy $U_0 < -0.5$V (higher than in the unbiased structure - compare dotted line in Fig. 7) is not fully understood at the moment. A step-like feature seen in the CV characteristics at $U_0 = -0.3$ V is related to QD charging [28]. The lack of separate peaks in CV due to charging with subsequent electrons [33-34] results from a relatively shallow electron confinement potential. An increase of the PL energy in the QDs charging region (-0.3 V $< U_0 < 0$V) can be explained in terms of selective tunnelling of carriers from QDs of sizes [35]. The tunnelling rate depends on the height of the triangular GaAs barrier, which forms in large electric field. Therefore it is larger in the case of QDs with higher

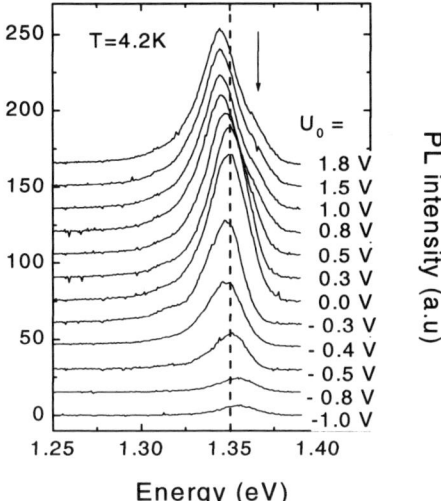

Figure 6. Photoluminescence in sample A at *T*=4.2K as a function of bias. The PL peak position (*E*=1.35 eV) in the unbiased sample is also shown for comparison.

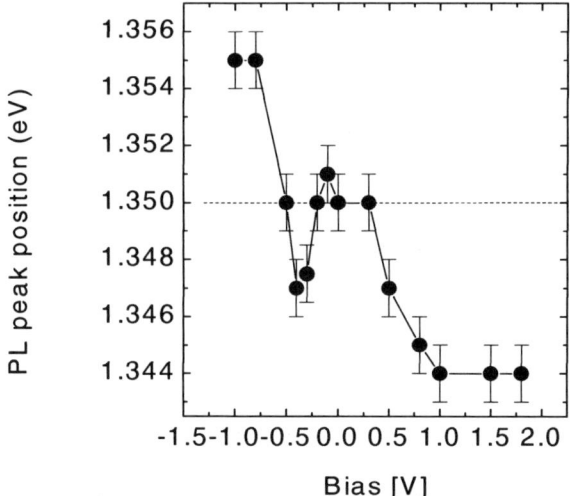

Figure 7. The bias dependence of the PL peak in sample A measured at *T*=4.2K. The PL peak position (*E*=1.35 eV) in the unbiased sample is also shown for comparison.

electron energies as compared to the QDs with lower electron energies. As a result the radiative efficiency in the QDs with higher electron energies becomes lower than in the QDs with lower electron energies. This leads to an apparent redshift of theinhomogeneously broadened PL peak. This effect was not observed in sample A at room temperature because the carrier evaporation dominates the tunnelling at higher temperatures. Another possible explanation of the PL energy increase would be the QCSE, however it was not observed in sample A at room temperature.

The PL energy decreases with decreasing electric field at $U_0 > 0.3$ V. This results from charging with subsequent electrons [36]. The optical transition energy in QDs changes as a results of QD charging. Additional energy comprises a repulsive energy term of electron-electron interaction, an attractive energy term of electron-hole interaction and the exchange energy term. To a good approximation the first two terms are equal and the latter influences the transition energy within the QDs. It was theoretically established that this additional exchange energy term is negative and a redshift should accompany the QDs charging with subsequent electrons [36]. The redshift of the PL energy with an increasing QDs electron occupancy was reported in InAs/GaAs QDs in field-effect structure [17] and in modulation doped InAs/GaAs QDs [37]. The additional PL feature, denoted with an arrow, results most likely from a transition involving an excited electron state to the ground hole state [17, 36].

5. Conclusions

In conclusion the PL from self-organised InGaAs/GaAs QDs in field-effect structures has been studied. It has been found that the QD PL can be completely quenched in a high electric field at room temperature. The electrically modulated PL technique was used to detect the spectral dependence of the PL quenching. The PL quenching was accompanied by a weak PL redshift in the sample with QDs of 9 ± 2 nm height, which was explained in terms of the QCSE.

At low temperature both the effect of field-induced PL quenching and the effect of QDs charging on the PL was observed. The PL redshift with increasing electric field was attributed to selective carrier tunnelling from the QDs. The PL redshift with increasing QDs electron occupation was explained in terms of charged excitons in the QDs.

Acknowledgements. The samples investigated in this work were grown by Dr R. Leon at the Australian National University, Canberra, Australia. Support from Dr C. Jagadish during the author's stay with Electronic Materials Engineering Department of ANU is gratefully acknowledged.

This work was supported in part by the Polish Committee for Scientific Research Grant no. PBZ-28.11.

References

1. Fafard S., Hinzer K., Raymond S., Dion M., McCaffrey J., Feng Y., and Charbonneau S. (1996) Red-emitting semiconductor dot lasers, *Science* **274**, 1350-1353,
2. Bimberg D., Ledentsov N. N., Grundmann M., Kirstaedter N., Schmidt O. G., Mao M. H., Ustinov V. M., Egorov A. Y., Zhukov A. E., Kopev P. S., Alferov Z. I., Ruvimov S. S., Goesele U., and Heydenreich J. (1996) InAs-GaAs quantum pyramid lasers: in situ growth, radiative lifetimes and polarization properties, *Jpn. J. Appl. Phys.* **35**, 1311-1319
3. Ashoori R.C. (1996) Electrons in artificial atoms, *Nature* **379**, 413-419
4. Fafard S., Leon R., Leonard D., Merz J. L., and Petroff P. M. (1994) Visible photoluminescence from *N*-dot ensembles and the linewidth of ultrasmall AlInAs/AlGaAs quantum dots, *Phys. Rev. B* **50**, 8086-8089
5. Leon R, Petroff P. M., Leonard D., and Fafard S. (1995) Spatially resolved visible qluminescence of self-assembled semiconductor quantum dots, *Science* **267**, 1966-1969
6. Leonard D., Krishnamurthy M., Reaves C.M., Denbaars S.P., and Petroff P.M. (1993) *Appl. Phys, Lett.* **63**, 3203-3205,
7. Leonard D., Pond K., and Petroff P.M. (1994) Critical layer thickness for self-assembled InAs island on GaAs, *Phys. Rev B* **50**, 11687-11692;
8. Leon R., Fafard S., Leonard D., Merz J.L., and Petroff P.M., (1995) Visible luminescence from semicondutor quantum dots in large ensembles, *Appl. Phys. Lett.* **67**, 521-523.
9. Mendez E.E., Bastard G., Chang L.L., Esaki L, Morkoc H., and Fisher R. (1982) Effect of an electric field on the luminescence of GaAs quantum wells, *Phys. Rev B.* **26**, 7101-7104
10. Koehler K., Polland H.-J., Schultheis L., and Tu C.W. (1988) Photoluminescence of two-dimensional excitons in an electric field: Lifetime enhancement and field ionization in GaAs quantum wells, *Phys. Rev. B* **38**, 5496-5503.
11. Empedocles S.A. and Bawendi M.G., (1997) Quantum-confined Stark effect in single CdSe nanocrystallite Quantum dots, *Science*, **278**, 2114-2117
12. Empedocles S.A., Norris D.J., and Bawendi M.G. (1996) Photoluminescence spectroscopy of single CdSe nanocrystallie quantum dots, *Phys. Rev. Lett*, **77**, 3873-3876
13. Heller W., Bockelmann U., and Abstreiter G. (1998) Electric-field effects on excitons in quantum dots, *Phys. Rev. B* **57**, 6270-6273
14. Bockelmann U., Roussignol Ph., Filoramo A., Heller W., Abstreiter G., Brunner K., Bohm G., and Weimann G. (1996) Time resolved spectroscopy of single quantum dots: Fermi Gas of Excitons?, *Phys. Rev. Lett.* **76**, 3622-3625
15. Pistol M.-E., Hessman D., Lindahl J., Montelius L., and Samuelson L (1996) STM-based luminescence spectroscopy on single quantum dots, *Mater. Sc. Eng B* **42**, 82-87
16. Raymond S., Reynolds J.P., Merz J.L., Fafard S., Feng Y., and Charbonneau, S., (1998) Asymmetric Stark shift in AlInAs/AlGaAs self-assembled dots, *Phys. Rev. B* **58**, R13415- R13418
17. Schmidt K.H., Medeiros-Ribeiro, Petroff P.M. (1998) Photoluminescence of charged InAs self-assembled quantum dots, *Phys. Rev. B* **58**, 3597-3600
18. Lobo C (1997) unpublished
19. Lobo C. and Leon R (1998) InGaAs island shapes and adatom migration behavoiur on (100), (110), (111), and (311) GaAs surfaces, *J. Appl. Phys.* **83**, 4168-4172
20. Schmidt K.H., Medeiros-Ribeiro, Garcia J., and Petroff P.M. (1997) Size quantization effects in InAs self-assembled quantum dots, *Appl. Phys. Lett.* **70**, 1727-1729
21. Leon R., Kim Y., Jagadish C., Gal M., Zou J., and Cockayne D. J. H. (1996) Effects of interdiffusion on the luminescence of InGaAs/GaAs quantum dots, *Appl. Phys. Lett.* **69**, 1888-1890
22. Malik S., Roberts C., Murray R., and Pate M. (1997) Tuning self-assembled InAs quantum dots by rapid thermal annealing, *Appl. Phys. Lett* **71**, 1987-1989
23. Mo Q. W., Fan T. W., Gong Q., Wu J., Wang Z. G., and Bai Y.Q. (1998) Effects of annealing on self-organized InAs quantum islands on GaAs (100), *Appl. Phys. Lett.* **73**, 3518-3520

24. Leon R., Kim Y., Jagadish C., Gal M., Zou J., and Cockayne D. J. H. (1996) Effects of interdiffusion on the luminescence of InGaAs/GaAs quantum dots, *Appl. Phys. Lett.* **69**, 1888-1890

25. Lobo C., Leon R., Fafard S., and Piva P. G. (1998) Intermixing induced changes in the radiative emission from III-V quantum dots, *Appl. Phys. Lett.* **72**, 2850-2852

26. Leon R., Fafard S., Piva P. G., Ruvimov S., and Liliental-Weber Z.(1998) Tunable intersublevel transitions in self-forming semiconductor quantum dots, *Phys. Rev. B* **58**, R4262-R4265

27. Leon R., Williams D. R. M., Krueger J., Weber E. R., and Melloch M. R. (1997) Diffusivity transients and radiative recombination in intermixed $In_{0.5}Ga_{0.5}As$/GaAs quantum structures, *Phys. Rev B* **56**, R4336-R4339

28. Babiński A, Wysmołek A., Tomaszewicz T., Baranowski J.M., Leon R., Lobo C, and Jagadish C. (1998) Electrically modulated photoluminescence in self-organized InGaAs/GaAs quantum dots, *Appl. Phys. Lett.* **72**, 2811-2813

29. Brounkov P.N., Polimeni A., Stoddart S.T., Henini M., Eaves L., Main P.C., Kovsh A.R., Musikhin Y.G., and Konnikov S.G. (1998) Electronic structure of self-assembled InAs quantum dots in GaAs matrix, *Appl. Phys. Lett.* **73**, 1092-1094

30. Bimberg D., Ledentsov N. N., Grundmann M., Kirstaedter N., Schmidt O. G., Mao M. H., Ustinov V. M., Egorov A. Y., Zhukov A. E., Kopev P. S., Alferov Z. I., Ruvimov S. S., Goesele U., and Heydenreich J. (1996) InAs-GaAs quantum dots: from growth to lasers, *phys. stat. sol (b)* **194**, 159-173

31. Babiński A, Tomaszewicz T., Wysmołek A., Baranowski J.M., Lobo C., Leon R., and Jagadish C. will be published in (1999*) Microcrystalline and Nanocrystalline Semiconductors - 1998. Symposium Mater. Res. Soc. 1998,*

32. Korona, K.P, et al. to be published

33. Drexler H., Leonard D., Hansen W., Kotthaus J.P., and Petroff P.M. (1994) Spectroscopy of quantum levels in charge-tunable InGaAs quantum dots, *Phys. Rev. Lett.* **73**, 2252-2255

34. Medeiros-Ribeiro G., Leonard D., and Petroff P. M. (1995) Electron and hole energy levels in InAs self-assembled quantum dots, *Appl. Phys. Lett.* **66**, 1767-1769

35. Skolnick M., this conference

36. Wójs A. and Hawrylak P.(1997) Theory of photoluminescence from modulation doped self-assembled quantum dots in a magnetic field, *Phys. Rev. B* **55**, 13 066- 13071

37. Lee J,I, Lee H.G., Shin E., Yu S., Kim D., and Ihm G. (1997) Many-body effect on modulation-doped InAs/GaAs quantum dots, *Appl. Phys. Lett* **70**, 2885-2887

HOMOGENEOUS LINEWIDTH OF OPTICAL TRANSITIONS AND ELECTRONIC ENERGY RELAXATION IN QUANTUM DOTS

K. KRÁL AND Z. KHÁS

*Institute of Physics, Academy of Sciences of Czech Republic,
Na Slovance 2,
CZ-18221 Prague 8, Czech Republic*

1. Introduction

There is an increasing interest in the properties of nanostructures, in which the carrier motion is confined in all three dimensions, as in self-assembled quantum dots and nanocrystallite quantum dots [1]. The reason for this interest lies, for example, in the properies of the 0D (zero-dimensional) structures, which promise efficient low-treshhold current lasers for opto-electronic applications. One of the practical problems in this respect is the question of the rapidity of the relaxation of the excited electronic subsystem and the question of the width of the optical line. It is the aim of theoretical and experimental studies to answer the question of which properties of 0D objects are inherent properties of these structures, and which are determined or influenced by the technological process.

Semiconductor materials of the type of GaAs have been studied extensively. The Fröhlich coupling of the charge carriers with the longitudinal optical (LO) lattice vibrations is known to play a very important role in the process of energy transfer from the excited carrier system to the subsystem of lattice vibrations. In quantum dots based on GaAs, of the lateral size of the order of 100 Å, the structure of the optical vibrations of a sample with a quantum dot may be approximated by the phonons of the GaAs crystal lattice. The carriers inside the dot may then be expected to be coupled to those LO modes which are near the Brillouin zone center, where LO phonons are practically dispersionless. This property of the optical phonons, together with the discrete structure of the density of states in a quantum dot with a rigid lattice, allows to expect that, with the exclusion of the exact resonance between the electronic inter-level separation and the LO

M.L. Sadowski et al. (eds.), Optical Properties of Semiconductor Nanostructures, 405–420.

phonon energy, the otherwise strong Fröhlich coupling will not provide an effective relaxation channel of electronic energy.

Quantum dots have an obvious similarity to an atomic nucleus. It appears that there are theoretical concepts, developed for many decades, which can be transferred from the atomic nucleus to the theory of quantum dots. The reader is refered to the work [2] and to the references cited therein.

It is the purpose of this presentation to briefly summarise the present state of studies of the role of Fröhlich coupling in electron energy relaxation phenomena, in both theory and experiment. Recent optical linewidth measurements are briefly reviewed and new theoretical results are presented, showing that not only the electronic relaxation, but also the optical line shape appears to be seriously influenced by the interaction of electrons with polar phonons.

2. Experiments on electronic relaxation

In earlier experiments the luminescence of 0D structures was found to decrease with decreasing lateral size of the quantum dots [3, 4, 5]. Besides attributing this decrease of luminescence to technological difficulties in the process of realising the lateral confinement for electrons and holes, a different explanation was suggested [6], ascribing it to the so called 'phonon−bottleneck effect' (see ref. [7]). The phonon-bottleneck hypothesis was supported by an earlier detailed analysis of the electron-phonon interaction in quantum dots, based on the Born approximation to electron-phonon scattering [8, 9, 10, 11].

In experiments, the relaxation of electronic energy was frequently measured on the scale of picoseconds [12, 13, 14, 15, 16, 17, 18, 19, 20, 21], being therefore fast enough to cause doubts about the existence of the bottleneck effect. A significant dependence of the electronic relaxation rate on the quantum-dot size was not reported in papers [12, 13, 14, 15, 16, 17, 18, 19, 20, 21] and the relaxation efficiency was reported to be independent of the relation between the electronic energy-level separations and the optical-phonon energy [18]. Basing on the experimental observations, it has been pointed out that the relaxation process of electrons in quantum dots should be considered as a multiphonon process [9, 13, 17, 19, 20, 21], and that multiphonon transitions should be expected to be the main mechanism providing electron energy relaxation.

The issue of the existence of the phonon bottleneck effect in quantum 0D structures does not appear to be clarified as yet. Although Empedocles *et al.* [22] conclude on the basis of their photoluminescence measurements that the phonon bottleneck is absent from the process of electronic relaxation

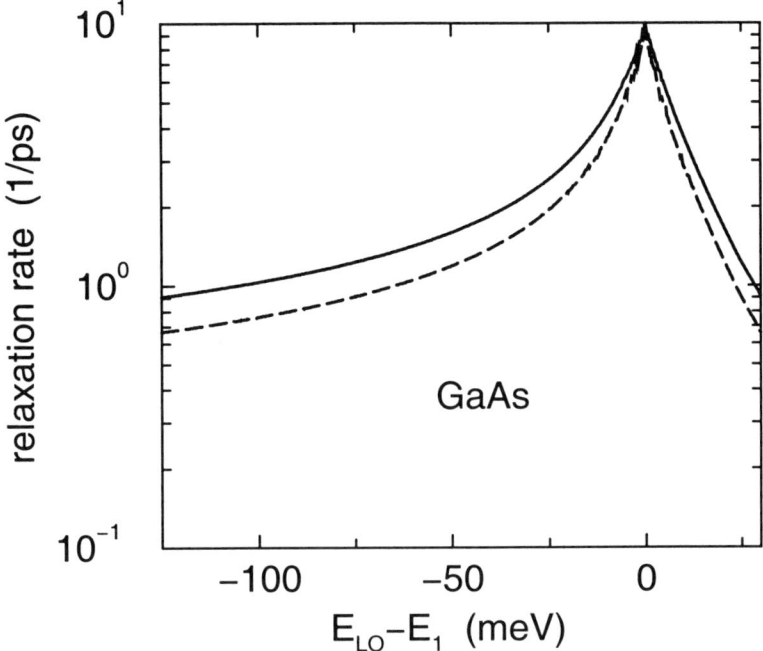

Figure 1. The relaxation rate $| dN_1/dt |$ in a GaAs quantum dot, as a function of the detuning $E_{LO} - E_1$ at room temperature (full curve) and at 77 K (dashed line).

from excited states in CdSe 30-50 Å large nanocrystallites, Murdin [23] interpret their decay rate measurements in Landau quantised InAs quantum dots as evidence for a suppression of the LO phonon scattering and for a phonon bottleneck between the lowest dot states. To a certain extent, the bottleneck issue also includes the matter of defining the concept of the phonon bottleneck in quantum dots.

The experimentally measured rapidity of the electronic relaxation in quantum dots appears to be greater than expected on the basis of simple theoretical estimates of, for example, the role of electron-phonon scattering [9], Auger effect [24] or defect related processes [25].

3. Measurements of the optical linewidth

There is a large amount of literature on the subject of the optical linewidth in quantum dots. The key effect of these measurements was shown to be inhomogeneous broadening, due to the spread of lateral dot size measured simultaneously in certain experiments. In order to avoid this inhomogeneous broadening feature in the spectra, experimental techniques were used allowing to measure the optical response of individual dots.

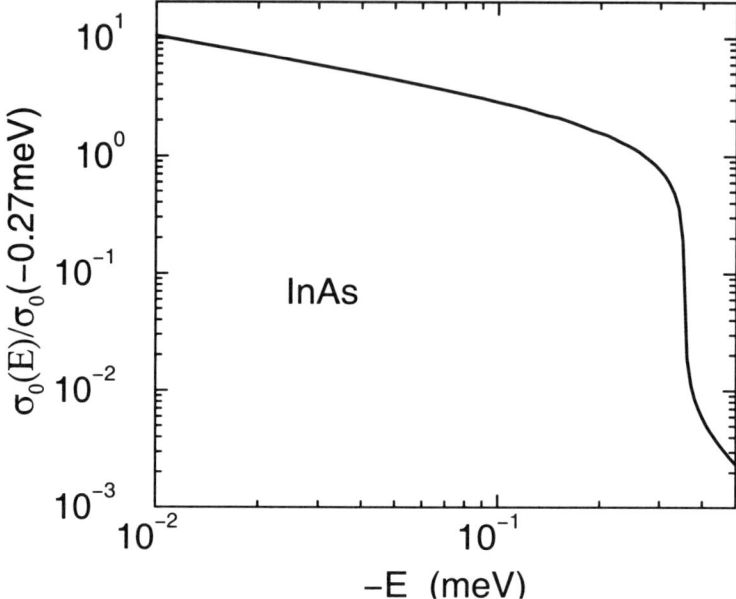

Figure 2. Log-log plot of the low-energy shoulder of the theoretical spectral density $\sigma_0(E)$ in an InAs quantum dot with $E_1 = \frac{3}{2}E_{LO}$ at a temperature of 20 K.

Empedocles [22] measured the photoluminescence of single CdSe nano-crystalline quantum dots with a lateral size of 40-50 Å. With a spectral resolution of the measuring apparatus of 0.12 meV they found that at a temperature of 10 K the spectral linewidth was 0.12 to 0.245 meV. The spectral width was shown to depend significantly on the excitation intensity of the light. The width of the optical lines was shown to increase up to about 4.6 meV at the highest excitation intensity used. At high excitation intensities the lineshape of the optical line appears to be well pronounced, having the approximate shape of a right-angled triangle with a shoulder on the low energy side. Valenta [26] studied CuBr nanocrystals embedded in a glass matrix. In the transient four-wave mixing experiment they found the dephasing times to depend on the excitation intensity and therefore on the sample temperature. They found that the dephasing time decreased from a value of about 6 ps at low intensity down to about half a picosecond at high excitation intensity, thus confirming the common trend of the increase of the optical linewidth with sample temperature. Grundmann [27] measured cathodoluminescence in a single-dot setup. The lateral size of the InAs quantum dots in the GaAs matrix was about 120 Å. They found the FWHM (full width at half maximum) of the spectral lines to be 0.15 meV. They regard this value as being limited from above by the spectral resolution of the experiment (0.15 meV). The measured carrier lifetime of 340 ps is interpreted in that paper as providing a lower limit of 0.012 meV to the FWHM.

Notomi [28] performed excitation spectroscopy of individual InGaAs quantum dots of lateral size of 400 Å and a height of 30 Å at temperatures from 5 to 50 K. Measuring the photoluminescence they found that the FWHM of the optical transitions in the dots is about 0.4 meV, being constant in the temperature interval from 5 to 50 K. Using photoluminescence excitation spectroscopy they found that the FWHM of the lines was about 0.5 to 1 meV in the interval of inter-level energy separations in the dot from 25 meV to about 60 meV. At these low temperatures, they did not find any sign of a resonant behaviour at the inter-subband energy separation equal the energy of the optical phonon (34 meV).

There are experiments in which the optical line shape is found to be quite large, from 5 or 10 meV at temperatures below 50 K up to several tens of meV at room temperature. Banin [29] measured 29 Å large CdSe and InP nanocrystals in an organic film, while Schoenlein measured the linewidth in CdSe 11 Å nanocrystals [30]. The linewidths in the very small quantum dots have a tendency to be much larger than the linewidths in larger dots [29].

4. Theory of electronic relaxation

The quantum dot system appears to be relatively simple because the electronic structure of the bound states, unperturbed by the lattice motion, may consist of only a few discrete energy levels. It is well known that the electron-phonon Hamiltonian of the quantum dot can be exactly diagonalised when the coupling of the electrons to the lattice vibrations is restricted to include only the transverse coupling (see below) terms [31].

Another simple and exactly solvable model, with no dissipation of the electronic energy, is the system with two electronic energy levels coupled to a single mode of the lattice vibrations [32], with the Hamiltonian formally equivalent to the Jaynes Cummings model [33]. A similar non-dissipating electron-phonon system was studied in a one-dimensional system [34]. Although we are aware of several simplified systems in which the electronic subsystem does not relax the energy, it is not yet clear which of their properties can be generalised to slightly more complicated realistic systems. Systems which we are likely meet in the real quantum dots are those in which the carriers, confined in the dots, interact with a large number of the lattice vibrational modes, and in which the electron-phonon coupling is more general than that of the Jaynes Cummings model. Unless further progress is made in the area of solving the electronic relaxation problem in

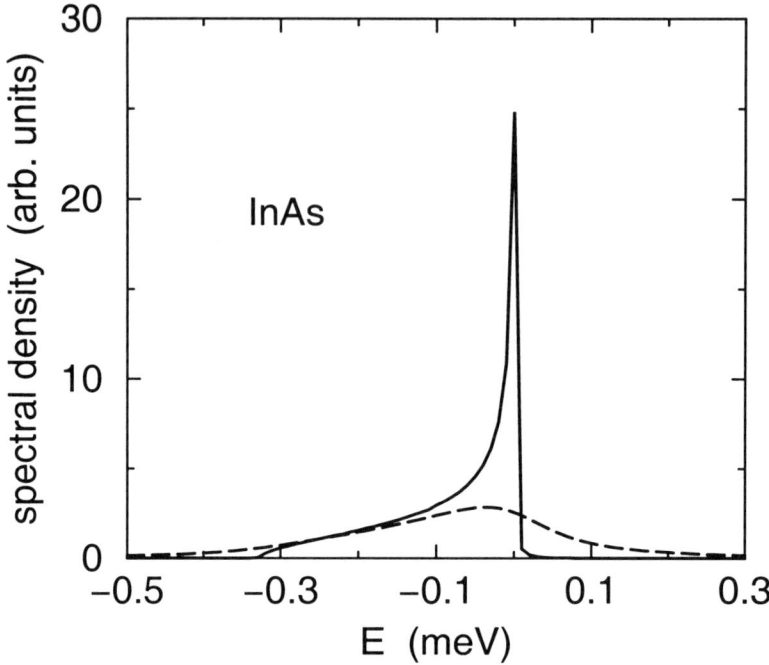

Figure 3. Lineshape of $\sigma_0(E)$ without spectral resolution ($\Delta = 0.001\,\text{meV}$) (full line) and with the spectral resolution $2\Delta = 0.15$ meV (dashed line). Temperature $T = 20\,\text{K}$, $E_1 = 2.5 \times E_{LO}$.

the zero-dimensional structures by exact methods, approximative theoretical approaches are in order.

The role of multiphonon states in the formulation of the electron transport problem in quantum dots was emphasised in papers [7, 35, 36, 37, 38, 39]. The need for a self-consistent treatment of the effect of "collision broadening" of the electronic energies in quantum dots was emphasised in references [36, 37]. These two requirements were recently taken into account in a study of the electronic spectral density [7, 38], where very sharp spectral density peaks were obtained. This sharpness was interpreted as an indication of a very long electronic lifetime, without paying sufficient attention to the lineshape of the spectral density features. In our previous paper [40] it was shown that the spectral density peaks may diverge to infinity as the inverse of the square root of the energy variable. Although such maxima in the electronic spectral density function are integrable, the relation of the "width" of such spectral features to the electronic lifetime may not be as simple as in the case of a Lorentzian peak shape. The relation of the electronic spectral densities to the rate of relaxation of the electronic energy, reported preliminarily in references [40, 41, 42], is treated in the

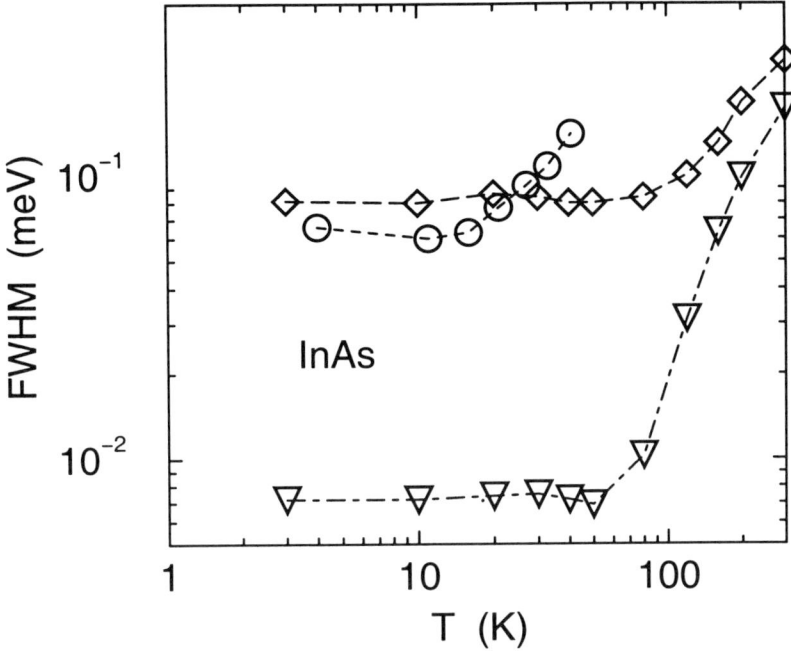

Figure 4. The linewidth (FWHM) of $\sigma_0(E)$ calculated in an InAs quantum dot (diamonds) at $E_{LO} - E_1$=-43.54 meV, with the experimental spectral resolution 2Δ equal to 0.03 meV. Circles are experimental data of Ota *et al.* [56]. Triangles - calculated with 2Δ=0.002 meV.

present work.

Besides the electron-LO-phonon coupling, other mechanisms were considered recently as possible candidates for explaining the experimental data on the fast electronic relaxation in quantum dots. Thus, the finite lifetime of the optical phonon was shown to provide an efficient mechanism of relaxing the severe restrictions imposed by the energy conservation in the Born approximation upon the electron-LO-phonon scattering [43, 44]. Also, the ultrafast electron energy relaxation in quantum dots has been recently suggested to be explained by the interaction of carriers with defect states in the quantum dot, taking into account the lattice relaxation mechanism [45].

It has been shown recently that electronic scattering in low-dimensional structures should be treated with caution. Namely, the Born approximation to the electronic scattering, giving good results in Monte Carlo semiclassical simulations of the electronic transport properties of bulk semiconductors, appears sometimes to be rather insufficient in quasi-two and quasi-one dimensional structures such as quantum wells and quantum wires [46, 47]. These observations provide additional arguments in favour of going beyond

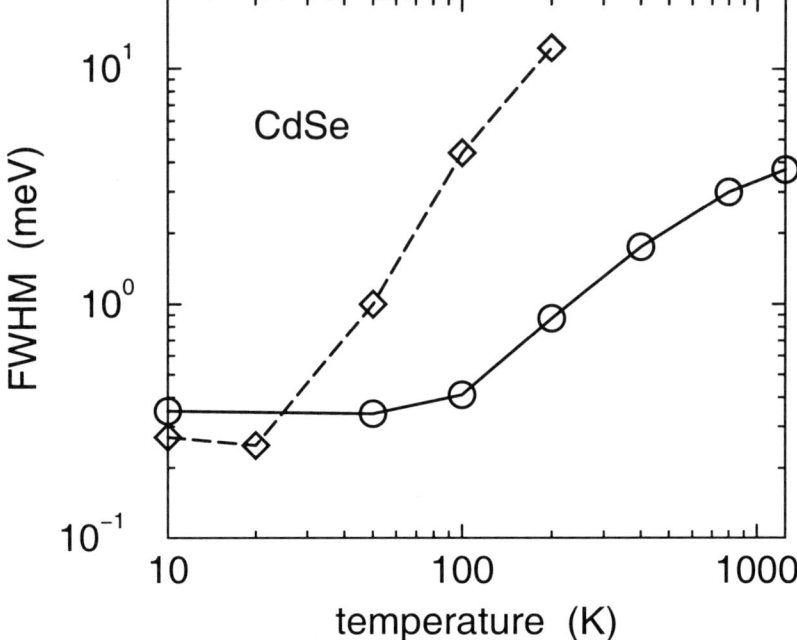

Figure 5. Temperature dependence of the width (FWHM) of the main peak of $\sigma_0(E)$ ($N_0 = 1$) computed with the finite experimental spectral resolution $2\Delta = 0.12$ meV in CdSe. Detuning $E_{LO} - E_1 = -65$ meV. Full line - calculated within the present TDLF approximation (see text). Dashed line - calculated in the TD approximation (see Ref. [40]).

the Born approximation in the electron-phonon scattering in quantum dots. Understanding the failure of the Born approximation as an implication of the multiple reflections of the charge carrier from the boundaries of the low-dimensional structure, it can be expected that such dimensionality effects should be present in zero-dimensional structures such as quantum dots. In such a case, the multiple scattering of electrons should be taken into account and the multiphonon states of the electron-phonon system of the quantum dot should be of importance.

In this work we wish to present the latest development following the theoretical approach submitted for publication recently [40, 48]. In the references [40, 48] a mechanism was considered for the electron energy relaxation in quantum dots. The quantum dot was considered to be a quantum dot of electron states, forgetting completely about the holes in the valence band states. Only the two lowest lying electronic states (ignoring spin) were considered for the single electron system, namely the ground state and the lowest excited state of a cubic dot with infinitely high potential walls. The material inside the dot is characterized by an effective mass and the static κ_0 and high-frequency κ_∞ dielectric constants. In the representation of these two electronic states the Hamiltonian of the electron (c_m) interacting with the dispersionless longitudinal optical (LO) phonons ($b_{\mathbf{q}}$) of the bulk material reads:

$$H = H_0 + H_1, \tag{1}$$

where

$$H_0 = \sum_{n=0}^{1} E_n c_n^+ c_n + \sum_{\mathbf{q}} E_{LO} b_{\mathbf{q}}^+ b_{\mathbf{q}} \tag{2}$$

and H_1 is the Fröhlich coupling [40]

$$H_1 = \sum_{m,n=0}^{1} \sum_{\mathbf{q}} A_q \Phi(n,m,\mathbf{q})(b_{\mathbf{q}} - b_{-\mathbf{q}}^+) c_n^+ c_m . \tag{3}$$

In the latter equation $q = |\,\mathbf{q}\,|$ while $A_q = (-ie/q)[E_{LO}(\kappa_\infty^{-1} - \kappa_0^{-1})]^{1/2}(2\varepsilon_0 V)^{-1/2}$. The electron-phonon coupling depends on the quantum dot size via the form-factor

$$\Phi(n,m,\mathbf{q}) = \int d^3\mathbf{r} \psi_n^*(\mathbf{r}) e^{i\mathbf{q}\mathbf{r}} \psi_m(\mathbf{r}). \tag{4}$$

In the operator H_1 there are two kinds of terms: the ones which are diagonal in the electronic index (with $m = n$, adiabatic or transverse terms) and the other terms with $m \neq n$ (longitudinal terms). The adiabatic terms can be eliminated by a canonical transformation [49, 31]. Let us note that Takagahara [50] considered a Hamiltonian similar to (1), using deformation potential and piezoelectric coupling of carriers to acoustic phonons, with the longitudinal terms absent, and calculated the homogeneous linewidth of an exciton, interpreting the linewidth as a fluctuation of the exciton energy due to the fluctuation of the number of phonons. His theoretical estimate of the optical linewidth agrees rather well with experiment in ultrasmall (several nanometers) quantum dots.

In contrast to Takagahara's model, in which a single state of the exciton is coupled to acoustic phonons via only the transverse (adiabatic) coupling, here we turn our attention to the coupling of the electron to the dispersionless LO phonons. The electron can occupy two orbital states in the dot and both adiabatic and longitudinal coupling terms are included.

The Hamiltonian (1) was used to the study of the rate of electron-energy relaxation in [40]. The theory of the nonequilibrium statistical operator (NSO) was used [51]. The relaxation rate was expressed in the form

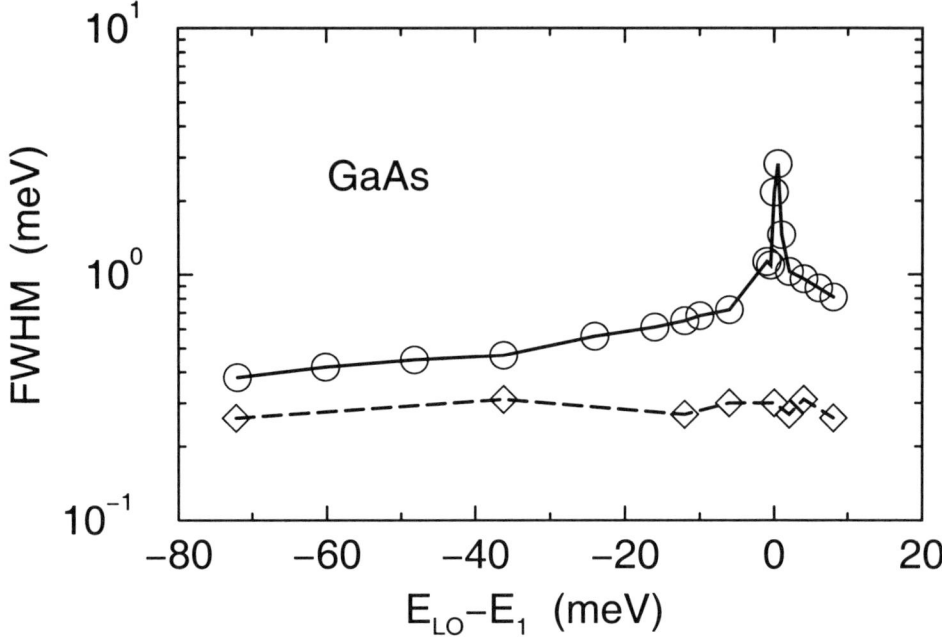

Figure 6. The linewidth (FWHM) of $\sigma_0(E)$ calculated in a GaAs quantum dot at two temperatures: 300 K - fulll line, 30 K - dashed line. Experimental spectral resolution assumed to be $2\Delta = 0.12\,\text{meV}$.

of a convolution of two electronic spectral densities corresponding to the electronic ground and excited states $(n = 0, 1)$. The numerical results of calculations of the electronic relaxation rate were presented recently [42]. The electronic spectral densities were calculated in the Tamm-Dancoff (TD) (or self-consistent Hartree-Fock) approximation to the electronic self-energy. The numerical results [42] show that in the present model the relaxation rate can be about $(1 \div 10)/\text{ps}$ in quantum dots of InAs, GaAs, or InP.

The quantum dot is a rather singular object in the sense that the spectral density of the electron is expected to have a shape not very far from the delta function. From the first-order perturbation theory one can expect that when the electron energy-level separation is in resonance with the optical phonon energy, threshold effects may become significant. Generally, threshold effects bring the importance of higher order corrections to interaction vertexes [52]. This is why we will now consider the Lang-Firsov (LF) transformation and eliminate in a simple way the transverse terms in H_1. The meaning of this transformation can be interpreted as follows: (i) the terms which do not contribute to the relaxation of the electronic energy in the lowest order of perturbation theory are eliminated, (ii) in this way part of the vertex corrections are performed in both the Dyson equation for the self-energy and in the relaxation rate formula. As a matter of fact, by removing the transverse terms we remove terms proportional to the first power of $\Phi(n, n, \mathbf{q})$, but the transverse coupling reappears in the modified longitudinal terms. Approximating these higher order terms by the average, assumed to be taken at a temperature T, we arrive at the new effective electron-LO-phonon operator. Instead of H_1 we now have $V_\parallel$:

$$V_\parallel = \sum_{\mathbf{q}} A_q \Phi(0, 1, \mathbf{q})(\mathbf{q}) e^{-\Omega} c_0^+ (b_{\mathbf{q}} - b_{-\mathbf{q}}^+) c_1 + c.c., \qquad (5)$$

where, denoting $v_{mn}(\mathbf{q}) = A_q \Phi(m, n, \mathbf{q})$,

$$\Omega = \sum_{\mathbf{q}} E_{LO}^{-2} \left[| v_{11}(\mathbf{q}) |^2 + | v_{22}(\mathbf{q}) |^2 - 2 v_{11}(\mathbf{q}) v_{22}^*(\mathbf{q}) \right] \left(\nu_{\mathbf{q}} + \frac{1}{2} \right). \qquad (6)$$

The rate of change of the population of the excited state, dN_1/dt, is (N_0 is the population of the electron ground state):

$$\frac{dN_1}{dt} = -\frac{2\pi}{\hbar} \alpha_{01} e^{-2\Omega} \qquad (7)$$

$$\left[N_1(1 - N_0) \left((1 + \nu_{LO}) \int_{-\infty}^{\infty} dE\, \sigma_1(E)\sigma_0(E - E_{LO}) \right.\right.$$

$$\left. + \nu_{LO} \int_{-\infty}^{\infty} dE \sigma_1(E)\sigma_0(E + E_{LO}) \right)$$

$$- N_0(1 - N_1) \left((1 + \nu_{LO}) \int_{-\infty}^{\infty} dE \sigma_0(E)\sigma_1(E - E_{LO}) \right.$$

$$\left.\left. + \nu_{LO} \int_{-\infty}^{\infty} dE \sigma_0(E)\sigma_1(E + E_{LO}) \right) \right],$$

where $\alpha_{mn} = \sum_{\mathbf{q}} | v_{mn}(\mathbf{q}) |^2$, σ_0 and σ_1 are electronic spectral densities and ν_{LO} is the Bose-Einstein distribution of LO-phonons at a lattice temperature T. The electronic spectral densities and the self-energy are (as already stated above) calculated from the Dyson equation, thus allowing the perturbation contributions to be summed up to infinite order. This amounts to the inclusion of multiphonon states. Such a multiphonon nature of the states of the system is consistent with the presence of coherent LO phonon excitations detected experimentally in in quantum dots (see e. g. Ref.[53]). The equation for the electronic self-energy in the presently used Tamm-Dancoff approximation [40], with the coupling operator transformed using the Lang-Firsov transformation, is now (TDLF approximation):

$$M_n(E) = \sum_{m \neq n} e^{-2\Omega} \alpha_{nm} \tag{8}$$

$$\times \quad \{ \quad \frac{1 - N_m + \nu_{LO}}{E - E_m - E_{LO} - M_m(E - E_{LO}) + i0_+}$$

$$+ \quad \frac{N_m + \nu_{LO}}{E - E_m + E_{LO} - M_m(E + E_{LO}) + i0_+} \}.$$

In this work we neglect the correction to the electronic energies and the phonon-less direct coupling between the electronic states resulting from the LF transformation. The numerical results of the relaxation rate are given in Fig. 1. The relaxation rate is calculated with $N_1 = 1$ (relaxation from the electronic excited state). Comparing this result to the relaxation rate calculated in the previous TD approximation (see Ref. [42]) we see that the relaxation rate near the resonance between the electronic excitation energy and the LO phonon energy is nearly the same. At nonzero values of detuning the difference between these two approximations to the self-energy is minor. A remarkable difference is the absence of the resonance spikes at such detunings, when E_1 equals a multiple of E_{LO} (see Ref. [41, 42]. These maxima obtained in the earlier calculations appear thus to be connected with the presence of the transverse (adiabatic) terms in the electron-LO-phonon coupling. The absence of the adiabatic terms in $V_\parallel$ is therefore reflected by the absence of the resonances in the rate at $E_1 - E_{LO} = nE_{LO}$, with $n > 1$. On the other hand, the magnitude of the resonance spike at $n = 1$, and its width, correspond to a recent experiment [23], in which the electron-energy relaxation rate in the magnetic-field confined quantum dots was measured. Their result can be interpreted as a confirmation of the absence of an LO phonon bottleneck. It is also seen that in materials such as GaAs the "vertex corrections" considered here do not have a strong impact on the numerical values of the computed relaxation rate. Let us also remark that the experiment of Murdin [23] does not allow us to decide whether the resonances at $E_1 - E_{LO} = nE_{LO}$, with $n > 1$, should be present in the theoretical relaxation rate or not.

5. Theoretical interpretation of the linewidth

Out of various mechanisms which could possibly contribute to the width of the optical lines [53] only the electron-LO-phonon coupling (Fröhlich) will be considered here. The contribution of acoustic phonons [50] will be omitted in this work. The line shape of the optical transition of the electron between the conduction band states in the dot and the valence band states is assumed to be proportional to the electronic spectral density $\sigma_0(E)$ of

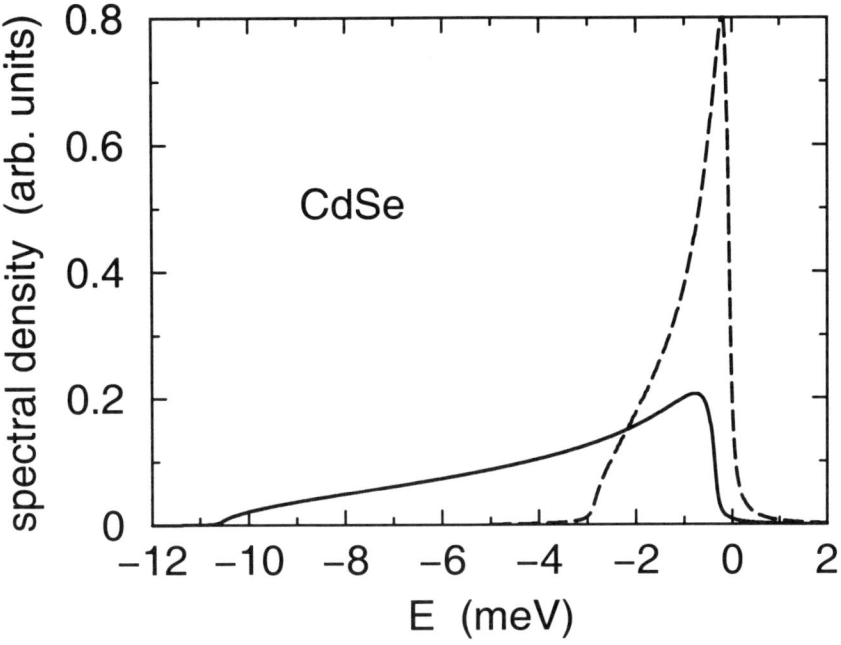

Figure 7. Triangle shape of the electronic spectral density in a CdSe quantum dot with detuning $E_{LO} - E_1 = -65$ meV. The spectral resolution is $2\Delta = 0.12$ meV. Temperatures: 1250 K - full line, 200 K - dashed line

only the electron in the lowest energy state ($n = 0$). The spectral density $\sigma_0(E)$, given by the solution of the Dyson equation (8), is thus calculated in the state with the occupation $N_0 = 1$.

Using simple analytical argumantation it was shown earlier [40] that the main optical lines depend on energy as $1/\sqrt{E}$, while there are some satellite maxima which remain finite for any value of the energy variable E. Figure 2 shows the log-log plot of the low-energy side of the main spectral feature of the spectral density $\sigma_0(E)$ (see Fig. 3) at $E = 0$ in InAs, confirming the $1/\sqrt{E}$ functional dependence in this case. Therefore, the $1/\sqrt{E}$ shape of the line does not allow us to speak about the 'full width at the half of maximum' (FWHM) of the spectral feature. In experiments the energy resolution is usually limited from below by the experimental apparatus properties. In order to take this fact into account, a finite positive number Δ is added to the imaginary part $| ImM_m |$ of the self-energy, chosen so that 2Δ equals the experimental spectral resolution. Fig. 3 shows the difference between the theoretical lineshape in InAs and the lineshape calculated with the nonzero Δ, as it is expected to be seen in experiment. The full line in Fig. 3 is in fact computed with $\Delta = 0.001$ meV in order to improve the convergence of the iterations. This value of the broadening also expresses the natural width of the corresponding optical transition in luminescence measurement. According to the calculation (full line in Fig. 3 and also Fig. 4) an experiment performed in an InAs quantum dot at 20 K, with the experimental spectral resolution of about 1 μeV, would give a FWHM of about 7 μeV. This value of the experimental linewidth could then be considered as a lower limit of the observed data. The spectral resolution usually used in experiments is about 0.1 meV [54, 55]. Recently a better value of the spectral resolution (about 30 μeV) was used in an experiment measuring microscopic photoluminescence spectra in quantum dots [56]. In this work the FWHM was found to have a plateau in the temperature interval from 3 K to about 20 K, where it has a value of about $(45 \div 65)\mu$eV. This experiment was performe on island-shaped InAs, AlInAs and GaAs quantum dots, several monolayers thick and having a lateral size of about $300 \div 600$Å. The comparison of the experiment on InAs dots [56] with the calculated data is presented in Fig. 4. Clearly, the experiments show that the measured FWHM follows closely the spectral resolution of the experimental apparatus in the case of narrow widths.

In the following the linewidths are computed with a finite value of the experimental spectral resolution 2Δ taken into account. The exception is of course the relaxation rate in Fig. 1), computed from the spectral densities with no experimental resolution included. The temperature dependence of the $\sigma_0(E)$ linewidth in CdSe is presented in Fig. 5, giving a comparison of the width computed using the present approximation (TDLF) with the result of a calculation employing the approximation (TD) of Refs. [40, 41, 42], as a function of sample temperature for a single size of the dot, so that $E_{LO} - E_1 = -65$ meV. The low temperature regions contain a plateau, giving a finite value of the FWHM in the limit of $T \to 0$ in both approximations. The curve calculated in the TD approximation (diamonds) is not displayed in the region of $T > 200$ K, because the calculated spectral density curve no longer consists of separated peaks. The low temperature values of the FWHM correspond to the experimental data [22].

Murdin [23] used a phenomenological approach to express theoretically the relaxation rate of electrons in magnetic field confined InAs-AlSb quantum dots. In order to fit the theoretical formula to their experimental data they needed to assume a homogeneous broadening of the electronic states. They assumed that the broadening is 0.3 meV at temperatures from 4 K to 80 K. In our model of an InAs dot, the lineshape, calculated (without and with the experimental resolution) at 20 K, is shown in Fig. 3 for the case of $E_1 = 2.5 \times E_{LO}$. Although we obtain a line with an obviously nonzero width, we cannot speak about width in the usual sense. The FWHM (here about 0.24 meV), calculated with the experimental resolution included, has however little in common with the linewidth assumed by Murdin.

The dependence of the FWHM on the size of the dot (or on the detuning between the phonon energy E_{LO} and the electronic energy level separation $E_1 - E_0$ (E_0 here is put equal to zero)) is shown in Fig. 6. This Figure shows a tendency of the FWHM to be peaked near zero detuning, at low temperatures.

The FWHM (Fig. 5), calculated in a CdSe dot, corresponds to experimental data [22] also with respect to the shape of the spectral density $\sigma_0(E)$. The triangle-like shape of $\sigma_0(E)$ (Fig. 7), computed for the temperature near the melting point of CdSe, reproduces the shape of the optical line in the experiment [22] under the condition of very intensive excitation by the light pulse.

6. Summary

The properties of the electronic spectral density, having at low temperatures a very narrow width due to a nearly $1/\sqrt{(E)}$ dependence of $\sigma_0(E)$ on the energy variable, give nevertheless a finite relaxation rate in the two approximations considered here. Because of the characteristic shape of the optical features, the experimental information about the lineshape thus appears to be resolution determined. The Fröhlich coupling of electrons and LO phonons gives the numerical results of the linewidth which are, taking into account the simple model considered, quite close to the observed data in not very small quantum dots.

References

1. N. N. Ledentsov, Fiz. Tekh. Poluprovod. **32**, 385 (1998).
2. W. D. Weiss and R. G. Nazmitdinov, PRB **55**, 16310 (1997).
3. M. Notomi et al., Appl. Phys. Lett. **58**, 720(1991).
4. H. E. G. Arnot et al., Superlatt. Microstruct. **5**, 459 (1989).
5. M. Kohl et al., Phys. Rev. Lett. **63**, 2124 (1989).
6. H. Benisty et al., Phys. Rev. B **44**, 10945 (1991).
7. T. Inoshita et al., Physica B **227**, 373 (1996).
8. U. Bockelmann et al., Phys. Rev. B **42**, 8947 (1990).
9. T. Inoshita et al., Phys. Rev. B **46**, 7260 (1992).
10. U. Bockelmann, Semicond. Sci. Technol. **9**, 865 (1994).
11. H. Benisty, Phys. Rev. B **51**, 13281 (1995).
12. F. Daiminger et al., Semicond. Sci. Technol. **9**, 896 (1994).
13. B. Ohnesorge et al., Phys. Rev. B **54**, 11532 (1996).
14. U. Bockelmann et al., Phys. Rev. Lett. **76**, 3622 (1996).
15. S. Raymond et al., Phys. Rev. B **54**, 11548 (1996).
16. K. L. Schumacher et al., Semicond. Sci. Technol. **11**, 1173 (1996).
17. M. J. Steer et al., Phys. Rev. B **54**, 17738 (1996).
18. U. Woggon et al., Phys. Rev. B **54**, 17681 (1996).
19. P. D. Wang et al., Superlatt. Microstruct. **21**, 259 (1997).
20. R. Heitz et al., Phys. Rev. B **56**, 10435 (1997).

420

21. R. Heitz *et al.*, Appl. Phys. Lett. **68**, 361 (1996).
22. S. A. Empedocles , Phys. Rev. Lett. **77**, 3873 (1996).
23. B. N. Murdin , Phys. Rev. B **59**, R7817 (1999).
24. U. Bockelmann , Phys. Rev. B **46**, 15574 (1992).
25. D. F. Schroeter , Phys. Rev. B **54**, 1486 (1996).
26. J. Valenta , Phys. Rev. B **57**, 1774 (1998).
27. M. Grundmann , Phys. Rev. Lett. **74**, 4043 (1995).
28. M. Notomi , Phys. Rev. B **53**, 15743 (1996).
29. U. Banin , Phys. Rev. B **55**, 7059 (1997).
30. R. W. Schoenlein , Phys. Rev. Lett. **70**, 1014 (1993).
31. G. D. Mahan, *Many-Particle Physics*, Second Edition. (Plenum Press, New York and London, 1993).
32. R. Zimmermann , J. Lumin. **58**, 271 (1994).
33. E. T. Jaynes and F. W. Cummings, Proc. IEEE **41**, 89 (1963).
34. V. Meden , Phys. Rev. B **52**, 5624 (1995).
35. V. A. Kovarskii *et al.*, Sov. Phys. Semicond. **26**, 1025 (1992).
36. I. Vurgaftman , Appl. Phys. Lett. **64**, 232 (1994).
37. I. Vurgaftman , Phys. Rev. B **50**, 14309 (1994).
38. T. Inoshita , Phys. Rev. B **56**, R4355 (1997).
39. J. A. Kenrow , Phys. Rev. Lett. **78**, 4873 (1997).
40. K. Král , Phys. Rev. B **57**, R2061 (1998).
41. K. Král and Z. Khás, Phys. Stat. Sol. (b) **208**, R5 (1998).
42. K. Král and Z. Khás, in *Proceedings of the 24th International Conference on Physics of Semiconductors* (World Scientific Publishing, Singapore, to be published.)
43. X.-Q. Li *et al.*, in *Proceedings of the 24th International Conference on Physics of Semiconductors* (World Scientific Publishing, Singapore, to be published.)
44. X.-Q. Li and Y. Arakawa, Phys. Rev. B **57**, 12285 (1998).
45. X.-Q. Li and Y. Arakawa, Phys. Rev. B **56**, 10423 (1997).
46. M. Moško *et al.*, Phys. Rev. B **31** 16860 (1995).
47. P. Wagner *et al.*, Carrier impurity collisions in a narrow quantum wire: Born approximation versus exact solutions, *Proceedings of the Workshop on Heterostructure Epitaxy and Devices*, NATO ASI Series, Kluwer Academic Publishers, 1998.
48. K. Král and Z. Khás, submitted for publication.
49. I. G. Lang and Yu. A. Firsov, Zhurn. Exp. Teor. Fiz. **43**, 1843 (1962).
50. T. Takagahara, Phys. Rev. Lett. **71**, 3577 (1993).
51. D. N. Zubarev, *Neravnovesnaya Statisticheskaya Termodinamika*(Nauka, Moscow, 1971).
52. I. V. Levinson and E. I. Rashba, Usp. Fiz. Nauk **111**, 683 (1973).
53. D. M. Mittleman , Phys. Rev. B **49**, 14435 (1994).
54. V. Zwiller , Phys. Rev. B **59**, 5021 (1999).
55. L. Landin , Science **280**, 262 (1999).
56. K. Ota , Physica E **2**, 573 (1998).

RAMAN SPECTROSCOPY OF EXCITON-ACOUSTIC PHONON COUPLING IN SEMICONDUCTOR NANOCRYSTALS

S.V. GOUPALOV, A.I. EKIMOV, O.G. LUBLINSKAYA, AND
I.A. MERKULOV
A.F. Ioffe Physico-Technical Institute, Politechnicheskaya
26,
194021 St. Petersburg, Russia

While electron and hole size quantisation in spherical semiconductor nanocrystals embedded in glass matrices has been much studied in the last 15 years, the question of acoustic phonon confinement remains open. Only simple models of either bulk-like phonons [1] or vibrational modes of an elastic sphere with free or fixed surface [2-5] were applied to describe the acoustic phonon energy spectrum in nanocrystals. In these two limiting cases the acoustic phonons either do not sense the interface between the nanocrystal and the glass matrix or can not penetrate through it, i.e. the probabilities of either phonon reflection or transmission are neglected. However the low-frequency Raman spectra line shapes are expected to be very sensitive to these probabilities when the excitation is far from the excitonic resonance and thus a more accurate consideration of acoustic phonons in nanocrystals is required.

In our work acoustic phonons are considered as vibrational modes of an elastic sphere surrounded by an elastic continuum. These modes are classified by the acoustic phonon total angular momentum, F, and come in two types [3, 4, 6]. In the first type are torsional modes, which are purely transversal. The modes of the second type, called spheroidal, have a mixed character of longitudinal and transversal modes. There is however one special case of the spheroidal modes where $F = 0$ and the mode is purely longitudinal. The displacement under such a vibration has only the radial component (so-called "breathing" mode). While calculating the acoustic phonon energy spectrum we use the boundary conditions assuming continuity of the displacement and radial strain at the nanocrystal surface. This treatment allows us to take into account both the probabilities of acoustic phonon reflection and transmission at the nanocrystal-matrix interface.

M.L. Sadowski et al. (eds.), Optical Properties of Semiconductor Nanostructures, 421–424.

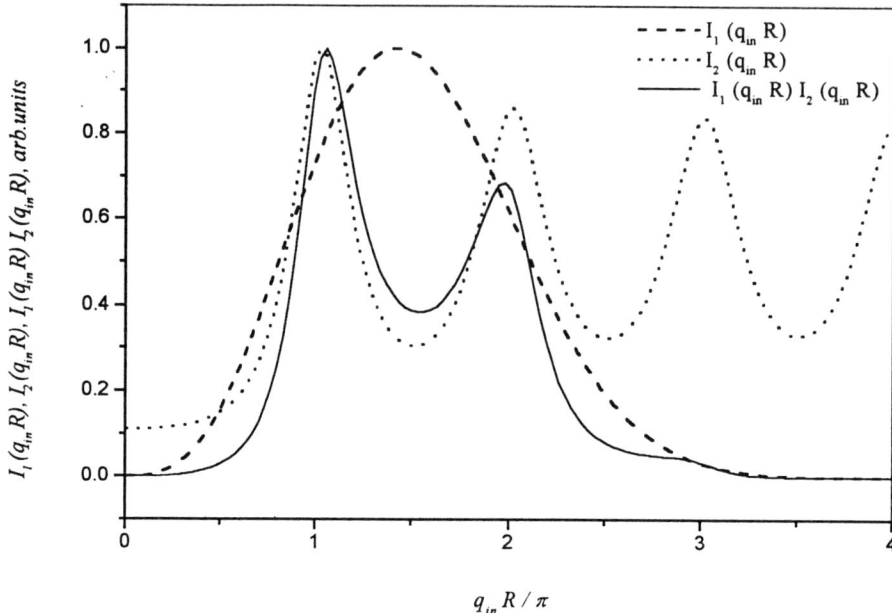

Figure 1. Exciton, $I_1(q_{in}R)$ (dashed line) and phonon, $I_2(q_{in}R)$ (dotted line), contributions to the Raman spectral line plotted as functions of the product of the phonon wave number inside the nanocrystal, q_{in}, and the nanocrystal radius, R. The calculation is made for CdS nanocrystals in the silica oxide glass. The product of the two contributions is shown as a solid line.

We consider exciton-acoustic phonon interaction due to the deformation potential coupling mechanism which was shown to predominate in semiconductor nanocrystals [3]. The exciton states are treated in the model of the isotropic cubic semiconductor nanocrystal, where the Luttinger Hamiltonian in the spherical approximation is used to describe the valence-band states. In this model the optically allowed exciton states are characterised by the total exciton angular momentum equal to one. From the angular momentum conservation law it follows that only the phonon modes with $F = 0, 1, 2$ can participate in the scattering process. Taking into account the parity conservation law and the symmetry under time inversion operation it is possible to show [6] that only spheroidal modes with $F = 0$ and $F = 2$ are Raman active.

We have measured low-frequency Raman spectra of CdS nanocrystals of different radii ($R = 15 \div 70$ Å) in the silica oxide glass matrix under

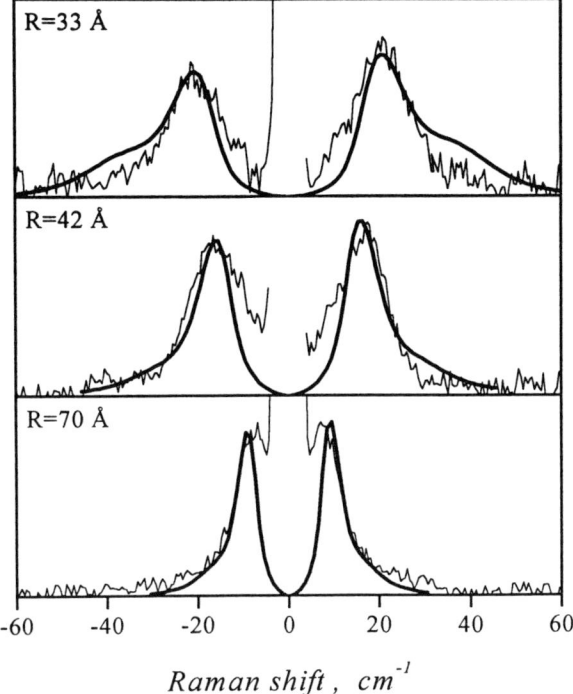

Figure 2. Low-frequency Raman spectra measured at room temperature together with calculated ones for samples with different nanocrystal mean radii.

an excitation far from the excitonic resonance. Raman lines fully polarised along with the excitation have been observed and attributed as coming from the scattering by the "breathing" vibrational modes of nanocrystals.

The developed theory allows an accurate description of the observed spectral line shapes and positions without any fitting parameters. The spectral shape is governed by two factors: (i) the acoustic phonon confinement and (ii) the exciton size quantisation. The phonon contribution is shown in Fig. 1 as a dotted line. The maxima correspond to the phonon resonant reflection. If the phonons were absolutely confined in the nanocrystal and could not tunnel to the matrix [2-5] the maxima would be infinitely narrow. In case of bulk-like phonons [1] there would be no maxima at all. Taking into account both the probabilities of acoustic phonon reflection and transmission at the nanocrystal-matrix interface we have a set of maxima with finite widths. The contribution corresponding to the exciton size quantisation is shown in Fig. 1 as a dashed line. In k-space the exciton envelope function is concentrated within a distance of the order of the reciprocal nanocrystal radius, and thus only acoustic phonons with the wave numbers lying in the same range can be effectively emitted or absorbed. The product of these two factors is shown in Fig. 1 as a solid line. One can conclude that the spectral line shape is mainly due to the phonon factor while the exciton size quantisation leads to a suppression of the maxima corresponding to the higher energy phonon resonances.

The observed room-temperature Raman spectra obtained after the rejection of a line corresponding to quasi-local oscillations of glass together with the calculated spectra are shown in Fig. 2 for samples with different mean nanocrystal radii $R = 70, 42, 33$ Å. A Gaussian nanocrystal size distribution with a dispersion of $\pm 20\%$ [7] has been taken into account in the calculations.

References

1. Sirenko, A.A., Belitsky, V.I., Ruf, T., Cardona, M., Ekimov, A.I., Trallero-Giner, C. (1998) Spin-flip and acoustic-phonon Raman scattering in CdS nanocrystals, *Phys. Rev. B* **58**, 2077-2087.
2. Tanaka, A., Onari, S., Arai, T. (1993) Low-frequency Raman scattering from CdS microcrystals embedded in a germanium dioxide glass matrix, *Phys. Rev. B* **47**, 1237-1243.
3. Takagahara, T. (1996) Electron-phonon interactions in semiconductor nanocrystals, *Journal of Luminescense* **70**, 129-143.
4. Saviot, L., Champagnon, B., Duval, E., Kudriavtsev, I.A., Ekimov, A.I. (1996) Size dependence of acoustic and optical vibrational modes of CdSe nanocrystals in glasses, *Journal of Non-Crystalline Solids* **197**, 238-246.
5. Saviot, L., Champagnon, B., Duval, E., Ekimov, A.I. (1998) Resonant low-frequency Raman scattering in CdS-doped glasses, *Phys. Rev. B* **57**, 341-346.
6. Goupalov, S.V. and Merkulov, I.A. (1999) Theory of Raman scattering from acoustic vibrations of nanocrystals, *Physics of Solid State* **41**, 1473-1483.
7. Lublinskaya, O., Gurevich, S., Ekimov, A., Kudryavtsev, I., Osinskii, A., Gandais, M., Wang, Y. (1998) CdS nanocrystal growth in thin silica films: evolution of size distribution function, *Journal of Crystal Growth* **184/185**, 360-364.

EXCITONS IN QUANTUM-DOT QUANTUM WELL STRUCTURES

W. JASKÓLSKI
  Instytut Fizyki UMK, Grudziadzka 5, Toruń, Poland

AND

GARNETT W. BRYANT
  National Institute of Standards and Technology, Gaithersburg, MD, USA

Abstract. The energy levels and excitation spectra of multi-shell nanocrystals composed of ZnS and CdS layers are investigated. The calculations are performed both in the 6-band envelope function approximation and in the tight-binding approach. Charge separation of pair states, with the electron and hole in different layers, is observed in a number of structures. For some structures, a dim-exciton ground state is observed.

Quantum-dot quantum well (QDQW) nanocrystals are composed of an internal spherical or tetrahedral semiconductor core which is coated with several shells of different semiconductors [1]. As in quantum well systems, barriers and wells can be defined by the ordering and composition of the layers. For example, the radial band profile of a QDQW with a ZnS core, a CdS shell and a ZnS cladding layer has an internal quantum well in both the conduction band (CB) and the valence band (VB). The band profile of a CdS/ZnS/CdS QDQW contains an internal barrier that can be used to control the coupling between the core and cladding layers. Transition energies and optical dynamics in these structures can be designed by changing the core size and thickness of each shell.

Multiple closely spaced excitations with different oscillator strengths have been observed [1]. This suggests that a detailed description of the band states, including heavy and light holes and band nonparabolicity, is needed. We previously developed a multiband effective mass theory for QDQWs [2]. However, one must worry whether effective mass theory is appropriate for QDQWs, especially when the layers are only a few monolayers thick.

M.L. Sadowski et al. (eds.), Optical Properties of Semiconductor Nanostructures, 425–430.

An atomistic approach may be necessary for such small structures. In this paper, we present two multiband models for QDQWs. One is our effective mass theory. The other is an empirical tight-binding model that should be more appropriate for the smallest structures. We compare results obtained from the two models to better assess the applicability of each. We consider CdS/ZnS/CdS QDQWs. CdS and ZnS are both wide gap semiconductors that should be well described by tight-binding theory and, near the band edges, by effective mass theory.

In our effective mass theory we use a 6-band spherical envelope function approximation (SEFA) [3] for the hole states. In the SEFA the hole states are described by three quantum numbers: nLF, where F is the total angular momentum $(F = J+L)$, L is the lowest quantum number corresponding to the envelope angular momentum, $J=3/2$ or $1/2$ is the Bloch VB-edge angular momentum and n is the main quantum number. The hole Hamiltonian is determined by two Luttinger parameters γ, γ_1 and the split-off band separation energy Δ. The electron states are found by solving a one-band effective-mass equation with CB nonparabolicity taken into account [3]. Transition rates are determined by summing over all pair states (of the same energy) and averaging over the linear polarisations of the dipole transition operator. The effective mass parameters we use yield electron, heavy-hole and light-hole effective masses that are close to literature values [4]. The VB offset between ZnS and CdS is set to 0.3 eV and the electron and hole barriers for tunnelling out of the QDQW are 3.5 eV.

We have developed an empirical tight-binding theory for QDQWs to test the multiband theory, especially for monolayer wells where the atomic character of the well and internal interfaces should be essential. We use an sp^3s^* model with nearest neighbour coupling but no spin-orbit splitting [5]. Calculations for spherical QDQWs are done for the dot centre midway between a neighbouring cation and anion. We assume that the lattice constant of the barrier layer relaxes to the lattice constant of the core and the cladding layers to define the size of the structure. However, the bulk tight-binding parameters [5] are used for all three layers. Using these tight-binding parameters, we obtain heavy hole masses smaller than those obtained in effective mass theory for the chosen Luttinger parameters. Because the heavy holes have small confinement energies, these differences in the SEFA and TB heavy hole masses are not essential.

To make the comparison between SEFA and TB, we consider a series of CdS/ZnS/CdS nanocrystals consisting of a 2.4-nm CdS core covered by different layers of ZnS and CdS. We start with a nanocrystal that has a 2.4-nm CdS core capped by a 1.2-nm wide ZnS barrier-shell and we add an external CdS cladding, layer by layer: 0.3-nm, 0.6-nm, 0.9-nm and 1.2-nm. The energies of several lowest electron and hole states, for a sequence

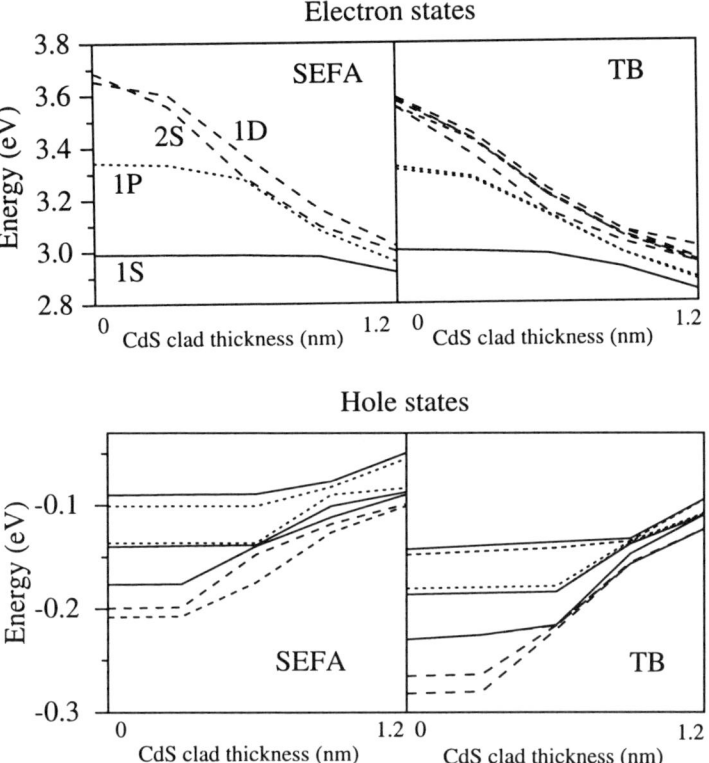

Figure 1. SEFA and TB energies of several lowest electron and hole states for the sequence of investigated CdS/ZnS/CdS QDQWs with the CdS clad thickness varying from 0 to 1.2 nm.

of the investigated QDQWs, are shown in Fig. 1. These lowest levels all lie below the ZnS band edge and are trapped mostly in the core or the cladding. There is a clear correspondence between SEFA and TB electron energy levels with similar spacing and dependence on cladding thickness. The SEFA electron states are easily labelled and ordered by their spherical symmetry. The lowest two TB electron states have the degeneracy appropriate for the spherical symmetry of the corresponding SEFA states. Higher TB electron states overlap in energy and it is difficult to assign the approximate spherical symmetries to these states. The SEFA and TB hole states are shifted, otherwise they also show similar spacing and dependence on cladding thickness. The differences are due in part to the different hole masses obtained with the two models. Spherical symmetries of the SEFA holes states change with cladding thickness. When there is no CdS cladding, the hole ground state is 1S3/2 while the first excited state is 1P3/2. When

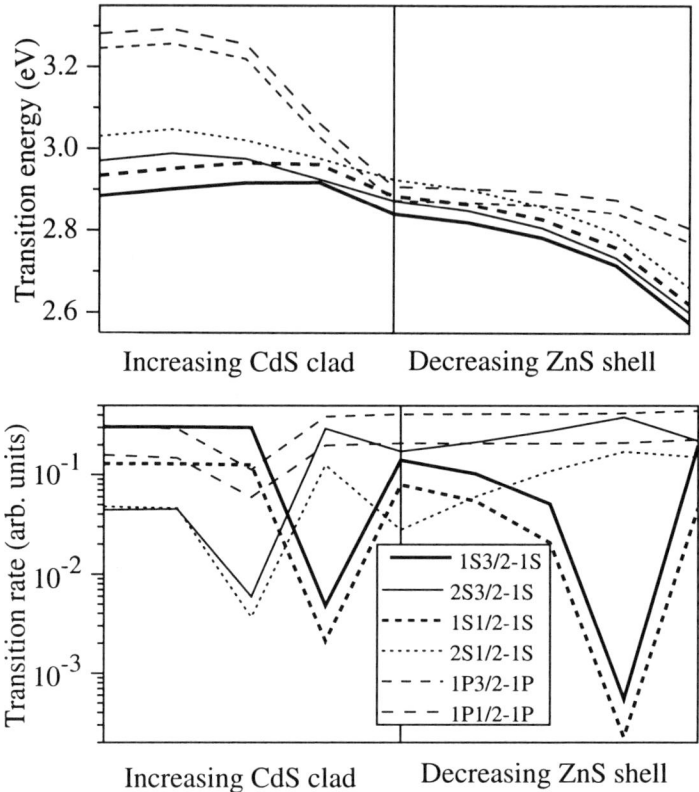

Figure 2. SEFA and TB radial charge densities of the electron and hole ground states for CdS/ZnS/CdS QDQWs with different widths of the CdS clad: 0.6 nm (solid lines) and 1.2 nm (dotted lines). The rapid variation in the TB densities results from the discreteness of the lattice.

the outer cladding layer is thick enough the 1P3/2 state becomes the hole ground state. The ordering of the other levels has a more complicated dependence on cladding thickness. No spherical symmetries can be assigned to the TB hole states because these symmetries are broken by the hole mixing on the lattice. Ground state electron and hole charge densities are shown in Fig. 2. A clear crossover from confinement in the core to confinement in the cladding layer occurs as the thickness of the cladding layer is increased. This crossover is reflected in the energies and symmetries of the states and can be used to tailor transition energies, transitions rates, and charge separation for pair states.

Transition energies and rates calculated in the SEFA for the lowest energy transitions are shown in Fig. 3 for a sequence of CdS/ZnS/CdS QDQWs with a 2.4-nm CdS core. Starting from a CdS/ZnS QDQW with

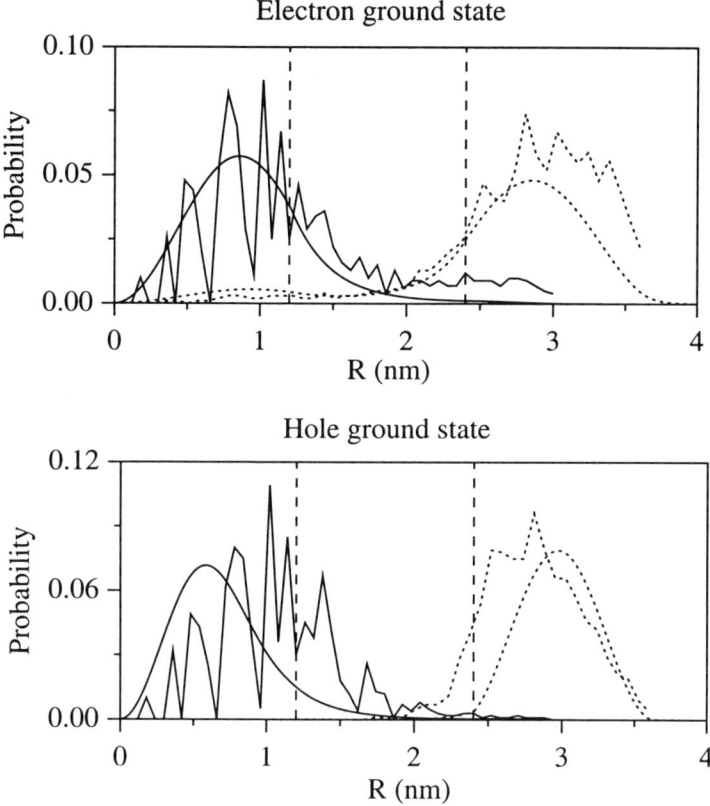

Figure 3. Transition energies and transition rates (calculated in SEFA) of several pair states for the sequence of investigated CdS/ZnS/CdS nanocrystals.

no cladding layer, the CdS cladding layer is increased to 1.2 nm in the first half of the sequence. In the second half of the sequence, the ZnS barrier is removed while the cladding layer is kept fixed. The transition energy is the energy separation of the pair state corrected by the Brus estimate for pair binding energy. No pair correlation is included in these calculations. When the CdS cladding layer is added to the CdS/ZnS QDQW, the effect of charge separation is clearly seen in the large reduction in transition rates for different pair states at different clad thicknesses. For large cladding thickness, the electron and hole are again in the same layer and the transition rates increase. Transition energies decrease for thick cladding due to the reduction in confinement within the entire nanocrystal. For the largest cladding thickness considered there is considerable overlap of states with different symmetries. When the barrier thickness is reduced, transition energies are further reduced, due to the reduction in confinement energy when the states can be in both the core and the cladding. Large reductions in the transition rates again indicate charge separation in the pair states that arises when different states tunnel between the core and the cladding.

Quantum dot quantum wells provide a new class of *quantum dot* heteronanostructures with tailorable optical properties. Modelling these structures, especially for monolayer shells and cladding, is a challenge. A comparison of a solid state model, such as the spherical envelop function approximation, and an atomistic, tight-binding approach shows good agreement for electron states even for the smallest structures. For hole states, the two models also predict similar results with the same dependence on layer thickness. A detailed comparison of the results for holes must be completed to fully understand the transition from the continuum to the atomic limit. Just as in the case of coupled multiple quantum well systems, transition energies, rates and charge overlap can be controlled by the layer thicknesses and compositions. Bright and dim exciton states can be engineered by adjusting the spatial separation of electrons and holes.

References

1. Mews, A., *et al.* (1996) Structural and spectroscopic investigations of CdS/HgS/CdS quantum-dot quantum wells *Phys. Rev. B* **53** R13242-R13245.
2. Jaskólski, W. and Bryant, G.W. (1998) Multiband theory of quantum-dot quantum wells: Dim excitons, bright excitons, and charge separation in heteronanostructures, *Phys. Rev. B* **57**, R4237-R4240.
3. Ekimov, A.I., *et al.*(1993) Absorption and intensity-dependent photoluminescence measurements on CdSe quantum dots: Assignment of the first electronic transitions, *J. Opt. Soc. Am. B* **10**, 100-107.
4. Broser, I., Broser, R. and Rosencweig M. (1982) in Madelung, O. Landolt-Börnstein (eds.) *Numerical Data and Functional relationships in Science and Technology, New Series, Group III, Vol. 17* Springer-Verlag, Berlin.
5. Lippens, P.E. and Lannoo, M. (1989)Calculation of the band gap for small CdS and ZnS crystallites *Phys. Rev. B* **39**, 10935-10942.

FINE STRUCTURE OF EXCITONS AND THE OVERHAUSER EFFECT IN QUANTUM DOTS

V.L.KORENEV, R.I.DZHIOEV, M.V.LAZAREV, M.N.TKACHUK,
B.P.ZAKHARCHENYA
Ioffe Physical Technical Institute, 194021 S. -Petersburg, Russia

Abstract. We present an experimental and theoretical study of the exciton-nuclear spin system in n-InP quantum dots. The polarised luminescence of an exciton bound to a neutral donor complex reveals anisotropic exchange splitting of optically active and inactive excitons. Optically oriented donor electrons and electrons in excitons polarise the lattice nuclei by hyperfine interaction. In turn, the polarised nuclei create an effective nuclear magnetic field affecting the polarisation of excitons. All nuclei species are polarised in InP islands, with the In nuclei undergoing a strong quadrupole interaction.

1.

The ground state of an e1-hh1 (1s) exciton with a heavy hole in a quantum well in a zinc blend lattice is fourfold degenerate and is characterised by the projection of the angular momentum $M=s+j=\pm 1, \pm 2$ on the growth axis $z \| [001]$ of the structure (the electron spin $s=\pm 1/2$ and the hole angular momentum $j=\pm 3/2$). The exchange interaction splits this state into a radiative doublet $|\pm 1\rangle$ and two close-lying optically inactive singlets (the superposition of the states $|\pm 2\rangle$). The localisation of an exciton on an anisotropic island reduces the symmetry of the system and the radiative doublet splits into two sublevels, linearly polarised in two orthogonal directions to be determined by the symmetry of the quantum dot (QD) [1]. The radiative doublet splitting manifests itself in the polarised luminescence (PL) of quantum dots even in the absence of a spectral resolution [2]. The more difficult problem is to determine the spin splitting of optically inactive excitons (+2 and -2 states), because due to selection rules they do not contribute directly to polarised luminescence.

The main idea of this research is that the dark excitonic states manifest themselves directly in the polarised PL of QDs doped by donors [3]. In this case the optically forbidden excitons can recombine through the formation of an exciton bound to a neutral donor complex (or charged exciton - trion, if the modulation doping technique is used to grow the sample). The ground state of the complex consists of two electrons with antiparallel spins and a hole. Only the hole is polarised in the complex, providing information about the spin state of both the optically active and inactive

431

M.L. Sadowski et al. (eds.), Optical Properties of Semiconductor Nanostructures, 431–434.

excitons before the complex formation. Thus, the spin splitting of both kinds of excitonic doublets is revealed in doped nanostructures. The hyperfine interaction of the electrons and crystal lattice nuclei has a large effect on the splitting of the exciton spin levels and polarisation of excitons [4]. Optically oriented electrons in donors and in excitons transfer their nonequilibrium spin to the nuclear spin system (the Overhauser effect) [5]. In turn, the polarised nuclei produce an effective magnetic field changing the splitting of the spin sublevels and the polarisation of excitons. Thus, a tight-binding exciton-nuclear spin system occurs in quantum dots.

2.

The structures were grown by MOVPE on GaAs substrates with (100)±30' orientation and contained a 500 nm thick InGaP buffer layer lattice-matched to the substrate, a layer of nanosized InP islands (the lateral size was 700 Å, the density of dots was $3 \times 10^9 \, cm^{-2}$) with a nominal thickness of three monolayers, and a 50 nm thick InGaP layer on top. The layers contained donor impurities $\approx 10^{15} \, cm^{-3}$. The samples were lowered into a cryostat with liquid helium and placed at the centre of an electromagnet. Photoexcitation was performed with a 10 W/cm^2 He-Ne laser beam ($h\nu$=1.96 eV) directed along the growth axis of the structure $z \| [001]$. The polarisation of the photoluminescence (PL) in the Faraday and Voigt geometry was measured at the maximum of the recombination emission band of the InP islands (λ=723 nm). The optical orientation and the measurement of the degree of circular polarisation of the recombination radiation were performed in two different regimes described below.

3.

In the first regime the sign of the circular polarisation of the exciting light was changed with a high frequency (26 kHz). In this case the nuclear spin does not follow the oscillating electron spin and there is no dynamic polarisation of the nuclei (the nuclei spin relaxation rate ~1 Hz). The polarisation of PL is analysed with a quarter-wave plate and a linear polariser. The effective degree of circular polarisation $\rho_c = \left(I_+^+ - I_+^- \right) / \left(I_+^+ + I_+^- \right)$ was measured. Here I_+^+ and I_+^- are the intensities of the σ^+ components of PL for σ^+ and σ^- excitation, respectively. Fig.1a shows the dependence $\rho_c(B)$ corresponding to the recombination of a D^0X complex. The dependence reflects the restoration of spin polarisation of optically inactive excitons and optically allowed states. For B=0 the excitons are unpolarised. The nonzero value of $\rho_c(0)$ is attributed to the spin polarisation of donor electrons. The nonmonotonic behaviour of $\rho_c(B)$ is due to the contributions of opposite signs of optically inactive and active excitons into polarised PL. A simple example illustrates this point. The σ^+-polarised light creates free optically active excitons with M=+1. If the hole spin in a free exciton relaxes quickly before trapping on the neutral donor, there are equal numbers of M=+1 excitons and M=-2 dark excitons. The trapping of an M=+1 exciton creates the D^0X complex with total momentum $M_T=+3/2$ (the electron spins are antiparallel). The recombination of such a complex leads to the σ^+ polarised PL. In the same way, the dark exciton forms the $M_T=-3/2$ complex, emitting a σ^- photon. Thus, the contributions of optically active and forbidden excitons are of opposite signs.

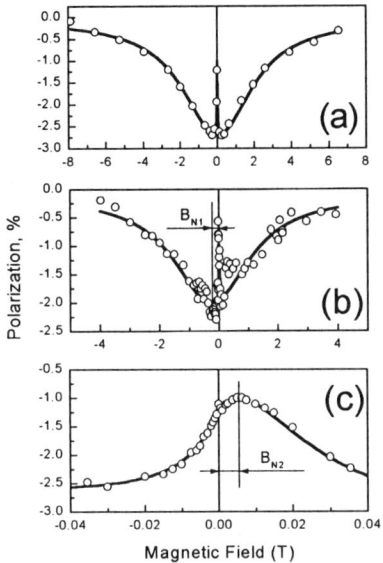

Figure 1. Optical orientation of excitons in magnetic field in the Faraday geometry in the absence (a) and in the presence (b), (c) of nuclear dynamic polarisation.

In zero magnetic field excitons are depolarised as a result of anisotropic exchange interaction mixing the +1 and -1 states (+2 and -2 states) of optically active (inactive) excitons. Low magnetic field (B~100 G) restores the orientation of optically inactive excitons and the degree of circular polarisation ρ_c becomes more negative (Fig.1a). The orientation of optically active excitons is restored at a higher magnetic field values (B~2 T) which causes ρ_c to increase (the anisotropic exchange splitting of optically allowed states is greater than that of dark states). Thus, the dependence $\rho_c(B)$ is nonmonotonic, with the orientation of optically active and dark excitons being restored at essentially different magnetic field values.

4.

The Overhauser effect in InP QDs arises under excitation by light of a constant helicity. A quartz modulator serves as the analyser of the PL polarisation. In this case the degree of circular polarisation ρ_c is measured. Circles in Fig.1 (b,c) show the dependence $\rho_c(B)$ under σ^+-excitation. The maximum of $\rho_c(B)$ (dark excitons are unpolarised) is achieved at B=54 G (Fig.1c) rather than B=0. This means that dark excitons experience the effect of the total field $B+B_{N2}$, which is the sum of external and effective nuclear (B_{N2}) magnetic fields. The maximum corresponds to the compensation of the nuclear field by the external one, i.e. B $\approx -B_{N2}$. In the same way, the polarisation of optically active excitons is restored by the total field $B+B_{N1}$, where B_{N1} is the nuclear field acting upon the optically allowed states. Hence, the minimum of their polarisation (Fig.1b) is achieved at B $\approx -B_{N1} \approx -2$ kG. The difference between the values of B_{N1} and B_{N2} is mainly connected with that of the g-factors of optically active and inactive excitons. If the value of the nuclear field did not depend on the external magnetic field, the presence of nuclear polarisation would shift the dependence $\rho_c(B)$ for optically active (inactive) states by B_{N1} (B_{N2}). We ascribe the shifts to the contribution to the Overhauser effect from oriented donor electrons, whose polarisation does not depend on B. However, besides the shift there is a striking asymmetry of $\rho_c(B)$ with respect to the maximum (Fig.4c). This means that the nuclear field depends on B. We interpret this fact as a contribution to the nuclear polarisation by electrons in excitons whose polarisation depends on B. As a result, the exciton - nuclear spin system appears to be non-linear: the optically oriented donor and exciton electrons polarise nuclei. In turn, polarised nuclei create an effective magnetic field, changing the polarisation of excitons.

5.

The InP islands contain nuclei of three species: ^{31}P (100%, spin 1/2), ^{115}In (95.5 %, spin 9/2) and ^{113}In (4.5 %, spin 9/2). An NMR signal at the Larmor frequency of ^{31}P has been detected optically [4]. In contrast, the NMR signal at the Larmor frequency of In nuclei has not been observed. Nevertheless, the dependence of ρ_c on magnetic field in the Voigt geometry (the Hanle effect) indicates that the In nuclei are mainly polarised in the InP islands. The In experiences a strong quadrupole interaction, changing the NMR frequency. The main axis of the interaction is parallel to the growth direction. This causes the nuclear field of In to be pinned to the [001] axis. The pinning suppresses the Hanle effect of donor electrons, which was observed experimentally [6].

References

1. Ivchenko E.L., and Pikus G.E. (1995) Superlattices and other heterostructures. *Symmetry and optical phenomena*, Springer-Verlag, Berlin.
2. Dzhioev R.I., Zakharchenya B.P., Ivchenko E.L., Korenev V.L., Kusraev Yu.G., Ledentsov N.N., Ustinov V.M., Zhukov A.E., and Tsatsulnikov A.F. (1997) Fine structure of excitonic levels in quantum dots, *JETP Lett.* **65**, 804-808.
3. Dzhioev R.I., Zakharchenya B.P., Korenev V.L., Pak P.E., Vinokurov D.A., Kovalenkov O.V., and Tarasov I.S. (1998) Optical orientation of donor-bound excitons in nanosized InP/InGaP islands, *Phys.Solid State* **40**, 1587-1593.
4. Dzhioev R.I., Zakharchenya B.P., Korenev V.L., Pak P.E., Tkachuk M.N., Vinokurov D.A., and Tarasov I.S. (1998) Dynamic polarisation of nuclei in a self-organized ensemble of quantum-size n-InP/InGaP islands, *JETP Lett.* **68**, 745-749.
5. Dyakonov M.I., and Perel V.I. (1984) Theory of optical spin orientation of electrons and nuclei in semiconductors, in *Cond. Matter Sciences* **8**, North-Holland, 11-73.
6. Dzhioev R.I., Korenev V.L., Lasarev M.V., and Zakharchenya B.P. (1999) Effect of nuclear dynamic polarisation on fine structure of excitons in a self-organized ensemble of nanosized islands n-InP/InGaP, *Phys.Solid State*, to be published.

NATO ARW PARTICIPANTS

Adamowski Janusz
Faculty of Physics and Nuclear Technics,
Technical University (AGH),
Al. Mickiewicza 30, 30-059 Krakow,
POLAND

Babiński Adam
Institute of Experimental Physics,
Warsaw University, ul Hoza 69,
00-681 Warszawa, POLAND

Bardyszewski Witold
Institute of Theoretical Physics,
Warsaw University, ul. Hoza 69,
00-681 Warszawa, POLAND

Bar-Joseph Israel
Department of Condensed Matter
Physics, Weizmann Institute of Science,
Rehovot 76100, ISRAEL

Brazis Romuald
Laboratory of Plasma Phenomena,
Semiconductor Physics Institute,
A.Goštauto 11, 2600 Vilnius,
LITHUANIA

Bugajski Maciej
Dept. Phys. and Technol. of Low
Dimensional Structures, Inst. of Electron
Technology, Al. Lotnikow 32/46,
02-668 Warszawa, POLAND

Ciulin Victoria
Inst. of Micro- and Optoelectronics,
Physics Departments, Swiss Federal
Institute of Technology,
CH-1015 Lausanne, SWITZERLAND

Dudziak Eugeniusz
Institute of Physics, Wroclaw University
of Technology, ul. Wyspianskiego 27,
50-370 Wroclaw, POLAND

Dyakonov Michel
A.F. Ioffe Physico-Technical Institute,
Politechnicheskaya 26,
194021 St. Petersburg, RUSSIA

Finley Jonathan J.
University of Sheffield, Department of
Physics, Hounsfield Road, Sheffield,
S3 7RH, UNITED KINGDOM

Forchel Alfred
Technische Physik, Experimentelle
Physik III, Universität Würzburg,
Am Hubland, D-97074 Würzburg,
GERMANY

Furdyna Jacek K..
Department of Physics and Electrical
Engineering, University of Notre Dame,
Notre Dame, IN 46556, USA

Gaj Jan
Institute of Experimental Physics,
Warsaw University, ul Hoza 69,
00-681 Warszawa, POLAND

Goupalov Sergei
A.F. Ioffe Physico-Technical Institute,
Politechnicheskaya 26,
194021 St. Petersburg, RUSSIA

Grynberg Marian
Institute of Experimental Physics,
Warsaw University, ul Hoza 69,
00-681 Warszawa, POLAND

Hawrylak Paweł
Institute for Microstructural Sciences,
National Research Council, Ottawa,
Ontario K1A OR6, CANADA

Jaskólski Włodzimierz
Instytut Fizyki, M. Kopernik University,
ul. Grudziadzka 5,
87 100 Torun, POLAND

Karrai Khaled
Ludwig-Maximilian-Universität, Sektion
Physik, Geschwister-Scholl-Platz 1,
D-80539 München, GERMANY

Kavaliauskas Julius
Semiconductor Physics Institute,
A. Goštauto 11,
2600 Vilnius, LITHUANIA

Kavokin Kiril
A.F. Ioffe Physico-Technical Institute,
Politechnicheskaya 26,
194021 St. Petersburg, RUSSIA

Kochereshko Vladimir
A.F. Ioffe Physico-Technical Institute,
Politechnicheskaya 26,
194021 St. Petersburg, RUSSIA

Korenev Vladimir L.
A.F. Ioffe Physico-Technical Institute,
Politechnicheskaya 26,
194021 St. Petersburg, RUSSIA

Kossacki Piotr
Inst. of Micro- and Optoelectronics,
Physics Departments, Swiss Federal
Institute of Technology, CH-1015
Lausannes, SWITZERLAND

Kotthaus Jörg P.
Ludwig-Maximilian-Universität, Sektion
Physik, Geschwister-Scholl-Platz 1, D-
80539 München, GERMANY

Koutselas Ioannis
Theoretical and Physical Chemistry
Institute, National Hellenic Research
Foundation, 48 Vas. Constantine Ave.,
11635 Athens, GREECE

Kral Karel
Institute of Physics, Academy of
Sciences of Czech Republic,
Na Slovance 2, 18221 Prague 8,
CZECH REPUBLIC

Kusraev Yuri
A.F. Ioffe Physico-Technical Institute,
Politechnicheskaya 26,
194021 St. Petersburg, RUSSIA

Kutrowski Mirosław
Institute of Physics, Polish Academy of
Sciences, Al. Lotnikow 32/46,
02 668 Warszawa, POLAND

Littlewood Peter
Cavendish Laboratory, University of
Cambridge, Madingley Road,
Cambridge, CB3 0HE,
UNITED KINGDOM

MacDonald Allan H.
Department of Physics,
Indiana University,
Bloomington, IN 47405, USA

Martinez Gerard
Grenoble High Magnetic Field
Laboratory, MPI/FKF and CNRS,
F-38042 Grenoble, FRANCE

McCombe Bruce D.
Department of Physics, State University
of New York, Fronczak Hall,
Buffalo, N.Y. 14260, USA

Merkulov Igor
A.F. Ioffe Physico-Technical Institute,
Politechnicheskaya 26,
194021 St. Petersburg, RUSSIA

Misiewicz Jan
Institute of Physics, Wroclaw University
of Technology, ul. Wyspianskiego 27,
50-370 Wroclaw, POLAND

Nicholas Robin J.
Department of Physics, Clarendon
Laboratory, University of Oxford,
Parks Road, Oxford, OX1 3PU,
UNITED KINGDOM

Oszwałdowski Rafał
Instytut Fizyki, M. Kopernik Univeristy
ul. Grudziadzka 5,
87 100 Torun, POLAND

Podor Balint
Hungarian Academy of Sciences,
Research Institute for Technical,
Physics and Material Science,
H-1525 Budapest, HUNGARY

Potemski Marek
Grenoble High Magnetic Field
Laboratory, MPI/FKF and CNRS,
F-38042 Grenoble, FRANCE

Quinn John J.
University of Tennessee,
Knoxville, TN 37906, USA

Rogatch Andrey
Belarussian State University,
220050 Minsk, BELARUS

Ryan John F.
Department of Physics, Clarendon
Laboratory, University of Oxford,
Parks Road, Oxford, OX1 3PU,
UNITED KINGDOM

Sadowski Marcin
Groupe d'Etude des Semiconducteurs,
Université Montpellier 2,
Pl. E. Bataillon,
F-34095 Montpellier, France

Skolnick Maurice S.
Department of Physics, University of
Sheffield, Sheffield, S1 3JD,
UNITED KINGDOM

Stępniewski Roman
Institute of Experimental Physics,
Warsaw University, ul Hoza 69,
00-681 Warszawa, POLAND

Tejedor Carlos
Dept. de Fisica de la Materia
Condensada, Universidad Autonoma de
Madrid, Cantoblanco,
E-28049 Madrid, SPAIN

Teran Francisco
Grenoble High Magnetic Field
Laboratory, MPI/FKF and CNRS,
BP 166, F-38042 Grenoble, FRANCE

Viña Luis
Depto. Fisica de Materiales (C-IV),
Universidad Autonoma de Madrid,
Cantoblanco, E-28049 Madrid, SPAIN

Warburton Richard
Ludwig-Maximilian-Universität, Sektion
Physik, Geschwister-Scholl-Platz 1,
D-80539 München, GERMANY

Załużny Mirosław
Institute of Physics,
M. Curie-Sklodowska University,
pl. M. Curie-Sklodowskiej 1,
20-031 Lublin, POLAND

AUTHOR INDEX

SUBJECT INDEX

443